*PROGRESS IN*

# Molecular Biology and Translational Science

*Volume 109*

*PROGRESS IN*

# Molecular Biology and Translational Science

## The Proteasomal System in Aging and Disease

*edited by*

Tilman Grune

*Institute of Nutrition, Friedrich Schiller University, Jena, Germany*

*Volume 109*

ELSEVIER

AMSTERDAM • BOSTON • HEIDELBERG • LONDON
NEW YORK • OXFORD • PARIS • SAN DIEGO
SAN FRANCISCO • SINGAPORE • SYDNEY • TOKYO
Academic Press is an imprint of Elsevier

Academic Press is an imprint of Elsevier
32 Jamestown Road, London, NW1 7BY, UK
Radarweg 29, PO Box 211, 1000 AE Amsterdam, The Netherlands
225 Wyman Street, Waltham, MA 02451, USA
525 B Street, Suite 1900, San Diego, CA 92101-4495, USA

This book is printed on acid-free paper. ∞



**Library of Congress Cataloging-in-Publication Data**
A catalog record for this book is available from the Library of Congress

**British Library Cataloguing in Publication Data**
A catalogue record for this book is available from the British Library

ISBN: 978-0-12-397863-9
ISSN: 1877-1173

For information on all Academic Press publications
visit our website at elsevierdirect.com

Printed and bound by CPI Group (UK) Ltd, Croydon, CR0 4YY

# Contents

Contributors ........ ix
Preface ........ xi

## Structure of the Proteasome ........ 1

Tobias Jung and Tilman Grune

I. Introduction ........ 1
II. The 20S Proteasome ........ 2
III. Regulation of the 20S Proteasome ........ 15
IV. Conclusion ........ 25
References ........ 26

## Structure and Function of the Ubiquitin–Proteasome System: Modulation of Components ........ 41

Niki Chondrogianni and Efstathios S. Gonos

I. The Ubiquitin–Proteasome System ........ 41
II. Modulation of the UPS Components ........ 51
III. Conclusion ........ 62
Acknowledgments ........ 62
References ........ 63

## Immunoproteasomes: Structure, Function, and Antigen Presentation ........ 75

Deborah A. Ferrington and Dale S. Gregerson

I. Introduction ........ 75
II. Immunoproteasome Structure ........ 77
III. Regulation of Gene Expression ........ 79
IV. Assembly of the 20S Core ........ 82
V. Enzymatic Activity ........ 84
VI. Immunoproteasome Knockout Mice ........ 86
VII. Immunoproteasome Function ........ 88
VIII. Mutations and Linkage to Human Disease ........ 101
References ........ 105

Interactions of the Proteasomal System with Chaperones: Protein Triage and Protein Quality control. . . . . . . . . . . . 113

Marc Kästle and Tilman Grune

I. Introduction .......... 114
II. Cytosolic Chaperones: The Hsp Family .......... 115
III. Refolding or Degradation? .......... 123
IV. Heat Shock Response After Proteasome Inhibition .......... 127
V. Inhibition of Hsps: Influence on Proteasomal Degradation .......... 131
VI. Hsps and the Proteasomal System During Oxidative Stress and the Aging Process .......... 133
VII. Neurodegenerative Diseases: Connection to the Hsp Family and the Proteasome .......... 136
VIII. Early-Stage Protein Quality Control and ERAD .......... 139
IX. Conclusion .......... 150
References .......... 150

Proteasome Inhibitors . . . . . . . . . . . . . . . . . . . . . . . . . 161

Boris Cvek

I. Short History of Bortezomib .......... 161
II. Bortezomib in Multiple Myeloma .......... 163
III. Bortezomib in Mantle Cell Lymphoma .......... 166
IV. Bortezomib in Other Cancers .......... 167
V. Bortezomib in Diseases Different from Cancer .......... 170
VI. The Adverse Effects of Bortezomib .......... 171
VII. Second-Generation of 20S Proteasome Inhibitors .......... 173
VIII. Early Drugs as Proteasome Inhibitors .......... 205
IX. Immunoproteasome Inhibitors .......... 207
X. Beyond the 20S Proteasome .......... 208
Acknowledgment .......... 209
References .......... 209

Degradation of Damaged Proteins: The Main Function of the 20S Proteasome . . . . . . . . . . . . . . . . . . . . . . . 227

Andrew M. Pickering and Kelvin J. A. Davies

I. Introduction .......... 228
II. Formation of Oxygen Radicals and Toxic Oxidants .......... 228
III. Removal of Oxygen Radicals and Toxic Oxidants by the Cell .......... 231
IV. The Role of the 20S Proteasome Versus the 26S Proteasome in Degrading Oxidized Proteins .......... 232
V. Mechanism of Selective Degradation of Damaged Proteins by the 20S Proteasome .......... 234

VI. The 20S Proteasome in Aging and the Removal of Age-related Protein Damage ........ 236
VII. The 20S Proteasome in Adaptation to Protein Damage ........ 238
VIII. Removal of Damaged Proteins by the 20S Proteasome in the Nucleus.... 239
IX. Alternative Forms of the 20S Proteasome in the Cytoplasm and Other Non-nuclear Compartments and Their Role(s) in the Removal of Damaged Proteins........ 241
X. Summary........ 243
References ........ 244

## Changes of the Proteasomal System During the Aging Process ........ 249

Martin A. Baraibar and Bertrand Friguet

I. Introduction ........ 249
II. Age-Related Decline of Proteasome Function in Organs and Tissues.... 252
III. Proteasome Dysfunction During Cellular Aging........ 255
IV. Modulation of Proteasome Function in Model Organisms ........ 257
V. Mechanisms Involved in the Age-Related Decline of Proteasome Function........ 260
VI. Rejuvenating Proteasome Function as an Antiaging Strategy........ 264
Acknowledgments........ 268
References ........ 268

## Proteasome and Cancer ........ 277

Betul Catalgol

## The Ubiquitin–Proteasome System and Cardiovascular Disease........ 295

Saul R. Powell, Joerg Herrmann, Amir Lerman, Cam Patterson, and Xuejun Wang

I. Introduction ........ 296
II. Regulation of the Cardiac Proteasome ........ 297
III. Involvement of the UPS in atherosclerosis........ 300
IV. The Role of Cardiac E3 Ligases........ 312
V. The UPS in PQC and Cardiac Proteinopathy ........ 319
VI. The UPS in Heart Failure ........ 325
VII. The UPS in Myocardial Ischemia/Reperfusion Injury ........ 327
VIII. Concluding Statement ........ 333
Acknowledgments........ 334
References ........ 334

## Role of the Ubiquitin–Proteasome in Protein Quality Control and Signaling: Implication in the Pathogenesis of Eye Diseases . . . . . . . . . . . . . . . . . . . . . . . . . . . . . 347

Fu Shang and Allen Taylor

I. Introduction ........ 348
II. Role of UPP in Lens Protein Quality Control ........ 350
III. The Role of UPP in Regulating Lens Cell Proliferation ........ 355
IV. The Role of UPP in Regulating Lens Fiber Differentiation and Lens Development ........ 357
V. UPP Functions in Retinal Development ........ 360
VI. The UPP Regulates Signal Transduction: Implication in the Pathogenesis of Age-Related Macular Degeneration ........ 363
VII. UPP Function in Regulating the NF-κB Signaling Pathway ........ 363
VIII. UPP Function in Regulating Expression of Angiogenic Factors ........ 365
IX. The Role of UPP in Visual Signaling ........ 366
X. Redox Regulation of UPP Activity in the Eye ........ 367
XI. Interaction Between the UPP and Cellular Antioxidant Systems ........ 369
XII. Impairment or Dysregulation of the UPP and Age-Related Eye Diseases ........ 370
XIII. Interaction Between the UPP and ALP in RPE: Implications Regarding the Pathogenesis of AMD ........ 373
XIV. Protecting the UPP from Oxidative Stress May Be a Valid Strategy for Prolonging the Vision Function ........ 376
XV. Conclusions and Future Perspectives ........ 376
Acknowledgments ........ 377
References ........ 377

## Proteasome and Neurodegenerative Diseases. . . . . . . . 397

Betul Catalgol and Tilman Grune

I. Introduction ........ 397
II. Proteasome in Alzheimer's Disease ........ 399
III. Parkinson's Disease ........ 402
IV. Huntington's Disease ........ 403
V. Amyotrophic Lateral Sclerosis ........ 404
VI. Friedreich Ataxia ........ 405
References ........ 406

Index ........ 415

# Contributors

*Numbers in parentheses indicate the pages on which the authors' contributions begin.*

**Martin A. Baraibar,** Laboratoire de Biologie Cellulaire du Vieillissement, UR4-IFR83, Université Pierre et Marie Curie–Paris 6, 4 place Jussieu, Paris Cedex 05, France (249)

**Betul Catalgol,** Department of Biochemistry, Faculty of Medicine, Genetic and Metabolic Diseases Research Center (GEMHAM), Marmara University, Haydarpasa, Istanbul, Turkey (277, 397)

**Niki Chondrogianni,** National Hellenic Research Foundation, Institute of Biology, Medicinal Chemistry and Biotechnology, Athens, Greece (41)

**Boris Cvek,** Department of Cell Biology & Genetics, Palacky University, Olomouc, Czech Republic (161)

**Kelvin J. A. Davies,** Ethel Percy Andrus Gerontology Center of the Davis School of Gerontology and Division of Molecular & Computational Biology, Department of Biological Sciences, Dornsife College of Letters, Arts & Sciences: The University of Southern California, Los Angeles, California, USA (227)

**Deborah A. Ferrington,** Department of Ophthalmology, University of Minnesota, Minneapolis, Minnesota, USA (75)

**Bertrand Friguet,** Laboratoire de Biologie Cellulaire du Vieillissement, UR4-IFR83, Université Pierre et Marie Curie–Paris 6, 4 place Jussieu, Paris Cedex 05, France (249)

**Efstathios S. Gonos,** National Hellenic Research Foundation, Institute of Biology, Medicinal Chemistry and Biotechnology, Athens, Greece (41)

**Dale S. Gregerson,** Department of Ophthalmology, University of Minnesota, Minneapolis, Minnesota, USA (75)

**Tilman Grune,** Institute of Nutrition, Friedrich Schiller University, Jena, Germany, (1, 113, 397)

**Joerg Herrmann,** Department of Internal Medicine, Division of Cardiovascular Diseases, Mayo Clinic, Rochester, Minnesota, USA (295)

**Tobias Jung,** Institute of Nutrition, Friedrich Schiller University, Jena, Germany (1)

**Marc Kästle,** Institute of Nutrition, Friedrich Schiller University, Jena, Germany (113)

**Amir Lerman,** Department of Internal Medicine, Division of Cardiovascular Diseases, Mayo Clinic, Rochester, Minnesota, USA (295)

**Cam Patterson,** Department of Medicine; Department of Pharmacology, and Department of Cell and Developmental Biology, The McAllister Heart Institute, University of North Carolina Chapel Hill, North Carolina, USA (295)

**Andrew M. Pickering,** Ethel Percy Andrus Gerontology Center of the Davis School of Gerontology and Division of Molecular & Computational Biology, Department of Biological Sciences, Dornsife College of Letters, Arts & Sciences: The University of Southern California, Los Angeles, California, USA (227)

**Saul R. Powell,** Center for Heart and Lung Research, The Feinstein Institute for Medical Research, Manhasset, New York, USA (295)

**Fu Shang,** Laboratory for Nutrition and Vision Research, USDA Human Nutrition Research Center on Aging, Boston, Massachusetts, USA (347)

**Allen Taylor,** Laboratory for Nutrition and Vision Research, USDA Human Nutrition Research Center on Aging, Boston, Massachusetts, USA (347)

**Xuejun Wang,** Protein Quality Control and Degradation Research Center, Division of Basic Biomedical Sciences, Sanford School of Medicine, University of South Dakota, Vermillion, South Dakota, USA (295)

# Preface

The proteasomes, together with the proteasome-associated proteins and cellular systems, are among the most complex cellular structures. Literally, there is no cellular process without involvement of proteins, which need to be degraded in a controlled manner or removed if they are damaged and this function is largely taken over by the proteasome. This leading role of the proteasome among the intracellular proteases leads, therefore, to an involvement of the proteasomal system in almost all cellular processes.

During evolution, several systems developed with the aim to maintain the integrity of the cellular protein pool. The idea that proteins are not stable within the body but undergo a permanent turnover was first raised by Schönheimer and colleagues in the late 1930s.[1] However, how these pathways were realized remained unknown until Christian de Duve discovered in the early 1950s the lysosomes, leading to the discovery of their enzymatic activities, including the proteolytic enzymes.[2] It was accepted for some time that lysosomes were responsible for the degradation of most cellular structures, including proteins. The proteasome was discovered as a structure in 1968/1969 by Harris *et al.*[3] It took some time to associate this particle with the process of protein degradation and only the description of the ubiquitination process led to the discovery of the real functions of the proteasome. One of the key articles was published in 1978 by Ciechanover, Hod, and Hershko describing the involvement of several fractions of reticulate lysate in the ATP-dependent degradation of globin.[4] In 2004, this pioneering research was acknowledged with awarding the Nobel Prize in chemistry for Ciechanover, Hershko, and Rose.

Ever since then, the research on the proteasome became wider and wider and led to the discovery of the role of the proteasomal system in the pathophysiology of several human diseases and in the physiology of the aging process itself.

This volume focuses on some major roles of the proteasome in human health and disease. The volume is divided into two parts: the first part (Chapters 1–5) describes the current knowledge of the structure, function, and regulation of the proteasomal system, whereas the second part (Chapters 6–11) describes the role of the proteasome in aging and disease.

The structure of the proteasome is described in a chapter by Jung and Grune, referring to today's knowledge on the structure of the 20S proteasome and of some of the proteasomal regulators. The next chapter by Chondrogianni and Gonos focuses on the function of this system with special attention to the ubiquitin system. In the next chapter, Ferrington and Gregerson describe a special form of the proteasomal system—the immunoproteasome. This proteasomal form was found to be induced by immunostimulatory cytokines. However, today's understanding attributes functions beyond the immune response to the immunoproteasome. A relatively new field is the understanding of the interaction of the proteasomal system with chaperons and the interactions of both systems as a response to the accumulation of unfolded proteins as well in the ER as in the cytosol. This is reviewed in the chapter by Kästle and Grune. The last chapter of the first part of the volume, written by Cvek, describes the wealth of proteasomal inhibitors known today.

The second part of the volume describes the role of the proteasomal system in physiology and pathophysiology of disease and aging. One of the first functions attributed to the proteasome already in the mid-1980s is the selective degradation of oxidatively modified proteins.[5] Ever since then our understanding of this process is becoming better, as reviewed by Pickering and Davies. However, the proteasomal system itself is also affected by oxidative stress and aging. The role of the increasingly failing proteasomal function during aging is described by Baraibar and Friguet. Moreover, the proteasomal function is also affected by diseases. This refers to the proteasomal malfunction in cardiovascular diseases, neurodegeneration, and eye diseases, as reviewed by Powell *et al.*, Shang/Taylor, and Catalgol/Grune. Special attention was given to the possibilities of treatment of cancer cells with proteasomal inhibitors, as shown in the chapter of Catalgol.

As you will see, by studying this volume, one can hardly overestimate the importance of the UPS for normal cellular metabolism. By realizing that this system is conserved during evolution from archaebacteria through mammals, its importance is further underlined. During its evolution, components of the proteasomal system diverged in order to meet the ever more challenging tasks in more complex organisms. The proteasomal system interacts with an increasing number of other cellular systems. It is expected that the research on the proteasomal system will discover more functions in near future. Furthermore, as already demonstrated by the usage of proteasomal inhibitors in cancer, more therapeutic approaches or strategies are expected to be developed in the future—either for the proteasome directly or for the proteasome–ubiquitin system.

Tilman Grune
Volume Editor

## References

1. Schönheimer R, Rittenberg D, Foster GL, Keston AS, Ratner S. The application of the nitrogen isotope N15 for the study of protein metabolism. *Science* 1938;**88**:599–600.

2. de Duve C, Gianetto R, Appelmans F, Wattiaux R. Enzymic content of the mitochondria fraction. *Nature* 1953;**172**:1143–4.
3. Harris JR. The isolation and purification of a macromolecular protein component from the human erythrocyte ghost. *Biochim Biophys Acta* 1969;**188**:31–42.
4. Ciechanover A, Hod Y, Hershko A. A heat-stable polypeptide component on an ATP-dependent proteolytic system from reticulocytes. *Biochem Biophys Res Commun* 1978;**81**:1100–5.
5. Davies KJ. Free radicals and protein degradation in human red blood cells. *Prog Clin Biol Res* 1985;**195**:15–27.

# Structure of the Proteasome

Tobias Jung and Tilman Grune

*Institute of Nutrition, Friedrich Schiller University, Jena, Germany*

I. Introduction ........ 1
II. The 20S Proteasome ........ 2
A. The Proteasomal α-Subunits ........ 7
B. The Proteasomal β-Subunits ........ 10
C. Intracellular Assembly of the 20S Proteasome ........ 12
D. Modeling of the 20S Proteasomal Proteolysis ........ 15
III. Regulation of the 20S Proteasome ........ 15
A. The 19S Regulator ........ 16
B. The Immunoproteasome ........ 16
C. The Thymus-Specific Proteasome (Thymoproteasome) ........ 19
D. The 11S Regulator ........ 20
E. The Hybrid Proteasome (PA28–20S–PA700) ........ 21
F. The PA200 Regulator Protein ........ 23
IV. Conclusion ........ 25
References ........ 26

The ubiquitin-proteasomal system is an essential element of the protein quality control machinery in cells. The central part of this system is the 20S proteasome. The proteasome is a barrel-shaped multienzyme complex, containing several active centers hidden at the inner surface of the hollow cylinder. So, the regulation of the substrate entry toward the inner proteasomal surface is a key control mechanism of the activity of this protease.

This chapter outlines the knowledge on the structure of the subunits of the 20S proteasome, the binding and structure of some proteasomal regulators and inducible proteasomal subunits. Therefore, this chapter imparts the knowledge on proteasomal structure which is required for the understanding of the following chapters.

## I. Introduction

In order to maintain the functionality and the viability of a cell, most of the cellular proteins are subjected to a highly regulated turnover. To realize this, proteins that are misfolded, (oxidatively) damaged, or no longer required, have to be recognized and removed.[1–4] Removal of proteins is usually realized via proteolytic degradation. The most important proteolytic intracellular system of the cytosol is the proteasomal system, an evolutionarily very old and distributed machinery that was found to be present even in many of the oldest bacteria, as

Progress in Molecular Biology
and Translational Science, Vol. 109
DOI: 10.1016/B978-0-12-397863-9.00001-8

well as in plants and animals. The central part of the proteasomal system is the 20S "core" proteasome, a large multisubunit and multicatalytic protease, as well as several different regulators that can change the activity of the specificity of the "core" particle.

In the following sections, we describe the structure and function of the 20S proteasome and its regulators. For a better differentiation between the variations of the "proteasome," the 20S "core" particle is always referred to as "proteasome," and the other forms, according to the regulators that are attached to that "core."

## II. The 20S Proteasome

The 20S proteasome represents the catalytic part of the proteasomal system, a highly regulated group of proteins that perform degradation of damaged or misfolded proteins, regulation of their life spans,[1–6] and "quality control" of newly synthesized proteins[7–12] that are involved in regulation of the cell cycle,[5] gene expression,[13–17] immune responses,[18–23] responses to (oxidative) stress,[24–28] and carcinogenesis.[29–31] Furthermore, the nuclear protein is involved in the maintenance of chromatin and influences DNA repair.[32–34] So an evermore increasing spectrum of cellular functions are related to the proteasomal system.

The term "20S" results from the sedimentation constant of the proteasome "core" particle.[35] The mammalian form of this particle is of a cylindrical structure of about $100 \times 160$ Å that contains four homologous rings (two alpha ($\alpha$-) and two beta ($\beta$-)rings, arranged in the sequence $\alpha$-$\beta$-$\beta$-$\alpha$), which are built of seven different subunits each. The three-dimensional structure of the large protease of several organisms has been investigated extensively via X-ray crystallography.[36–41]

Two basic forms of the proteasome are known: the ancestral one that is found in Archaea bacteria like *Thermoplasma acidophilum* and the evolutionarily higher form of yeast, plants, and animals.

As the evolutionary higher form, the ancestral proteasome contains four heptameric rings, arranged in the common $\alpha$-$\beta$-$\beta$-$\alpha$ sequence, but in this ancestral proteasome each ring contains seven equal subunits, so only one $\alpha$- and $\beta$-subunit are present (see Fig. 1). It is obvious that from these simple forms of the proteasome a more complex one evolved via the divergence of the single subunits into several homologous ones. So the evolutionary higher form of the 20S proteasome contains 14 different subunits overall ($\alpha_1$–$\alpha_7$ and $\beta_1$–$\beta_7$), showing molecular masses between 20 and 30 kDa (see Table I), summarizing to a molecular weight of some 700 kDa. While the catalytic centers are located in the inner $\beta$-rings, the outer $\alpha$-rings of the proteasome

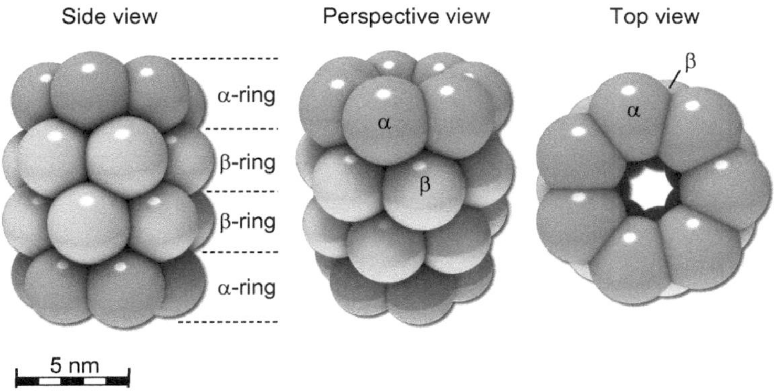


FIG. 1. The structure of the archaea 20S proteasome from *Thermoplasma acidophilum*. This figure shows a basic model of the archaea proteasomal structure. As shown on the left, the proteasome contains four homologous rings in the sequence α-β-β-α. Each ring contains seven identical subunits: the α-ring only α-, the β-ring only β-subunits, as shown in the central image. The right panel shows the arrangement of the α-subunits in a vertical view onto an α-ring. (See Color Insert.)

are responsible for the regulation of substrate entrance to the inner proteolytic chamber, as well as for recognition and binding of the substrates themselves. So the α-subunits are able to change both the activity and specificity of the proteasome. The proteolytic centers found in the inner rings are encoded by three different β-subunits ($\beta_1$, $\beta_2$, and $\beta_5$). Thus, due to the symmetric arrangement of the different rings, the inner chamber contains 6 different proteolytic centers, protected inside the proteasome in the evolutionary higher form of the proteasome, but 14 in the ancestral one. The inside of the proteasome is subdivided into two "ante chambers" (between the α- and β-rings) and one single "main chamber," found between the two β-rings (see Fig. 2). The "main chamber" is also the location of the catalytic centers.

As the proteasome is today referred to as a proteolytic system, several regulators are binding to the core proteasome, modulating the proteasomal activity. Today, a set of several different proteasomal regulators are known, all binding to the α-subunits of the outer proteasomal rings. The 11S regulator particle, in most organisms termed "PA28" or "REG," is formed of three different subunits (PA28α, PA28β, and PA28γ arranged in several diverse combinations). In *Trypanosoma brucei*, this ATP-independent regulator is called "PA26." Another important regulator is the ATP-dependent "19S," also known as "PA700" regulator; its analogue in archaea is termed "PAN." Several other regulators are known, including the nuclear regulator "PA200," which is known in three different isoforms (PA200i, PA200ii, and PA200iii) and

TABLE I

HERE, THE MOLECULAR MASSES (AFTER POSTTRANSCRIPTIONAL PROCESSING, AS FOUND IN THE ASSEMBLED WHOLE 20S "CORE" PROTEASOME) OF THE PROTEASOMAL SUBUNITS FROM BOTH HUMAN AND YEAST PROTEASOME ARE LISTED, AS WELL AS THE SUBUNITS OF TWO PROTEASOMAL REGULATOR CAPS (11S AND 19S)

20S "core" proteasome

| Systematic | *S. cerevisiae* | *Homo sapiens* | Mass [kDa] | Literature |
|---|---|---|---|---|
| $\alpha_1$ | C7/Prs2 | HsPROS27/HsIota | 27.5 | 42 |
| $\alpha_2$ | Y7 | HsC3 | 25.9 | 43,44 |
| $\alpha_3$ | Y13 | HsC9 | 29.5 | 45,46 |
| $\alpha_4$ | Pre6 | HsC6/XAPC7 | 27.9 | 47,48 |
| $\alpha_5$ | Pup2 | HsZeta | 26.4 | 49–51 |
| $\alpha_6$ | Pre5 | HsC2/HsPROS30 | 30.2 | 52 |
| $\alpha_7$ | C1/Prs1 | HsC8 | 28.4 | 53–56 |
| $\beta_1$ | Pre3 | HsDelta/Y | 25.3 (21.9) | 57–59 |
| $\beta_{1i}$ | – | Lmp2 | 23.2 (20.9) | 60–63 |
| $\beta_2$ | Pup1 | Z | 20.0 (24.5) | 64–66 |
| $\beta_{2i}$ | – | Mecl1 | 28.9 (23.8) | 67,68 |
| $\beta_3$ | Pup3 | HsC10-II | 22.9 | 69,70 |
| $\beta_4$ | C11 / Pre1 | HsC7-I | 22.8 | 69,71 |
| $\beta_5$ | Pre2 | X/MB1 | N/A (22.4) | 71 |
| $\beta_{5i}$ | – | Lmp7 | 30.4 (21.2) | 67,72–74 |
| $\beta_6$ | C5/Prs3 | HsC5 | 26.5 (23.3) | 75,76 |
| $\beta_7$ | Pre4 | HsN3/HsBPROS26 | 29.2 (24.4) | 69,77,78 |

11S (PA28) activator cap

| Systematic | Other names | Mass [kDa] | Literature |
|---|---|---|---|
| 11S subunit $\alpha$ | REG$\alpha$ or PA28$\alpha$ | 28.723 | 79,80 |
| 11S subunit $\beta$ | REG$\beta$ or PA28$\beta$ | 27.348 | 20 |
| 11S subunit $\gamma$ | REG$\gamma$ or PA28$\gamma$ | 30.886 | 81–84 |

19S (PA700) regulator cap

| Systematic | Other names | Mass [kDa] | Literature |
|---|---|---|---|
| *ATPase-subunits* | | | |
| Rpt1 | S7 or p48, Mss1, Yta3, Cim5 | 48.633 | 85 |
| Rpt2 | S4 or p56Yhs4, Yta5, Mts2 | 49.184 | 86–88 |
| Rpt3 | S6b or p48, Tbp7, Yta2, Ynt1, MS73 | 47.336 | 89,90 |
| Rpt4 | S10b or p42, Sug2, Pcs1, Crl13, CADp44 | 44.173 | 91,92 |
| Rpt5 | S6a or p50, Tbp1, Yta1 | 49.118 | 93,94 |
| Rpt6 | S8 or p45, Trip1, Sug1, Cim3, Crl3, Tby1, Tbp10 | 45.653 | 95–97 |
| *Non-ATPase-subunits* | | | |
| Rpn1 | S2 or p97, Trap2, Nas1, Hrd2, Rpd1, Mts4 | 100.199 | 98 |
| Rpn2 | S1 or p112, Sen3 | 105.866 | 99,100 |
| Rpn3 | S3 or p58, Sun2 | 61.005 | 101 |

(*Continues*)

TABLE I (*Continued*)

| 19S (PA700) regulator cap | | | |
|---|---|---|---|
| Systematic | Other names | Mass [kDa] | Literature |
| Rpn4 | Son1 or Ufd5 | 60.152 | 102–104 |
| Rpn5 | P55 or Nas5 | 52.904 | 105–107 |
| Rpn6 | S9 or p44.5 | 47.447 | 108–110 |
| Rpn7 | S10a or p44, HUMORF07 | 45.531 | 111–113 |
| Rpn8 | S12 or p40, Mov-34, Nas3 | 37.060 | 114 |
| Rpn9 | S11 or p40.5, Les1, Nas7 | 42.945 | 115–117 |
| Rpn10 | S5a or p54, ASF1, Sun1, Mcb1, Mbp1 | 40.736 | 118–120 |
| Rpn11 | S13 or Poh1, Mpr1, Par1 | 34.577 | 121–123 |
| Rpn12 | S14 or p31, Nin1, Mts3 | 30.004 | 98,124,125 |
| Rpn15 | DSS1 or SHFM1 (in human), SEM1 (in yeast) | 8.146 | 126,127 |

The molecular masses in brackets represent the peptide mass before posttranscriptional processing. According to their position ($\alpha$- or $\beta$-ring) in the mature proteasome, the systematic names of the subunits were defined. The "i" in the systematic names indicates an $\gamma$-interferon-"inducible" proteasomal subunit. The molecular masses of the proteasome are according to Coux *et al.*,[128] and the information about the regulator caps is according to Finley *et al.*[129] and Baumeister *et al.*.[130] The molecular weights of the single proteins were taken from the corresponding literature and checked using the site "www.wolframalpha.com." Molecular weights from "www.wolframalpha.com" were very similar to the weights taken from the literature, except for the subunits Rpt1 (57.199 kDa according to www.wolfram.com), Rpn4 (24.551 kDa), and Rpn12 (39.481 kDa).

contributes to spermatogenesis[131] and DNA repair.[132] However, only PA200i seems to bind to the proteasome, while the other two isoforms can be found in nuclear foci[133] without any proteasomal interaction.

Furthermore, PR39 and PI31, which work as cellular proteasomal inhibitors, are known. PR39 is a short peptide of only 39 amino acids, first extracted from porcine bone marrow, and functions as a noncompetitive inhibitor, both in yeast and mammals. Its mechanism of inhibition is unique: via binding to the $\alpha_7$-subunit of the proteasome, an allosteric change of the whole proteasomal structure is induced that decreases its proteolytic activity and affects the binding to the 19S-regulator.[134] PI31, a mammalian protein first discovered by DeMartino,[135] competes with the $\alpha$- and $\beta$-form of PA28 for binding of the proteasome.[136]

Several other proteins interact with the proteasomal system and are able to regulate the proteasomal proteolytic activities. The heat shock protein 90 (Hsp90) is another known cellular proteasomal regulator.[137–139] Furthermore, the proteasomal activity seems to be regulated by phosphorylation of its substrates or of components of the proteasomal system itself[53,95,140,141] and the nuclear proteasome by poly-ADP-ribose.[142–144]

As mentioned, the proteolytic activity is localized inside the proteasome in the main chamber. While only the subunits $\beta_1$, $\beta_2$, and $\beta_5$ show proteolytic activity, the others do not; furthermore, in some mammalian cells, the active

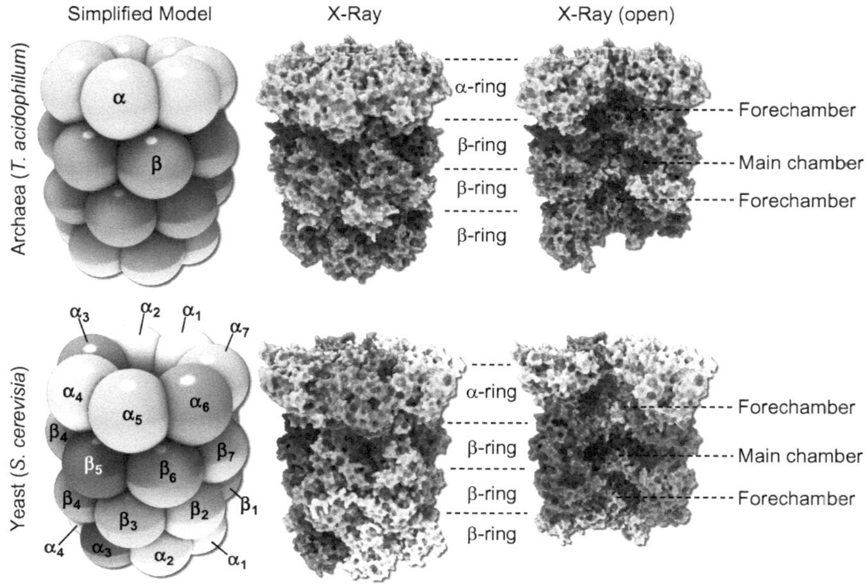


FIG. 2. Structures of the archaea and eukaryotic 20S proteasomes. Here, the structures of the archaea and the eukaryotic proteasomes are compared. The upper line of images shows a simple descriptive model (left), a highly detailed model calculated from X-ray structure analysis (middle), and the inside of the archaea 20S proteasome (from *Thermoplasma acidophilum*) after removal of several α- and β-subunits. The simple ball model shows the α-rings in pink and the β-rings in turquoise, while the more detailed one alternates those colors in order to accentuate the single rings. The right image shows the inner structure of the prokaryotic proteasome, subdivided into two fore chambers between the α- and β-rings, and the main chamber formed by the two β-rings. The bottom row of images shows the same for the eukaryotic 20S proteasome from *Saccharomyces cerevisiae*. This type of proteasome contains seven different α- and β-subunits, each arranged in two α- and β-rings. The single subunits are color coded in the same way in the simple ball model on the left and the more complex models from X-ray structure analysis (middle and right). The right image reveals the inner structure of the yeast 20S proteasome, showing the same subdivision into two fore and one main chamber, as found in the archaea proteasome too. (See Color Insert.)

subunits can be replaced by their γ-interferon (IFN-γ)-inducible isoforms $\beta_{1i}$, $\beta_{2i}$, and $\beta_{5i}$. However, in fact, that is not a replacement but a *de novo* synthesis of new proteasomes. The presence of the IFN-γ-inducible isoforms of the proteasome results in a change of the fragment length of the product of oligopeptides. It is surmised that the inducible proteasomal forms play a significant role in the antigen presentation of the adaptive immune response. Interestingly, in the thymus, a specific third variation of the $\beta_5$-subunit was recently found by Hirano, the replacement of the $\beta_5$-subunit by the so-called $\beta_{5t}$ one.[145] The β-subunits of the proteasome provide their own class of proteases that show no evolutionary relation to other known proteases but a very high relationship to each other, suggesting a common ancestor.[36] According to the arrangement of the proteasomal subunits as found in the archaeic form from *T. acidophilum*, the different subunits have been divided into two classes, the α- and β-subunits.[146] Normally, Eubacteria contain no 20S proteasome. Nevertheless, in a subgroup of Eubacteria, the so-called actinobacteria, proteasomal genes and even proteasomes were found. In those actinobacteria, HslVU (also known as ClpQY), an ATP-dependent hexameric protease[147,148] that shows structural similarities to the β-subunits of the 20S proteasome, was identified. It is a dimer containing both the proteasome-related protease HslIV and HslU, an ATPase.[149]

The proteasome is not the only proteolytic system of the cell, but one of the most important ones. Among the others are the lysosomal system,[150,151] containing many different cathepsins,[152–154] and the calpains associated with the cytoskeleton. While the main task of the lysosomal system includes the degradation of intracellular organelles, the proteasomal system is the most important one regarding the recognition and degradation of (damaged) proteins. It is assumed that between 70% and 90%[155,156] of the misfolded, (oxidatively) damaged, or no longer needed proteins are degraded via the proteasomal pathway. While the pure degradation of dysfunctional or misfolded proteins occurs in an ATP-independent way, the regulatory degradation of functional proteins is ATP-dependent and, moreover, the proteins targeted for degradation have to be labeled with a chain of ubiquitin molecules (polyubiquitination).

In mammalian cells, the amount of proteasome can be up to 1% of the whole protein pool (in liver and kidney cells).[157] The proteasome can be found both in the cytosolic and in the nuclear compartment of a cell, and can also be bound to the endoplasmic reticulum (ER) as well as being in association with the cytoskeleton.[158] The mammalian form of the 20S proteasome was first discovered and isolated from human red blood cells by Harris in 1968 and termed "cylindrin"[159,160] following the shape of the large protein complex. Other scientists have termed it "macroxyproteinase,"[161–163] "hollow cylinder" protein,[164,165] "multicatalytic proteinase complex,"[166–170] or "prosome."[158,171] Today, the term "proteasome" is used the most often.

## A. The Proteasomal α-Subunits

Both recognition and access of the substrate into the inner proteolytic chamber of the proteasome is regulated by the α-rings. After contact with a substrate, an (oxidatively) damaged/misfolded, and (partly) defolded protein, a conformational change of the α-rings is induced that virtually "unlocks" the gate they form to control and regulate substrate entrance to the inner chamber. That gate is formed by the N-terminal ends of the three subunits $\alpha_2$, $\alpha_3$, and $\alpha_4$.[172] The N-terminal ends of those three subunits are pointing in the direction of the proteasomal symmetric axis and block mechanically the entrance to the proteolytic centers formed by the two β-rings. Incubation of isolated proteasomes with low concentrations of sodium dodecylsulfate (SDS), an agent that induces a slight defolding of the proteasomal subunits, resulted in a significant increase of proteolytic activity, caused by an opening of the proteasomal gate that facilitates substrate entrance[128] (Fig. 3).

Similar effects could be induced by repeated freezing–thawing cycles as well as under conditions of low ionic strength. Those experimental results suggested an involvement of structural changes of the α-subunits that regulate proteasomal substrate entrance. A further possibility of substrate entrance modulation is the binding of a regulatory subunit (like 11S, 19S, or PA200) to the proteasomal α-rings. Such a binding can increase proteasomal activity up to 10-fold[173,174] and induce a change in substrate specificity, which is induced by the maximal opening of the substrate channel of the proteasome via a conformational rearrangement of the blocking N-terminal ends of some α-subunits.

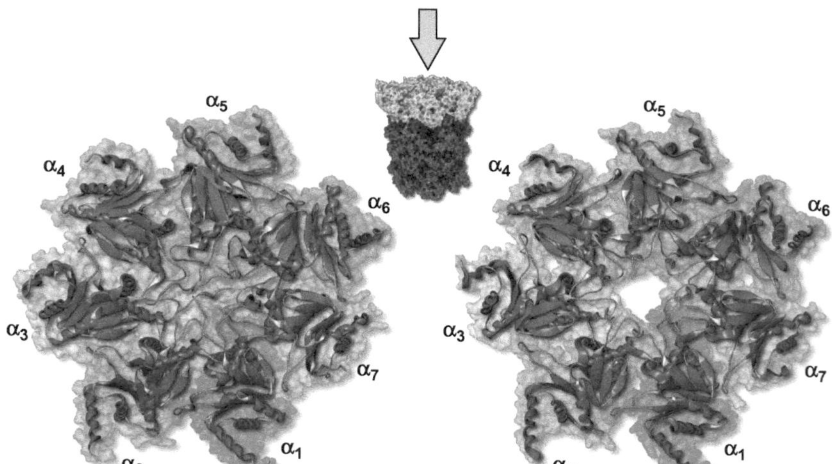


FIG. 3. Gating of the eukaryotic 20S proteasome. This image shows a structural rearrangement of the α-rings of the yeast 20S proteasome without and after activation. Activation can be induced by binding of a substrate protein, short oligopeptides, or regulating proteins (in this case, Blm10; see the text). After activation, the substrate accessibility of the proteasome increases, mainly due to an opening of the gate that is formed by the N-terminal structures of the α-subunits. Please note the massive reorganization of the N-terminus of $\alpha_3$ (yellow). After activation, the α-rings open a channel of about 13 Å to the inner proteolytic centers. The small icon in the middle shows the point of view (arrow) from which the reader is looking at the depicted α-rings. (See Color Insert.)

The "activation" of the proteasome seems to be mediated via the binding of the α-rings to hydrophobic amino acids (normally buried inside the correctly folded/native form of a globular and water-soluble proteins) that are exposed by (oxidatively) damaged or unfolded/misfolded proteins; that binding induces the conformational change that results in an opening of the proteasomal gate.[175]

The proteasomal gate shows a maximal diameter of about 13 Å (archaea proteasome) in its maximal opened state: this is sufficient to enable the entrance of a defolded substrate protein, usually represented by a single chain of amino acids. The individual N-terminal ends of the single gating subunits reveal unique three-dimensional structures that are essential for the gating of the proteasome. Some of the involved structures are highly conserved in eukaryotic cells. The so-called YDR-motif (Tyr8-Asp9-Arg10) can be found in every single α-subunit, as well as in archaea and eukaryotic cells, and may be working as a joint, bending the gating structures in order to modulate proteasomal activities.[172] The most important part seems to be played by the $\alpha_3$ subunit: $\alpha_3\Delta N$ mutant of yeast, which miss the last nine amino acids (GSRRYDSRT) that are found in the wild type, and show a permanently increased proteasomal activity that cannot be modulated/increased any further by SDS-exposure of the proteasome.[172] However, the characteristics of substrate binding or degradation are not or significantly less affected. In contrast, the $\alpha_7\Delta N$ mutant did not reveal any significant increase in proteolytic activity, while the $\alpha_3\alpha_7\Delta N$ mutant induced significantly more activity in casein degradation than either of the single deletions.[176] So, especially the YDR motive of the $\alpha_3$-subunits seems to be essential for stabilization of the gate, involving allosteric effects that affect also the subunits $\alpha_2$ and $\alpha_4$.

Interestingly, in the archaea proteasome (from *T. acidophilum*), where the α-rings are built of seven identical subunits, some oligopeptides revealed the ability to induce a conformational change in the N-termini of the α-subunits resulting in an opening of the gate. In this conformational change, the last 13 amino acids are involved as well as a slight turn (about 4°) of every single subunit.

The same gate-opening reconfiguration can be induced by the attachment of PAN in its ATP-bound confirmation, as cryo-electron microscopic experiments have revealed.[177] The oligopeptides mentioned above (seven or more amino acid residues) are the different C-terminal sequences of PAN subunits[178] that are able to bind the gaps between the single α-subunits of the proteasome. Those residues are termed the HbXY motifs. After binding of PAN or an oligopeptide, the gate of the archaea proteasome "opens" by a structural rearrangement. The diameter of the "closed" gate is usually about 9 Å and thus notably smaller than the channel directing to the inner proteolytic chamber of the proteasome with a diameter of about 23 Å. As mentioned earlier, after "activation" the diameter of the open gate increases to about 13–20 Å. A similar mechanism was revealed in studies of the PA26-binding: the gate structures were opened by the C-termini of PA26 and a so-called activation loop that induces both a movement of the N-terminal structures of the α-rings and a faint turning of every α-subunit.[177]

Usually, the 20S proteasome is found in the cell in its "inactive" state but can be "activated" by regulators, unfolded proteins, or proteasomal substrates. Furthermore, it has to be discriminated between the proteasomal peptidase and protease activities: while "peptidase activities" represent the degradation of small oligopeptides and are almost independent of the gate status, "protease activities" stand for the degradation of a whole unfolded protein and are considerably dependent on the gate status. This suggests that the gating α-subunits have only little interaction with small peptide fragments but a key function in the degradation of whole protein substrates.

## B. The Proteasomal β-Subunits

In contrast to the gating/regulating function of the proteasomal α-units, the main task of the β-units is the proteolytic process itself.

The ancient archaea proteasome contains seven identical β-subunits in one ring and thus seven proteolytic centers, too. In contrast, further developed yeast and the mammalian proteasomes contain only three different centers per β-subunit ring, localized on the subunits $\beta_1$, $\beta_2$, and $\beta_5$. In 2002, Unno *et al.* proposed a novel N-terminal nucleophile hydrolase activity,[40,41] formed by the Thr8 residue of subunit $\beta_7$, after the analysis of bovine proteasomes' X-ray crystal structure. The proteolytic centers of the active subunits are found in the inner chamber of the 20S proteasome complex. Figure 4 shows the catalytic centers of the archaea proteasome. Each of the three known proteolytic subunits shows a different preference for substrate binding:

- $\beta_1$ Shows a peptidyl–glutamyl–peptide hydrolysing activity (caspase-like activity, cleaving after *acidic* amino acids, thus also termed as "post-glutamyl-peptide hydrolytic" activity).[179]
- $\beta_2$ Shows a trypsin-like activity, cleaving after *basic* amino acids.
- β5 Shows a chymotrypsin-like activity and cleaves after *neutral* amino acids.[180]

In all three active subunits, the active center is formed by the N-terminus (Thr1) of the corresponding proteasomal β-subunits (Fig. 5). Noteworthy, the proteases that show similar specificities like trypsin[181] or chymotrypsin[182] are serine-proteases. Typical products of proteasomal degradation are oligo-peptides with lengths between 2 and 35 amino acids.[183] That distribution shows three different maxima: 2–3, 8–10, and 20–30 amino acids, while the average length is about 8–12 residues.[86]

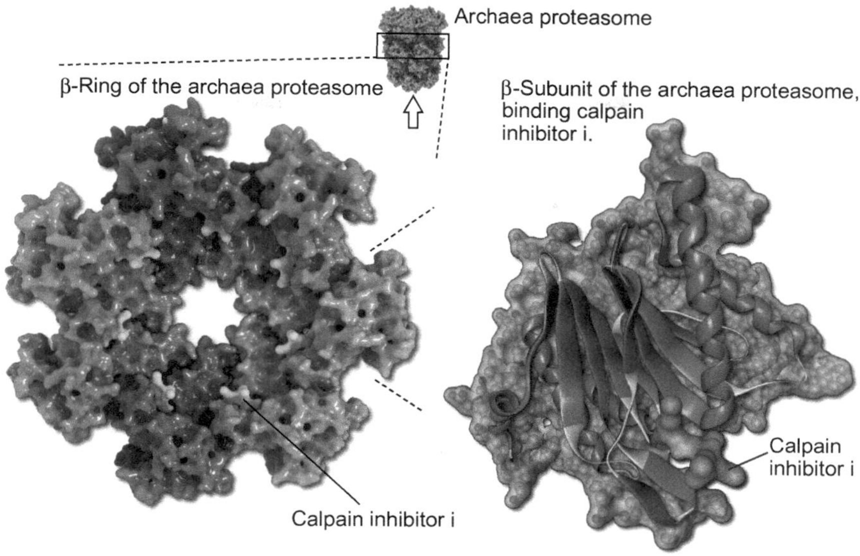


FIG. 4. Active centers of the archaea 20S proteasome. Here, the β-ring of the archaea proteasome from *Archaeoglobus fulgidus* is shown. The small icon in the upper left corner shows the angle of view (arrow) and the part of the proteasome the reader is looking at, seeing the side of the β-ring facing the proteolytic main chamber of the proteasome. The single identical β-subunits of the displayed archaea β-ring are color coded, and the small orange molecule bound to every single one is the calpain inhibitor i. The right image shows a single β-subunit in complex with calpain inhibitor i (orange). (See Color Insert.)

Besides Thr1, the amino residues Asp17 and Lys33 turned out to be essential for the functionality of all three active centers. At the same time, the residues Ser129/166/169, which are localized surrounding Thr1, seem to be important for stabilization of the three-dimensional structure of each active center (Fig. 6). Possible allosteric effects between the single active centers are still discussed: Some experimental outcome suggests such interactions,[184] but most of the proteasomal inhibitors bind to the $\beta_5$-subunit, resulting only in a significant reduction of the associated specificity (chymotrypsin like), but no change in the activities of the other two proteolytic subunits. In contradiction to these experimental results is the "bite and chew" model from Kisselev[185] that proposes indeed allosteric interactions between the active subunits; according to this model, substrates for the $\beta_5$-subunit like Suc-LLVY-MNA or Suc-FLF-MNA significantly increased the activity of $\beta_1$ (caspase-like). In the same way, inhibiting substrates for one activity indirectly had an effect on the other two activities. Thus, Kisselev concluded a cyclic mechanism in the degradation of a protein substrate. The chymotrypsin-like activity of $\beta_5$ initiates degradation of an amino acid chain and

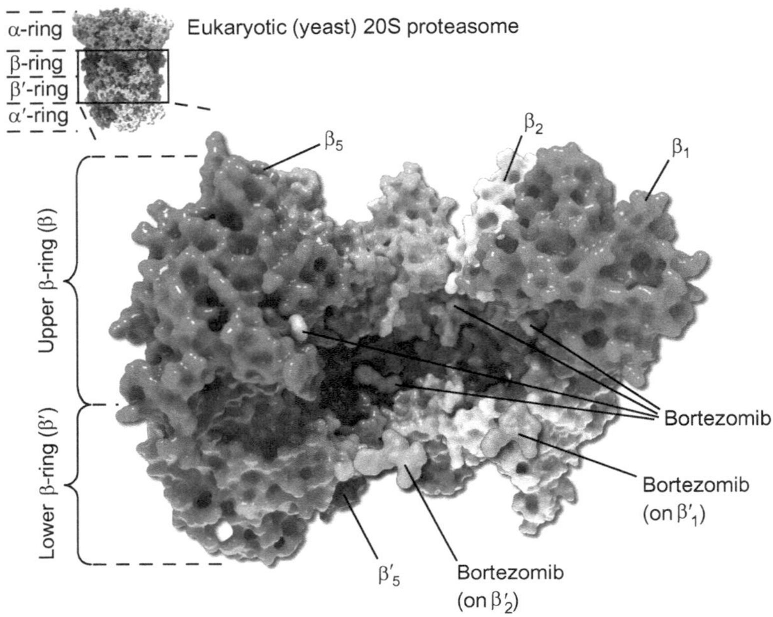


FIG. 5. Active centers of the eukaryotic 20S proteasome. (See Color Insert.)

triggers further cleavage by the $\beta_1$-subunit (showing peptidyl–glutamyl–peptide hydrolysing activity); during $\beta_1$-mediated substrate cleavage ("chewing"), $\beta_5$ activity is inhibited. If no further cleavage by $\beta_1$ is possible, then $\beta_5$ is "reactivated," starting the cycle over and over again. According to Kisselev's model, allosteric interactions between the active subunits are essential for substrate degradation.[185] However, knockout mutants of yeast revealed that active proteasomal subunits are important for cellular survival and functionality (cell division) and showed that the proteolytic capacities of the single subunits vary in their importance: $\beta_5 >> \beta_2 > \beta_1$. This results in the fact that double knockout mutants $\beta_1/\beta_5$ and $\beta_1/\beta_2$ are still viable, while $\beta_2/\beta_5$ are not.[186,187]

## C. Intracellular Assembly of the 20S Proteasome

The first step in proteasome assembly is the association of the α-ring. In order to prevent any unspecific oligomerization of the α-subunits, this process is guided by different chaperones. These chaperones are called proteasome assembly chaperones (PAC). Until now four different forms of these chaperones are known: PAC 1–4 (in humans)[188–190] and their equivalents from yeast,

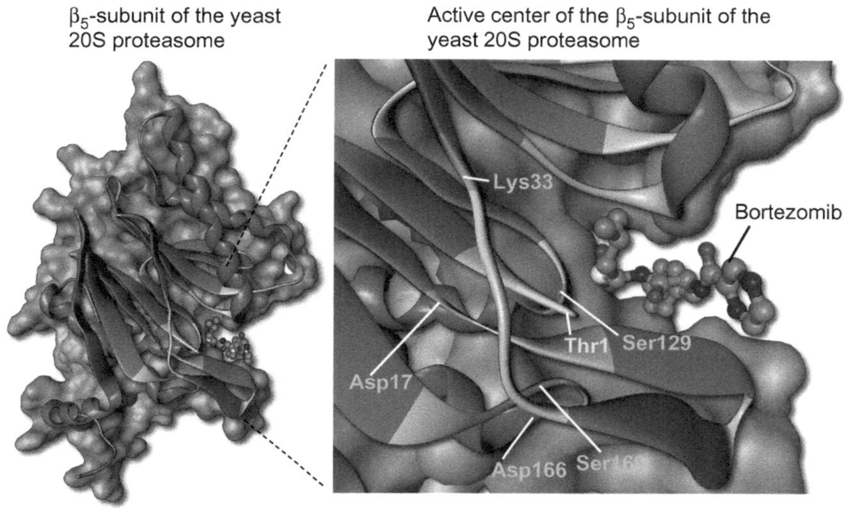


FIG. 6. Amino acids in the active center of the eukaryotic 20S proteasome. The small icon in the upper left corner shows the part of the whole 20S proteasome (from *Saccharomyces cerevisiae*) that is shown in the enlarged image: the two β-rings. Some of the subunits are removed in order to expose the main chamber with its six proteolytic centers overall. The upper β-ring still contains $\beta_1$ (red), $\beta_2$ (yellow), and $\beta_5$ (blue), and the lower ring only $\beta_5'$ (blue). Every single of those active subunits binds a bortezomib molecule (purple) to its active center. The two bortezomib molecules that are bound to the removed active subunits of the lower β-ring ($\beta_1'$ and $\beta_2'$) are shown in orange. Here, the $\beta_5$-subunit of a 20S proteasome (from *S. cerevisiae*) is shown. The left part of the image shows the whole structure, and the right part the active center and the most important amino acids that are involved. The center binds a single molecule of bortezomib. (See Color Insert.)

termed proteasome biogenesis-associated protein (Pba) 1–4.[191,192] However, such a terminology is not uniquely used and so sometimes in yeast, Pba1 is called POC1, Pba2 POC2 or PAC2, Pba3 PAC3 or Dmp2[188] or POC3, and finally Pba4 Dmp1 or POC4. These chaperones form two different heterodimers to become active: the first one is PAC1–PAC2 in humans and Pba1–Pba2 (POC1–POC2) in yeast; the second one is PAC3–PAC4 in humans and Pba3–Pba4 (Dmp1–Dmp2) in yeast.[189] PAC1–PAC2 are involved in the assembly of the α-ring. The PAC1–PAC2 heterodimer first binds to the subunits $\alpha_5$ and $\alpha_7$, followed by stepwise incorporation of $\alpha_6$ and $\alpha_1$ (both bind on the $\alpha_7$ side), then followed by $\alpha_2$ (binds to $\alpha_1$), $\alpha_3$ (binds $\alpha_2$), and $\alpha_4$ (binds both to $\alpha_3$ and $\alpha_5$), driven by mutual interactions.[193] This assembly is supplemented by the PAC3–PAC4 heterodimer, attaching to the $\alpha_2$-subunit.

After this, the PAC3–PAC4 heterodimer (in yeast Pba3–Pba4/Dmp1–Dmp2) provides incorporation of the single β-subunits one after the other. In a first step, PAC3–PAC4 binds and incorporates first $\beta_2$ and then $\beta_3$. In further steps, the subunits $\beta_4$, $\beta_5$, $\beta_6$, $\beta_1$, and $\beta_7$[145] are incorporated. At this stage, the active β-subunits ($\beta_1$, $\beta_2$, and $\beta_5$) are still in their inactive form (pro$\beta_1$, pro$\beta_2$, and pro$\beta_5$), due to a later removed prosequence. $\beta_6$ and $\beta_7$ are not proteolytic active, but nevertheless contain a prosequence. The PAC3–PAC4 heterodimer detaches from the complex after $\beta_3$ is incorporated. This ends up with the formation of a so-called half proteasome built with one α- and one β-ring. The corresponding intermediates of a complete α-ring binding different β-subunits including the so-called half-mer (a complete α-ring ring with all β-subunits incorporated except $\beta_7$) have already been identified, suggesting the sequential characteristic of that assembly.[194] Another identified intermediate is the so-called 13S (built with a complete α-ring and the subunits β2, β3, and β4).[195] Until now, the exact order of β-subunit incorporation in yeast is still unclear. In yeast, Ump1 binds after $\beta_2$, $\beta_3$, and $\beta_4$ are already recruited in the complete α-ring, while in humans it binds the α-ring together with $\beta_2$.[145]

So, two of those "half proteasomes" finally assemble at the holoproteasome, the functional 20S "core" particle. In yeast, assembly of the holoproteasome is mediated by the proteasome maturation factor Ump1[196,197] (also called "proteassemblin"[193,198] or "POMP",[199–201] for proteasome maturation protein). The human form of Ump1 is termed "hUmp1." One Ump1 binds to every single "half proteasome." After the last β-subunit ($\beta_7$) is incorporated into the "half-mer," the dimerization of two "half mers" is induced. That whole complex (also termed "16S intermediate") now contains a complete α- and β-ring, as well as one Ump1 and the PAC1–PAC2 heterodimer. In this process, the extended C-terminus of $\beta_7$ from one "half proteasome" interacts between the $\beta_1$- and $\beta_2$-subunit of the opposite β-ring from the other "half proteasome." Incorporation of $\beta_7$ seems to be a rate-limiting step in this process, since overexpression of $\beta_7$ massively increases 16S dimerization.[202] Assembly of the holoproteasome is followed by an autocatalytic "activation" of the proteolytic β-subunits: Their N-terminal ends are degraded (setting free the N-terminal Thr1 on $\beta_1$, $\beta_2$, and $\beta_3$), followed by degradation of both Ump1 molecules inside the proteolytic chamber as the first substrate of the functional 20S holoproteasome. After this, the two attached PAC1–PAC2 heterodimers are degraded, too.

Another protein involved in 16S dimerization is Hsc73: one is bound to each 16S, and detaches after assembly of both 16S. Since intracellular immunolocalization revealed a stable binding of Ump1 to the outer membrane of the ER,[199] it is supposed that the complete α-ring is bound to Ump1 and is not released before all of the β-subunits are recruited. As much as 75–88% of the proteasome maturation intermediates were colocalized with the ER membrane, while formation of the holoproteasome seems to take place in the cytosol of the cell.[203]

Interestingly, in contrast, assembly of the archaea proteasome (from *T. acidophilum*), which only contains a single type of both α- and β-subunits, is independent of any chaperone proteins: a coexpression of both subunits in

*Escherichia coli* resulted in the formation of functional archaea 20S proteasomes.[36,146] This auto-assembly is enabled by characteristic loops (L-loops) of the α-subunits that can assemble with other α-subunits. The β-subunits have no such structure and, thus, their assembly depends of a complete α-ring that functions as an assembly "platform."[204]

## D. Modeling of the 20S Proteasomal Proteolysis

In order to describe 20S proteasomal (with and without regulators) degradation, several mathematical models have been developed. For the prediction of antigens presented by the immune system, a model describing cleavage sites and fragment length of a given oligopeptide may be useful. For this purpose, two main strategies exist: one concentrates on the predictions of the fragments generated during proteolytic degradation of a substrate, though the results may become inaccurate if the produced fragments overlap[205]; the other strategy is focused on the prediction of potential cleavage sites, returning the probability of the occurrence of a specific fragment.[206] In this model, the produced fragments are normally determined by a potential cleavage site between two amino acids with respect to one or two residues neighboring that locus on each side. In contrast, other models try to simulate the process of degradation from a mechanistic point of view, considering kinetic rates in order to calculate the velocity of proteolytic degradation for a given substrate protein. The computed results have to be proven in an empiric way, while the models are based on "learning sets" that result from the fragments occurring after degradation of a known protein.

# III. Regulation of the 20S Proteasome

In order to prevent uncontrolled and unregulated proteolytic degradation in a cell, the proteasomal degradation has to be carefully regulated. Therefore, during evolution, a set of regulators has developed that are able to control both recognition and degradation of proteasomal substrates. The regulated entrance into the proteolytic inner chamber of the 20S core proteasome is realized via the gating α-rings that may be "activated" by binding to the exposed hydrophobic stretches of oxidatively damaged proteins. Usually, these hydrophobic structures are buried inside the soluble proteins, but after (oxidative) modification, these structures may be exposed, whereas native and correctly folded proteins have to be targeted (via polyubiquitination) for proteasomal degradation. The most important regulator for the recognition of ubiquitin-labeled proteins is the so-called 19S regulator complex that cooperates with another cellular machinery, the ubiquitination system.

## A. The 19S Regulator

The 19S regulator, also known as "PA700" or "proteasome activator 700 kDa"[207–210] is built up of two main structures: a ring-shaped base that binds to the α-rings of the 20S "core" proteasome and a lid that recognizes and binds polyubiquitinated proteins, thus regulating substrate entrance to the 20S proteasome.

The base ring contains at least 10 different subunits (Rpt1–Rpt6, Rpn1, Rpn2, Rpn10, and Rpn13). The lid contains nine subunits (Rpn3, Rpn5–Rpn9, Rpn11, Rpn12, and Rpn15) that are also termed "DSS1" or "SHFM1" in humans, and "SEM1" in yeast. The Rpt-subunits show an ATPase activity, while the Rpn-subunits do not. Furthermore, Rpn11 in the lid contains a $Zn^{2+}$-dependent proteolytic center that is able to catalyze the proteolytic degradation of polyubiquitin chains that label native substrate proteins for degradation; after this, the ubiquitin molecules are released for reuse of the polyubiquitination machinery.

The Rpt2 (in humans also called S4 or p56, and in yeast YTA5 or mts2), Rpt3 (human form termed as S6, Tbp7, or P48, and the yeast form as YTA2), and Rpt5 (S60 or Tbp1 in humans, and YTA1 in yeast) subunits of the base ring play a role in gate opening of the attached α-subunits of the 20S proteasome,[211] while Rpn10 (S5a or Mbp1 in humans, and SUN1, MCB1, or pus1 in yeast) and Rpn13 (ADRM1 in humans, and DAQ1 in yeast) function as polyubiquitin receptors.[211] The main role of the small protein ubiquitin is the labeling of native proteins for proteasomal degradation, in order to regulate their intracellular amount of life span.

One 19S regulator might attach to each of the α-rings of the 20S "core" proteasome, forming a large particle of about 2 MDa,[212] termed the "26S proteasome." In a mechanism that is known from other regulators of the proteasome, the 19S particle makes substrate access to the "core" particle easier by "opening" the gating α-rings. It has been shown in yeast that the Rpt2-ATPase of the base ring is involved in this process.[86] Until now, no data from X-ray crystallography of the 19S regulator are available[211] and only some of the interactions of the single subunits are known[213]; thus, in Fig. 7, only a hypothetical structure of the 19S regulator cap can be displayed.

## B. The Immunoproteasome

A special inducible form of the 20S proteasome is called the immunoproteasome (i20S). The i20S proteasome can be induced by IFN-γ, and on the site of the subunits $\beta_1$, $\beta_2$, and $\beta_5$ their inducible equivalents ($\beta_{1i}$, $\beta_{2i}$, and $\beta_{5i}$) are located.[140,214–219] To achieve this, proteasomes have to be synthetized de novo.[220,221] Further inducers are both tumor necrosis factor alpha (TNF-α)[222,223] and lipopolysaccharides.[224] In general, the three inducible β-subunits

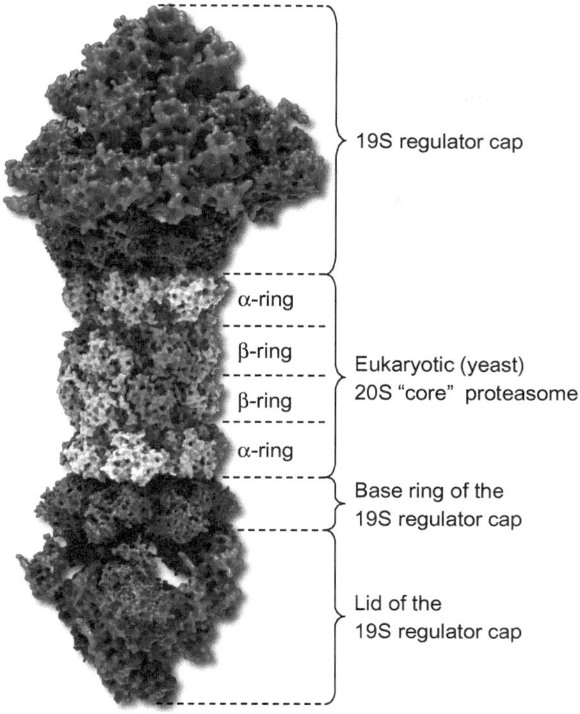


FIG. 7. A model of the structure of the eukaryotic 26S proteasome. This image shows a model of a eukaryotic 20S "core" proteasome (from *Saccharomyces cerevisiae*) bound to two 19S regulator caps. Since there are no data from X-ray structure analysis for the 19S regulator, this is just a model showing the very basic shape of 19S that is divided into a "base"-ring containing six subunits binding the α-ring of the 20S "core" proteasome and a "lid" containing nine subunits, responsible for recognition, binding, and unfolding (in an ATP-dependent way) of polyubiquitinated substrates, feeding them into the 20S proteasome for terminal degradation. Overall, the 19S regulator cap contains some 18 different subunits (see Table I), 10 for the whole "base" structure and nine for the "lid." (See Color Insert.)

are homologs of the constitutive ones and are indicated by an additional "i" (i20S); other names found in the literature for those inducible proteasomal subunits are low-molecular-weight protein 2 (LMP2, for $\beta_{1i}$), multicatalytic endopeptidase-like-complex-1 (MELC1, for $\beta_{2i}$), and LMP7 (for for $\beta_{5i}$).[225–227]

At the same time, the so-called 11S regulator (also known as PA28, PA26, or REG) is induced by the same cellular signal cascades. The major function of the immunoproteasome is the production of a specific short oligopeptide product pattern that can be presented by the major histocompatibility complex class I (MHC-I) on the cell's surface in immune response. Typical products of the immunoproteasome are short protein fragments, made up of about 8–10 amino acids that are optimized for MHC-I-presentation. Since i20S induction mainly depends on the amount of cytokines that are released in the tissue, it is suggested that immunoproteasomes mainly release new self-determinants that prevent autoimmune response in the surrounding uninfected cells.[228] On the other hand, Yewdell proposed in 2005[228] that the main functions of i20S are not the generation of MHC-I-presented antigens and that further research in this field is needed.

INF-$\gamma$ induces both the proteasome maturation factor Ump1 and the inducible forms of the proteolytic proteasomal subunits; paradoxically, the mRNA of Ump1 is increased, while the amount of free Ump1 decreases (as found in HeLa cells) and the half-life of that protein is reduced from 82 to only 21 min.[229] It turns out that this is caused by a massively increased proteasome formation that enhanced degradation of Ump1, which is the first proteasomal substrate (see above) after assembly of the holoproteasome. Considering this, the turnover of Ump1 can be used as an indicator for the formation rate of functional proteasomes. The formation of the immunoproteasome is turned into a highly dynamic process due to two main factors. The first one is the fact that the processed $\beta_{5i}$ has a higher affinity to Ump1 than the propeptide pro$\beta_{5i}$, which generates a higher rate of i20S formation compared to the one of c20S (this fact suggests two different binding sites).[229] The second factor is the much shorter half-life of i20S (about 27 h) compared to c20S (about 8–12 days).[230–232] The interplay of these two factors allows both a very quick expression and a fast removal of i20S. Seven days of continuous stimulation can completely replace c20S by i20S,[21] while after a shorter stimulus the ratio of c20S/i20S decreases.

However, after expression of the inducible subunits, the de novo assembled proteasomes do not always contain all six inducible catalytic subunits; many forms are found that only show between 1 and 5 of the inducible $\beta_i$-subunits, while the others are the constitutive ones. Whether there is a special need or function for such intermediate proteasomes is still unclear. Nevertheless, both $\beta_{5i}$ and Ump1 seem to be essential for the formation of i20S: cells that do not express $\beta_{5i}$ or Ump1 are not able to form i20S, even after IFN-$\gamma$ treatment.

In Ump1-knockdown cells, proteasomal-mediated proteolysis decreases rapidly to 60% after 24 h and to 40% after 48 h,[229] while the overall amount of cellular proteasomes is reduced significantly.[233] In the same way, an overexpression of Ump1 increases the proteasomal-mediated cellular proteolysis.[200] The number of $\beta_i$-subunits increases constantly over time in muscle tissue of aged rats compared to young ones (a three- to sixfold increase was found). The same results were detected for neurons, astrocytes, and endothelial cells in the hippocampus of elderly humans (about 70 years of age) compared to a younger control group (about 42 years of age).[234] Considering this, the inducible β-subunits seem to accumulate over time in cells and tissues that normally only contain the "housekeeping" form of the proteasome (c20S) and especially in cells prone to postmitotic aging, as neurons and muscle cells.

## C. The Thymus-Specific Proteasome (Thymoproteasome)

Another specific proteasomal subunit is the so-called $\beta_{5t}$, which was first discovered in mice, exclusively in cortical thymic epithelial cells (cTECs).[235] $\beta_{5t}$ plays an important role in the positive selection of thymocytes.[236] Accordingly, the term "thymoproteasome" was suggested for the "$\beta_{1i}$, $\beta_{2i}$, and $\beta_{5t}$-configuration" of the 20S proteasome. It turns out that the thymoproteasome is responsible for the generation of short antigenic oligopeptides presented on the cell surface, resulting in the positive selection of $CD8^+$ T-cells. The self-peptide production is dependent on the thymoproteasome and is essential for the development of an immune-competent repertoire of $CD8^+$ T-cells.[237]

Genomic analyses of the gene coding $\beta_{5t}$ (PSMB11) was performed by Sutoh *et al.*.[238] It turned out that teleost fish have two functional copies of PSMB11 (PSMB11a and PSMB11b), while chickens, turkeys, and zebra fish lost PSMB11, expressing neither thymo- nor immunoproteasomes. In mammals, reptiles, amphibians, and teleost fishes, PSMB11 is located close to PSMB5, which codes $\beta_5$ of the constitutive 20S proteasome. These results suggest that PSMB11 may originate from the older PSMB5 by tandem duplication. $\beta_{5t}$ shows a close relation to both $\beta_5$ and $\beta_{5i}$ and was found to be incorporated in about 20% of the thymic proteasomes.

In proteasomes containing $\beta_{5t}$, the inducible subunits $\beta_{1i}$ and $\beta_{2i}$ are preferentially incorporated compared to the constitutive ones $\beta_1$ and $\beta_2$. Antigens that are presented by the MHC-I[239–241] complexes show hydrophobic C-termini that function as an anchor in MHC-I binding[242] and that result from the characteristics of $\beta_{5t}$-mediated cleavage. In contrast to $\beta_5$ and $\beta_{5i}$, the proteolytic center of $\beta_{5t}$ contains hydrophobic amino residues that reduce the chymotrypsin-like proteasomal activity by 60–70%, without any effect on the other two activities.[235] The maximal velocity of proteolysis, as well as the Michaelis constant, is lower in $\beta_{5t}$ compared to both $\beta_5$ and $\beta_{5i}$. The result is a significantly decreased amount of oligopeptides released with a hydrophobic C-terminus that are preferably incorporated in the binding grove of MHC-I. Thus, $\beta_{5t}$ seems to reduce the amount of MHC-I-presentable antigens.

The result is both a lowered production and presentation of MHC-I-bound oligopeptides and thus a decreased interaction of cTECs with the αβ-T-cell antigen receptor, which causes a higher probability of positive selection of those cells.[235,236]

$\beta_{5i}$-deficient mouse models revealed an imperfect development of $CD8^+$ T-cells and a resulting decrease of those cells by about 80%,[243] suggesting that $\beta_{5i}$ may enhance the selection of $CD8^+$ T-cells. However, the amount of antigen-loaded MHC-I molecules presented on the surface of $\beta_{5t}$-deficient cells did not change. The lysosomal cathepsin S is a necessary factor in antigen presentation in most cells, but in contrast, this task is performed by cathepsin L in thymus cortical epithelial cells. After deletion of cathepsin L in those cells, the selection of $CD4^+$ was reduced without any influence on the amount of MHC-II. The presentation of different antigenic oligopeptides by MHC-I (in $CD8^+$ cells mediated by the proteasome) and MHC-II (in $CD4^+$ cells mediated by the lysosomal system/cathepsin L) decides the positive or negative selection of mature T-cells. T-cells showing a high affinity to self-antigens are sorted out, otherwise causing autoimmune reactions. In contrast, T-cells with a low affinity to MHC molecules will maturate, while mediocre affinity usually triggers positive selection.

## D. The 11S Regulator

The 11S regulator of the proteasome, also called "PA28," "REG," or "PA26" (in *T. brucei*), has different structures. Three different subunits of the 11S activator are known: PA28α, PA28β, and PA28γ. There are hexameric or heptameric structures described and, in addition to that, under defined conditions, various homo- or heteropolymerization products of the individual subunits are formed. Results indicated first that the 11S regulator has an $\alpha_3\beta_3$ structure, where both subunits were arranged alternatively.[244,245] However, later, an $\alpha_3\beta_4$ complex was detected. This particle interestingly contains a β–β dimer, but no α-subunit dimer.[246] There seem to exist several 11S forms in cells, as PA28$\alpha_3\beta_3$, PA28$\alpha_4\beta_3$, PA28$\alpha_3\beta_4$ (in each case with alternating arrangement of the α- and β-subunits), and PA28$\gamma_7$.[247] If PA28α and PA28β subunits are mixed in vitro in a ratio of α to β of 1.2, both PA$\alpha_3\beta_4$ and PA$\alpha_4\beta_3$ can be detected.[246] The PA28α-heptamer is instable, but can be formed in vitro, whereas the PA28β-heptamer cannot. However, PA28γ forms a stable heptamer.[248,249]

The base diameter of the 11S regulator is some 90 Å and the complex is about 60 Å in height; with a central cavity 20–30 Å wide. In general, the various 11S regulators are able to bind to the outer α-rings of the proteasome and change substrate degradation properties. However, like the degradation by the 20S proteasome, the degradation of substrates by the 11S–20S complex is ATP-independent (Fig. 8), suggesting that only unfolded proteins are substrates.[247] 11S binding increases the $\beta_2$-catalyzed cleavage about 10-fold and the $\beta_1$- and $\beta_5$-catalyzed cleavages by about 50-fold.[250,251] The PA28γ-isoform activates only $\beta_2$.[248] Binding of PA28 changes the conformation of the proteasome, thereby making it more efficient in proteolytic activity.[252]

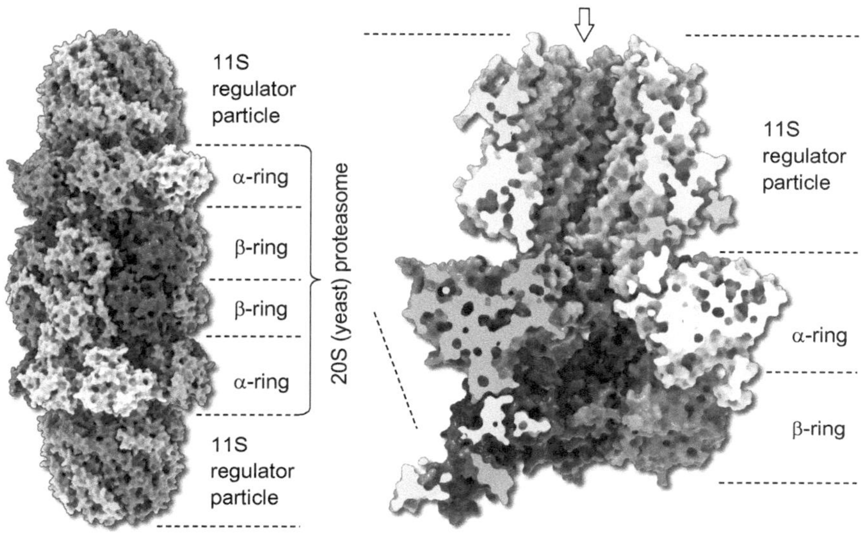


FIG. 8. The immunoproteasome and the 11S regulator. Here, the structure of the so-called immunoproteasome, a eukaryotic 20S "core" proteasome capped with two 11S regulators, is shown on the left side of the image. The right part shows a cross-section of the 11S regulator cap and an α- and β-ring of the 20S proteasome (a half proteasome). Please note that the gate of the "core" proteasome is "opened" and a channel through the regulator cap (arrow above the structure) enables substrate access to the main chamber of 20S and the proteolytic centers within. (See Color Insert.)

Interestingly, PA28α and PA28β are located only in the cytosol, but all three PA28 isoforms can be found in the nucleus.[253,254] The PA28γ isoform activates only $\beta_2$.[248] The PA28α,β regulators seem to be involved in the generation of oligopeptides in the immune response (see also immunoproteasome). So, PA28β knockout animals have a reduced immune function.[255] Interestingly, PA28γ-knockout animals show malfunctions in cell cycle regulation and apoptosis,[255,256] due to the role of PA28$\gamma_7$ in the degradation of nuclear lysine-free proteins.[257,258]

## E. The Hybrid Proteasome (PA28–20S–PA700)

The hybrid proteasome, thus termed by Tanahashi *et al.*,[259] contains both a PA700 (19S) and a PA28 (11S) regulator cap. Each of the regulators is bound to either end of the 20S proteasome. Tanahashi *et al.* determined the relative amounts of the different possible hybrid proteasomes (see Fig. 9). The 11S regulator is present in its hexameric form $(\alpha\beta)_3$ and the two different heptameric versions $(\alpha\beta)_3\alpha$ and $(\alpha\beta)_3\beta$ in hybrid proteasomes, whereas the

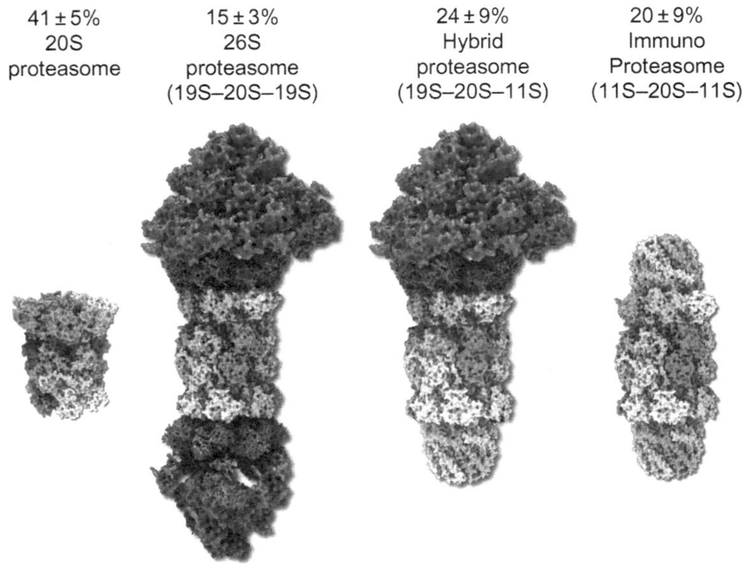


FIG. 9. The relative amount of proteasome types as found in the cytosol of HeLa cells. Here, the relative amount of proteasomal types are shown, starting with the uncapped 20S (on the left, "20S proteasome") and the amounts of the different fractions like 26S proteasome (19S–20S–19S, second from left), hybrid proteasome (19S–20S–11S, second from right), and the immunoproteasome (11S–20S–11S, on the right). (See Color Insert.)

heptameric PA28γ ($\gamma_7$) is not. Via immunoprecipitation it is possible to isolate the hybrid proteasomal forms that can be induced in cells in an IFN-γ-mediated way.[260] ATP is needed for attachment of PA28 to the 20S core proteasome, as well as in the ATP-dependent protein degradation of the PA28–20S–PA700 complex.[259] Considering the ATP-dependence of the regulator "core" particle attachment, the formation of PA28–20S–PA700 is very similar to that of PA700–20S–PA700 (the 26S proteasome).

Though the exact cellular function of those hybrid proteasomes is still unknown, it might be possible that the proteolytic specificities of the core proteasome bound to 19S change by binding of an additional PA28 regulator,[260] causing a different set of oligopeptide products produced during proteolytic degradation of a substrate. The proteolytic activity of 26S was found to be higher than that shown by the hybrid forms. However, a cooperation of immunoproteasome and hybrid forms of the proteasome in antigen processing might be possible, since both proteasomal forms can be induced via IFN-γ. So, it was suggested that most of the proteolysis required for MHC-I antigen presentation is performed via the 26S proteasome or the so-called hybrid proteasomes (PA28–20S–19S). Since the PA28 activator cap is not able to mediate the degradation of natively folded proteins, it appears that the substrate protein has to be unfolded first in an ATP-dependent way that is mediated by the 19S-regulator cap.

Whether the short oligopeptide products that are released after 19S–20S-mediated proteolytic degradation are further processed by another proteasome, perhaps containing a PA28 regulator protein, is still unclear. According to the "molecular coupling hypothesis", a hybrid proteasome is attached to a TAP1–TAP2 complex (TAP, transporter associated with antigen processing) protein channel in the ER membrane via the PA28 regulator cap. So, a polyubiquitinated and natively folded antigenic protein is recognized and unfolded by the 19S regulator cap and then guided into the core proteasome (in this special case, an immunoproteasome), where it is degraded. Fragment lengths as well as the fragment characteristics are influenced by the PA28 proteasomal regulator cap that delivers antigenic oligopeptide fragments directly into the TAP1–TAP2 protein complex. The advantage of such a direct transport is the protection of the fragments from cytosolic proteases. Further processing of the antigenic fragments (like N-terminal trimming) is done via ER-resident proteases as endoplasmic reticulum aminopeptidase associated with antigen processing and endoplasmic reticulum aminopeptidase 1 and 2. This proposed mechanism was inspired by the identification of the highly conserved "KEKE"-motifs at the distal side of PA28 that are not involved in binding or activation of the 20S "core" proteasome. KEKE motifs may be involved in protein–protein interactions and have been found in four subunits of the 20S proteasome and five subunits of the 19S proteasomal regulator cap. Furthermore, they have been found in both HSP90 and calnexin, two other proteins that play a role in epitope loading of MHC-I proteins. Another hypothesis gaining from those results suggests that heat shock proteins may be involved in immune response.[261,262] However, this idea has been disproven in 2006 using the SIINFEKL epitope of ovalbumin: experiments revealed that there is no promotion of its MHC-I presentation.[263]

## F. The PA200 Regulator Protein

The PA200 proteasomal activator cap is exclusively found in the nucleus of mammalian cells. The yeast homolog (from *Saccharomyces cerevisiae*), known as Blm10, shows a sequence homology of about 20%[264] to the mammalian form. First investigations of the PA200 structure and its binding to the 20S core proteasome were done using electron microscopy. Three-dimensional reconstructions of the gained data showed a slightly asymmetric dome structure (100 Å in diameter, as the 20S proteasomal α-ring, and about 60 Å in high) with an inner cavity that sits on one or both the α-rings like a cap[265] (Fig. 10). Differently from the other proteasomal regulators, PA200 is a monomeric

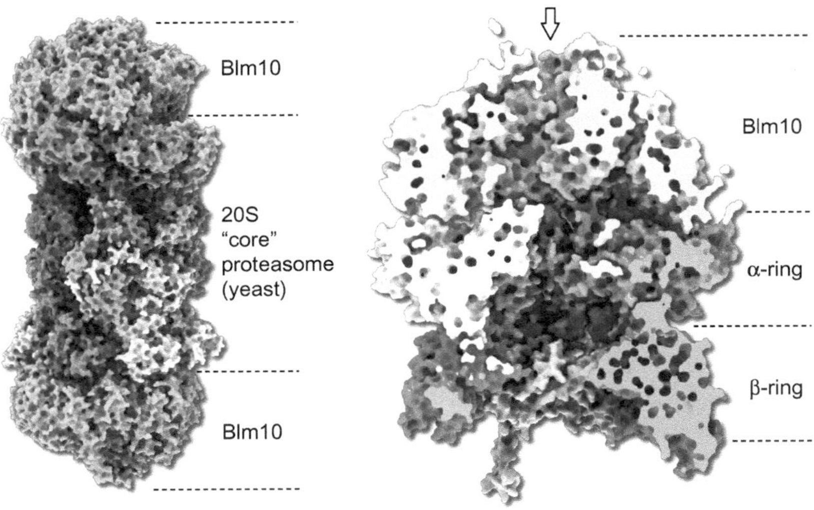


FIG. 10. The Blm10 regulator cap. The left part of the image shows a 20S "core" proteasome (from *Saccharomyces cerevisiae*) with two Blm10 regulator caps (the light blue dome-shaped structure on both ends of 20S) attached. The right part of the image shows a more detailed cross-section of a single Blm10 cap and one α- and β-ring of the corresponding 20S proteasome. The cross-section shows both an activated α-ring opening a channel to the inside of the proteasome and a channel in the Blm10 regulator cap (arrow on the right image) that enables substrate access, significantly increasing substrate turnover. (See Color Insert.)

structure of about 200 kDa. PA200 binding to the 20S core proteasome is mediated by a structure that contains several HEAT-repeats and that contacts almost every α-subunit, except $\alpha_7$. The yeast form (Blm10) binds to every single α-subunit.[264] Normally, the entrance to the inner proteolytic chamber is regulated by the N-termini of the α-subunits. This suggests that the PA200 regulator changes its structure in a way that enables enhanced substrate access to the "core" particle. This rearrangement turns out to be the main mechanism of activation by reconstructions from electron microscopic data.[265]

In the same way as the 11S regulator, PA200 increases the degradation of small protein fragments and releases fluorescent degradation products, but is not able to process a natively folded protein. A ratio of 20S/PA200–20S/PA200–20S–PA200 of 50:40:10 was measured using electron microscopic investigation of isolated proteasomes and PA200–proteasome complexes from bovine testes.[265] About the function of PA200, the current knowledge is very limited. Interestingly, experiments of gene deletion or overexpression did not induce significant change in the phenotype.[264] Only a role of PA200 in DNA repair after exposure to ionizing radiation or oxidizing agents seems to be ensured.[132]

So, in response to ionizing radiation, PA200 is expressed and accumulates in its hybrid form on chromatin.[133] After PA200 knockdown, cells showed genomic instabilities and reduced survival. The genome-stabilizing functions of PA200 seem to be a result of its ability to enhance the proteasomal $\beta_1$-mediated peptidyl—glutamyl-like proteolysis.[266] However, deletion of Blm10 in the yeast A364a-strain did not result in any effect on growth or viability after treatment of the cells with DNA-damaging agents like bleomycin or phleomycin.[264] Furthermore, no increased susceptibility of A364a to UV- or $\gamma$-irradiation, methyl methane sulfonate, camptothecin, or hydroxyurea could be detected. However, overexpression of Blm10 resulted in a reduced growth, but this can be an effect of increased binding to the 20S proteasome, thus detracting activity from other cellular functions of that protease. The presence of PA200 in the nucleus enables the formation of another proteasome complex: in yeast, Blm10 is able to form PA200–20S–19S.[132,267] After HeLa treatment with ionizing irradiation, immunoprecipitation revealed a coprecipitation of PA200 in complex with 20S–19S, even if the amounts of 20S and 19S were not increased. Thus, irradiation seems to induce the formation of the PA200–20S–19S complex in a DNA-damage response-mediated way.[266] Twenty-four hours after irradiation, the PA200–20S–19S complex showed an accumulation on chromatin. The trypsin-like (mediated by the $\beta_2$-subunit of the proteasome) activity associated with the chromatin showed a sixfold increase, and the peptidyl–glutamyl-like one (mediated by the $\beta_1$-subunit) up to a 19-fold increase, accompanied by a five- to eightfold increased amount of 20S on the chromatin.[268] That accumulation seems to be independent of ATM (a PI3-like kinase),[269,270] which starts the signal cascade after irradiation-mediated stress via triggering of the tumor suppressor p53.[271,272] One important function of PA200 might be an enhancement of the $\beta_1$ activity of the proteasome that is essential for the cellular survival after exposure to ionizing irradiation.[266]

## IV. Conclusion

As shown in this chapter, the proteasomal system is complex and far from being well understood. We referred in this chapter only to the principal structures of the proteasome, not describing the interaction with the ubiquitination machinery, or other proteasomal regulators; it becomes clear that one can hardly imagine any aspect of cellular life not related to the function of the ubiquitin–proteasome system. The importance of this system is further underlined by the fact that it appeared early in evolution, as in Archae bacteria, and evolved to a more complex structure with new regulators and specialized subunits. The evolution toward function and organ-specific isoforms underlines this chapter.

In the further chapters of this book, the functions of the proteasome will be described in more detail, and the role of the proteasome in a selection of diseases and cellular stress situations will be highlighted, always on the basis of the principal structural features of the proteasome described above.

## References

1. Dick LR, Moomaw CR, DeMartino GN, Slaughter CA. Degradation of oxidized insulin B chain by the multiproteinase complex macropain (proteasome). *Biochemistry* 1991;**30**:2725–34.
2. Sitte N, Merker K, Grune T. Proteasome-dependent degradation of oxidized proteins in MRC-5 fibroblasts. *FEBS Lett* 1998;**440**:399–402.
3. Grune T, Reinheckel T, Davies KJ. Degradation of oxidized proteins in K562 human hematopoietic cells by proteasome. *J Biol Chem* 1996;**271**:15504–9.
4. Friguet B, Bulteau AL, Chondrogianni N, Conconi M, Petropoulos I. Protein degradation by the proteasome and its implications in aging. *Ann N Y Acad Sci* 2000;**908**:143–54.
5. Takeuchi J, Toh-e A. Regulation of cell cycle by proteasome in yeast. *Tanpakushitsu Kakusan Koso* 1997;**42**:2247–54.
6. Davies KJ. Degradation of oxidized proteins by the 20S proteasome. *Biochimie* 2001;**83**: 301–10.
7. Schmitz A, Herzog V. Endoplasmic reticulum-associated degradation: exceptions to the rule. *Eur J Cell Biol* 2004;**83**:501–9.
8. Lord JM, Davey J, Frigerio L, Roberts LM. Endoplasmic reticulum-associated protein degradation. *Semin Cell Dev Biol* 2000;**11**:159–64.
9. Plemper RK, Wolf DH. Retrograde protein translocation: ERADication of secretory proteins in health and disease. *Trends Biochem Sci* 1999;**24**:266–70.
10. Plemper RK, Wolf DH. Endoplasmic reticulum degradation. Reverse protein transport and its end in the proteasome. *Mol Biol Rep* 1999;**26**:125–30.
11. Hampton RY. ER-associated degradation in protein quality control and cellular regulation. *Curr Opin Cell Biol* 2002;**14**:476–82.
12. Brodsky JL, McCracken AA. ER protein quality control and proteasome-mediated protein degradation. *Semin Cell Dev Biol* 1999;**10**:507–13.
13. Zimmermann J, Lamerant N, Grossenbacher R, Furst P. Proteasome- and p38-dependent regulation of ERK3 expression. *J Biol Chem* 2001;**276**:10759–66.
14. Wu Y, Luo H, Kanaan N, Wu J. The proteasome controls the expression of a proliferation-associated nuclear antigen Ki-67. *J Cell Biochem* 2000;**76**:596–604.
15. Blagosklonny MV, Wu GS, Omura S, El-Deiry WS. Proteasome-dependent regulation of p21WAF1/CIP1 expression. *Biochem Biophys Res Commun* 1996;**227**:564–9.
16. Dembla-Rajpal N, Seipelt R, Wang Q, Rymond BC. Proteasome inhibition alters the transcription of multiple yeast genes. *Biochim Biophys Acta* 2004;**1680**:34–45.
17. Stitzel ML, Durso R, Reese JC. The proteasome regulates the UV-induced activation of the AP-1-like transcription factor Gcn4. *Genes Dev* 2001;**15**:128–33.
18. Preckel T, Fung-Leung WP, Cai Z, Vitiello A, Salter-Cid L, Winqvist O, et al. Impaired immunoproteasome assembly and immune responses in PA28-/- mice. *Science* 1999;**286**: 2162–5.
19. Kloetzel PM, Soza A, Stohwasser R. The role of the proteasome system and the proteasome activator PA28 complex in the cellular immune response. *Biol Chem* 1999;**380**:293–7.

20. Schwarz K, Eggers M, Soza A, Koszinowski UH, Kloetzel PM, Groettrup M. The proteasome regulator PA28alpha/beta can enhance antigen presentation without affecting 20S proteasome subunit composition. *Eur J Immunol* 2000;**30**:3672–9.
21. Khan S, van den Broek M, Schwarz K, de Giuli R, Diener PA, Groettrup M. Immunoproteasomes largely replace constitutive proteasomes during an antiviral and antibacterial immune response in the liver. *J Immunol* 2001;**167**:6859–68.
22. Osterloh P, Linkemann K, Tenzer S, Rammensee HG, Radsak MP, Busch DH, et al. Proteasomes shape the repertoire of T cells participating in antigen-specific immune responses. *Proc Natl Acad Sci USA* 2006;**103**:5042–7.
23. Borissenko L, Groll M. Diversity of proteasomal missions: fine tuning of the immune response. *Biol Chem* 2007;**388**:947–55.
24. Mathew A, Mathur SK, Morimoto RI. Heat shock response and protein degradation: regulation of HSF2 by the ubiquitin-proteasome pathway. *Mol Cell Biol* 1998;**18**:5091–8.
25. Sulahian R, Sikder D, Johnston SA, Kodadek T. The proteasomal ATPase complex is required for stress-induced transcription in yeast. *Nucleic Acids Res* 2006;**34**:1351–7.
26. Kahn NW, Rea SL, Moyle S, Kell A, Johnson TE. Proteasomal dysfunction activates the transcription factor SKN-1 and produces a selective oxidative-stress response in Caenorhabditis elegans. *Biochem J* 2008;**409**:205–13.
27. Hahn JS, Neef DW, Thiele DJ. A stress regulatory network for co-ordinated activation of proteasome expression mediated by yeast heat shock transcription factor. *Mol Microbiol* 2006;**60**:240–51.
28. Bregegere F, Milner Y, Friguet B. The ubiquitin-proteasome system at the crossroads of stress-response and ageing pathways: a handle for skin care? *Ageing Res Rev* 2006;**5**:60–90.
29. Clawson GA. Protease inhibitors and carcinogenesis: a review. *Cancer Invest* 1996;**14**: 597–608.
30. Scheffner M, Whitaker NJ. Human papillomavirus-induced carcinogenesis and the ubiquitin-proteasome system. *Semin Cancer Biol* 2003;**13**:59–67.
31. Moriishi K, Mochizuki R, Moriya K, Miyamoto H, Mori Y, Abe T, et al. Critical role of PA28gamma in hepatitis C virus-associated steatogenesis and hepatocarcinogenesis. *Proc Natl Acad Sci USA* 2007;**104**:1661–6.
32. Walters KJ, Lech PJ, Goh AM, Wang Q, Howley PM. DNA-repair protein hHR23a alters its protein structure upon binding proteasomal subunit S5a. *Proc Natl Acad Sci USA* 2003;**100**:12694–9.
33. Bergink S, Severijnen LA, Wijgers N, Sugasawa K, Yousaf H, Kros JM, et al. The DNA repair-ubiquitin-associated HR23 proteins are constituents of neuronal inclusions in specific neurodegenerative disorders without hampering DNA repair. *Neurobiol Dis* 2006;**23**:708–16.
34. Krogan NJ, Lam MH, Fillingham J, Keogh MC, Gebbia M, Li J, et al. Proteasome involvement in the repair of DNA double-strand breaks. *Mol Cell* 2004;**16**:1027–34.
35. Hough R, Pratt G, Rechsteiner M. Purification of two high molecular weight proteases from rabbit reticulocyte lysate. *J Biol Chem* 1987;**262**:8303–13.
36. Lowe J, Stock D, Jap B, Zwickl P, Baumeister W, Huber R. Crystal structure of the 20S proteasome from the archaeon T. acidophilum at 3.4 A resolution. *Science* 1995;**268**:533–9.
37. Stock D, Ditzel L, Baumeister W, Huber R, Lowe J. Catalytic mechanism of the 20S proteasome of Thermoplasma acidophilum revealed by X-ray crystallography. *Cold Spring Harb Symp Quant Biol* 1995;**60**:525–32.
38. Groll M, Ditzel L, Lowe J, Stock D, Bochtler M, Bartunik HD, et al. Structure of 20S proteasome from yeast at 24 A resolution. *Nature* 1997;**386**:463–71.
39. Tomisugi Y, Unno M, Mizushima T, Morimoto Y, Tanahashi N, Tanaka K, et al. New crystal forms and low resolution structure analysis of 20S proteasomes from bovine liver. *J Biochem* 2000;**127**:941–3.

40. Unno M, Mizushima T, Morimoto Y, Tomisugi Y, Tanaka K, Yasuoka N, et al. The structure of the mammalian 20S proteasome at 2.75 A resolution. *Structure* 2002;**10**:609–18.
41. Unno M, Mizushima T, Morimoto Y, Tomisugi Y, Tanaka K, Yasuoka N, et al. Structure determination of the constitutive 20S proteasome from bovine liver at 2.75 A resolution. *J Biochem* 2002;**131**:171–3.
42. Petit F, Jarrousse AS, Dahlmann B, Sobek A, Hendil KB, Buri J, et al. Involvement of proteasomal subunits zeta and iota in RNA degradation. *Biochem J* 1997;**326**(Pt 1):93–8.
43. Tokunaga F, Aruga R, Iwanaga S, Tanaka K, Ichihara A, Takao T, et al. The NH2-terminal residues of rat liver proteasome (multicatalytic proteinase complex) subunits, C2, C3 and C8, are N alpha-acetylated. *FEBS Lett* 1990;**263**:373–5.
44. Du J, Mitch WE, Wang X, Price SR. Glucocorticoids induce proteasome C3 subunit expression in L6 muscle cells by opposing the suppression of its transcription by NF-kappa B. *J Biol Chem* 2000;**275**:19661–6.
45. Feist E, Dorner T, Kuckelkorn U, Schmidtke G, Micheel B, Hiepe F, et al. Proteasome alpha-type subunit C9 is a primary target of autoantibodies in sera of patients with myositis and systemic lupus erythematosus. *J Exp Med* 1996;**184**:1313–8.
46. Castano JG, Mahillo E, Arizti P, Arribas J. Phosphorylation of C8 and C9 subunits of the multicatalytic proteinase by casein kinase II and identification of the C8 phosphorylation sites by direct mutagenesis. *Biochemistry* 1996;**35**:3782–9.
47. Dong J, Chen W, Welford A, Wandinger-Ness A. The proteasome alpha-subunit XAPC7 interacts specifically with Rab7 and late endosomes. *J Biol Chem* 2004;**279**:21334–42.
48. Mukherjee S, Dong J, Heincelman C, Lenhart M, Welford A, Wandinger-Ness A. Functional analyses and interaction of the XAPC7 proteasome subunit with Rab7. *Methods Enzymol* 2005;**403**:650–63.
49. Mayau V, Baron B, Buttin G, Debatisse M. Twelve genes, including the unassigned proteasome zeta subunit gene, ordered within the human 1p13 region. *Mamm Genome* 1998;**9**:331–3.
50. Van GG, De JC, Pype S, Van CW, Julliams A, Vanderhoeven I, et al. Alzheimer's disease associated presenilin 1 interacts with HC5 and ZETA, subunits of the catalytic 20S proteasome. *Neurobiol Dis* 1999;**6**:376–91.
51. Jorgensen L, Hendil KB. Proteasome subunit zeta, a putative ribonuclease, is also found as a free monomer. *Mol Biol Rep* 1999;**26**:119–23.
52. Kania MA, DeMartino GN, Baumeister W, Goldberg AL. The proteasome subunit, C2, contains an important site for binding of the PA28 (11S) activator. *Eur J Biochem* 1996;**236**:510–6.
53. Bose S, Stratford FL, Broadfoot KI, Mason GG, Rivett AJ. Phosphorylation of 20S proteasome alpha subunit C8 (alpha7) stabilizes the 26S proteasome and plays a role in the regulation of proteasome complexes by gamma-interferon. *Biochem J* 2004;**378**:177–84.
54. Gerards WL, Enzlin J, Haner M, Hendriks IL, Aebi U, Bloemendal H, et al. The human alpha-type proteasomal subunit HsC8 forms a double ringlike structure, but does not assemble into proteasome-like particles with the beta-type subunits HsDelta or HsBPROS26. *J Biol Chem* 1997;**272**:10080–6.
55. Gerards WL, de Jong WW, Bloemendal H, Boelens W. The human proteasomal subunit HsC8 induces ring formation of other alpha-type subunits. *J Mol Biol* 1998;**275**:113–21.
56. Shu F, Guo S, Dang Y, Qi M, Zhou G, Guo Z, et al. Human aurora-B binds to a proteasome alpha-subunit HC8 and undergoes degradation in a proteasome-dependent manner. *Mol Cell Biochem* 2003;**254**:157–62.
57. Chung PA, Johnson J, Khramtsov NV, Upton SJ. Cloning and molecular characterization of a gene encoding a Cryptosporidium parvum putative 20S proteasome beta1-type subunit. *DNA Seq* 2000;**11**:309–14.

58. Lequeu J, Simon-Plas F, Fromentin J, Etienne P, Petitot AS, Blein JP, et al. Proteasome comprising a beta1 inducible subunit acts as a negative regulator of NADPH oxidase during elicitation of plant defense reactions. *FEBS Lett* 2005;**579**:4879–86.
59. Madding LS, Michel JK, Shockley KR, Conners SB, Epting KL, Johnson MR, et al. Role of the beta1 subunit in the function and stability of the 20S proteasome in the hyperthermophilic archaeon Pyrococcus furiosus. *J Bacteriol* 2007;**189**:583–90.
60. Dissemond J, Goette P, Moers J, Lindeke A, Goos M, Ferrone S, et al. Immunoproteasome subunits LMP2 and LMP7 downregulation in primary malignant melanoma lesions: association with lack of spontaneous regression. *Melanoma Res* 2003;**13**:371–7.
61. Wang HX, Wang HM, Lin HY, Yang Q, Zhang H, Tsang BK, et al. Proteasome subunit LMP2 is required for matrix metalloproteinase-2 and -9 expression and activities in human invasive extravillous trophoblast cell line. *J Cell Physiol* 2006;**206**:616–23.
62. Mishto M, Bellavista E, Santoro A, Stolzing A, Ligorio C, Nacmias B, et al. Immunoproteasome and LMP2 polymorphism in aged and Alzheimer's disease brains. *Neurobiol Aging* 2006;**27**:54–66.
63. Krause S, Kuckelkorn U, Dorner T, Burmester GR, Feist E, Kloetzel PM. Immunoproteasome subunit LMP2 expression is deregulated in Sjogren's syndrome but not in other autoimmune disorders. *Ann Rheum Dis* 2006;**65**:1021–7.
64. Schweisguth F. Dominant-negative mutation in the beta2 and beta6 proteasome subunit genes affect alternative cell fate decisions in the Drosophila sense organ lineage. *Proc Natl Acad Sci USA* 1999;**96**:11382–6.
65. Ramos PC, Marques AJ, London MK, Dohmen RJ. Role of C-terminal extensions of subunits beta2 and beta7 in assembly and activity of eukaryotic proteasomes. *J Biol Chem* 2004;**279**: 14323–30.
66. Baldisserotto A, Marastoni M, Trapella C, Gavioli R, Ferretti V, Pretto L, et al. Glutamine vinyl ester proteasome inhibitors selective for trypsin-like (beta2) subunit. *Eur J Med Chem* 2007;**42**:586–92.
67. Scheffler S, Kuckelkorn U, Egerer K, Dorner T, Reiter K, Soza A, et al. Autoimmune reactivity against the 20S-proteasome includes immunosubunits LMP2 (beta1i), MECL1 (beta2i) and LMP7 (beta5i). *Rheumatology (Oxford)* 2008;**47**:622–6.
68. Eleuteri AM, Kohanski RA, Cardozo C, Orlowski M. Bovine spleen multicatalytic proteinase complex (proteasome). Replacement of X, Y, and Z subunits by LMP7, LMP2, and MECL1 and changes in properties and specificity. *J Biol Chem* 1997;**272**:11824–31.
69. Nothwang HG, Tamura T, Tanaka K, Ichihara A. Sequence analyses and inter-species comparisons of three novel human proteasomal subunits, HsN3, HsC7-I and HsC10-II, confine potential proteolytic active-site residues. *Biochim Biophys Acta* 1994;**1219**:361–8.
70. Nishimura C, Tamura T, Akioka H, Tokunaga F, Tanaka K, Ichihara A. cDNA cloning of rat proteasome subunit RC10-II, assumed to be responsible for trypsin-like catalytic activity. *FEBS Lett* 1993;**336**:462–6.
71. Reidlinger J, Pike AM, Savory PJ, Murray RZ, Rivett AJ. Catalytic properties of 26 S and 20 S proteasomes and radiolabeling of MB1, LMP7, and C7 subunits associated with trypsin-like and chymotrypsin-like activities. *J Biol Chem* 1997;**272**:24899–905.
72. Egerer T, Martinez-Gamboa L, Dankof A, Stuhlmuller B, Dorner T, Krenn V, et al. Tissue-specific up-regulation of the proteasome subunit beta5i (LMP7) in Sjogren's syndrome. *Arthritis Rheum* 2006;**54**:1501–8.
73. Caudill CM, Jayarapu K, Elenich L, Monaco JJ, Colbert RA, Griffin TA. T cells lacking immunoproteasome subunits MECL-1 and LMP7 hyperproliferate in response to polyclonal mitogens. *J Immunol* 2006;**176**:4075–82.

74. Heink S, Fricke B, Ludwig D, Kloetzel PM, Kruger E. Tumor cell lines expressing the proteasome subunit isoform LMP7E1 exhibit immunoproteasome deficiency. *Cancer Res* 2006;**66**:649–52.
75. Trachtulec Z, Hamvas RM, Forejt J, Lehrach HR, Vincek V, Klein J. Linkage of TATA-binding protein and proteasome subunit C5 genes in mice and humans reveals synteny conserved between mammals and invertebrates. *Genomics* 1997;**44**:1–7.
76. Rodriguez-Vilarino S, Arribas J, Arizti P, Castano JG. Proteolytic processing and assembly of the C5 subunit into the proteasome complex. *J Biol Chem* 2000;**275**:6592–9.
77. Kopp F, Kristensen P, Hendil KB, Johnsen A, Sobek A, Dahlmann B. The human proteasome subunit HsN3 is located in the inner rings of the complex dimer. *J Mol Biol* 1995;**248**:264–72.
78. Thomson S, Rivett AJ. Processing of N3, a mammalian proteasome beta-type subunit. *Biochem J* 1996;**315**(Pt 3):733–8.
79. Knowlton JR, Johnston SC, Whitby FG, Realini C, Zhang Z, Rechsteiner M, et al. Structure of the proteasome activator REGalpha (PA28alpha). *Nature* 1997;**390**:639–43.
80. Zhang Z, Clawson A, Realini C, Jensen CC, Knowlton JR, Hill CP, et al. Identification of an activation region in the proteasome activator REGalpha. *Proc Natl Acad Sci USA* 1998;**95**:2807–11.
81. Zhang Z, Zhang R. Proteasome activator PA28 gamma regulates p53 by enhancing its MDM2-mediated degradation. *EMBO J* 2008;**27**:852–64.
82. Kanai K, Aramata S, Katakami S, Yasuda K, Kataoka K. Proteasome activator PA28gamma stimulates degradation of GSK3-phosphorylated insulin transcription activator MAFA. *J Mol Endocrinol* 2011;**47**:119–27.
83. Pratt G, Rechsteiner M. Proteasomes cleave at multiple sites within polyglutamine tracts: activation by PA28gamma(K188E). *J Biol Chem* 2008;**283**:12919–25.
84. Zannini L, Buscemi G, Fontanella E, Lisanti S, Delia D. REGgamma/PA28gamma proteasome activator interacts with PML and Chk2 and affects PML nuclear bodies number. *Cell Cycle* 2009;**8**:2399–407.
85. Szabo A, Pal M, Deak P, Kiss P, Ujfaludi Z, Pankotai T, et al. Molecular characterization of the Rpt1/p48B ATPase subunit of the Drosophila melanogaster 26S proteasome. *Mol Genet Genomics* 2007;**278**:17–29.
86. Kohler A, Cascio P, Leggett DS, Woo KM, Goldberg AL, Finley D. The axial channel of the proteasome core particle is gated by the Rpt2 ATPase and controls both substrate entry and product release. *Mol Cell* 2001;**7**:1143–52.
87. Ueda M, Matsui K, Ishiguro S, Kato T, Tabata S, Kobayashi M, et al. Arabidopsis RPT2a encoding the 26S proteasome subunit is required for various aspects of root meristem maintenance, and regulates gametogenesis redundantly with its homolog, RPT2b. *Plant Cell Physiol* 2011;**52**:1628–40.
88. Lee KH, Minami A, Marshall RS, Book AJ, Farmer LM, Walker JM, et al. The RPT2 Subunit of the 26S Proteasome Directs Complex Assembly, Histone Dynamics, and Gametophyte and Sporophyte Development in Arabidopsis. *Plant Cell* 2011;**23**:4298–317.
89. Kumar B, Kim YC, DeMartino GN. The C terminus of Rpt3, an ATPase subunit of PA700 (19 S) regulatory complex, is essential for 26 S proteasome assembly but not for activation. *J Biol Chem* 2010;**285**:39523–35.
90. Amoroso MR, Matassa DS, Laudiero G, Egorova AV, Polishchuk RS, Maddalena F, et al. TRAP1 and the proteasome regulatory particle TBP7/Rpt3 interact in the endoplasmic reticulum and control cellular ubiquitination of specific mitochondrial proteins. *Cell Death Differ* 2011;**19**:592–604.
91. Lim HS, Cai D, Archer CT, Kodadek T. Periodate-triggered cross-linking reveals Sug2/Rpt4 as the molecular target of a peptoid inhibitor of the 19S proteasome regulatory particle. *J Am Chem Soc* 2007;**129**:12936–7.

92. Han Y, Cao H, Jiang J, Xu Y, Du J, Wang X, et al. Rice ROOT ARCHITECTURE ASSOCIATED1 binds the proteasome subunit RPT4 and is degraded in a D-box and proteasome-dependent manner. *Plant Physiol* 2008;**148**:843–55.
93. Gallois JL, Guyon-Debast A, Lecureuil A, Vezon D, Carpentier V, Bonhomme S, et al. The Arabidopsis proteasome RPT5 subunits are essential for gametophyte development and show accession-dependent redundancy. *Plant Cell* 2009;**21**:442–59.
94. Lee SY, Mota-Peynado A, Roelofs J. Loss of Rpt5 protein interactions with the core particle and Nas2 protein causes the formation of faulty proteasomes that are inhibited by Ecm29 protein. *J Biol Chem* 2011;**286**:36641–51.
95. Zhang F, Hu Y, Huang P, Toleman CA, Paterson AJ, Kudlow JE. Proteasome function is regulated by cyclic AMP-dependent protein kinase through phosphorylation of Rpt6. *J Biol Chem* 2007;**282**:22460–71.
96. Satoh K, Sasajima H, Nyoumura KI, Yokosawa H, Sawada H. Assembly of the 26S proteasome is regulated by phosphorylation of the p45/Rpt6 ATPase subunit. *Biochemistry* 2001;**40**:314–9.
97. Chouduri AU, Tokumoto T, Dohra H, Ushimaru T, Yamada S. Functional and biochemical characterization of the 20S proteasome in a yeast temperature-sensitive mutant, rpt6-1. *BMC Biochem* 2008;**9**:20.
98. Tomko Jr. RJ, Hochstrasser M. Incorporation of the rpn12 subunit couples completion of proteasome regulatory particle lid assembly to lid-base joining. *Mol Cell* 2011;**44**:907–17.
99. Zmijewski JW, Banerjee S, Abraham E. S-glutathionylation of the Rpn2 regulatory subunit inhibits 26 S proteasomal function. *J Biol Chem* 2009;**284**:22213–21.
100. Arlt A, Minkenberg J, Kruse ML, Grohmann F, Folsch UR, Schafer H. Immediate early gene-X1 interferes with 26 S proteasome activity by attenuating expression of the 19 S proteasomal components S5a/Rpn10 and S1/Rpn2. *Biochem J* 2007;**402**:367–75.
101. Joshi KK, Chen L, Torres N, Tournier V, Madura K. A proteasome assembly defect in rpn3 mutants is associated with Rpn11 instability and increased sensitivity to stress. *J Mol Biol* 2011;**410**:383–99.
102. Wang X, Xu H, Ha SW, Ju D, Xie Y. Proteasomal degradation of Rpn4 in Saccharomyces cerevisiae is critical for cell viability under stressed conditions. *Genetics* 2010;**184**:335–42.
103. Dohmen RJ, Willers I, Marques AJ. Biting the hand that feeds: Rpn4-dependent feedback regulation of proteasome function. *Biochim Biophys Acta* 2007;**1773**:1599–604.
104. Wang L, Mao X, Ju D, Xie Y. Rpn4 is a physiological substrate of the Ubr2 ubiquitin ligase. *J Biol Chem* 2004;**279**:55218–23.
105. Yu Z, Kleifeld O, Lande-Atir A, Bsoul M, Kleiman M, Krutauz D, et al. Dual function of Rpn5 in two PCI complexes, the 26S proteasome and COP9 signalosome. *Mol Biol Cell* 2011;**22**:911–20.
106. Book AJ, Smalle J, Lee KH, Yang P, Walker JM, Casper S, et al. The RPN5 subunit of the 26s proteasome is essential for gametogenesis, sporophyte development, and complex assembly in Arabidopsis. *Plant Cell* 2009;**21**:460–78.
107. Yen HC, Espiritu C, Chang EC. Rpn5 is a conserved proteasome subunit and required for proper proteasome localization and assembly. *J Biol Chem* 2003;**278**:30669–76.
108. Isono E, Saito N, Kamata N, Saeki Y, Toh E. Functional analysis of Rpn6p, a lid component of the 26 S proteasome, using temperature-sensitive rpn6 mutants of the yeast Saccharomyces cerevisiae. *J Biol Chem* 2005;**280**:6537–47.
109. Santamaria PG, Finley D, Ballesta JP, Remacha M. Rpn6p, a proteasome subunit from Saccharomyces cerevisiae, is essential for the assembly and activity of the 26 S proteasome. *J Biol Chem* 2003;**278**:6687–95.
110. Lier S, Paululat A. The proteasome regulatory particle subunit Rpn6 is required for Drosophila development and interacts physically with signalosome subunit Alien/CSN2. *Gene* 2002;**298**:109–19.

111. Sha Z, Yen HC, Scheel H, Suo J, Hofmann K, Chang EC. Isolation of the Schizosaccharomyces pombe proteasome subunit Rpn7 and a structure-function study of the proteasome-COP9-initiation factor domain. *J Biol Chem* 2007;**282**:32414–23.
112. Lee BJ, Kwon SJ, Kim SK, Kim KJ, Park CJ, Kim YJ, et al. Functional study of hot pepper 26S proteasome subunit RPN7 induced by Tobacco mosaic virus from nuclear proteome analysis. *Biochem Biophys Res Commun* 2006;**351**:405–11.
113. Isono E, Saeki Y, Yokosawa H, Toh-e A. Rpn7 Is required for the structural integrity of the 26 S proteasome of Saccharomyces cerevisiae. *J Biol Chem* 2004;**279**:27168–76.
114. Thompson HG, Harris JW, Brody JP. Post-translationally modified S12, absent in transformed breast epithelial cells, is not associated with the 26S proteasome and is induced by proteasome inhibitor. *Int J Cancer* 2004;**111**:338–47.
115. Jin H, Li S, Villegas Jr. A. Down-regulation of the 26S proteasome subunit RPN9 inhibits viral systemic transport and alters plant vascular development. *Plant Physiol* 2006;**142**:651–61.
116. Takeuchi J, Toh-e A. Genetic dissection of the yeast 26S proteasome: cell cycle defects caused by the Deltarpn9 mutation. *Biochimie* 2001;**83**:333–40.
117. Takeuchi J, Fujimuro M, Yokosawa H, Tanaka K, Toh-e A. Rpn9 is required for efficient assembly of the yeast 26S proteasome. *Mol Cell Biol* 1999;**19**:6575–84.
118. Riedinger C, Boehringer J, Trempe JF, Lowe ED, Brown NR, Gehring K, et al. Structure of Rpn10 and its interactions with polyubiquitin chains and the proteasome subunit Rpn12. *J Biol Chem* 2010;**285**:33992–4003.
119. Zhang D, Chen T, Ziv I, Rosenzweig R, Matiuhin Y, Bronner V, et al. Together, Rpn10 and Dsk2 can serve as a polyubiquitin chain-length sensor. *Mol Cell* 2009;**36**:1018–33.
120. Walters KJ, Zhang N. Rpn10 protects the proteasome from Dsk2. *Mol Cell* 2008;**32**:459–60.
121. Guterman A, Glickman MH. Complementary roles for Rpn11 and Ubp6 in deubiquitination and proteolysis by the proteasome. *J Biol Chem* 2004;**279**:1729–38.
122. Spataro V, Simmen K, Realini CA. The essential 26S proteasome subunit Rpn11 confers multidrug resistance to mammalian cells. *Anticancer Res* 2002;**22**:3905–9.
123. Verma R, Aravind L, Oania R, McDonald WH, Yates III JR, Koonin EV, et al. Role of Rpn11 metalloprotease in deubiquitination and degradation by the 26S proteasome. *Science* 2002;**298**:611–5.
124. Smalle J, Kurepa J, Yang P, Babiychuk E, Kushnir S, Durski A, et al. Cytokinin growth responses in Arabidopsis involve the 26S proteasome subunit RPN12. *Plant Cell* 2002;**14**:17–32.
125. Takeuchi J, Toh-e A. Genetic evidence for interaction between components of the yeast 26S proteasome: combination of a mutation in RPN12 (a lid component gene) with mutations in RPT1 (an ATPase gene) causes synthetic lethality. *Mol Gen Genet* 1999;**262**:145–53.
126. Josse L, Harley ME, Pires IM, Hughes DA. Fission yeast Dss1 associates with the proteasome and is required for efficient ubiquitin-dependent proteolysis. *Biochem J* 2006;**393**:303–9.
127. Wei SJ, Williams JG, Dang H, Darden TA, Betz BL, Humble MM, et al. Identification of a specific motif of the DSS1 protein required for proteasome interaction and p53 protein degradation. *J Mol Biol* 2008;**383**:693–712.
128. Coux O, Tanaka K, Goldberg AL. Structure and functions of the 20S and 26S proteasomes. *Annu Rev Biochem* 1996;**65**:801–47.
129. Finley D, Tanaka K, Mann C, Feldmann H, Hochstrasser M, Vierstra R, et al. Unified nomenclature for subunits of the Saccharomyces cerevisiae proteasome regulatory particle. *Trends Biochem Sci* 1998;**23**:244–5.
130. Baumeister W, Walz J, Zuhl F, Seemuller E. The proteasome: paradigm of a self-compartmentalizing protease. *Cell* 1998;**92**:367–80.
131. Khor B, Bredemeyer AL, Huang CY, Turnbull IR, Evans R, Maggi Jr. LB, et al. Proteasome activator PA200 is required for normal spermatogenesis. *Mol Cell Biol* 2006;**26**:2999–3007.

132. Ustrell V, Hoffman L, Pratt G, Rechsteiner M. PA200, a nuclear proteasome activator involved in DNA repair. *EMBO J* 2002;**21**:3516–25.
133. Blickwedehl J, McEvoy S, Wong I, Kousis P, Clements J, Elliott R, et al. Proteasomes and proteasome activator 200 kDa (PA200) accumulate on chromatin in response to ionizing radiation. *Radiat Res* 2007;**167**:663–74.
134. Gaczynska M, Osmulski PA, Gao Y, Post MJ, Simons M. Proline- and arginine-rich peptides constitute a novel class of allosteric inhibitors of proteasome activity. *Biochemistry* 2003;**42**:8663–70.
135. Ma CP, Slaughter CA, DeMartino GN. Identification, purification, and characterization of a protein activator (PA28) of the 20 S proteasome (macropain). *J Biol Chem* 1992;**267**:10515–23.
136. Zaiss DM, Standera S, Holzhutter H, Kloetzel P, Sijts AJ. The proteasome inhibitor PI31 competes with PA28 for binding to 20S proteasomes. *FEBS Lett* 1999;**457**:333–8.
137. Wagner BJ, Margolis JW. Age-dependent association of isolated bovine lens multicatalytic proteinase complex (proteasome) with heat-shock protein 90, an endogenous inhibitor. *Arch Biochem Biophys* 1995;**323**:455–62.
138. Conconi M, Friguet B. Proteasome inactivation upon aging and on oxidation-effect of HSP 90. *Mol Biol Rep* 1997;**24**:45–50.
139. Imai J, Maruya M, Yashiroda H, Yahara I, Tanaka K. The molecular chaperone Hsp90 plays a role in the assembly and maintenance of the 26S proteasome. *EMBO J* 2003;**22**:3557–67.
140. Rivett AJ, Bose S, Brooks P, Broadfoot KI. Regulation of proteasome complexes by gamma-interferon and phosphorylation. *Biochimie* 2001;**83**:363–6.
141. Poppek D, Keck S, Ermak G, Jung T, Stolzing A, Ullrich O, et al. Phosphorylation inhibits turnover of the tau protein by the proteasome: influence of RCAN1 and oxidative stress. *Biochem J* 2006;**400**:511–20.
142. Ullrich O, Ciftci O, Hass R. Proteasome activation by poly-ADP-ribose-polymerase in human myelomonocytic cells after oxidative stress. *Free Radic Biol Med* 2000;**29**:995–1004.
143. Mayer-Kuckuk P, Ullrich O, Ziegler M, Grune T, Schweiger M. Functional interaction of poly (ADP-ribose) with the 20S proteasome in vitro. *Biochem Biophys Res Commun* 1999;**259**: 576–81.
144. Ullrich O, Diestel A, Bechmann I, Homberg M, Grune T, Hass R, et al. Turnover of oxidatively damaged nuclear proteins in BV-2 microglial cells is linked to their activation state by poly-ADP-ribose polymerase. *FASEB J* 2001;**15**:1460–2.
145. Hirano Y, Kaneko T, Okamoto K, Bai M, Yashiroda H, Furuyama K, et al. Dissecting beta-ring assembly pathway of the mammalian 20S proteasome. *EMBO J* 2008;**27**:2204–13.
146. Zwickl P, Lottspeich F, Baumeister W. Expression of functional Thermoplasma acidophilum proteasomes in Escherichia coli. *FEBS Lett* 1992;**312**:157–60.
147. Rohrwild M, Pfeifer G, Santarius U, Muller SA, Huang HC, Engel A, et al. The ATP-dependent HslVU protease from Escherichia coli is a four-ring structure resembling the proteasome. *Nat Struct Biol* 1997;**4**:133–9.
148. Song HK, Bochtler M, Azim MK, Hartmann C, Huber R, Ramachandran R. Isolation and characterization of the prokaryotic proteasome homolog HslVU (ClpQY) from Thermotoga maritima and the crystal structure of HslV. *Biophys Chem* 2003;**100**:437–52.
149. Li Z, Lindsay ME, Motyka SA, Englund PT, Wang CC. Identification of a bacterial-like HslVU protease in the mitochondria of Trypanosoma brucei and its role in mitochondrial DNA replication. *PLoS Pathog* 2008;**4**:e1000048.
150. de Duve C. The lysosome turns fifty. *Med Sci (Paris)* 2005;**21**:12–5.
151. Kurz T, Terman A, Gustafsson B, Brunk UT. Lysosomes and oxidative stress in aging and apoptosis. *Biochim Biophys Acta* 2008;**1780**:1291–303.
152. Turk D, Guncar G. Lysosomal cysteine proteases (cathepsins): promising drug targets. *Acta Crystallogr D Biol Crystallogr* 2003;**59**:203–13.

153. Turk V, Turk B, Turk D. Lysosomal cysteine proteases: facts and opportunities. *EMBO J* 2001;**20**:4629–33.
154. Roberts R. Lysosomal cysteine proteases: structure, function and inhibition of cathepsins. *Drug News Perspect* 2005;**18**:605–14.
155. Rock KL, Gramm C, Rothstein L, Clark K, Stein R, Dick L, et al. Inhibitors of the proteasome block the degradation of most cell proteins and the generation of peptides presented on MHC class I molecules. *Cell* 1994;**78**:761–71.
156. Lee DH, Goldberg AL. Proteasome inhibitors: valuable new tools for cell biologists. *Trends Cell Biol* 1998;**8**:397–403.
157. Peters JM. Proteasomes: protein degradation machines of the cell. *Trends Biochem Sci* 1994;**19**:377–82.
158. Scherrer K, Bey F. The prosomes (multicatalytic proteinases; proteasomes) and their relationship to the untranslated messenger ribonucleoproteins, the cytoskeleton, and cell differentiation. *Prog Nucleic Acid Res Mol Biol* 1994;**49**:1–64.
159. Harris JR. Release of a macromolecular protein component from human erythrocyte ghosts. *Biochim Biophys Acta* 1968;**150**:534–7.
160. Harris JR. The isolation and purification of a macromolecular protein component from the human erythrocyte ghost. *Biochim Biophys Acta* 1969;**188**:31–42.
161. Pacifici RE, Salo DC, Davies KJ. Macroxyproteinase (M.O.P.): a 670 kDa proteinase complex that degrades oxidatively denatured proteins in red blood cells. *Free Radic Biol Med* 1989;**7**:521–36.
162. Pacifici RE, Davies KJ. Protein degradation as an index of oxidative stress. *Methods Enzymol* 1990;**186**:485–502.
163. Salo DC, Pacifici RE, Lin SW, Giulivi C, Davies KJ. Superoxide dismutase undergoes proteolysis and fragmentation following oxidative modification and inactivation. *J Biol Chem* 1990;**265**:11919–27.
164. Malech HL, Marchesi VT. Hollow cylinder protein in the cytoplasm of human erythrocytes. *Biochim Biophys Acta* 1981;**670**:385–92.
165. White MD, Ralston GB. The 'hollow cylinder' protein of erythrocyte membranes. *Biochim Biophys Acta* 1979;**554**:469–78.
166. Rivett AJ. The multicatalytic proteinase. Multiple proteolytic activities. *J Biol Chem* 1989;**264**:12215–9.
167. Rivett AJ. The multicatalytic proteinase complex. *Revis Biol Celular* 1989;**20**:113–23.
168. Rivett AJ. The multicatalytic proteinase of mammalian cells. *Arch Biochem Biophys* 1989;**268**:1–8.
169. Orlowski M. The multicatalytic proteinase complex, a major extralysosomal proteolytic system. *Biochemistry* 1990;**29**:10289–97.
170. Rivett AJ, Sweeney ST. Properties of subunits of the multicatalytic proteinase complex revealed by the use of subunit-specific antibodies. *Biochem J* 1991;**278**(Pt 1):171–7.
171. Nothwang HG, Coux O, Bey F, Scherrer K. Prosomes and their multicatalytic proteinase activity. *Eur J Biochem* 1992;**207**:621–30.
172. Groll M, Bajorek M, Kohler A, Moroder L, Rubin DM, Huber R, et al. A gated channel into the proteasome core particle. *Nat Struct Biol* 2000;**7**:1062–7.
173. Adams GM, Crotchett B, Slaughter CA, DeMartino GN, Gogol EP. Formation of proteasome-PA700 complexes directly correlates with activation of peptidase activity. *Biochemistry* 1998;**37**:12927–32.
174. Glickman MH, Rubin DM, Fried VA, Finley D. The regulatory particle of the Saccharomyces cerevisiae proteasome. *Mol Cell Biol* 1998;**18**:3149–62.
175. Liu CW, Corboy MJ, DeMartino GN, Thomas PJ. Endoproteolytic activity of the proteasome. *Science* 2003;**299**:408–11.

176. Bajorek M, Finley D, Glickman MH. Proteasome disassembly and downregulation is correlated with viability during stationary phase. *Curr Biol* 2003;**13**:1140–4.
177. Rabl J, Smith DM, Yu Y, Chang SC, Goldberg AL, Cheng Y. Mechanism of gate opening in the 20S proteasome by the proteasomal ATPases. *Mol Cell* 2008;**30**:360–8.
178. Smith DM, Chang SC, Park S, Finley D, Cheng Y, Goldberg AL. Docking of the proteasomal ATPases' carboxyl termini in the 20S proteasome's alpha ring opens the gate for substrate entry. *Mol Cell* 2007;**27**:731–44.
179. Loidl G, Groll M, Musiol HJ, Ditzel L, Huber R, Moroder L. Bifunctional inhibitors of the trypsin-like activity of eukaryotic proteasomes. *Chem Biol* 1999;**6**:197–204.
180. Groll M, Huber R. Inhibitors of the eukaryotic 20S proteasome core particle: a structural approach. *Biochim Biophys Acta* 2004;**1695**:33–44.
181. Ofosu FA, Freedman J, Dewar L, Song Y, Fenton JW. A trypsin-like platelet protease propagates protease-activated receptor-1 cleavage and platelet activation. *Biochem J* 1998;**336**(Pt 2):283–5.
182. Klegeris A, McGeer PL. Chymotrypsin-like proteases contribute to human monocytic THP-1 cell as well as human microglial neurotoxicity. *Glia* 2005;**51**:56–64.
183. Luciani F, Kesmir C, Mishto M, Or-Guil M, de Boer RJ. A mathematical model of protein degradation by the proteasome. *Biophys J* 2005;**88**:2422–32.
184. Orlowski M, Wilk S. Catalytic activities of the 20 S proteasome, a multicatalytic proteinase complex. *Arch Biochem Biophys* 2000;**383**:1–16.
185. Kisselev AF, Akopian TN, Castillo V, Goldberg AL. Proteasome active sites allosterically regulate each other, suggesting a cyclical bite-chew mechanism for protein breakdown. *Mol Cell* 1999;**4**:395–402.
186. Heinemeyer W, Fischer M, Krimmer T, Stachon U, Wolf DH. The active sites of the eukaryotic 20 S proteasome and their involvement in subunit precursor processing. *J Biol Chem* 1997;**272**:25200–9.
187. Jager S, Groll M, Huber R, Wolf DH, Heinemeyer W. Proteasome beta-type subunits: unequal roles of propeptides in core particle maturation and a hierarchy of active site function. *J Mol Biol* 1999;**291**:997–1013.
188. Yashiroda H, Mizushima T, Okamoto K, Kameyama T, Hayashi H, Kishimoto T, et al. Crystal structure of a chaperone complex that contributes to the assembly of yeast 20S proteasomes. *Nat Struct Mol Biol* 2008;**15**:228–36.
189. Kusmierczyk AR, Hochstrasser M. Some assembly required: dedicated chaperones in eukaryotic proteasome biogenesis. *Biol Chem* 2008;**389**:1143–51.
190. Le TB, Barrault MB, Courbeyrette R, Guerois R, Marsolier-Kergoat MC, Peyroche A. 20S proteasome assembly is orchestrated by two distinct pairs of chaperones in yeast and in mammals. *Mol Cell* 2007;**27**:660–74.
191. Hirano Y, Hendil KB, Yashiroda H, Iemura S, Nagane R, Hioki Y, et al. A heterodimeric complex that promotes the assembly of mammalian 20S proteasomes. *Nature* 2005;**437**:1381–5.
192. Li X, Kusmierczyk AR, Wong P, Emili A, Hochstrasser M. beta-Subunit appendages promote 20S proteasome assembly by overcoming an Ump1-dependent checkpoint. *EMBO J* 2007;**26**:2339–49.
193. Jayarapu K, Griffin TA. Protein-protein interactions among human 20S proteasome subunits and proteassemblin. *Biochem Biophys Res Commun* 2004;**314**:523–8.
194. Xie Y. Structure, assembly and homeostatic regulation of the 26S proteasome. *J Mol Cell Biol* 2010;**2**:308–17.
195. Schmidt M, Schmidtke G, Kloetzel PM. Structure and structure formation of the 20S proteasome. *Mol Biol Rep* 1997;**24**:103–12.

196. McIntyre J, Baranowska H, Skoneczna A, Halas A, Sledziewska-Gojska E. The spectrum of spontaneous mutations caused by deficiency in proteasome maturase Ump1 in Saccharomyces cerevisiae. *Curr Genet* 2007;**52**:221–8.
197. Chen Q, Thorpe J, Dohmen JR, Li F, Keller JN. Ump1 extends yeast lifespan and enhances viability during oxidative stress: central role for the proteasome? *Free Radic Biol Med* 2006;**40**:120–6.
198. Griffin TA, Slack JP, McCluskey TS, Monaco JJ, Colbert RA. Identification of proteassemblin, a mammalian homologue of the yeast protein, Ump1p, that is required for normal proteasome assembly. *Mol Cell Biol Res Commun* 2000;**3**:212–7.
199. Fricke B, Heink S, Steffen J, Kloetzel PM, Kruger E. The proteasome maturation protein POMP facilitates major steps of 20S proteasome formation at the endoplasmic reticulum. *EMBO Rep* 2007;**8**:1170–5.
200. Chondrogianni N, Gonos ES. Overexpression of hUMP1/POMP proteasome accessory protein enhances proteasome-mediated antioxidant defence. *Exp Gerontol* 2007;**42**:899–903.
201. Hoefer MM, Boneberg EM, Grotegut S, Kusch J, Illges H. Possible tetramerisation of the proteasome maturation factor POMP/proteassemblin/hUmp1 and its subcellular localisation. *Int J Biol Macromol* 2006;**38**:259–67.
202. Marques AJ, Glanemann C, Ramos PC, Dohmen RJ. The C-terminal extension of the beta7 subunit and activator complexes stabilize nascent 20 S proteasomes and promote their maturation. *J Biol Chem* 2007;**282**:34869–76.
203. Schmidtke G, Schmidt M, Kloetzel PM. Maturation of mammalian 20 S proteasome: purification and characterization of 13 S and 16 S proteasome precursor complexes. *J Mol Biol* 1997;**268**:95–106.
204. Gallastegui N, Groll M. The 26S proteasome: assembly and function of a destructive machine. *Trends Biochem Sci* 2010;**35**:634–42.
205. Holzhutter HG, Frommel C, Kloetzel PM. A theoretical approach towards the identification of cleavage-determining amino acid motifs of the 20 S proteasome. *J Mol Biol* 1999;**286**: 1251–65.
206. Ginodi I, Vider-Shalit T, Tsaban L, Louzoun Y. Precise score for the prediction of peptides cleaved by the proteasome. *Bioinformatics* 2008;**24**:477–83.
207. DeMartino GN, Moomaw CR, Zagnitko OP, Proske RJ, Chu-Ping M, Afendis SJ, et al. PA700, an ATP-dependent activator of the 20 S proteasome, is an ATPase containing multiple members of a nucleotide-binding protein family. *J Biol Chem* 1994;**269**:20878–84.
208. Lam YA, DeMartino GN, Pickart CM, Cohen RE. Specificity of the ubiquitin isopeptidase in the PA700 regulatory complex of 26 S proteasomes. *J Biol Chem* 1997;**272**:28438–46.
209. Strickland E, Hakala K, Thomas PJ, DeMartino GN. Recognition of misfolding proteins by PA700, the regulatory subcomplex of the 26 S proteasome. *J Biol Chem* 2000;**275**:5565–72.
210. Liu CW, Millen L, Roman TB, Xiong H, Gilbert HF, Noiva R, et al. Conformational remodeling of proteasomal substrates by PA700, the 19 S regulatory complex of the 26 S proteasome. *J Biol Chem* 2002;**277**:26815–20.
211. Tanaka K. The proteasome: overview of structure and functions. *Proc Jpn Acad Ser B Phys Biol Sci* 2009;**85**:12–36.
212. Glickman MH, Ciechanover A. The ubiquitin-proteasome proteolytic pathway: destruction for the sake of construction. *Physiol Rev* 2002;**82**:373–428.
213. Chen C, Huang C, Chen S, Liang J, Lin W, Ke G, et al. Subunit-subunit interactions in the human 26S proteasome. *Proteomics* 2008;**8**:508–20.
214. Ortiz-Navarrete V, Seelig A, Gernold M, Frentzel S, Kloetzel PM, Hammerling GJ. Subunit of the '20S' proteasome (multicatalytic proteinase) encoded by the major histocompatibility complex. *Nature* 1991;**353**:662–4.

215. Aki M, Shimbara N, Takashina M, Akiyama K, Kagawa S, Tamura T, et al. Interferon-gamma induces different subunit organizations and functional diversity of proteasomes. *J Biochem* 1994;**115**:257–69.
216. Groettrup M, Khan S, Schwarz K, Schmidtke G. Interferon-gamma inducible exchanges of 20S proteasome active site subunits: why? *Biochimie* 2001;**83**:367–72.
217. Piccinini M, Mostert M, Croce S, Baldovino S, Papotti M, Rinaudo MT. Interferon-gamma-inducible subunits are incorporated in human brain 20S proteasome. *J Neuroimmunol* 2003;**135**:135–40.
218. Stratford FL, Chondrogianni N, Trougakos IP, Gonos ES, Rivett AJ. Proteasome response to interferon-gamma is altered in senescent human fibroblasts. *FEBS Lett* 2006;**580**:3989–94.
219. Stohwasser R, Giesebrecht J, Kraft R, Muller EC, Hausler KG, Kettenmann H, et al. Biochemical analysis of proteasomes from mouse microglia: induction of immunoproteasomes by interferon-gamma and lipopolysaccharide. *Glia* 2000;**29**:355–65.
220. Kloetzel PM. Antigen processing by the proteasome. *Nat Rev Mol Cell Biol* 2001;**2**:179–87.
221. Kruger E, Kuckelkorn U, Sijts A, Kloetzel PM. The components of the proteasome system and their role in MHC class I antigen processing. *Rev Physiol Biochem Pharmacol* 2003;**148**:81–104.
222. Llovera M, Garcia-Martinez C, Agell N, Lopez-Soriano FJ, Argiles JM. TNF can directly induce the expression of ubiquitin-dependent proteolytic system in rat soleus muscles. *Biochem Biophys Res Commun* 1997;**230**:238–41.
223. Lin SY, Chen WY, Lee FY, Huang CJ, Sheu WH. Activation of ubiquitin-proteasome pathway is involved in skeletal muscle wasting in a rat model with biliary cirrhosis: potential role of TNF-alpha. *Am J Physiol Endocrinol Metab* 2005;**288**:E493–501.
224. Nelson JE, Loukissa A, Altschuller-Felberg C, Monaco JJ, Fallon JT, Cardozo C. Up-regulation of the proteasome subunit LMP7 in tissues of endotoxemic rats. *J Lab Clin Med* 2000; **135**:324–31.
225. Driscoll J, Brown MG, Finley D, Monaco JJ. MHC-linked LMP gene products specifically alter peptidase activities of the proteasome. *Nature* 1993;**365**:262–4.
226. Glynne R, Powis SH, Beck S, Kelly A, Kerr LA, Trowsdale J. A proteasome-related gene between the two ABC transporter loci in the class II region of the human MHC. *Nature* 1991;**353**:357–60.
227. Cerundolo V, Kelly A, Elliott T, Trowsdale J, Townsend A. Genes encoded in the major histocompatibility complex affecting the generation of peptides for TAP transport. *Eur J Immunol* 1995;**25**:554–62.
228. Yewdell JW. Immunoproteasomes: regulating the regulator. *Proc Natl Acad Sci USA* 2005;**102**:9089–90.
229. Heink S, Ludwig D, Kloetzel PM, Kruger E. IFN-gamma-induced immune adaptation of the proteasome system is an accelerated and transient response. *Proc Natl Acad Sci USA* 2005;**102**:9241–6.
230. Tanaka K, Ichihara A. Half-life of proteasomes (multiprotease complexes) in rat liver. *Biochem Biophys Res Commun* 1989;**159**:1309–15.
231. Tanaka K, Ichihara A. Autodegradation of rat liver proteasomes (large multicatalytic proteinase complexes). *Biochem Biophys Res Commun* 1989;**158**:548–54.
232. Ichihara A, Tanaka K. Roles of proteasomes in cell growth. *Mol Biol Rep* 1995;**21**:49–52.
233. Witt E, Zantopf D, Schmidt M, Kraft R, Kloetzel PM, Kruger E. Characterisation of the newly identified human Ump1 homologue POMP and analysis of LMP7(beta 5i) incorporation into 20 S proteasomes. *J Mol Biol* 2000;**301**:1–9.
234. Mishto M, Santoro A, Bellavista E, Sessions R, Textoris-Taube K, Dal PF, et al. A structural model of 20S immunoproteasomes: effect of LMP2 codon 60 polymorphism on expression, activity, intracellular localisation and insight into the regulatory mechanisms. *Biol Chem* 2006;**387**:417–29.

235. Murata S, Sasaki K, Kishimoto T, Niwa S, Hayashi H, Takahama Y, et al. Regulation of $CD8^+$ T cell development by thymus-specific proteasomes. *Science* 2007;**316**:1349–53.
236. Murata S, Takahama Y, Tanaka K. Thymoproteasome: probable role in generating positively selecting peptides. *Curr Opin Immunol* 2008;**20**:192–6.
237. Nitta T, Murata S, Sasaki K, Fujii H, Ripen AM, Ishimaru N, et al. Thymoproteasome shapes immunocompetent repertoire of $CD8^+$ T cells. *Immunity* 2010;**32**:29–40.
238. Sutoh Y, Kondo M, Ohta Y, Ota T, Tomaru U, Flajnik MF, et al. Comparative genomic analysis of the proteasome beta5t subunit gene: implications for the origin and evolution of thymoproteasomes. *Immunogenetics* 2012;**64**:49–58.
239. Fruh K, Yang Y. Antigen presentation by MHC class I and its regulation by interferon gamma. *Curr Opin Immunol* 1999;**11**:76–81.
240. York IA, Rock KL. Antigen processing and presentation by the class I major histocompatibility complex. *Annu Rev Immunol* 1996;**14**:369–96.
241. Pamer E, Cresswell P. Mechanisms of MHC class I—restricted antigen processing. *Annu Rev Immunol* 1998;**16**:323–58.
242. Young AC, Nathenson SG, Sacchettini JC. Structural studies of class I major histocompatibility complex proteins: insights into antigen presentation. *FASEB J* 1995;**9**:26–36.
243. Bevan MJ. Immunology. The cutting edge of T cell selection. *Science* 2007;**316**:1291–2.
244. Song X, Mott JD, von KJ, Pramanik B, Slaughter CA, et al. A model for the quaternary structure of the proteasome activator PA28. *J Biol Chem* 1996;**271**:26410–7.
245. Ahn JY, Tanahashi N, Akiyama K, Hisamatsu H, Noda C, Tanaka K, et al. Primary structures of two homologous subunits of PA28, a gamma-interferon-inducible protein activator of the 20S proteasome. *FEBS Lett* 1995;**366**:37–42.
246. Zhang Z, Krutchinsky A, Endicott S, Realini C, Rechsteiner M, Standing KG. Proteasome activator 11S REG or PA28: recombinant REG alpha/REG beta hetero-oligomers are heptamers. *Biochemistry* 1999;**38**:5651–8.
247. Rechsteiner M, Realini C, Ustrell V. The proteasome activator 11 S REG (PA28) and class I antigen presentation. *Biochem J* 2000;**345**(Pt 1):1–15.
248. Realini C, Jensen CC, Zhang Z, Johnston SC, Knowlton JR, Hill CP, et al. Characterization of recombinant REGalpha, REGbeta, and REGgamma proteasome activators. *J Biol Chem* 1997;**272**:25483–92.
249. Li J, Rechsteiner M. Molecular dissection of the 11S REG (PA28) proteasome activators. *Biochimie* 2001;**83**:373–83.
250. Di CD. Human erythrocyte contains a factor that stimulates the peptidase activities of multicatalytic proteinase complex. *Ital J Biochem* 1992;**41**:213–24.
251. Kuehn L, Dahlmann B. Proteasome activator PA28 and its interaction with 20 S proteasomes. *Arch Biochem Biophys* 1996;**329**:87–96.
252. Dick TP, Ruppert T, Groettrup M, Kloetzel PM, Kuehn L, Koszinowski UH, et al. Coordinated dual cleavages induced by the proteasome regulator PA28 lead to dominant MHC ligands. *Cell* 1996;**86**:253–62.
253. Soza A, Knuehl C, Groettrup M, Henklein P, Tanaka K, Kloetzel PM. Expression and subcellular localization of mouse 20S proteasome activator complex PA28. *FEBS Lett* 1997;**413**:27–34.
254. Wojcik C, Tanaka K, Paweletz N, Naab U, Wilk S. Proteasome activator (PA28) subunits, alpha, beta and gamma (Ki antigen) in NT2 neuronal precursor cells and HeLa S3 cells. *Eur J Cell Biol* 1998;**77**:151–60.
255. Murata S, Kawahara H, Tohma S, Yamamoto K, Kasahara M, Nabeshima Y, et al. Growth retardation in mice lacking the proteasome activator PA28gamma. *J Biol Chem* 1999;**274**:38211–5.

256. Barton LF, Runnels HA, Schell TD, Cho Y, Gibbons R, Tevethia SS, et al. Immune defects in 28-kDa proteasome activator gamma-deficient mice. *J Immunol* 2004;**172**:3948–54.
257. Chen X, Barton LF, Chi Y, Clurman BE, Roberts JM. Ubiquitin-independent degradation of cell-cycle inhibitors by the REGgamma proteasome. *Mol Cell* 2007;**26**:843–52.
258. Li X, Lonard DM, Jung SY, Malovannaya A, Feng Q, Qin J, et al. The SRC-3/AIB1 coactivator is degraded in a ubiquitin- and ATP-independent manner by the REGgamma proteasome. *Cell* 2006;**124**:381–92.
259. Tanahashi N, Murakami Y, Minami Y, Shimbara N, Hendil KB, Tanaka K. Hybrid proteasomes. Induction by interferon-gamma and contribution to ATP-dependent proteolysis. *J Biol Chem* 2000;**275**:14336–45.
260. Hendil KB, Khan S, Tanaka K. Simultaneous binding of PA28 and PA700 activators to 20 S proteasomes. *Biochem J* 1998;**332**(Pt 3):749–54.
261. Wells AD, Rai SK, Salvato MS, Band H, Malkovsky M. Hsp72-mediated augmentation of MHC class I surface expression and endogenous antigen presentation. *Int Immunol* 1998;**10**:609–17.
262. Binder RJ, Blachere NE, Srivastava PK. Heat shock protein-chaperoned peptides but not free peptides introduced into the cytosol are presented efficiently by major histocompatibility complex I molecules. *J Biol Chem* 2001;**276**:17163–71.
263. Gonciarz-Swiatek M, Rechsteiner M. Proteasomes and antigen presentation: evidence that a KEKE motif does not promote presentation of the class I epitope SIINFEKL. *Mol Immunol* 2006;**43**:1993–2001.
264. Iwanczyk J, Sadre-Bazzaz K, Ferrell K, Kondrashkina E, Formosa T, Hill CP, et al. Structure of the Blm10-20 S proteasome complex by cryo-electron microscopy. Insights into the mechanism of activation of mature yeast proteasomes. *J Mol Biol* 2006;**363**:648–59.
265. Ortega J, Heymann JB, Kajava AV, Ustrell V, Rechsteiner M, Steven AC. The axial channel of the 20S proteasome opens upon binding of the PA200 activator. *J Mol Biol* 2005;**346**:1221–7.
266. Blickwedehl J, Agarwal M, Seong C, Pandita RK, Melendy T, Sung P, et al. Role for proteasome activator PA200 and postglutamyl proteasome activity in genomic stability. *Proc Natl Acad Sci USA* 2008;**105**:16165–70.
267. Schmidt M, Haas W, Crosas B, Santamaria PG, Gygi SP, Walz T, et al. The HEAT repeat protein Blm10 regulates the yeast proteasome by capping the core particle. *Nat Struct Mol Biol* 2005;**12**:294–303.
268. Rapic-Otrin V, McLenigan MP, Bisi DC, Gonzalez M, Levine AS. Sequential binding of UV DNA damage binding factor and degradation of the p48 subunit as early events after UV irradiation. *Nucleic Acids Res* 2002;**30**:2588–98.
269. Vantler M, Huntgeburth M, Caglayan E, Ten FH, Schnabel P, Rosenkranz S. PI3-kinase/Akt-dependent antiapoptotic signaling by the PDGF alpha receptor is negatively regulated by Src family kinases. *FEBS Lett* 2006;**580**:6769–76.
270. Zaiss DM, Standera S, Kloetzel PM, Sijts AJ. PI31 is a modulator of proteasome formation and antigen processing. *Proc Natl Acad Sci USA* 2002;**99**:14344–9.
271. Das GC, Holiday D, Gallardo R, Haas C. Taxol-induced cell cycle arrest and apoptosis: dose-response relationship in lung cancer cells of different wild-type p53 status and under isogenic condition. *Cancer Lett* 2001;**165**:147–53.
272. Essmann F, Pohlmann S, Gillissen B, Daniel PT, Schulze-Osthoff K, Janicke RU. Irradiation-induced translocation of p53 to mitochondria in the absence of apoptosis. *J Biol Chem* 2005;**280**:37169–77.

# Structure and Function of the Ubiquitin–Proteasome System: Modulation of Components

Niki Chondrogianni and
Efstathios S. Gonos

*National Hellenic Research Foundation, Institute of Biology, Medicinal Chemistry and Biotechnology, Athens, Greece*

I. The Ubiquitin–Proteasome System .......... 41
A. The Ubiquitin System .......... 42
B. The Proteasome .......... 47
II. Modulation of the UPS Components .......... 51
A. Ubiquitin System Modulation .......... 51
B. Proteasome System Modulation .......... 54
C. The Modulation of UPS System in Cancer and Therapy .......... 59
III. Conclusion .......... 62
References .......... 63

The ubiquitin–proteasome system (UPS) is directly or indirectly involved in all aspects of the cellular metabolic networks linked to either normal or pathologic processes. The highly orchestrated coordination of the many different parts that constitute the UPS augments the levels of difficulty in elucidating the unknown regulatory mechanisms of the system. At the same time, this increased complexity endows the system with the possibility of interfering in its different parts, depending on the surrounding cellular conditions and still specifically modulating the whole pathway. In this chapter, the main features of the UPS are summarized, with special attention given to the current data regarding the modulation of the different parts of this major system, either via genetic intervention or via treatment with naturally occurring or synthetic compounds.

## I. The Ubiquitin–Proteasome System

In eukaryotic cells, the majority of intracellular proteins are degraded by the ubiquitin–proteasome system (UPS).[1,2] This system relies initially on the coordination of three enzymes, termed E1, E2, and E3 ligases, to conjugate a small protein ubiquitin to the specific protein substrate that will thus be targeted to degradation.[2] On a second level, the system relies on the 26S proteasome, a large multicatalytic proteolytic particle that will execute the actual degradation of

Progress in Molecular Biology
and Translational Science, Vol. 109
DOI: 10.1016/B978-0-12-397863-9.00002-X

the tagged protein.[1] Following recognition of the tagged protein, the ubiquitin chains are released with the help of proteasome-associated deubiquitinases (DUBs)[3] to be recycled and the substrate is translocated in the inner proteolytic cavity where it is digested into short peptides that can be either displayed on the cell surface of the relative cells for immunosurveillance[4] or further chopped to free amino acids by different aminopeptidases and thus conclude the cycle of the amino acids.

## A. The Ubiquitin System

Ubiquitin conjugation is a multistep process highly significant for the cellular metabolism as it has been implicated so far with several functions including DNA repair,[5] transcriptional regulation,[6] endocytosis,[7] intracellular signaling,[8] and protein trafficking.[9] However, its best characterized function is related to the UPS and the intracellular protein turnover.[10,11]

### 1. Ubiquitin

Ubiquitin is the founding member of a family of structurally conserved proteins that regulate several processes in eukaryotic cells. Ubiquitin and its relatives carry out their functions through covalent attachment to other cellular proteins, thereby changing the stability, localization, or activity of the target protein.[12]

Ubiquitin is a small, essential protein of 76 residues that is ubiquitously found in all eukaryotic cells.[13] Its structure is compact, with a five-stranded antiparallel β-sheet traversed by a single helix.[14] Several important sites are identified in ubiquitin with the C-terminal carboxyl group of G76 being the major player in the process of ubiquitination, along with the primary amino group in each of the seven lysine residues (K48, K63, K6, K11, K27, K29, and K33) that exist in the molecule. Proteasomal degradation has thus far been shown to be mediated mainly by polyubiquitination based on Lys48 as the conjugated residue,[15] although chains based on other lysines have also been implicated in targeting proteins to the proteasome.[16,17]

Protein labeling by covalent attachment of ubiquitin to protein substrates is a multistep process orchestrated by the sequential action of three types of enzymes: E1, E2, and E3[18] (Fig. 1). The reaction begins with the ubiquitin-activating enzyme (E1), which initially adenylates the C-terminal glycine of ubiquitin and then forms a thioester bond between the activated glycine residue and a cysteine residue on the E1 catalytic site, through an ATP-dependent mechanism yielding a reactive E1–ubiquitin thioester intermediate. The activated ubiquitin is subsequently passed to one of a number of distinct ubiquitin-conjugating enzymes (E2s) by transthiolation to a conserved E2

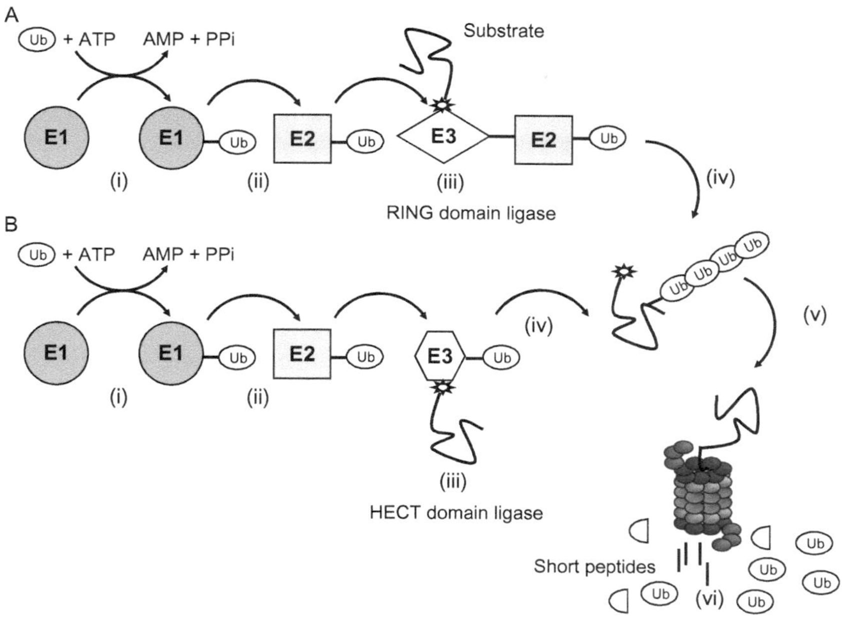


FIG. 1. The ubiquitin–proteasome system. (Ai and Bi) ATP-dependent activation of ubiquitin is catalyzed by E1 enzyme, the ubiquitin-activating enzyme. (Aii and Bii) Transfer of the activated ubiquitin to E2, a ubiquitin-carrier protein (ubiquitin-conjugating enzyme). (Aiii) In the case of a RING domain ligase, the ubiquitin-charged E2 binds to the E3 ligase that carries the substrate protein and (Aiv) transfers the activated ubiquitin moiety directly to the substrate. (Biii) In the case of an HECT domain ligase, ubiquitin is transferred from the ubiquitin-charged E2 to a conserved Cys residue in the E3 ligase and then to the substrate (Biv). The described steps are repeated in order to produce the polyubiquitin chain. (A, Bv and A, Bvi) The ubiquitinated substrate is degraded to short peptides by the 26S proteasome with release of free and reusable ubiquitin. Some of the ubiquitin molecules are degraded in this process along with the substrate. (For color version of this figure, the reader is referred to the online version of this chapter.)

active-site cysteine. The third step involves a specific ubiquitin–protein ligase (E3) that recruits the target protein and mediates the transfer of the activated ubiquitin from the E2 enzyme to the substrate. In most cases, an ε-NH2 group of a lysine residue on the substrate attacks the thioester bond between the ubiquitin and E2, and an isopeptide bond is formed, linking the activated C-terminal glycine of ubiquitin to the amino group in the attacking lysine of the target substrate.[12] However, when the substrate does not contain any accessible lysine residues, ubiquitin is instead conjugated to the α-amino group of the N-terminus of the target protein.[19] According to the sequential addition model (the standard model), this cycle of ubiquitination is successively repeated, thus producing polyubiquitin chains on the target protein. Several additional models have been proposed to explain the polyubiquitin chain assembly like the "hit and run" model, the "indexation" model, the "seesaw" model, and the "hubrid model" (reviewed in Ref. 18). Although the sequential addition model is the most accepted, several of the above-mentioned variations have been shown to occur *in vivo*. In the context of the sequential addition model, another type of enzymes, the E4s (chain elongation factors[20]), have been also proposed to act via binding to the ubiquitin moieties of preformed conjugates and the downstream catalysis of further ubiquitination. These elongation factors still require an E3 ligase of the homologous to E6-associated protein (E6AP) carboxyl terminus (HECT) class (see below) for chain assembly. The proteins proposed to be E4s have been shown to function as E3s in other assays.[21] They contain a U-box motif with a tertiary structure very similar to that of the really interesting new gene (RING) domain.[22] Therefore, it is suggested that the E4 elongation factors are actually E3s that can use the ubiquitin thioester-linked HECT domain of their cognate E3s in much the same way that other E3s use a cognate ubiquitin–E2 thioester for chain elongation. Despite the mechanism of the ubiquitin chain elongation, the length of the chain can vary largely. However, in the shade of protein degradation, at least four ubiquitin residues are needed for efficient recognition and processing by the 26S proteasome.[23,24] Finally, it is necessary to report that the ubiquitin signal is not always related to a polyubiquitin chain. Ubiquitin can also be conjugated on a protein as a single moiety in a process termed monoubiquitination or as several ubiquitin moieties in a process termed multiple monoubiquitination, thus generating structural diversity to control distinct biological processes.[25] However, these processes are out of the scope of this chapter.

## 2. E3 Ubiquitin–Protein Ligases

In the human genome, one can find few E1s[26] and several E2s but, in general, the specificity of ubiquitination is mediated by the specific E3s. There are two main structural families of E3 ligases, known as RING-finger containing E3 enzymes and HECT domain-containing E3 enzymes. Smaller families of ligases (U-box, plant homology domain, and zinc finger) have also been described.[27] In humans, RING-based E3s are specified by over 600 human genes,[28] whereas the HECT family counts about 30 members.[29] An E3 ligase in combination with an E2 ubiquitin-conjugating enzyme catalyzes the final step of ubiquitination of a target protein ending up in producing an isopeptide bond. Upon polyubiquitination on specific lysines of the target protein, the E3 ubiquitin ligase targets specific protein substrates for degradation by the proteasome. Therefore, E3s primarily dictate the specificity of the ubiquitination system.

*a. RING-Based E3 Ligases.* RING ligases serve as scaffolds that facilitate direct transfer of ubiquitin from the E2 enzymes to lysine residues on the target protein (Fig. 1A). The RING domain mediates the interaction with the appropriate E2 enzyme (Fig. 2A) and consists of a short motif, rich in cysteine and histidine residues.[30] Although in many E3s the substrate-binding site resides in the same polypeptide as the RING domain, thus functioning as a single subunit enzyme, certain RING domain proteins belong to complexes where substrate recognition is mediated by a separate subunit like in the cullin–RING complexes. The most known case of such a RING-based E3 ligase is the Skp1/cullin/F-box (SCF) complex.[31–34] SCF complex is composed of a modular E3 core containing CUL1 and Rbx1 (also called Roc1), and a substrate specificity module composed of Skp1 and a member of the F-box family of proteins.[35] The CUL1/Rbx1 complex functions as a scaffold to assemble the E2 ubiquitin-conjugating enzyme with the substrate specificity module.[36] CUL1 interacts with Rbx1 through its C terminus and with Skp1 through its N-terminus. The interaction of F-box proteins with Skp1 occurs through the F-box motif, an ~40-amino acid motif first identified in budding yeast Cdc4p and human cyclin F, the latter giving the name to the entire family.[37] In addition to the F-box motif, all F-box proteins have a protein–protein interaction domain for substrate binding and recognition that falls into the following three classes: (1) WD40 repeat-containing F-Box proteins (FBXWs), such as β-TRCP and FBXW7; (2) leucine-rich repeats-containing F-Box proteins (FBXLs), including Skp2; and (3) F-Box proteins containing other diverse domains (FBXOs), which include proteins with other domain types in the C-terminal region.[38] The SCF complex is responsible for the degradation of various proteins including cell cycle components,[39] transcription factors,[40] and components related to the maintenance of genome stability.[41]

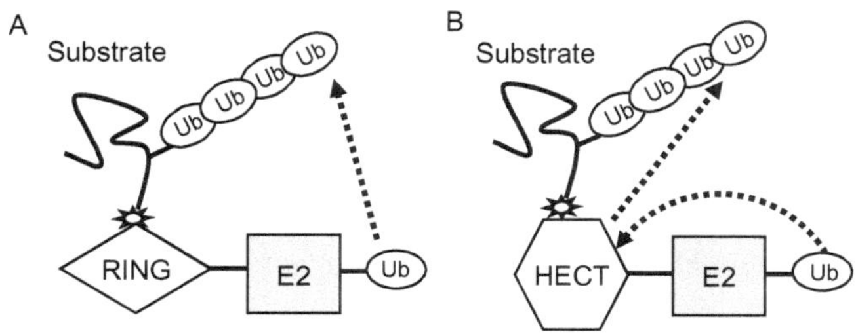


FIG. 2. RING and HECT domain E3 ligases. (A) The RING domain that carries the protein substrate mediates the interaction with the appropriate ubiquitin-charged E2 enzyme. Direct transfer of ubiquitin from the E2 to the target substrate occurs. (B) In the case of an HECT domain ligase, ubiquitin is primarily transferred from the appropriate ubiquitin-charged E2 enzyme to a conserved Cys residue of E3 ligase, forming a thioester intermediate and then to the protein substrate.

*b. HECT-Based E3 Ligases.* A key feature of HECT E3s is a conserved Cys residue to which ubiquitin binds prior to its transfer to the substrate (Figs. 1B and 2B). HECT E3 ligases form a thioester intermediate with ubiquitin before transferring it to the target protein.[29] The HECT domain consists of ~350 amino acids that are found at the C terminus of proteins. It was first described in human papillomavirus (HPV) E6AP and, therefore, E6AP E3 ligase is the founding member of the HECT E3 family.[42] Its most characterized role is the p53 degradation in cells infected with oncogenic strains of HPV.[43]

HECT E3s can either function on their own or in conjunction with accessory or adaptor proteins. Based on N-terminal domain architecture, the 28 human HECT E3s can be divided into three groups: the Nedd4 family (9 members), the HERC family (6 members), and other HECTs (13 members).[29] These ligases play pivotal roles in regulating cell signaling, protein trafficking, and have an equally important role in normal physiology and in human diseases such as cancer.[29]

3. Deubiquitinases

Ubiquitination is a reversible posttranslational modification with key roles in various signal transduction cascades and in determining protein stability. The key enzymes that perform this reverse reaction are called DUBs. In humans, there are approximately 100 DUBs and can be subdivided into five families: ubiquitin C-terminal hydrolases (UCHs), ubiquitin-specific proteases (USPs), ovarian tumor proteases (OTUs), Josephins, and JAB1/MPN/MOV34 metalloenzymes (JAMMs; also known as MPN+). The UCH, USP, OTU, and Josephin families are Cys proteases, whereas the JAMM/MPN+ family members are zinc metalloproteases.[3] DUBs play a pivotal role in three major processes: ubiquitin processing, reversal of ubiquitin signaling or protein degradation, and ubiquitin recycling. Regarding ubiquitin processing, ubiquitin can be transcribed from several genes as a linear fusion protein of multiple ubiquitin molecules or fused to the N-terminus of certain ribosomal proteins. Therefore, DUB activity is needed to acquire free ubiquitin moieties. Second, DUBs can remove ubiquitin chains from posttranslationally modified proteins, leading to reversal of ubiquitin signaling or to protein stabilization by rescue from either proteasomal or lysosomal degradation. Even if the degradation is committed to occur, DUB activity of proteasome-associated DUBs is necessary for ubiquitin recycling and maintenance of ubiquitin homeostasis. It has been shown that when ubiquitin levels are low in yeast, the cells sense the ubiquitin stress and lead to the induction of the proteasome-associated DUB Ubp6. This enhances loading of proteasomes with Ubp6, thereby altering proteasome function and consequently maintaining the ubiquitin pool.[44] Finally, DUBs can be used to edit the form of ubiquitin modification by trimming ubiquitin chains.[3]

DUB activity is highly specific and regulated at multiple levels to distinguish between the many ubiquitin-like molecules, isopeptides (using an ε-amino group) and linear peptides (using an α-amino group), and between different types of ubiquitin linkage and chain structure.[3] It has also been shown that substrate-specific DUBs exist. For example, it was found that Usp29 binds and cleaves polyubiquitin chains from p53 upon oxidative stress, thus coordinating the molecular and cellular response to oxidative stress.[45] Accordingly, it has been also shown that inhibition of a Usp14 DUB by a small-molecule inhibitor specifically enhances degradation of several proteasome substrates.[46,47]

## B. The Proteasome

The 26S proteasome is a highly conserved multicatalytic enzyme that can be dissociated into two distinct subcomplexes: the 20S core proteasome and the regulatory 19S complexes that bind either to one or both ends of the 20S core (Fig. 3). It is a large complex of 2.4 MDa that is found both in the cytoplasm and nucleus.[48] Controlled protein degradation is predominately catalyzed by the proteasome. Therefore, this complex is responsible for cell clearance of abnormal, denatured, or in general damaged proteins as well as for the regulated degradation of short-lived proteins.[49,50]

### 1. The 20S Proteasome

The proteolytic core, the 20S proteasome, is a 700-kDa multisubunit enzyme complex composed of seven different α- and seven different β-subunits arranged as a barrel-shape stack of four heptameric rings (Fig. 3). The two

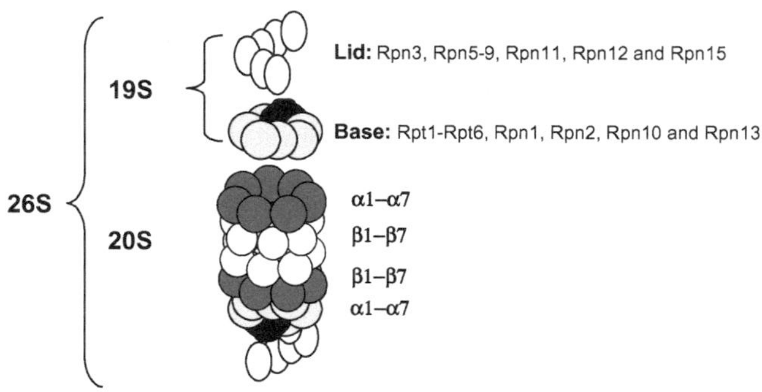


Fig. 3. Schematic representation of the 20S proteasome and the 19S regulatory particle. The 26S proteasome is composed of the 20S core complex and one or two 19S regulatory complexes that are attached at one or two ends of the 20S complex, respectively. The 20S proteasome is a multisubunit enzyme complex composed of seven different α- and seven different β-subunits arranged as a barrel-shape stack of four heptameric rings. The 19S complex is composed of two subcomplexes, the base and the lid.

outer α-subunits rings ($\alpha_{1-7}$) embrace two central head-to-head oriented rings containing β-subunits ($\beta_{1-7}$), thus acquiring the final $\alpha_{1-7}\ \beta_{1-7}\ \beta_{1-7}\ \alpha_{1-7}$ structure. In the α-rings, there is a narrow entry port (11–15 Å) that is blocked by a net made of the N-terminal domains of the individual α-subunits.[51] The proteolytic active sites are hosted in the internal chamber that is composed of β-subunits at the center of the complex.[52,53] Three of the β-subunits, $\beta_1$, $\beta_2$, and $\beta_5$, are responsible for the proteasome-hydrolyzing activities that cleave peptide bonds on the carboxyl site of acidic (peptidylglutamylpeptide-hydrolyzing or caspase-like activity, PGPH or C-L), basic (trypsin-like activity, T-L), and hydrophobic (chymotrypsin-like activity, CT-L) amino acids respectively.[50] The catalytic activities of these active β-type subunits are associated with their N-terminal threonine residue that acts as a nucleophile in hydrolysis.

Proteasome biogenesis and assembly is a complex, accurately ordered multistep event. It involves the *de novo* biosynthesis of α- and β-subunits and the following interaction with proteasome assembling chaperones, 1, 2, 3, and 4 (PAC1–4).[54,55] The PAC1–PAC2 dimer assists the assembly of one α-ring, followed by the PAC3–PAC4-assisted stepwise incorporation of precursor β-subunits that result in the formation of a single half proteasome.[56] Two of these half proteasomes with the parallel help of the UMP1/POMP accessory protein give rise to the pre-holeoproteasome. In this pre-complex, the active β-subunits (β1, β2, and β5) still contain N-terminal propeptides and they are therefore inactive. The proteolytic β-subunits are then activated with the cleavage of their N-terminal precursor by autocatalytic degradation, leading to mature 20S proteasome. UMP1/POMP chaperone is the first substrate of the newly assembled and active proteasome.[57] After this cleavage, the still attached PAC1–PAC2 dimers are also degraded[56] and the 20S mature proteasome is complete.

There are also special types of proteasomes that can be found in specific cells or under specific conditions. The immunoproteasomes can be found in cells of the hematopoietic lineage and cells exposed to (pro)inflammatory cytokines.[58] In this proteasome type, the constitutively expressed $\beta_1$, $\beta_2$, and $\beta_5$ subunits are substituted during *de novo* proteasome biosynthesis by $\beta_{1i}$, $\beta_{2i}$, and $\beta_{5i}$ subunits respectively. As a result, although these particles digest proteins at rates similar to those for constitutive proteasomes, they generate a higher fraction of peptides with the appropriate C-termini and length for optimal binding of peptides to MHC-I molecules. Specifically, they show reduced cleavage following acidic residues and enhanced cleavage following hydrophobic residues.[59]

Finally, additional tissues may also have their own type of proteasomes as it was shown for cortical thymic epithelial cells where the $\beta_{5i}$ subunit is substituted by $\beta_{5t}$, thus giving rise to the thymoproteasomes, which are thought to function in the positive selection of MHC class I-restricted T cells because they have reduced ability to cleave after hydrophobic residues. Therefore, they produce peptides having weak affinity for MHC-I molecules.[60] The presence of cell-type-specific proteasome subunits has also been found in *Drosophila* where testis-specific proteasome subunits have been identified to be necessary for spermatogenesis.[61,62]

## 2. The 19S Complex

The 19S regulatory particle (also known as RP/PA700) is a large complex of ~1 MDa that consists of at least 19 different subunits. It is attached on either one or both ends of the 20S core, thus producing the 26S proteasome which is central for the ATP/ubiquitin-dependent protein degradation.[63]

Following treatment with high salt concentrations, the 19S particle can be dissociated into two subcomplexes, namely the lid and the base (Fig. 3). The lid covers the base and consists of nine subunits (Rpn3, Rpn5–9, Rpn11, Rpn12, and Rpn15) and it is involved in the substrate uptake and deubiquitination. The Rpn11 subunit contains a highly conserved Jab1/MPN domain-associated metalloisopeptidase (JAMM) motif-EX(n)HXHX(10)D, thus coupling the degradation with deubiquitination.[64] Proteasomal base biogenesis requires multiple, functionally overlapping chaperones and follows a model in which base subunits form specific subcomplexes that then assemble into the base.[65,66] Likewise, the lid forms several lid intermediates and then Rpn12 bridges the lid and base, mediating both stable binding to the final lid intermediate and lid–base joining.[67]

The base consists of six ATPase subunits (Rpt1–Rpt6) as well as four non-ATPase subunits (Rpn1, Rpn2, Rpn10, and Rpn13) involved in the recognition, unfolding, and further translocation of the substrate.[68] More specifically, although these ATPases share high sequence homology, differences can be found mainly at their N-terminal domains, indicating possible distinct roles. Elegant electron microscopy and biochemical studies have shown that the six homologous ATPases are associated with the α-rings of the 20S particle, thus performing the opening of the pore of the α-gated channel.[69–74] Nevertheless, the pore is still narrow and, therefore, ATPase subunits also mediate the unfolding and translocation of the substrates into the 20S core.[75] Rpt2 plays a role in opening the gated α-ring to facilitate substrate entry,[76] while Rpt5 is implicated in the recognition of the substrate-linked polyubiquitin chain.[77] Rpn10[77] and Rpn13[78] have also been shown to interact with polyubiquitin chains, thus attributing a recognition role to these subunits as well. Intriguingly, Rpn10 and Rpn13 are not essential for yeast viability,[78,79] thus suggesting that additional, still unidentified ubiquitin/polyubiquitin-binding protein(s) might exist that circumvent the absence of the already known ubiquitin receptors. Rpn1 is also known to interact with various ubiquitin chain receptors that are not integral components of the proteasome but associate with it like Rad23 and Dsk2 in yeast and HHR23 and ubiquitin/protein-linking IAP with cytoskeleton

(PLIC) in humans.[80,81] These cofactors may also recruit substrates to the 26S proteasomes and, therefore, it is also possible that polyubiquitinated non-proteasomal shuttling proteins and ubiquitin-binding proteasomal subunits act in parallel, and that some of their functions are redundant.

### 3. Other Proteasome Activators

There are also additional regulators of the proteasome, with the 11S complex (also known as PA28/REG/PA26) being the best characterized example.[63] These factors are less broadly conserved than the ATP-dependent activators, and their substrates and biological functions are less clear. Nevertheless, the used mechanisms to activate proteasomes have been better characterized. Given that ubiquitin/polyubiquitin-binding protein properties have not yet been attributed to any of these subunits, it is most likely that these complexes are involved in ubiquitin-independent degradation pathways. Moreover, given that these activators have not been shown to possess ATPase activity, it is suggested that their substrates are small peptides or proteins with unstructured regions and, therefore, active unfolding is not needed.

There are three 11S isoforms called *PA28α, β, γ* (otherwise called REG α, β, γ).[82] *PA28α* and *PA28β*, which can be found only in vertebrates, preferentially form a heteroheptamer, while *PA28γ*, which is found in higher eukaryotes, is a homoheptamer. It is generally reported that 11S activators stimulate the hydrolysis of model peptide substrates instead of proteins. Nevertheless, PA28γ has been implicated in the degradation of specific natively unfolded proteins.[83–85] *PA28α* and *PA28β* can be induced by interferon and have been implicated in the production of MHC class I ligands, although the mechanistic basis for this function remains elusive.[59] *PA28γ* is not induced by interferon and it has been implicated mainly in the degradation of nuclear proteins like certain cyclin-dependent kinase inhibitors (p21, p16, p19).[86]

Blm10/PA200 (*Saccharomyces cerevisiae*/human) is another known activator that does not need ATP to function and is supposed to stimulate the hydrolysis of peptides but not proteins. Blm10/PA200 has been proposed to function in a surprisingly broad variety of processes, including 20S proteasome assembly,[87] DNA repair,[88,89] genomic stability,[90] and proteasome inhibition,[91] among others.

Finally, there is also an additional type of supra-proteasome complex, the hybrid proteasomes that are built of one 20S core proteasome with both a PA700 and a PA28 regulator particles attached at the two edges (PA28–20S–PA700).[92] The exact function of this proteasome is unknown, but a possible action speculates that the substrate protein is recognized and bound by the 19S regulator and degraded by the core particle, which changes its proteolytic specificities upon PA28 binding.[93] It is known that most of the proteolysis for MHC-I antigen presentation is performed by both the 26S proteasome and the so-called hybrid proteasome (PA28–20S–19S).

## II. Modulation of the UPS Components

There is no doubt that the UPS is absolutely necessary for cell physiology and survival.[94] It is also well established that under certain conditions, the UPS or at least some of its components are inhibited, thus generating homeostatic instability that leads to dysfunction and finally cell death. The most characteristic example is the linkage of the impaired proteasome function with neurodegenerative diseases like Alzheimer's and Parkinson's disease and the characteristic accumulation of undegradable material.[95,96] It has been suggested that induction of the protein catabolism can represent a powerful defensive or even therapeutic strategy against these age-related diseases.[95] On top of that, UPS impairment has not only been linked with diseases but also with naturally occurring conditions like aging and the related oxidative stress.[97,98] Therefore, it has been suggested that proteasome activation could serve as a possible novel antiaging strategy.[99] We report here on studies concerning the modulation of the UPS components in relation to enhancement of proteasome activities and protein turnover in situations where protein degradation is declining such as aging, oxidative stress, and age-related diseases.[94,97,100] Given that there are numerous studies that have associated the ubiquitin system modulation with cancer progression and therapy, we are going to refer to some of these studies in a separate section.

### A. Ubiquitin System Modulation

Transcriptional control of ubiquitin is tightly regulated. In *S. cerevisiae*, under normal growth conditions, the bulk of cellular ubiquitin is provided by *UBI1-3* genes. However, under stress conditions, induction of *UBI4* occurs.[101,102] The complexity of regulation of ubiquitin biosynthesis probably corresponds to the multiple cellular roles of ubiquitin and, therefore, the need for precise control of the ubiquitin levels to maintain cellular homeostasis. Moreover, a novel pathway of cellular response to ubiquitin depletion was revealed.[44] Ubp6 is a proteasome-associated deubiquitinating enzyme. Upon low levels of intracellular ubiquitin, Ubp6 abundance increases. Once ubiquitin levels are restored, Ubp6 levels return to baseline. Ubiquitin-dependent induction of *Ubp6* results in enhanced *Ubp6* levels on the proteasomes and thus in altered proteasome composition, thereby achieving higher rates of ubiquitin recycling and finally restoration of the intracellular ubiquitin levels.[44]

Another deubiquitinating enzyme, ATX-3 (the *Caenorhabditis elegans* orthologue for ataxin-3), has also been shown to mediate enhanced life span and stress compensation.[103] More specifically, worms deficient for *atx-3* and *cdc-48.1* (a chaperone-like AAA–ATPase that interact with different DUBs[104]), demonstrated extended life span by up to 50%, mediated through the insulin-like growth factor 1 signaling pathway (IIS). In contrast, longevity was significantly reduced by transgenic expression of *atx-3*. This study identified a novel link between protein degradation, longevity, and ubiquitin chain editing.[103]

Mammalian proteasomes have been found so far to be associated with three DUBs: Rpn11, Uch37, and Usp14.[64,105,106] Uch37 and Usp14 associate reversibly with the proteasome, whereas Rpn11 is a stoichiometric subunit. It has been shown that inhibition of Usp14 function through a small-molecule inhibitor enhances proteasome activity and specific degradation of oxidized proteins.[46] Recently, it was shown that transgenic complementation of ataxia mice with neuronally expressed ubiquitin prevents early postnatal lethality, restores muscle mass, and corrects developmental and functional deficits resulting from the loss of *Usp14*. Therefore, this study attributes a critical role to ubiquitin homeostasis in synaptic development and function while it also reveals a potential therapy through induction of ubiquitin.[107] Likewise, *Rpn11* overexpression has also a positive result on proteasome activity and the progression of aging (see Section II.B.2).

Despite the important role of ubiquitin and DUBs manipulation in the modulation of the ubiquitin system, the main enzymes that have been more extensively studied are the E3-specific ligases. Their modulation has been shown to affect many aspects of biological function, including aging, stress resistance, and longevity by specifically altering the turnover of target substrates. It is also worth mentioning that in most of the studies, modulation of E3 ligases refers to silencing experiments and not to induction through ectopic overexpression. A few examples will be presented below.

The Skp1–Cul1–F-box E3 ligase complex was among the first to be identified as a key component for the extended life span of *C. elegans* IIS mutants.[108] This study showed that depletion of (a) some of the *C. elegans* cullin scaffolding proteins (with CUL-1 being the most effective), (b) two Skp1-related proteins (SKR-1 and SKR-1), and (c) four F-box proteins (lin-23, fbxa-121, phi-3, and F59B2.8), reduced the extension of life span in IIS mutants without however producing any effects on other long-lived mutants or wild-type worms. Moreover, Ghazi *et al.*[108] have suggested that this complex extends organismal life span at least in part by promoting the transcriptional activity of the FOXO transcription factor DAF-16, possibly by targeting co-factors that affect its ability to activate longevity-promoting genes. It is known that upon its induction, DAF-16 accumulates in the nucleus,[109–111] where it activates (or represses) individual antioxidant, antimicrobial, metabolic, and other genes whose combined activities promote significant alterations in life span.[112,113] A similar study in human cells has revealed the role of *Skp2*, the F-box protein of the Skp2$^{SCF}$ ubiquitin E3 ligase[38] in cellular senescence. Genetic inactivation of *Skp2* alone does not induce cellular senescence response. However, the combined inactivation of *Skp2* and tumor suppressors PTEN or ARF results in cellular senescence both *in vivo* and *in vitro*.[114]

In another study, Li *et al.*[115] have also revealed an E3 ligase, RLE-1 (regulation of longevity by E3), which regulates aging in *C. elegans*. In contrast to the SCF CUL-1 case, disruption of RLE-1 expression increases the life span of *C. elegans*. Further experimentation revealed that RLE-1 is an E3 ubiquitin ligase for DAF-16, thus suggesting that RLE-1 opposes longevity by targeting DAF-16 for polyubiquitination and, consequently, for degradation.[115] The results of Li *et al.*[115] and Ghazi *et al.*[108] show that ubiquitin-mediated proteolytic regulation targets both DAF-16/FOXO to limit life span and additional factors to extend life span.

The beneficial role of another E3 ligase, *WWP-1*, was also revealed in relation to diet restriction (DR) and the relevant life span extension.[116] Overexpression of this HECT E3 ligase extended the life span of wild-type worms by 20% under conditions of *ad libitum* feeding. This extension was shown to be dependent upon FoxA transcription factor and independent of the FoxO transcription factor, DAF-16. Reduction of the ligase completely suppressed the extended longevity of diet-restricted animals. Moreover, it was also shown that the E2 ubiquitin-conjugating enzyme UBC-18 interacts with *WWP-1* and is required for its ubiquitin ligase activity and the extended longevity exhibited by the overexpressor worms. The authors suggested that these enzymes function to ubiquitinate substrates that regulate DR-induced longevity.[116] In accordance to this report in *C. elegans*, it was recently shown that overexpression of *WWP-1* delayed cellular senescence in human fibroblasts as well, whereas its knockdown led to premature senescence. The WWP-1-mediated delay of senescence was related to the promotion of p27$^{Kip1}$ degradation.[117]

The WD40 repeat protein *WDR-23* (in *C. elegans*, *WDR23* in humans) has been shown to function with the CUL4/DDB1 ubiquitin ligase to regulate nuclear abundance and activity of SKN-1 transcription factor (SKN-1 in *C. elegans*, Nrf2 in humans).[118] *WDR-23* recruits SKN-1 to CUL-4/DDB-1 ubiquitin ligase complex and targets it to degradation. Genetic targeting of *WDR-23* increased oxidative stress resistance and extended life span in worms.[118] A more recent study suggests that XREP-1 (*WDR-23*) is likely to be a functional equivalent of mammalian Keap1, the main regulator of Nrf2.[119] Accordingly, in *Drosophila*, male *keap1* heterozygous flies showed significantly increased survival rates after exposure to the free-radical generator paraquat, and they also lived significantly longer than controls under normal growth conditions.[120]

The von Hippel–Lindau (VHL) tumor suppressor homolog VHL-1 is a cullin E3 ubiquitin ligase that negatively regulates the hypoxic response by promoting ubiquitination and degradation of the hypoxic response transcription factor HIF-1 (hypoxia-inducible factor 1).[121] Loss of VHL-1 in *C. elegans* significantly increases life span and enhances resistance to polyglutamine and amyloid beta toxicity under normoxic conditions.[122] This effect is due at least in part to the increased stability of HIF-1 following loss of its E3 ligase. However, in human cells, it has been shown that VHL loss actuates an HIF-independent senescence program mediated by Rb and p400,[123] thus pinpointing to additional roles of VHL that do not include HIF-1.

The former examples of E3 ligase modulation reveal the extremely wide potential of therapeutic interventions by interfering with the ubiquitin system. Moreover, modulation of E3 ligases assures a high specificity regarding the protein target that can save researchers, medical doctors, and patients from the nightmare of side effects that accompany an intervention on a molecule that is involved in more than one cellular pathway.

## B. Proteasome System Modulation

### 1. 20S Complex Induction

Data in the literature are very limited regarding the activation/upregulation of the proteasome. The first data that appeared concerned proteasome activation through genetic manipulation of subunits of the main proteasome core. Goldberg and colleagues have succeeded the enhancement of hydrophobic (CT-L) and basic (T-L) proteasome activities following transfection of lymphoblasts and HeLa cells with $\beta_{5i}$ subunit, along with the increase of basic activity (T-L) following transfection of the same cell lines with the $\beta_{1i}$ subunit.[124] The same group has also shown the stimulation of acidic activity (PGPH) after overexpression of the $\beta_1$ subunit in HeLa cells.[125] Following these important preliminary findings, we have taken a comprehensive approach to fully decode the molecular and biochemical means of proteasome activation. Specifically, we have found that transfection of the $\beta_5$ subunit in WI38/T and HL60 cells resulted in upregulation of other β-type subunits, thus implying a common regulatory loop, and the recruitment of "free" α-type subunits to produce new and functional proteasomes. The developed "proteasome-activated cell lines" exhibited increased rates of proteolysis following treatment with various oxidants that were followed by enhanced resistance and cell survival. Importantly, when the $\beta_5$ subunit overexpression was performed in primary human embryonic fibroblasts (IMR90 cells), an extension of life span by approximately 15–20% was achieved while, most importantly, the cells maintained their "young" phenotype for longer.[126] The observed coregulation of β-type subunits has also been confirmed in other cell types, like the lens epithelial cells[127] or the murine neuroblastoma cells,[128] following overexpression of the $\beta_5$ subunit. Moreover, it was also shown that restoration of the normal levels of proteasome subunits through genetic manipulation in aged human fibroblasts reduces the levels of various aging biomarkers, thus further revealing the positive effects of proteasome activation on the retardation of senescence.[129]

UMP1 and POMP proteins in yeast and humans respectively have been identified as key molecules in the process of proteasome assembly.[130–133] We have achieved proteasomal upregulation via overexpression of the hUMP1/POMP protein in human fibroblasts. Upon its overexpression, the established clones exhibited increased rates of proteasome assembly and elevated proteasome activities. Upon administration of oxidative stressors, this enhanced proteasome function led to better recovery ability.[134] *UMP1* overexpression in yeast has been also shown to enhance viability during oxidative stress mainly through the enhanced preservation of proteasome-mediated protein degradation which, in turn, extends yeast stationary phase survival.[135] Moreover, silencing of *UMP1* by RNA interference has been shown to induce proteasome inhibition in yeast that eventually promotes a robust increase of protein oxidation resulting in autophagic death.[136] In contrast, deletion of *UMP1* has a neutral effect on cellular viability under normal growth conditions, but impairs the ability of yeast cells to survive under stationary phase conditions.[137]

*Rpn4* is another critical player that has been identified in yeast. It is a transcription factor responsible for the regulation of most of the yeast proteasome genes.[138] It is very tightly regulated and it is involved in the feedback circuit that governs the homeostatic regulation of the yeast proteasome.[139] *Rpn4* has been also involved in the coordinated activation of proteasome expression under stress conditions.[140] Investigation of Rpn4 has revealed that its induction is necessary for yeast survival under stress conditions.[141] However, its uncontrollable hyperactivation can be detrimental,[142] thus highlighting the need for highly controlled Rpn4 induction and downstream proteasome activation. Recently, it was shown that elevated proteasome capacity through *Rpn4* manipulation extends also the replicative life span of *S. cerevisiae*. Moreover, this elevated capacity results in improved clearance of toxic Huntington fragments in the relative yeast model for neurodegenerative diseases.[143] The observed life span extension has been suggested to originate from prolonged elimination of damaged protein in the old mother cells. Regardless of its importance, an *Rpn4* homologue in humans has not been identified.

The above-described studies revealed that proteasome activation is feasible through genetic manipulation of a specific 20S proteasome subunit or a proteasome-associated factor. Nevertheless, due to technical and ethical restraints of a genetic manipulation, the interest in the field is in parallel turned to the identification of natural or synthetic compounds that may enhance proteasome activities. Recently, a study showed that methylene blue (MB) reduces

Aβ levels in 3xTg-AD mice and rescues early cognitive deficit while at the same time the authors reported an increase of two out of the three proteasome activities (CT-L and T-L).[144] However, the underlying mechanism of proteasome activation was not elucidated. The same improvement in cognitive functions was also reported for patients on MB compared with patients in placebo[145] but it still remains uninvestigated whether proteasome activation is also observed in these human subjects.

It is well established that compound-mediated proteasome activation may occur via conformational changes of the proteasome structure per se by promoting the "opening" of its closed structure, thus leading to increased proteasome amount and enhanced function. More specifically, structural alterations of proteasome subunits have been shown to affect proteasome activity through changes of the 20S barrel conformation from the latent form to the active one. More specifically, SDS and some fatty acids such as olein, linoleic, and linolenic acids have been shown to stimulate proteasome activities in the test tube,[146,147] while potassium chloride had a negative effect[148] by favoring the open or closed conformation of the proteasome, respectively, and thus by partially denaturing the complex. Proteasome-activating hydrophobic peptides have been shown to be bound as modifiers at noncatalytic sites, thus mimicking the effect of 11S complex by opening the α-gated pore.[149] Synthetic peptidyl alcohols, nitriles, *p*-nitroanilides, and esters were also reported to activate the proteasome through binding to the PA28 binding site[150] along with cellular lipid components such as ceramides, lysophosphatidylinositol, and cardiolipin.[151–153] Based on these studies, we have isolated oleuropein, the most abundant of the phenolic compounds in *Olea europaea* leaf extract, olive oil, and olives,[154] and we have demonstrated that it has a stimulatory impact on proteasome activities *in vitro*. The increased activities promoted cellular resistance to oxidants while constant exposure of human primary fibroblasts to oleuropein conferred cellular life span extension.[155] Our results suggested that oleuropein most likely acts through structural changes of the 20S α-gated channel conformation in a similar manner to SDS, albeit its effects are considerably stronger. A similar effect was described for a lipid algae extract (*Phaeodactylum tricornutum*), which stimulated 20S proteasome peptidase activities of human keratinocytes following UVA and UVB irradiation.[156] Betulinic acid, a pentacyclic triterpene, constitutes another proteasome activator that preferably enhances the CT-L activity without any significant effects on the other two proteasome activities.[157]

Despite the importance of the proteasomal mechanism, its regulation is not yet entirely elucidated, especially in higher eukaryotes. There are reports in mice indicating that the expression of proteasome subunits and the relative proteasome activities can be modulated by exogenous stimuli. Specifically, Kensler and coworkers have reported that the genes forming the 26S proteasome complex are coordinately regulated by nuclear factor erythroid-derived 2-related factor 2 (Nrf2) in response to antioxidants such as 3H-1,2-dithiole-3-thione (D3T)[158] or sulforafane,[128] resulting in increased protection against various oxidants. Nrf2 belongs to the family of Cap'n'collar (Cnc) transcription factors and it is the main factor that mediates adaptive responses to cellular stress. Nrf2 orchestrates the transcriptional response of cells to oxidative stressors and electrophilic xenobiotics and it is the central mediator of a prominent antioxidant response system. It activates the transcription of multiple phase 2 genes encoding for proteins that protect cells against oxidative and electrophilic stress, such as glutathione S-transferases, γ-glutamylcysteine ligases, heme-oxygenase 1 (HO-1), and NADPH quinone oxidoreductase (Nqo1)[159] through a common DNA regulatory cisacting element, called antioxidant response element (ARE) or electrophile responsive element (EpRE; 5′-TGA[C/T]NNNGC-3′).[160] Activation of the Nrf2 signaling pathway is regulated by Kelch-like ECH-associated protein 1 (Keap1) according to changes in the intracellular redox state. Under nonstimulated conditions, Nrf2 is sequestered in the cytoplasm, where it is associated with Keap1.[161] Keap1 does not just passively sequester Nrf2 in the cytoplasm but plays an active role in targeting Nrf2 for ubiquitination and proteasomal degradation.[161] Depending on the type of stimulus, Nrf2 may become phosphorylated and/or Keap1 may be modified, resulting in the disruption of the Keap1–Nrf2 complex and the nuclear translocation of Nrf2.[162] Once in the nucleus, Nrf2 heterodimerizes with a small musculo-aponeurotic fibrosarcoma (Maf) protein and binds to AREs/EpREs, thereby promoting transcription of various cytoprotective genes.[163,164] In rodent models, it has been shown that signaling by Nrf2 defends against oxidative stress and aging-associated brain disorders.[165–168] Decreased abundance of Nrf2 has been also associated with neurodegeneration in humans (Friedreich's ataxia,[169] Alzheimer's disease[170]). Moreover, in humans, polymorphisms that decrease Nrf2 abundance have been associated with various pathologies of the skin,[171] respiratory system,[172] and digestive tract.[173–175] On top of that, CnC factors have life span-extending functions in invertebrates[176] further exhibiting the role of these factors in the progression of aging and the regulation of longevity.

It has been reported that several chemopreventive agents and natural compounds can induce Nrf2 function at low doses. Examples of potent plant-derived Nrf2 inducers include H-1,2-dithiole-3-thione (D3T), carnosol, curcumin, epigallocatechin-3-gallate, lycopene, quercetin, resveratrol, sulforaphane, and wasabi.[177,178] However, only few of them have been correlated to proteasome regulation via Nrf2 in mice and murine cells[128,158] and we have recently identified hederagenin and 18α-glycyrrhetinic acid, two triterpenoids, as potential Nrf2 inducers. Extensive analysis of 18α-glycyrrhetinic acid revealed that cell treatment with this compound induces *Nrf2* transcription and promotes its activation and translocation into the nucleus of human primary fibroblasts, resulting in elevated levels of proteasome expression and activity and increased cellular survival. Inhibition of the induction of proteasome activities abrogated the enhanced cell survival following administration of the Nrf2 inducer in parallel to the oxidant supplementation.[179] More importantly, continuous treatment of the cells with this compound throughout their replicative life span led to its extension.[179]

Finally, we have also shown that CT-L proteasome activity and proteasome content are induced upon quercetin treatment in HFL-1 human primary cells.[180] Since quercetin is a known enhancer of Nrf2 transcription factor[181] and Nrf2 has been shown to regulate genes forming the 26S proteasome complex,[158] the quercetin-mediated proteasome activation that we report could be mediated via the Nrf2 pathway. Quercetin-treated cells challenged with oxidants exhibited increased survival ability and the proteasome activation was found to be at least, in part, the responsible mechanism.[180] The same life span-extending results of quercetin were also observed in *C. elegans*[182,183] but these studies have not investigated the possibility of a potential link with proteasome activation.

The related transcription factor, nuclear factor erythroid-derived 2-related factor 1 (Nrf1), also known as NFE2L1/LCRF1/TCF11, is also a member of the CNC subfamily of basic-leucine zipper (bZIP) transcription factors (CNC-bZIP).[184] The respective *NFE2L1* gene encodes two main isoforms[185]: the longer one is known as TCF11 (772 aa) and the shorter one as Nrf1 (741 aa). Similar to Nrf2, TCF11/Nrf1 has been shown to induce the transcription of cytoprotective genes.[186] It was recently shown that TCF11, rather than Nrf2, is responsible for the upregulation of the proteasome system following proteasome inhibition via an ERAD-dependent feedback loop.[187,188]

NF-κB is another transcription factor that has been involved in the induction of proteasome expression under specific conditions. More specifically, it has been suggested that increased expression of the ubiquitin–proteasome pathway in murine myotubes by proteolysis-inducing factor (PIF) is associated with activation of the transcription factor NF-κB.[189–191] PIF is a tumor product that has the ability to enhance proteasome activities, the expression of the 20S proteasome core subunits, and several 19S regulatory subunits and, therefore, the proteasome-mediated protein degradation.[189,192]

### 2. 19S and 11S Complex Induction

Overexpression of various subunits of the regulatory proteasome complexes (19S and 11S complexes) has been also shown to act against the progression of senescence and age-related diseases. Overexpression of *PA28γ* was shown to enhance the survival of Huntington's disease neuronal model cells while *Rpn10* overexpression did not confer the same results.[193] In *Drosophila*, *Rpn11* overexpression suppressed the age-related reduction of the 26S proteasome activity, resulting in the extension of flies' life spans with suppression of the age-dependent accumulation of ubiquitinated proteins.[194] Moreover, *Rpn11* overexpression delayed the age-related polyglutamine-induced progressive neurodegeneration, thus further suggesting that maintaining the 26S proteasome with age could extend the life span and suppress the age-related progression of neurodegenerative diseases.[194] In *C. elegans*, it was shown that AIP-1, the worm homologue of mammalian arsenic-inducible proteasomal 19S regulatory particle-associated protein (AIRAP) is required for normal life span and resistance to the consequences of protein misfolding. Its silencing sensitized worms to acute toxic effects of exposure to sodium arsenite while animals raised in *aip-1* RNAi exhibited a shorter median life span.[195] Moreover, the overexpression of AIP-1 protected animals against Aβ toxicity. This protection coincided with reduced accumulation of Aβ, which is consistent with the AIP-1-enhancing protein degradation.[196] In accordance to these observations, AIRAP has been shown to protect mammalian cells from arsenite toxicity.[197] Later, it was delineated that AIRAP overexpression adapts the human cell's core protein degradation machinery to counteract proteotoxicity induced by an environmental toxin like arsenite.[198] More specifically, AIRAP-containing proteasomes were more active than conventional particles and thus more efficient to cope with the arsenite-induced proteotoxicity.[198]

The above examples indicate how by modulating one of the subunits of the 19S and 11S complexes, proteasome function can be induced, thereby exhibiting its positive effects under conditions of proteolysis malfunction. It is, however, noteworthy that for several of these 19S subunits, a proteasome-independent role has been discovered, thus further complicating the elucidation of the proteasome subunit induction and the downstream effects. For example, Rpt6 and Rpt4 have been reported to facilitate the conversion of soluble mutant Htt amino-terminal fragments to their aggregated state independent of the proteasome. More specifically, overexpression of *Rpt4* or *Rpt6* in primary neurons facilitated aggregation of mutant huntingtin and ataxin-3 without affecting proteasomal degradation. Conversely, reducing *Rpt6* or Rpt4 levels decreased the number of inclusions in the neurons, indicating that endogenous *Rpt4* and *Rpt6* facilitate inclusion formation.[199] Therefore, special attention should be given to the translation of the results regarding proteasome subunits modulation.

## C. The Modulation of UPS System in Cancer and Therapy

Cancer cells exhibit higher levels of proteasome activities.[200,201] This property might be linked to their intense needs due to increased proliferation rates, their intracellular enhanced levels of oxidative stress,[202] or their constant exposure to multiple cytokines and growth factors that affects proteasome structure and function.[203,204] On top of that, many proteins encoded by oncogenes and tumor suppressor genes lie among the proteasomal substrates,[205,206] with p53[207] and p27$^{Kip1}$ inhibitor of cyclin-dependent kinases (cdk)[208] being the most studied examples. For the aforementioned reasons in combination with the high specificity of target protein selection by E3 ubiquitin ligases,[209] the researchers suggested the UPS system as a target in cancer therapy.[210]

Proteasome inhibitor bortezomib is currently used as a novel anticancer drug that has shown promise in the treatment of various hematological malignancies (for more details, see last paragraph of the current section). However, its clinical efficacy has been hampered by the emergence of drug-resistance phenomena. Study of the molecular mechanism of bortezomib resistance in the resistant human myelomonocytic THP1 revealed (1) an Ala49Thr mutation residing in a highly conserved bortezomib-binding pocket in the $\beta_5$ proteasome subunit protein; (2) a dramatic overexpression (up to 60-fold) of $\beta_5$ protein but not of other proteasome subunits including $\beta_1$, $\beta_2$, and $\alpha_4$; (3) high levels of cross-resistance to $\beta_5$ subunit-targeted cytotoxic peptides 4A6, MG132, MG262, and ALLN, but not to a broad spectrum of chemotherapeutic drugs; (4) no marked changes in CT-L proteasome activity; and (5) restoration of bortezomib sensitivity in bortezomib-resistant cells by siRNA-mediated silencing of $\beta_5$ gene expression.[211] Analogous mutated $\beta_5$ subunit overexpression was also found in nonsmall cell lung cancer[212] and in leukemia cells.[213] On top of that, overexpression of the wild-type $\beta_5$ proteasome subunit as well as amplification of the $\beta_5$ gene was also reported in T-lymphoblastic lymphoma/leukemia cells derived from Jurkat line.[214] These findings establish a novel mechanism of bortezomib resistance associated with the selective overexpression of a mutant or wild-type $\beta_5$ protein and verify the important role of proteasome subunit modulation in cancer cell survival, resistance, and consequently in cancer therapy.

E3 ligases that target oncogene products or tumor suppressor gene products are involved in the maintenance of a normal level of these products. Therefore, inactivation of E3 ligases that target oncogene products or overexpression of E3 ligases that target tumor suppressor proteins can play a role in carcinogenesis or malignant progression. Selected examples of E3 ligases that are modulated during cancer progression will be presented.

The SCF complex controls many cancer-associated proteins including both tumor suppressor proteins and oncogene products.[215] Among the different F-box proteins, *Skp2* and *Fbw7* have been associated with control of many human cancer-associated proteins.[216] Tumor suppressor proteins like cdk inhibitors and cell cycle regulators are the main targets of *Skp2*, with p27$^{Kip1}$ being the best characterized one.[217] Therefore, overexpression of *Skp2* has been reported in many cancers, such as prostate cancer,[218] esophageal squamous cell carcinoma,[219] melanoma,[220] ovarian adenocarcinoma,[221] colorectal cancer,[222] and breast cancer,[223] among others. Moreover, gene amplifications of the *Skp2* gene were found in human gastric cancers.[224] Consequently, *Skp2* is suggested to be an oncogene.

*Fbw7* mainly targets the degradation of oncogenic proteins such as Cyclin E, c-Myc, c-Jun, c-Myb, Notch, and mTOR.[225] Moreover, deletion and mutation of *Fbw7* is found in human cancers.[226] Therefore, as reduced *Fbw7* expression level and loss-of-function mutations are found in a wide range of human cancers, *Fbw7* is considered a tumor suppressor gene.

HECT-type E3 ligases control the TGFβ-Smad pathway[227,228] whereas some HECT-type E3 ligases target the p53 family of proteins for degradation, that is, E6AP–E6,[43] ARF-BP1 (Huwe1) for p53,[229] Itch for p63 and p73,[230,231] and *WWP1* for p63.[232] Because these HECT-type E3 ligases are often overexpressed in human cancers, they may be associated with tumor cell growth. E3 ligases with an intrinsic single RING-finger domain such as Mdm2, Pirh2, and COP1 also target p53.[233] Mdm2 is the most investigated one, found overexpressed in several types of tumors exhibiting variable predictive significance.[234] Nevertheless, overexpression of the other p53 ligases has been also reported; for Pirh2 in lung,[235] prostate,[236] and head and neck cancers[237] and for Cop1 in breast and ovarian adenocarcinomas.[238] Regarding Mdm2, it has also been shown that it regulates the ubiquitin degradation of retinoblastoma protein (pRb), another tumor suppressor gene product.[239] An inverse correlation of pRb and Mdm2 has been shown in human lung cancer.[240] Therefore, Mdm2 plays a pivotal regulatory role on the main tumor suppressor proteins.

Accumulation of HIF-1 is observed in most types of solid tumor and is frequently associated with poor prognosis and cancer progression.[241] The protein stability of the catalytic subunit HIFα is regulated by the VHL E3 ligase and is considered a tumor suppressor E3 ligase.[121] VHL aberrations have been identified in familial and sporadic clear-cell kidney carcinomas, hemangioblastomas of the retina and central nervous system, and pheochromocytomas.[242]

The numerous cases of E3 ligase aberration or activation in different types of cancers leading to UPS modulation further justify the increased value of the system in cancer therapy. The therapeutic strategy can be subdivided into two categories: (a) modulation of specific components of the UPS system like the E3-specific ligase of an oncogenic protein or a tumor suppressor gene product, a regulatory component of the proteasome, or the ubiquitination system itself; and (b) inhibition of the proteasome activities and function. The latter category is the one that has already been applied. There are multiple naturally occurring and synthetic proteasome inhibitors[94] and one of the synthetic ones, dipeptidyl boronate bortezomib (Velcade™; formerly known as PS-341, LDP-341, and MLN341[243]), is the first specific proteasome inhibitor to enter clinical trials for cancer patients.[244] Bortezomib shows higher affinity for $\beta_5$ proteasome subunit followed by a weaker affinity for $\beta_1$ subunit, leaving the trypsin-like activity that is related to the $\beta_2$ subunit almost unaffected. Bortezomib has direct antitumor[210] as well as radiosensitizing effects[245] on cancer cells *in vitro* and *in vivo*. Bortezomib received accelerated Food and Drug Administration approval in May 2003 for the treatment of refractory multiple myeloma, in May 2005 for the treatment of patients with multiple myeloma who had received at least one prior therapy, in December 2006 for the treatment of patients with mantle cell lymphoma who have received at least one prior therapy, and in June 2008 for the treatment of patients with multiple myeloma. There are currently several phase I and phase II clinical trials in different types of cancers using this potential drug.[246] Although the results gained so far by this therapeutic strategy are promising, manipulation/targeting of more specific components of the UPS will definitely be beneficial.

## III. Conclusion

The UPS system is related to almost all cellular processes, making it initially a very unlikely target for therapeutic approaches and biomedical applications. However, the results collected so far for bortezomib refute the investigators' concerns and pinpoint to future directions that can turn out to be even more successful. Modulation of specific parts of the UPS might narrow down the adverse effects of manipulation of a more general target like the 20S/26S complex. For example, manipulation of a specific E3 ligase to either suppress or induce the degradation of a certain protein might be a perfect therapeutic strategy for the treatment of diseases like cancer by specific induction or inhibition of one responsible cancer-associated protein. Nevertheless, the activation of the proteasome or the UPS can also find myriad applications in conditions where progressive decrease of the UPS function occurs. Aging and/or longevity and the aging-related pathologies lie among these conditions. For example, initial results implicating proteasome activation on the treatment of age-related pathologies are very promising.[144,193,196] Consequently, there is intense need to further elucidate the mechanisms that regulate the gene and protein expression and the assembly of the different parts of the UPS in order to seek specific modulators (naturally occurring or synthetic) that will restore the protein turnover equilibrium.

### Acknowledgments

Cited work from the laboratory has been funded by the EU projects Protage, Functionage, and Proteomage to E. S. G. We also thank CMST COST Action CM1001 for the support.

## References


1. Glickman MH, Ciechanover A. The ubiquitin-proteasome proteolytic pathway: destruction for the sake of construction. *Physiol Rev* 2002;**82**:373–428.
2. Finley D. Recognition and processing of ubiquitin-protein conjugates by the proteasome. *Annu Rev Biochem* 2009;**78**:477–513.
3. Komander D, Clague MJ, Urbé S. Breaking the chains: structure and function of the deubiquitinases. *Nat Rev Mol Cell Biol* 2009;**10**:550–63.
4. Strehl B, Seifert U, Krüger E, Heink S, Kuckelkorn U, Kloetzel PM. Interferon-gamma, the functional plasticity of the ubiquitin-proteasome system, and MHC class I antigen processing. *Immunol Rev* 2005;**207**:19–30.
5. Pickart CM. DNA repair: right on target with ubiquitin. *Nature* 2002;**419**:120–1.
6. Kodadek T. No Splicing, no dicing: non-proteolytic roles of the ubiquitin-proteasome system in transcription. *J Biol Chem* 2010;**285**:2221–6.
7. Haglund K, Di Fiore PP, Dikic I. Distinct monoubiquitin signals in receptor endocytosis. *Trends Biochem Sci* 2003;**28**:598–603.
8. Tokunaga F, Sakata S, Saeki Y, Satomi Y, Kirisako T, Kamei K, et al. Involvement of linear polyubiquitylation of NEMO in NF-kappaB activation. *Nat Cell Biol* 2009;**11**:123–32.
9. d'Azzo A, Bongiovanni A, Nastasi T. E3 ubiquitin ligases as regulators of membrane protein trafficking and degradation. *Traffic* 2005;**6**:429–41.
10. Pickart CM. Targeting of substrates to the 26S proteasome. *FASEB J* 1997;**11**:1055–66.
11. Navon A, Ciechanover A. The 26 S proteasome: from basic mechanisms to drug targeting. *J Biol Chem* 2009;**284**:33713–8.
12. Pickart CM, Eddins MJ. Ubiquitin: structures, functions, mechanisms. *Biochim Biophys Acta* 2004;**1695**:55–72.
13. Pickart CM. Mechanisms underlying ubiquitination. *Annu Rev Biochem* 2001;**70**:503–33.
14. Vijay-Kumar S, Bugg CE, Wilkinson KD, Cook WJ. Three-dimensional structure of ubiquitin at 2.8 A resolution. *Proc Natl Acad Sci USA* 1985;**82**:3582–5.
15. Pickart CM. Ubiquitin in chains. *Trends Biochem Sci* 2000;**25**:544–8.
16. Saeki Y, Kudo T, Sone T, Kikuchi Y, Yokosawa H, Toh-e A, et al. Lysine 63-linked polyubiquitin chain may serve as a targeting signal for the 26S proteasome. *EMBO J* 2009;**28**:359–71.
17. Xu P, Duong DM, Seyfried NT, Cheng D, Xie Y, Robert J, et al. Quantitative proteomics reveals the function of unconventional ubiquitin chains in proteasomal degradation. *Cell* 2009;**137**:133–45.
18. Hochstrasser M. Lingering mysteries of ubiquitin-chain assembly. *Cell* 2006;**124**:27–34.
19. Ravid T, Hochstrasser M. Diversity of degradation signals in the ubiquitin-proteasome system. *Nat Rev Mol Cell Biol* 2008;**9**:679–90.
20. Koegl M, Hoppe T, Schlenker S, Ulrich HD, Mayer TU, Jentsch S. A novel ubiquitination factor, E4, is involved in multiubiquitin chain assembly. *Cell* 1999;**96**:635–44.
21. Hatakeyama S, Yada M, Matsumoto M, Ishida N, Nakayama KI. U box proteins as a new family of ubiquitin-protein ligases. *J Biol Chem* 2001;**276**:33111–20.
22. Ohi MD, Vander Kooi CW, Rosenberg JA, Chazin WJ, Gould KL. Structural insights into the U-box, a domain associated with multi-ubiquitination. *Nat Struct Biol* 2003;**10**:250–5.
23. Thrower JS, Hoffman L, Rechsteiner M, Pickart CM. Recognition of the polyubiquitin proteolytic signal. *EMBO J* 2000;**19**:94–102.
24. Pickart CM, Fushman D. Polyubiquitin chains: polymeric protein signals. *Curr Opin Chem Biol* 2004;**8**:610–6.
25. Sadowski M, Suryadinata R, Tan AR, Roesley SN, Sarcevic B. Protein monoubiquitination and polyubiquitination generate structural diversity to control distinct biological processes. *IUBMB Life* 2011;**64**:136–42.

26. Groettrup M, Pelzer C, Schmidtke G, Hofmann K. Activating the ubiquitin family: UBA6 challenges the field. *Trends Biochem Sci* 2008;**33**:230–7.
27. de Bie P, Ciechanover A. Ubiquitination of E3 ligases: self-regulation of the ubiquitin system via proteolytic and non-proteolytic mechanisms. *Cell Death Differ* 2011;**18**:1393–402.
28. Deshaies RJ, Joazeiro CA. RING domain E3 ubiquitin ligases. *Annu Rev Biochem* 2009;**78**:399–434.
29. Rotin D, Kumar S. Physiological functions of the HECT family of ubiquitin ligases. *Nat Rev Mol Cell Biol* 2009;**10**:398–409.
30. Freemont PS, Hanson IM, Trowsdale J. A novel cysteine-rich sequence motif. *Cell* 1991;**64**:483–94.
31. Kamura T, Koepp DM, Conrad MN, Skowyra D, Moreland RJ, Iliopoulos O, et al. Rbx1, a component of the VHL tumor suppressor complex and SCF ubiquitin ligase. *Science* 1999;**284**:657–61.
32. Ohta T, Michel JJ, Schottelius AJ, Xiong Y. ROC1, a homolog of APC11, represents a family of cullin partners with an associated ubiquitin ligase activity. *Mol Cell* 1999;**3**:535–41.
33. Tan P, Fuchs SY, Chen A, Wu K, Gomez C, Ronai Z, et al. Recruitment of a ROC1-CUL1 ubiquitin ligase by Skp1 and HOS to catalyze the ubiquitination of I kappa B alpha. *Mol Cell* 1999;**3**:527–33.
34. Seol JH, Feldman RM, Zachariae W, Shevchenko A, Correll CC, Lyapina S, et al. Cdc53/cullin and the essential Hrt1 RING-H2 subunit of SCF define a ubiquitin ligase module that activates the E2 enzyme Cdc34. *Genes Dev* 1999;**13**:1614–26.
35. Cardozo T, Pagano M. The SCF ubiquitin ligase: insights into a molecular machine. *Nat Rev Mol Cell Biol* 2004;**5**:739–53.
36. Zheng N, Schulman BA, Song L, Miller JJ, Jeffrey PD, Wang P, et al. Structure of the Cul1–Rbx1–Skp1–F box Skp2 SCF ubiquitin ligase complex. *Nature* 2002;**416**:703–9.
37. Bai C, Sen P, Hofmann K, Ma L, Goebl M, Harper JW, et al. SKP1 connects cell cycle regulators to the ubiquitin proteolysis machinery through a novel motif, the F-box. *Cell* 1996;**86**:263–74.
38. Jin J, Cardozo T, Lovering RC, Elledge SJ, Pagano M, Harper JW. Systematic analysis and nomenclature of mammalian F-box proteins. *Genes Dev* 2004;**18**:2573–80.
39. Skaar JR, Pagano M. Control of cell growth by the SCF and APC/C ubiquitin ligases. *Curr Opin Cell Biol* 2009;**21**:816–24.
40. Irniger S, Braus GH. Controlling transcription by destruction: the regulation of yeast Gcn4p stability. *Curr Genet* 2003;**44**:8–18.
41. Silverman JS, Skaar JR, Pagano M. SCF ubiquitin ligases in the maintenance of genome stability. *Trends Biochem Sci* 2011;**37**:66–73.
42. Huibregtse JM, Scheffner M, Beaudenon S, Howley PM. A family of proteins structurally and functionally related to the E6-AP ubiquitin–protein ligase. *Proc Natl Acad Sci USA* 1995;**92**:2563–7.
43. Scheffner M, Huibregtse JM, Vierstra RD, Howly PM. The HPV-16 E6 and E6-AP complex functions as a ubiquitin-protein ligase in the ubiquitination of p53. *Cell* 1993;**75**:495–505.
44. Hanna J, Meides A, Zhang DP, Finley D. A ubiquitin stress response induces altered proteasome composition. *Cell* 2007;**129**:747–59.
45. Liu J, Chung HJ, Vogt M, Jin Y, Malide D, He L, et al. JTV1 co-activates FBP to induce USP29 transcription and stabilize p53 in response to oxidative stress. *EMBO J* 2011;**30**:846–58.
46. Lee BH, Lee MJ, Park S, Oh DC, Elsasser S, Chen PC, et al. Enhancement of proteasome activity by a small-molecule inhibitor of USP14. *Nature* 2010;**467**:179–84.

47. Lee MJ, Lee BH, Hanna J, King RW, Finley D. Trimming of ubiquitin chains by proteasome-associated deubiquitinating enzymes. *Mol Cell Proteomics* 2011;**10**:R110.003871.
48. Brooks P, Fuertes G, Murray RZ, Bose S, Knecht E, Rechsteiner MC, et al. Subcellular localization of proteasomes and their regulatory complexes in mammalian cells. *Biochem J* 2000;**346**:155–61.
49. Ciechanover A. Intracellular protein degradation: from a vague idea through the lysosome and the ubiquitin-proteasome system and onto human diseases and drug targeting. *Cell Death Differ* 2005;**12**:1178–90.
50. Goldberg AL. Functions of the proteasome: from protein degradation and immune surveillance to cancer therapy. *Biochem Soc Trans* 2007;**35**:12–7.
51. Groll M, Bajorek M, Kohler A, Moroder L, Rubin DM, Huber R, et al. A gated channel into the proteasome core particle. *Nat Struct Biol* 2000;**7**:1062–7.
52. Löwe J, Stock D, Jap B, Zwickl P, Baumeister W, Huber R. Crystal structure of the 20S proteasome from the archaeon T. acidophilum at 3.4 A resolution. *Science* 1995;**268**:533–9.
53. Groll M, Ditzel L, Löwe J, Stock D, Bochtler M, Bartunik HD, et al. Structure of 20S proteasome from yeast at 2.4 A resolution. *Nature* 1997;**386**:463–71.
54. Hirano Y, Hendil KB, Yashiroda H, Iemura S, Nagane R, Hioki Y, et al. A heterodimeric complex that promotes the assembly of mammalian 20S proteasomes. *Nature* 2005;**437**:1381–5.
55. Hirano Y, Hayashi H, Iemura S, Hendil KB, Niwa S, Kishimoto T, et al. Cooperation of multiple chaperones required for the assembly of mammalian 20S proteasomes. *Mol Cell* 2006;**24**:977–84.
56. Hirano Y, Kaneko T, Okamoto K, Bai M, Yashiroda H, Furuyama K, et al. Dissecting β-ring assembly pathway of the mammalian 20S proteasome. *EMBO J* 2008;**27**:2204–13.
57. Heinemeyer W, Ramos PC, Dohmen RJ. The ultimate nanoscale mincer: assembly, structure and active sites of the 20S proteasome core. *Cell Mol Life Sci* 2004;**61**:1562–78.
58. Sijts EJ, Kloetzel PM. The role of the proteasome in the generation of MHC class I ligands and immune responses. *Cell Mol Life Sci* 2011;**68**:1491–502.
59. Groettrup M, Kirk CJ, Basler M. Proteasomes in immune cells: more than peptide producers? *Nat Rev Immunol* 2010;**10**:73–8.
60. Murata S, Takahama Y, Tanaka K. Thymoproteasome: probable role in generating positively selecting peptides. *Curr Opin Immunol* 2008;**20**:192–6.
61. Yuan X, Miller M, Belote JM. Duplicated proteasome subunit genes in Drosophila melanogaster encoding testes-specific isoforms. *Genetics* 1996;**144**:147–57.
62. Zhong L, Belote JM. The testis-specific proteasome subunit Prosalpha6T of D. melanogaster is required for individualization and nuclear maturation during spermatogenesis. *Development* 2007;**134**:3517–25.
63. Stadtmueller BM, Hill CP. Proteasome activators. *Mol Cell* 2011;**41**:8–19.
64. Verma R, Aravind L, Oania R, McDonald WH, Yates 3rd. JR, Koonin EV, et al. Role of Rpn11 metalloprotease in deubiquitination and degradation by the 26S proteasome. *Science* 2002;**298**:611–5.
65. Funakoshi M, Tomko Jr. RJ, Kobayashi H, Hochstrasser M. Multiple assembly chaperones govern biogenesis of the proteasome regulatory particle base. *Cell* 2009;**137**:887–99.
66. Tomko Jr. RJ, Funakoshi M, Schneider K, Wang J, Hochstrasser M. Heterohexameric ring arrangement of the eukaryotic proteasomal ATPases: implications for proteasome structure and assembly. *Mol Cell* 2010;**38**:393–403.
67. Tomko Jr. RJ, Hochstrasser M. Incorporation of the rpn12 subunit couples completion of proteasome regulatory particle lid assembly to lid-base joining. *Mol Cell* 2011;**44**:907–17.
68. Hanna J, Finley D. A proteasome for all occasions. *FEBS Lett* 2007;**581**:2854–61.

69. Walz J, Erdmann A, Kania M, Typke D, Koster AJ, Baumeister W. 26S proteasome structure revealed by three-dimensional electron microscopy. *J Struct Biol* 1998;**121**:19–29.
70. Baumeister W, Walz J, Zuhl F, Seemuller E. The proteasome: paradigm of a self-compartmentalizing protease. *Cell* 1998;**92**:367–80.
71. Nickell S, Mihalache O, Beck F, Hegerl R, Korinek A, Baumeister W. Structural analysis of the 26S proteasome by cryoelectron tomography. *Biochem Biophys Res Commun* 2007;**353**:115–20.
72. Nickell S, Beck F, Scheres SH, Korinek A, Förster F, Lasker K, et al. Insights into the molecular architecture of the 26S proteasome. *Proc Natl Acad Sci USA* 2009;**106**:11943–7.
73. Smith DM, Chang SC, Park S, Finley D, Cheng Y, Goldberg AL. Docking of the proteasomal ATPases' carboxyl termini in the 20S proteasome's alpha ring opens the gate for substrate entry. *Mol Cell* 2007;**27**:731–44.
74. Thompson D, Hakala K, DeMartino GN. Subcomplexes of PA700, the 19 S regulator of the 26 S proteasome, reveal relative roles of AAA subunits in 26 S proteasome assembly and activation and ATPase activity. *J Biol Chem* 2009;**284**:24891–903.
75. Braun BC, Glickman M, Kraft R, Dahlmann B, Kloetzel PM, Finley D, et al. The base of the proteasome regulatory particle exhibits chaperone-like activity. *Nat Cell Biol* 1999;**1**:221–6.
76. Köhler A, Cascio P, Leggett DS, Woo KM, Goldberg AL, Finley D. The axial channel of the proteasome core particle is gated by the Rpt2 ATPase and controls both substrate entry and product release. *Mol Cell* 2001;**7**:1143–52.
77. Lam YA, Lawson TG, Velayutham M, Zweier JL, Pickart CM. A proteasomal ATPase subunit recognizes the polyubiquitin degradation signal. *Nature* 2002;**416**:763–7.
78. Husnjak K, Elsasser S, Zhang N, Chen X, Randles L, Shi Y, et al. Proteasome subunit Rpn13 is a novel ubiquitin receptor. *Nature* 2008;**453**:481–8.
79. Wilkinson CR, Seeger M, Hartmann-Petersen R, Stone M, Wallace M, Semple C, et al. Proteins containing the UBA domain are able to bind to multi-ubiquitin chains. *Nat Cell Biol* 2001;**3**:939–43.
80. Elsasser S, Gali RR, Schwickart M, Larsen CN, Leggett DS, Muller B, et al. Proteasome subunit Rpn1 binds ubiquitin-like protein domains. *Nat Cell Biol* 2002;**4**:725–30.
81. Verma R, Oania R, Graumann J, Deshaies RJ. Multiubiquitin chain receptors define a layer of substrate selectivity in the ubiquitin-proteasome system. *Cell* 2004;**118**:99–110.
82. Rechsteiner M, Hill CP. Mobilizing the proteolytic machine: cell biological roles of proteasome activators and inhibitors. *Trends Cell Biol* 2005;**15**:27–33.
83. Mao I, Liu J, Li X, Luo H. REGgamma, a proteasome activator and beyond? *Cell Mol Life Sci* 2008;**65**:3971–80.
84. Nie J, Wu M, Wang J, Xing G, He F, Zhang L. REGgamma proteasome mediates degradation of the ubiquitin ligase Smurf1. *FEBS Lett* 2010;**584**:3021–7.
85. Suzuki R, Moriishi K, Fukuda K, Shirakura M, Ishii K, Shoji I, et al. Proteasomal turnover of hepatitis C virus core protein is regulated by two distinct mechanisms: a ubiquitin-dependent mechanism and a ubiquitin-independent but PA28gamma-dependent mechanism. *J Virol* 2009;**83**:2389–92.
86. Chen X, Barton LF, Chi Y, Clurman BE, Roberts JM. Ubiquitin-independent degradation of cell-cycle inhibitors by the REGgamma proteasome. *Mol Cell* 2007;**26**:843–52.
87. Fehlker M, Wendler P, Lehmann A, Enenkel C. Blm3 is part of nascent proteasomes and is involved in a late stage of nuclear proteasome assembly. *EMBO Rep* 2003;**4**:959–63.
88. Schmidt M, Haas W, Crosas B, Santamaria PG, Gygi SP, Walz T, et al. The HEAT repeat protein Blm10 regulates the yeast proteasome by capping the core particle. *Nat Struct Mol Biol* 2005;**12**:294–303.

89. Ustrell V, Hoffman L, Pratt G, Rechsteiner M. PA200, a nuclear proteasome activator involved in DNA repair. *EMBO J* 2002;**21**:3516–25.
90. Blickwedehl J, Agarwal M, Seong C, Pandita RK, Melendy T, Sung P, et al. Role for proteasome activator PA200 and postglutamyl proteasome activity in genomic stability. *Proc Natl Acad Sci USA* 2008;**105**:16165–70.
91. Lehmann A, Jechow K, Enenkel C. Blm10 binds to pre-activated proteasome core particles with open gate conformation. *EMBO Rep* 2008;**9**:1237–43.
92. Tanahashi N, Murakami Y, Minami Y, Shimbara N, Hendil KB, Tanaka K. Hybrid proteasomes. Induction by interferon-gamma and contribution to ATP-dependent proteolysis. *J Biol Chem* 2000;**275**:14336–45.
93. Hendil KB, Khan S, Tanaka K. Simultaneous binding of PA28 and PA700 activators to 20S proteasomes. *Biochem J* 1998;**332**:749–54.
94. Jung T, Catalgol B, Grune T. The proteasomal system. *Mol Aspects Med* 2009;**30**:191–296.
95. Paul S. Dysfunction of the ubiquitin-proteasome system in multiple disease conditions: therapeutic approaches. *Bioessays* 2008;**30**:1172–84.
96. Oddo S. The ubiquitin-proteasome system in Alzheimer's disease. *J Cell Mol Med* 2008;**12**:363–73.
97. Chondrogianni N, Gonos ES. Proteasome dysfunction and mammalian aging: involved steps and factors. *Exp Gerontol* 2005;**40**:931–8.
98. Kriegenburg F, Poulsen EG, Koch A, Krüger E, Hartmann-Petersen R. Redox control of the ubiquitin-proteasome system: from molecular mechanisms to functional significance. *Antioxid Redox Signal* 2011;**15**:2265–99.
99. Chondrogianni N, Gonos ES. Proteasome activation as a novel anti-aging strategy. *IUBMB Life* 2008;**60**:651–5.
100. Vernace VA, Arnaud L, Schmidt-Glenewinkel T, Figueiredo-Pereira ME. Aging perturbs 26S proteasome assembly in Drosophila melanogaster. *FASEB J* 2007;**21**:2672–82.
101. Ozkaynak E, Finley D, Solomon MJ, Varshavsky A. The yeast ubiquitin genes: a family of natural gene fusions. *EMBO J* 1987;**6**:1429–39.
102. Finley D, Ozkaynak E, Varshavsky A. The yeast polyubiquitin gene is essential for resistance to high temperatures, starvation, and other stresses. *Cell* 1987;**48**:1035–46.
103. Kuhlbrodt K, Janiesch PC, Kevei É, Segref A, Barikbin R, Hoppe T. The Machado-Joseph disease deubiquitylase ATX-3 couples longevity and proteostasis. *Nat Cell Biol* 2011;**13**:273–81.
104. Jentsch S, Rumpf S. Cdc48 (p97): a "molecular gearbox" in the ubiquitin pathway? *Trends Biochem Sci* 2007;**32**:6–11.
105. Borodovsky A, Kessler BM, Casagrande R, Overkleeft HS, Wilkinson KD, Ploegh HL. A novel active site-directed probe specific for deubiquitylating enzymes reveals proteasome association of USP14. *EMBO J* 2001;**20**:5187–96.
106. Yao T, Song L, Xu W, DeMartino GN, Florens L, Swanson SK, et al. Proteasome recruitment and activation of the Uch37 deubiquitinating enzyme by Adrm1. *Nat Cell Biol* 2006;**8**:994–1002.
107. Chen PC, Bhattacharyya BJ, Hanna J, Minkel H, Wilson JA, Finley D, et al. Ubiquitin homeostasis is critical for synaptic development and function. *J Neurosci* 2011;**31**:17505–13.
108. Ghazi A, Henis-Korenblit S, Kenyon C. Regulation of Caenorhabditis elegans lifespan by a proteasomal E3 ligase complex. *Proc Natl Acad Sci USA* 2007;**104**:5947–52.
109. Henderson ST, Johnson TE. daf-16 integrates developmental and environmental inputs to mediate aging in the nematode Caenorhabditis elegans. *Curr Biol* 2001;**11**:1975–80.
110. Lee RY, Hench J, Ruvkun G. Regulation of C. elegans DAF-16 and its human ortholog FKHRL1 by the daf-2 insulin-like signaling pathway. *Curr Biol* 2001;**11**:1950–7.

111. Lin K, Hsin H, Libina N, Kenyon C. Regulation of the Caenorhabditis elegans longevity protein DAF-16 by insulin/IGF-1 and germline signaling. *Nat Genet* 2001;**28**:139–45.
112. Murphy CT, McCarroll SA, Bargmann CI, Fraser A, Kamath RS, Ahringer J, et al. Genes that act downstream of DAF-16 to influence the lifespan of Caenorhabditis elegans. *Nature* 2003;**424**:277–83.
113. Lee SS, Kennedy S, Tolonen AC, Ruvkun G. DAF-16 target genes that control C. elegans life-span and metabolism. *Science* 2003;**300**:644–7.
114. Lin HK, Chen Z, Wang G, Nardella C, Lee SW, Chan CH, et al. Skp2 targeting suppresses tumorigenesis by Arf-p53-independent cellular senescence. *Nature* 2010;**464**:374–9.
115. Li W, Gao B, Lee SM, Bennett K, Fang D. RLE-1, an E3 ubiquitin ligase, regulates C. elegans aging by catalyzing DAF-16 polyubiquitination. *Dev Cell* 2007;**12**:235–46.
116. Carrano AC, Liu Z, Dillin A, Hunter T. A conserved ubiquitination pathway determines longevity in response to diet restriction. *Nature* 2009;**460**:396–9.
117. Cao X, Xue L, Han L, Ma L, Chen T, Tong T. WW domain-containing E3 ubiquitin protein ligase 1 (WWP1) delays cellular senescence by promoting p27(Kip1) degradation in human diploid fibroblasts. *J Biol Chem* 2011;**286**:33447–56.
118. Choe KP, Przybysz AJ, Strange K. The WD40 repeat protein WDR-23 functions with the CUL4/DDB1 ubiquitin ligase to regulate nuclear abundance and activity of SKN-1 in Caenorhabditis elegans. *Mol Cell Biol* 2009;**29**:2704–15.
119. Hasegawa K, Miwa J. Genetic and cellular characterization of Caenorhabditis elegans mutants abnormal in the regulation of many phase II enzymes. *PLoS One* 2010;**5**:e11194.
120. Sykiotis GP, Bohmann D. Keap1/Nrf2 signaling regulates oxidative stress tolerance and life-span in Drosophila. *Dev Cell* 2008;**14**:76–85.
121. Kim W, Kaelin Jr. WG. The von Hippel-Lindau tumor suppressor protein: new insights into oxygen sensing and cancer. *Curr Opin Genet Dev* 2003;**13**:55–60.
122. Mehta R, Steinkraus KA, Sutphin GL, Ramos FJ, Shamieh LS, Huh A, et al. Proteasomal regulation of the hypoxic response modulates aging in C. elegans. *Science* 2009; **324**:1196–8.
123. Young AP, Schlisio S, Minamishima YA, Zhang Q, Li L, Grisanzio C, et al. VHL loss actuates a HIF-independent senescence programme mediated by Rb and p400. *Nat Cell Biol* 2008;**10**:361–9.
124. Gaczynska M, Rock K, Spies T, Goldberg AL. Peptidase activities of proteasomes are differentially regulated by the major histocompatibility complex-encoded genes for LMP2 and LMP7. *Proc Natl Acad Sci USA* 1994;**91**:9213–7.
125. Gaczynska M, Goldberg AL, Tanaka K, Hendil KB, Rock K. Proteasome subunits X and Y alter peptidase activities in opposite ways to the interferon-gamma-induced subunits LMP2 and LMP7. *J Biol Chem* 1996;**271**:17275–80.
126. Chondrogianni N, Tzavelas C, Pemberton AJ, Nezis IP, Rivett AJ, Gonos ES. Overexpression of proteasome beta 5 subunit increases the amount of assembled proteasome and confers ameliorated response to oxidative stress and higher survival rates. *J Biol Chem* 2005;**280**:11840–50.
127. Liu Y, Liu X, Zhang T, Luna C, Liton PB, Gonzalez P. Cytoprotective effects of proteasome bet5 subunit overexpression in lens epithelial cells. *Mol Vis* 2007;**13**:31–8.
128. Kwak MK, Cho JM, Huang B, Shin S, Kensler TW. Role of increased expression of the proteasome in the protective effects of sulphoraphane against hydrogen peroxide-mediated cytotoxicity in murine neuroblastoma cells. *Free Radic Biol Med* 2007;**43**:809–17.
129. Hwang JS, Hwang JS, Chang I, Kim S. Age-associated decrease in proteasome content and activities in human dermal fibroblasts: restoration of normal level of proteasome subunits reduces aging markers in fibroblasts from elderly persons. *J Gerontol A Biol Sci Med Sci* 2007;**62**:490–9.

130. Ramos PC, Hockendorff J, Johnson ES, Varshavsky A, Dohmen RJ. Ump1p is required for proper maturation of the 20S proteasome and becomes its substrate upon completion of the assembly. *Cell* 1998;**92**:489–99.
131. Witt E, Zantopf D, Schmidt M, Kraft R, Kloetzel PM, Kruger E. Characterisation of the newly identified human Ump1 homologue POMP and analysis of LMP7 (beta 5i) incorporation into 20S proteasomes. *J Mol Biol* 2000;**301**:1–9.
132. Burri L, Hockendorff J, Boehm U, Klamp T, Dohmen RJ, Lévy F. Identification and characterization of a mammalian protein interacting with 20S proteasome precursors. *Proc Natl Acad Sci USA* 2000;**97**:10348–53.
133. Griffin TA, Slack JP, McCluskey TS, Monaco J, Colbert RA. Identification of proteassemblin, a mammalian homologue of the yeast protein, Ump1p, that is required for normal proteasome assembly. *Mol Cell Biol Res Commun* 2000;**3**:212–7.
134. Chondrogianni N, Gonos ES. Overexpression of hUMP1/POMP proteasome accessory protein enhances proteasome-mediated antioxidant defence. *Exp Gerontol* 2007;**42**:899–903.
135. Chen Q, Thorpe J, Dohmen JR, Li F, Keller JN. Ump1 extends yeast lifespan and enhances viability during oxidative stress: central role for the proteasome? *Free Radic Biol Med* 2006;**40**:120–6.
136. Chen Q, Ding Q, Thorpe J, Dohmen JR, Keller JN. RNA interference toward UMP1 induces proteasome inhibition in *Saccharomyces cerevisiae*: evidence for protein oxidation and autophagic cell death. *Free Radic Biol Med* 2005;**38**:226–34.
137. Chen Q, Thorpe J, Ding Q, El-Amouri IS, Keller JN. Proteasome synthesis and assembly are required for survival during stationary phase. *Free Radic Biol Med* 2004;**37**:859–68.
138. Mannhaupt G, Schnall R, Karpov V, Vetter I, Feldmann H. Rpn4p acts as a transcription factor by binding to PACE, a nonamer box found upstream of 26S proteasomal and other genes in yeast. *FEBS Lett* 1999;**450**:27–34.
139. Ju D, Wang L, Mao X, Xie Y. Homeostatic regulation of the proteasome via an Rpn4-dependent feedback circuit. *Biochem Biophys Res Commun* 2004;**321**:51–77.
140. Hahn JS, Neef DW, Thiele DJ. A stress regulatory network for co-ordinated activation of proteasome expression mediated by yeast heat shock transcription factor. *Mol Microbiol* 2006;**60**:240–51.
141. Wang X, Xu H, Ju D, Xie Y. Disruption of Rpn4-induced proteasome expression in *Saccharomyces cerevisiae* reduces cell viability under stressed conditions. *Genetics* 2008;**180**:1945–53.
142. Wang X, Xu H, Ha SW, Ju D, Xie Y. Proteasomal degradation of Rpn4 in *Saccharomyces cerevisiae* is critical for cell viability under stressed conditions. *Genetics* 2010;**184**:335–42.
143. Kruegel U, Robison B, Dange T, Kahlert G, Delaney JR, Kotireddy S, et al. Elevated proteasome capacity extends replicative lifespan in *Saccharomyces cerevisiae*. *PLoS Genet* 2011;**7**:e1002253.
144. Medina DX, Caccamo A, Oddo S. Methylene blue reduces aβ levels and rescues early cognitive deficit by increasing proteasome activity. *Brain Pathol* 2011;**21**:140–9.
145. Gura T. Hope in Alzheimer's fight emerges from unexpected places. *Nat Med* 2008;**14**:894.
146. Dahlmann B, Rutschmann M, Kuehn L, Reinauer H. Activation of the multicatalytic proteinase from rat skeletal muscle by fatty acids or sodium dodecyl sulphate. *Biochem J* 1985;**228**:171–7.
147. Watanabe N, Yamada S. Activation of the 20S proteasomes from spinach leaves by fatty acids. *Plant Cell Physiol* 1996;**37**:147–51.
148. Kohler A, Cascio P, Leggett DS, Woo KM, Goldberg AL, Finley D. The axial channel of the proteasome core particle is gated by the Rpt2 ATPase and controls both substrate entry and product release. *Mol Cell* 2001;**7**:1143–52.

149. Kisselev AF, Kaganovich D, Goldberg AL. Binding of hydrophobic peptides to several non-catalytic sites promotes peptide hydrolysis by all active sites of 20 S proteasomes. Evidence for peptide-induced channel opening in the alpha-rings. *J Biol Chem* 2002;**277**:22260–70.
150. Wilk S, Chen WE. Synthetic peptide-based activators of the proteasome. *Mol Biol Rep* 1997;**24**:119–24.
151. Ohkubo I, Gasa S, Namikawa C, Makita A, Sasaki M. Human erythrocyte multicatalytic proteinase: activation and binding to sulphated galacto- and lactosylceramides. *Biochem Biophys Res Commun* 1991;**174**:1133–40.
152. Matsumura K, Aketa K. Proteasome (multicatalytic proteinase) of sea urchin sperm and its possible participation in the acrosome reaction. *Mol Reprod Dev* 1991;**29**:189–99.
153. Ruiz de Mena I, Mahillo E, Arribas J, Castano JG. Kinetic mechanism of activation by cardiolipin (diphosphatidylglycerol) of the rat liver multicatalytic proteinase. *Biochem J* 1993;**296**:93–7.
154. Bonoli M, Bendini A. Qualitative and semiquantitative analysis of phenoloic compounds in extra virgin olive oils as a function of the ripening degree of olive fruits by different analytical techniques. *J Agric Food Chem* 2004;**52**:7026–32.
155. Katsiki M, Chondrogianni N, Chinou I, Rivett AJ, Gonos ES. The olive constituent oleuropein exhibits proteasome stimulatory properties in vitro and confers life span extension of human embryonic fibroblasts. *Rejuvenation Res* 2007;**10**:157–72.
156. Bulteau AL, Moreau M, Saunois A, Nizard C, Friguet B. Algae extract-mediated stimulation and protection of proteasome activity within human keratinocytes exposed to UVA and UVB irradiation. *Antioxid Redox Signal* 2006;**8**:136–43.
157. Huang L, Ho P, Chen CH. Activation and inhibition of the proteasome by betulinic acid and its derivatives. *FEBS Lett* 2007;**581**:4955–9.
158. Kwak MK, Wakabayashi N, Greenlaw JL, Yamamoto M, Kensler TW. Antioxidants enhance mammalian proteasome expression through the Keap1-Nrf2 signaling pathway. *Mol Cell Biol* 2003;**23**:8786–94.
159. Nguyen T, Yang CS, Pickett CB. The pathways and molecular mechanisms regulating Nrf2 activation in response to chemical stress. *Free Radic Biol Med* 2004;**37**:433–41.
160. Wasserman WW, Fahl WE. Functional antioxidant responsive elements. *Proc Natl Acad Sci USA* 1997;**94**:5361–6.
161. Furukawa M, Xiong Y. BTB protein Keap1 targets antioxidant transcription factor Nrf2 for ubiquitination by the Cullin 3-Roc1 ligase. *Mol Cell Biol* 2005;**25**:162–71.
162. Surh YJ, Kundu JK, Na HK. Nrf2 as a master redox switch in turning on the cellular signalling involved in the induction of cytoprotective genes by some chemopreventive phytochemicals. *Planta Med* 2008;**74**:1526–39.
163. Itoh K, Chiba T, Takahashi S, Ishii T, Igarashi K, Katoh Y, et al. An Nrf2/small Maf heterodimer mediates the induction of phase II detoxifying enzyme genes through antioxidant response elements. *Biochem Biophys Res Commun* 1997;**236**:313–22.
164. Venugopal R, Jaiswal AK. Nrf2 and Nrf1 in association with Jun proteins regulate antioxidant response element-mediated expression and coordinated induction of genes encoding detoxifying enzymes. *Oncogene* 1998;**17**:3145–56.
165. Zhao J, Kobori N, Aronowski J, Dash PK. Sulforaphane reduces infarct volume following focal cerebral ischemia in rodents. *Neurosci Lett* 2006;**393**:108–12.
166. Zhao J, Moore AN, Redell JB, Dash PK. Enhancing expression of Nrf2-driven genes protects the blood brain barrier after brain injury. *J Neurosci* 2007;**27**:10240–8.
167. Zhao X, Sun G, Zhang J, Strong R, Dash PK, Kan YW, et al. Transcription factor Nrf2 protects the brain from damage produced by intracerebral hemorrhage. *Stroke* 2007;**38**:3280–6.
168. Dash PK, Zhao J, Orsi SA, Zhang M, Moore AN. Sulforaphane improves cognitive function administered following traumatic brain injury. *Neurosci Lett* 2009;**460**:103–7.

169. Paupe V, Dassa EP, Goncalves S, Auchère F, Lönn M, Holmgren A, et al. Impaired nuclear Nrf2 translocation undermines the oxidative stress response in Friedreich ataxia. *PLoS One* 2009;**4**:e4253.
170. Ramsey CP, Glass CA, Montgomery MB, Lindl KA, Ritson GP, Chia LA, et al. Expression of Nrf2 in neurodegenerative diseases. *J Neuropathol Exp Neurol* 2007;**66**:75–85.
171. Guan CP, Zhou MN, Xu AE, Kang KF, Liu JF, Wei XD, et al. The susceptibility to vitiligo is associated with NF-E2-related factor2 (Nrf2) gene polymorphisms: a study on Chinese Han population. *Exp Dermatol* 2008;**17**:1059–62.
172. Marzec JM, Christie JD, Reddy SP, Jedlicka AE, Vuong H, Lanken PN, et al. Functional polymorphisms in the transcription factor NRF2 in humans increase the risk of acute lung injury. *FASEB J* 2007;**21**:2237–46.
173. Arisawa T, Tahara T, Shibata T, Nagasaka M, Nakamura M, Kamiya Y, et al. The relationship between Helicobacter pylori infection and promoter polymorphism of the Nrf2 gene in chronic gastritis. *Int J Mol Med* 2007;**19**:143–8.
174. Arisawa T, Tahara T, Shibata T, Nagasaka M, Nakamura M, Kamiya Y, et al. Association between promoter polymorphisms of nuclear factor-erythroid 2-related factor 2 gene and peptic ulcer diseases. *Int J Mol Med* 2007;**20**:849–53.
175. Arisawa T, Tahara T, Shibata T, Nagasaka M, Nakamura M, Kamiya Y, et al. Nrf2 gene promoter polymorphism is associated with ulcerative colitis in a Japanese population. *Hepatogastroenterology* 2008;**55**:394–7.
176. Sykiotis GP, Bohmann D. Stress-activated cap'n'collar transcription factors in aging and human disease. *Sci Signal* 2010;**3**:re3.
177. Eggler AL, Gay KA, Mesecar AD. Molecular mechanisms of natural products in chemoprevention: induction of cytoprotective enzymes by Nrf2. *Mol Nutr Food Res* 2008;**52**:S84–94.
178. Lau A, Villeneuve NF, Sun Z, Wong PK, Zhang DD. Dual roles of Nrf2 in cancer. *Pharmacol Res* 2008;**58**:262–70.
179. Kapeta S, Chondrogianni N, Gonos ES. Nuclear erythroid factor 2-mediated proteasome activation delays senescence in human fibroblasts. *J Biol Chem* 2010;**285**:8171–84.
180. Chondrogianni N, Kapeta S, Chinou I, Vassilatou K, Papassideri I, Gonos ES. Anti-ageing and rejuvenating effects of quercetin. *Exp Gerontol* 2010;**45**:763–71.
181. Tanigawa S, Fujii M, Hou DX. Action of Nrf2 and Keap1 in ARE-mediated NQO1 expression by quercetin. *Free Radic Biol Med* 2007;**42**:1690–703.
182. Kampkötter A, Timpel C, Zurawski RF, Ruhl S, Chovolou Y, Proksch P, et al. Increase of stress resistance and lifespan of *Caenorhabditis elegans* by quercetin. *Comp Biochem Physiol B Biochem Mol Biol* 2008;**149**:314–23.
183. Pietsch K, Saul N, Menzel R, Stürzenbaum SR, Steinberg CE. Quercetin mediated lifespan extension in *Caenorhabditis elegans* is modulated by age-1, daf-2, sek-1 and unc-43. *Biogerontology* 2009;**10**:565–78.
184. Chan JY, Han XL, Kan YW. Cloning of Nrf1, an NF-E2-related transcription factor, by genetic selection in yeast. *Proc Natl Acad Sci USA* 1993;**90**:11371–5.
185. Luna L, Johnsen O, Skartlien AH, Pedeutour F, Turc-Carel C, Prydz H, et al. Molecular cloning of a putative novel human bZIP transcription factor on chromosome 17q22. *Genomics* 1994;**22**:553–62.
186. Biswas M, Chan JY. Role of Nrf1 in antioxidant response element-mediated gene expression and beyond. *Toxicol Appl Pharmacol* 2010;**244**:16–20.
187. Radhakrishnan SK, Lee CS, Young P, Beskow A, Chan JY, Deshaies RJ. Transcription factor Nrf1 mediates the proteasome recovery pathway after proteasome inhibition in mammalian cells. *Mol Cell* 2010;**38**:17–28.
188. Steffen J, Seeger M, Koch A, Krüger E. Proteasomal degradation is transcriptionally controlled by TCF11 via an ERAD-dependent feedback loop. *Mol Cell* 2010;**40**:147–58.

189. Whitehouse AS, Tisdale MJ. Increased expression of the ubiquitin-proteasome pathway in murine myotubes by proteolysis-inducing factor (PIF) is associated with activation of the transcription factor NF-kappaB. *Br J Cancer* 2003;**89**:1116–22.
190. Smith HJ, Wyke SM, Tisdale MJ. Role of protein kinase C and NF-kappaB in proteolysis-inducing factor-induced proteasome expression in C(2)C(12) myotubes. *Br J Cancer* 2004;**90**:1850–7.
191. Wyke SM, Russell ST, Tisdale MJ. Induction of proteasome expression in skeletal muscle is attenuated by inhibitors of NF-kappaB activation. *Br J Cancer* 2004;**91**:1742–50.
192. Lorite MJ, Smith HJ, Arnold JA, Morris A, Thompson MG, Tisdale MJ. Activation of ATP-ubiquitin-dependent proteolysis in skeletal muscle in vivo and murine myoblasts in vitro by a proteolysis-inducing factor (PIF). *Br J Cancer* 2001;**85**:297–302.
193. Seo H, Sonntag KC, Kim W, Cattaneo E, Isacson O. Proteasome activator enhances survival of huntington's disease neuronal model cells. *PLoS One* 2007;**2**:e238.
194. Tonoki A, Kuranaga E, Tomioka T, Hamazaki J, Murata S, Tanaka K, et al. Genetic evidence linking age-dependent attenuation of the 26S proteasome with the aging process. *Mol Cell Biol* 2009;**29**:1095–106.
195. Yun C, Stanhill A, Yang Y, Zhang Y, Haynes CM, Xu CF, et al. Proteasomal adaptation to environmental stress links resistance to proteotoxicity with longevity in Caenorhabditis elegans. *Proc Natl Acad Sci USA* 2008;**105**:7094–9.
196. Hassan WM, Merin DA, Fonte V, Link CD. AIP-1 ameliorates beta-amyloid peptide toxicity in a *Caenorhabditis elegans* Alzheimer's disease model. *Hum Mol Genet* 2009;**18**:2739–47.
197. Sok J, Calfon M, Lu J, Lichtlen P, Clark SG, Ron D. Arsenite-inducible RNA-associated protein (AIRAP) protects cells from arsenite toxicity. *Cell Stress Chaperones* 2001;**6**:6–15.
198. Stanhill A, Haynes CM, Zhang Y, Min G, Steele MC, Kalinina J, et al. An arsenite-inducible 19S regulatory particle-associated protein adapts proteasomes to proteotoxicity. *Mol Cell* 2006;**23**:875–85.
199. Rousseau E, Kojima R, Hoffner G, Djian P, Bertolotti A. Misfolding of proteins with a polyglutamine expansion is facilitated by proteasomal chaperones. *J Biol Chem* 2009;**284**:1917–29.
200. Kumatori A, Tanaka K, Inamura N, Sone S, Ogura T, Matsumoto T, et al. Abnormally high expression of proteasomes in human leukemic cells. *Proc Natl Acad Sci USA* 1990;**87**:7071–5.
201. Kanayama H, Tanaka K, Aki M, Kagawa S, Miyaji H, Satoh M, et al. Changes in expressions of proteasome and ubiquitin genes in human renal cancer cells. *Cancer Res* 1991;**51**:6677–85.
202. Montero AJ, Jassem J. Cellular redox pathways as a therapeutic target in the treatment of cancer. *Drugs* 2011;**71**:1385–96.
203. Rock KL, York IA, Saric T, Goldberg AL. Protein degradation and the generation of MHC class I-presented peptides. *Adv Immunol* 2002;**80**:1–70.
204. Llovera M, García-Martínez C, Agell N, López-Soriano FJ, Argilés JM. TNF can directly induce the expression of ubiquitin-dependent proteolysis in rat soleus muscles. *Biochem Biophys Res Commun* 1997;**230**:238–41.
205. Hershko A. The ubiquitin system for protein degradation and some of its roles in the control of the cell division cycle. *Cell Death Differ* 2005;**12**:1191–7.
206. Kitagawa K, Kotake Y, Kitagawa M. Ubiquitin-mediated control of oncogene and tumor suppressor gene products. *Cancer Sci* 2009;**100**:1374–81.
207. Maki CG, Huibregtse JM, Howley PM. In vivo ubiquitination and proteasome-mediated degradation of p53. *Cancer Res* 1996;**56**:2649–54.
208. Pagano M, Tam SW, Theodoras AM, Beer-Romero P, Del Sal G, Chau V, et al. Role of the ubiquitin-proteasome pathway in regulating abundance of the cyclin-dependent kinase inhibitor p27. *Science* 1995;**269**:682–95.

209. Hershko A, Ciechanover A. The ubiquitin system. *Annu Rev Biochem* 1998;**67**:425–79.
210. Adams J, Palombella VJ, Sausville EA, Johnson J, Destree A, Lazarus DD, et al. Proteasome inhibitors: a novel class of potent and effective antitumor agents. *Cancer Res* 1999;**59**:2615–22.
211. Oerlemans R, Franke NE, Assaraf YG, Cloos J, van Zantwijk I, Berkers CR, et al. Molecular basis of bortezomib resistance: proteasome subunit beta5 (PSMB5) gene mutation and overexpression of PSMB5 protein. *Blood* 2008;**112**:2489–99.
212. de Wilt LH, Jansen G, Assaraf YG, van Meerloo J, Cloos J, Schimmer AD, et al. Proteasome-based mechanisms of intrinsic and acquired bortezomib resistance in non-small cell lung cancer. *Biochem Pharmacol* 2012;**83**:207–17.
213. Franke NE, Niewerth D, Assaraf YG, van Meerloo J, Vojtekova K, van Zantwijk CH, et al. Impaired bortezomib binding to mutant β5 subunit of the proteasome is the underlying basis for bortezomib resistance in leukemia cells. *Leukemia* 2012;**26**:757–68.
214. Lü S, Chen Z, Yang J, Chen L, Gong S, Zhou H, et al. Overexpression of the PSMB5 gene contributes to bortezomib resistance in T-lymphoblastic lymphoma/leukemia cells derived from Jurkat line. *Exp Hematol* 2008;**36**:1278–84.
215. Nakayama K, Nakayama KI. Ubiquitin ligases: cell-cycle control and cancer. *Nat Rev Cancer* 2006;**6**:369–81.
216. Kipreos ET, Pagano M. The F-box protein family. *Genome Biol* 2000;**1**:1–7.
217. Carrano AC, Eytan E, Hershko A, Pagano M. SKP2 is required for ubiquitin-mediated degradation of the CDK inhibitor p27. *Nat Cell Biol* 1999;**1**:193–9.
218. Ben-Izhak O, Lahav-Baratz S, Meretyk S, Ben-Eliezer S, Sabo E, Dirnfeld M, et al. Inverse relationship between Skp2 ubiquitin ligase and the cyclin dependent kinase inhibitor p27Kip1 in prostate cancer. *J Urol* 2003;**170**:241–5.
219. Fukuchi M, Masuda N, Nakajima M, Fukai Y, Miyazaki T, Kato H, et al. Inverse correlation between expression levels of p27 and the ubiquitin ligase subunit Skp2 in early esophageal squamous cell carcinoma. *Anticancer Res* 2004;**24**:777–83.
220. Li Q, Murphy M, Ross J, Sheehan C, Carlson JA. Skp2 and p27kip1 expression in melanocytic nevi and melanoma: an inverse relationship. *J Cutan Pathol* 2004;**31**:633–42.
221. Shigemasa K, Gu L, O'Brien TJ, Ohama K. Skp2 overexpression is a prognostic factor in patients with ovarian adenocarcinoma. *Clin Cancer Res* 2003;**9**:1756–63.
222. Hershko D, Bornstein G, Ben-Izhak O, Carrano A, Pagano M, Krausz MM, et al. Inverse relation between levels of p27(Kip1) and of its ubiquitin ligase subunit Skp2 in colorectal carcinomas. *Cancer* 2001;**91**:1745–51.
223. Radke S, Pirkmaier A, Germain D. Differential expression of the F-box proteins Skp2 and Skp2B in breast cancer. *Oncogene* 2005;**24**:3448–58.
224. Yokoi S, Yasui K, Saito-Ohara F, Koshikawa K, Iizasa T, Fujisawa T, et al. A novel target gene, SKP2, within the 5p13 amplicon that is frequently detected in small cell lung cancers. *Am J Pathol* 2002;**161**:207–16.
225. Welcker M, Clurman BE. FBW7 ubiquitin ligase: a tumour suppressor at the crossroads of cell division, growth and differentiation. *Nat Rev Cancer* 2008;**8**:83–93.
226. Thompson BJ, Buonamici S, Sulis ML, Palomero T, Vilimas T, Basso G, et al. The SCFFBW7 ubiquitin ligase complex as a tumor suppressor in T cell leukemia. *J Exp Med* 2007;**204**:1825–35.
227. Bernassola F, Karin M, Ciechanover A, Melino G. The HECT family of E3 ubiquitin ligases: multiple players in cancer development. *Cancer Cell* 2008;**14**:10–21.
228. Inoue Y, Imamura T. Regulation of TGF-β family signaling by E3 ubiquitin ligases. *Cancer Sci* 2008;**99**:2107–12.
229. Chen D, Kon N, Li M, Zhang W, Qin J, Gu W. ARF-BP1/Mule is a critical mediator of the ARF tumor suppressor. *Cell* 2005;**121**:1071–83.

230. Rossi M, De Laurenzi V, Munarriz E, Green DR, Liu YC, Vousden KH, et al. The ubiquitin-protein ligase Itch regulates p73 stability. *EMBO J* 2005;**24**:836–48.
231. Melino G, Knight RA, Cesareni G. Degradation of p63 by Itch. *Cell Cycle* 2006;**5**:1735–9.
232. Li Y, Zhou Z, Chen C. WW domain-containing E3 ubiquitin protein ligase 1 targets p63 transcription factor for ubiquitin-mediated proteasomal degradation and regulates apoptosis. *Cell Death Differ* 2008;**15**:1941–51.
233. Brooks CL, Gu W. p53 ubiquitination: Mdm2 and beyond. *Mol Cell* 2006;**21**:307–15.
234. Onel K, Cordon-Cardo C. MDM2 and prognosis. *Mol Cancer Res* 2004;**2**:1–8.
235. Duan W, Gao L, Druhan LJ, Zhu WG, Morrison C, Otterson GA, et al. Expression of Pirh2, a newly identified ubiquitin protein ligase, in lung cancer. *J Natl Cancer Inst* 2004;**96**:1718–21.
236. Logan IR, Gaughan L, McCracken SR, Sapountzi V, Leung HY, Robson CN. Human PIRH2 enhances androgen receptor signaling through inhibition of histone deacetylase 1 and is overexpressed in prostate cancer. *Mol Cell Biol* 2006;**26**:6502–10.
237. Shimada M, Kitagawa K, Dobashi Y, Isobe T, Hattori T, Uchida C, et al. High expression of Pirh2, an E3 ligase for p27, is associated with low expression of p27 and poor prognosis in head and neck cancers. *Cancer Sci* 2009;**100**:866–72.
238. Dornan D, Bheddah S, Newton K, Ince W, Frantz GD, Dowd P, et al. COP1, the negative regulator of p53, is overexpressed in breast and ovarian adenocarcinomas. *Cancer Res* 2004;**64**:7226–30.
239. Uchida C, Miwa S, Kitagawa K, Hattori T, Isobe T, Otani S, et al. Enhanced Mdm2 activity inhibits pRB function via ubiquitin-dependent degradation. *EMBO J* 2005;**24**:160–9.
240. Miwa S, Uchida C, Kitagawa K, Hattori T, Oda T, Sugimura H, et al. Mdm2-mediated pRB downregulation is involved in carcinogenesis in a p53-independent manner. *Biochem Biophys Res Commun* 2006;**340**:54–61.
241. Ohh M. Ubiquitin pathway in VHL cancer syndrome. *Neoplasia* 2006;**8**:623–9.
242. Kapitsinou PP, Haase VH. The VHL tumor suppressor and HIF: insights from genetic studies in mice. *Cell Death Differ* 2008;**15**:650–9.
243. Adams J, Behnke M, Chen S, Cruickshank AA, Dick LR, Grenier L, et al. Potent and selective inhibitors of the proteasome: dipeptidyl boronic acids. *Bioorg Med Chem Lett* 1998;**8**:333–8.
244. Adams J. Development of the proteasome inhibitor PS-341. *Oncologist* 2002;**7**:9–16.
245. Pervan M, Pajonk F, Sun JR, Withers HR, McBride WH. Molecular pathways that modify tumor radiation response. *Am J Clin Oncol* 2001;**24**:481–5.
246. Cvek B, Dvorak Z. The ubiquitin-proteasome system (UPS) and the mechanism of action of bortezomib. *Curr Pharm Des* 2011;**17**:1483–99.

# Immunoproteasomes: Structure, Function, and Antigen Presentation

Deborah A. Ferrington and Dale S. Gregerson

*Department of Ophthalmology, University of Minnesota, Minneapolis, Minnesota, USA*

I. Introduction .......... 75
II. Immunoproteasome Structure .......... 77
III. Regulation of Gene Expression .......... 79
IV. Assembly of the 20S Core .......... 82
V. Enzymatic Activity .......... 84
VI. Immunoproteasome Knockout Mice .......... 86
VII. Immunoproteasome Function .......... 88
A. Antigen Processing and the Immune Response .......... 90
B. Nonimmune Roles for I-proteasome .......... 95
VIII. Mutations and Linkage to Human Disease .......... 101
References .......... 105

Immunoproteasomes contain replacements for the three catalytic subunits of standard proteasomes. In most cells, oxidative stress and proinflammatory cytokines are stimuli that lead to elevated production of immunoproteasomes. Immune system cells, especially antigen-presenting cells, express a higher basal level of immunoproteasomes. A well-described function of immunoproteasomes is to generate peptides with a hydrophobic C terminus that can be processed to fit in the groove of MHC class I molecules. This display of peptides on the cell surface allows surveillance by CD8 T cells of the adaptive immune system for pathogen-infected cells. Functions of immunoproteasomes, other than generating peptides for antigen presentation, are emerging from studies in immunoproteasome-deficient mice, and are complemented by recently described diseases linked to mutations or single-nucleotide polymorphisms in immunoproteasome subunits. Thus, this growing body of literature suggests a more pleiotropic role in cell function for the immunoproteasome.

**Abbreviations:** IFN-γ, interferon-gamma; LMP, low-molecular-weight protein; MECL-1, Multicatalytic endopeptidase complex subunit 1; MHC class I, major histocompatibility complex class I

## I. Introduction

While the first descriptions of the proteasome were published in the 1970s,[1,2] it was not until the 1990s that the scientific community became aware of a specialized form of the proteasome[3–6] that was later called the

Progress in Molecular Biology
and Translational Science, Vol. 109
DOI: 10.1016/B978-0-12-397863-9.00003-1

immunoproteasome (i-proteasome).[7] The name i-proteasome was first given to 20S cores that had incorporated the low-molecular-weight proteins (LMPs) LMP2 and LMP7. These proteins were significantly upregulated in response to the major immunomodulatory cytokine interferon-gamma (IFN-$\gamma$). An additional factor influencing the choice of name was that the genes encoding LMP2 and LMP7 were located within the major histocompatibility complex (MHC) class II region. The third i-proteasome subunit, the multicatalytic endopeptidase complex subunit 1 (MECL-1), was discovered about a decade later.[8–10] This subunit also responds to IFN-$\gamma$ stimulation; however, the gene that encodes this protein lies outside the MHC class II region.

Proteasome subtypes are defined by their catalytic subunits. The standard proteasome catalytic subunits include β1, β2, and β5, which are constitutively expressed in all cells. The i-proteasome catalytic subunits, also known as the inducible subunits, are LMP2, MECL-1, and LMP7. An intermediate proteasome containing a mixture of both the standard and i-proteasome subunits has also been described. The most recent proteasome subtype to be discovered is the thymus-specific proteasome, which substitutes the β5 subunit with an alternate protein (β5t).[11]

The nomenclature for the various proteasome subunits is quite confusing because multiple names have been assigned to each subunit. This is due in part to contributions from multiple laboratories working on different organisms. Table I provides a list of names for the different catalytic subunits and should be a useful reference for future reading, as the nomenclature is not completely standardized. The UniProtKB nomenclature and gene names are also provided to aid in accessing online information. For the purpose of this chapter, we will

TABLE I

NOMENCLATURE AND ACTIVITY FOR PROTEASOME CATALYTIC SUBUNITS

| Proteasome subunit[a] | Common name[b] | Alternative names[c] | Gene | Activity |
|---|---|---|---|---|
| Beta type-6 | β1 | Y, Delta, LMP19, *Pre3* | *PSMB6* | Caspase-like |
| Beta type-9 | β1i | <u>Lmp2</u>, RING12, MC7 | *PSMB9* | Chymotrypsin-like |
| Beta type-7 | β2 | Z, MC14, Lmp9, *Pup1* | *PSMB7* | Trypsin-like |
| Beta type-10 | β2i | <u>MECL-1</u>, Lmp10 | *PSMB10* | Trypsin-like |
| Beta type-5 | β5 | X, epsilon, Lmp17, MB1, *Pre2*, Doa3, Prgl | *PSMB5* | Chymotrypsin-like |
| Beta type-8 | β5i | <u>Lmp7</u>, RING10, Y2, C13 | *PSMB8* | Chymotrypsin-like |
| Beta type-11 | β5t | Thymus-specific β5 | *PSMB11* | Chymotrypsin-like |

[a]UniProtKB nomenclature (www.uniprot.org).
[b]Baumeister,[12] names in italics are for yeast subunits.
[c]Underlined alternative names will be used for the i-proteasome subunits.

use the common name for the standard and thymoproteasome subunits (β1, β2, β5, β5t) and the alternate names for the i-proteasome subunits (LMP2, LMP7, MECL-1).

## II. Immunoproteasome Structure

The basic structure of all proteasome subtypes is essentially the same. Each 20S core particle is composed of four stacked rings of seven subunits each. The two outer rings contain the constitutively expressed α-subunits, which interact with regulatory complexes such as PA28 and PA700. The two inner rings contain the β-subunits. Three of the β-subunits in each ring perform distinct proteolytic activities. In the standard proteasome, activity of the catalytic subunits β1, β2, and β5 have been classified as caspase-like, trypsin-like, and chymotrypsin-like for cleavage after acidic, basic, and hydrophobic amino acids, respectively (Table I). The standard catalytic subunits β1, β2, and β5 can be replaced in nascent proteasome cores by the inducible subunits LMP2, MECL-1, and LMP7, respectively. While the MECL-1 and LMP7 subunits perform the same type of activities as the β2 and β5, the LMP2 subunit performs chymotrypsin-like activity and cleaves after hydrophobic amino acids. The structural basis for this change is discussed in a later section. It has been suggested that the altered activity of the LMP2 subunit facilitates the generation of peptides for antigen presentation, which requires peptides with hydrophobic amino acids in the C-terminal position.

Each catalytic subunit is expressed with a propeptide that ranges in size from 20 to 72 amino acids (Table II). Cleavage of the propeptide is essential for maturation of the 20S core and activation of the catalytic threonine residue. Thus, the difference in the molecular mass of the unprocessed and the mature protein allows one to distinguish these two species on a high-percentage sodium dodecyl sulfate gel. A comparison of the amino acids and molecular mass of the unprocessed and mature proteins for humans and mice is provided in Table II. For most of the proteins, there is good agreement between the two species, except for the presence of the two isoforms of the β5 and LMP7 subunits that are unique to humans.

There is high sequence homology between the standard catalytic subunits and their i-proteasome correlates (Fig. 1A). The percentage of amino acid sequence identities is between 60% and 70%. When considering the conservative substitutions in each subunit, the amino acid sequence similarities range from 76% to 83%. The amino acid sequence for each subunit is also highly conserved between species. Sequence comparison for humans and mice shows that the primary sequence is 83–96% identical and 90–100% similar (Fig. 1B). Thus, it appears that there is strong evolutionary pressure to maintain the primary and secondary structure for all of the proteasome catalytic subunits.

TABLE II

PROTEASOME CATALYTIC SUBUNIT SIZE

| Subunit name | Primary Acc. (#)[a] | | Amino acids (#)[b] | | Molecular weight (Da)[b] | | Pro-pep (# AA) |
|---|---|---|---|---|---|---|---|
| | Human | Mouse | Human | Mouse | Human | Mouse | |
| β1 | P28072 | Q60692 | 239/205 | 238/205 | 25358/21904 | 25379/21997 | 33/32 |
| LMP2 | P28065 | P28076 | 219/199 | 219/199 | 23264/21276 | 23397/21326 | 20 |
| β2 | Q99436 | P70195 | 277/234 | 277/234 | 29965/25218 | 29891/25252 | 43 |
| MECL-1 | P40306 | O35955 | 273/234 | 273/234 | 28936/24648 | 29063/24804 | 39 |
| β5[1c] | P28074 | O55234 | 263/204 | 264/205 | 28480/22458 | 28532/22529 | 59 |
| β5[2c] | P28074 | | 203/144 | | 21845/15822 | | 9 |
| LMP7[1c] | P28062 | P28063 | 276/204 | 276/204 | 30354/22660 | 30260/22650 | 72 |
| LMP7[2c] | P28062 | | 272/204 | | 29770/22660 | | 68 |
| β5t | A5LHX3 | Q8BG41 | 300/251 | 302/253 | 32530/27232 | 33220/27851 | 49 |

[a]UniProtKB (www.uniprot.org).
[b]Unprocessed protein/mature protein.
[c]Two isoforms, indicated as 1 and 2, are known for β5 and β5i in humans.

A

| | β1 | LMP2 | β2 | MECL | β5 | LMP7 | β5t |
|---|---|---|---|---|---|---|---|
| β1 | | 78 | 51 | 46 | 48 | 51 | 47 |
| LMP2 | 62 | | 51 | 51 | 51 | 53 | 45 |
| β2 | 30 | 29 | | 76 | 50 | 49 | 49 |
| MECL | 27 | 32 | 59 | | 51 | 49 | 45 |
| β5 | 31 | 31 | 29 | 31 | | 83 | 67 |
| LMP7 | 33 | 32 | 28 | 29 | 71 | | 67 |
| β5t | 32 | 30 | 29 | 28 | 50 | 50 | |

B

| | Identical | Similar |
|---|---|---|
| β1 | 89 | 98 |
| LMP2 | 93 | 98 |
| β2 | 96 | 100 |
| MECL | 87 | 95 |
| β5 | 94 | 99 |
| LMP7 | 88 | 95 |
| β5t | 83 | 90 |

FIG. 1. Sequence homology for proteasome catalytic subunits. Percentage of the amino acid sequence that is identical (boxed in gray) or similar (boxed in white) for proteasome catalytic subunits. (A) Comparison of the primary sequence for catalytic subunits from mouse is shown. (B) Comparison of the primary sequence for each catalytic subunit comparing mouse versus human is shown. Percentages were obtained using the ClustalW algorithm from the UniProt Consortium.

The 20S is the predominant proteasome form in most cells and, in fact, may be the primary mechanism for the ATP-independent degradation of oxidized proteins following an oxidative insult.[13,14] The core particle, containing all standard or i-proteasome catalytic subunits, or a mixture of both standard and i-proteasome subunits, has been described.[15–17] Considering the various combinations of subunits, it is theoretically possible to generate 36 different 20S core complexes that differ in their proteolytic capacity and specificity. This is a dizzying number of proteasome subtypes when one also includes the different regulatory complexes that can associate with each end to form cores with PA700 or PA28 bound, or one molecule each of PA700 and PA28 to form the hybrid proteasome. Of particular importance to the i-proteasome is PA28. This regulatory complex contains seven proteins. In the cytosol, the α- and β-subunits form this complex, but the γ-subunit is present in the nucleus. Expression of PA28 is upregulated by IFN-γ, suggesting a role in regulating i-proteasome function. When this regulatory molecule binds to the outer ring of the α-subunits, it causes structural rearrangement of the N-terminal portion of the α-subunits that opens the central chamber and facilitates access of proteins to the catalytic core.[18]

## III. Regulation of Gene Expression

The two genes that encode LMP2 and LMP7 were first discovered in the early 1980s by Monaco and McDevitt.[19,20] These two genes were subsequently mapped to the MHC class II region, where they are clustered with the TAP-1 and TAP-2 genes. Like LMP2 and LMP7, the expression of TAP proteins is also upregulated by IFN-γ. The TAP proteins are transport molecules embedded in the endoplasmic reticulum (ER), where they shuttle peptides destined for MHC class I antigen presentation from the cytoplasm into the lumen of the ER. The MHC class II region is located on chromosome 6 in humans and chromosome 17 in mice.

The genomic organization of the gene for LMP2 is very similar for both humans and mice. The gene is approximately 2.3 Kb in size and contains six exons of identical length.[21,22] The LMP2 gene is unusual in that it contains a bidirectional promoter that is shared with the TAP-1 gene.[23] Both genes are divergently transcribed from a central promoter region of 539 base pairs that has no TATA box, but rather has several GC boxes that are the likely transcriptional start sites. Transcription is initiated at multiple sites for each gene, but remarkably, there is no overlap in transcripts between genes. The promoter region also contains binding sites for multiple transcription factors, including the interferon consensus sequence-2 and gamma interferon activated sequence elements that bind the dimer signal transducers and activators of transcription-1 (Stat-1) and interferon regulatory factor-1 (IRF-1).[24] The Stat-1/IRF-1 dimers are the major transcription factors involved in IFN-γ signal transduction. Gene expression of LMP2 and LMP7 is strongly upregulated by IFN-γ.[25] The promoter also contains additional binding sites for transcription factors, including 10 GC-rich regions, two cAMP responsive elements, two AP-2 elements, four Egr response elements, and three regions of consensus sequence for NF-κB binding.[21,26] Transcription factors that have been associated with LMP2 regulation include NF-κB, Sp1, AP-1, cAMP responsive element binding protein (CREB), and, more recently, Zif268, also known as Egr1.[21,23,26] Both LMP2 and LMP7 were recently identified as target genes for Zif268, which suppresses transcription of these genes. It has been suggested that Zif268 may modulate the transcriptional response of Sp1 and NF-κB in the LMP2 and LMP7 genes since the binding sites of these three transcription factors lie in close proximity to each other.[26]

There are a few differences in the genomic organization of the LMP7 gene when comparing the human and mouse genomes.[27] In the mouse, there are six exons, while the human gene contains seven exons. The difference occurs in the first human exon, which is modified by numerous insertions and deletions. This exon also contains two frame stop codons that results in the production of two different LMP7 isoforms in humans. In contrast, the mouse gene encodes for only one protein. The difference in amino acid sequence between the two LMP7 isoforms produced by the human gene is limited to the N-terminal portion of the protein, which is removed in the mature protein. Thus, while the catalytic activity is identical, there could be potential differences in assembly in the 20S core particle due to the altered amino acid composition of the N terminus. In both mouse and human, the 3′ end of the TAP-1 gene lies ~1 Kb upstream from the initial translational codon of LMP7. The promoter region contains a TATA box, two GC-rich regions, one cAMP regulatory element, four Sp1 sites, Egr response elements, and one region with the consensus sequence for NF-κB binding. Thus, similar transcription factors regulate both the LMP7 and LMP2 genes.[26,27]

Discovery of the third i-proteasome gene for the MECL-1 subunit (*PSMB10*) occurred much later than the two i-proteasome genes encoded in the MHC class II region. The initial studies identified the MECL-1 subunit as a component of the i-proteasome based on its structural similarities to LMP2 and LMP7 and its responsiveness to IFN-γ.[8,10,28] The gene for MECL-1 is found outside the MHC class II region and is expressed as a single copy.[9,29] In the mouse, it is located on chromosome 8 and in humans, on chromosome 16. Investigation of the genomic organization of the *PSMB10* gene showed similarities between the mouse and human genome. Each gene contained eight exons of equal size with introns inserted at exactly the same place. The mouse gene was slightly larger than the gene in humans (2.5 vs. 2.3 Kb) due to a long interspersed repetitive DNA sequence known as L! that is located upstream of the promoter region. The promoter regions contain no CAAT or TATA boxes and, like other genes missing these elements, MECL-1 has multiple transcriptional start sites. The promoter also contains two elements that confer responsiveness to IFN and multiple additional regions that bind transcription factors, including Sp1, AP-1/2, and NF-κB.

The gene for the β5t subunit of the thymoproteasome was discovered by Murata and colleagues[11] during a genome-wide database search for proteasome-related genes. The β5t gene is located adjacent to the β5 gene (*PSMB5*) and is encoded by a single exon. The β5 and β5t genes are transcribed in opposite directions and, in the mouse, are separated by 7.3 kb on chromosome 14. Based on Northern and Western blot analyses, expression of the β5t gene and gene product occurs exclusively in the thymus. The presence of the β5t gene was confirmed in both mice and humans.

Standard proteasomes are constitutively expressed in nearly all mammalian cells. In contrast, i-proteasome expression is generally lower under basal conditions, but can be significantly upregulated when cells are exposed to various factors, such as IFN-γ, or environmental stressors, such as oxidative stress. An exception to this generalization is the cells of the immune system, which can constitutively express i-proteasome at high levels. For example, cells in the spleen contain nearly all i-proteasome and low levels of the standard proteasome subunits.[30]

IFN-γ is the most well-studied factor used to investigate i-proteasome expression. In addition to affecting i-proteasome, IFN-γ also upregulated a host of other genes associated with generating peptides, including molecules from the MHC class I and II pathways, TAP and PA28. IFN-γ can also affect the population of proteasome subtypes present by causing dephosphorylation of the 20S, which favors dissociation with PA700 and association with PA28.[31] Induction of i-proteasome subunit expression by IFN-γ has been well established for both cultured immune cells[10] and cultured nonimmune cells, such as neurons[32] and epithelial cells of the retina.[33] This cytokine-induced expression results from binding of the Stat-1 and IRF-1 transcription factors to multiple IFN-γ consensus/activation sequences in the promoter region of the LMP2, LMP7, and MECL-1 genes.[24,29] Other cytokines, such as IFNα/β, lipopolysaccharide, and TNFα also elicit an inflammatory response that involves increased i-proteasome expression.

The presence of binding sites for multiple transcription factors, such as CREB and NF-κB on the promoter region of i-proteasome genes suggests that additional cytokine-independent mechanisms of regulation are possible.[26,27] For example, nitric oxide increases i-proteasome expression via the cGMP/cAMP pathway.[34] In the cascade of reactions that begins with elevated levels of nitric oxide, cGMP/cAMP activates the protein kinases A and G, which in turn phosphorylate CREB. Phosphorylation triggers the nuclear translocation of

CREB and subsequent transcription of the i-proteasome subunits. This mechanism has been proposed to explain high basal i-proteasome expression in endothelial cells, which are exposed to constitutively high levels of nitric oxide.[34]

The ability of the cells to produce different subtypes of the proteasome, each possessing different catalytic properties and substrate preference, has several advantages. First, this plasticity allows the cell to respond to changing environmental conditions and adjust the proteolytic capacity to match each new challenge. Second, the presence of different subtypes increases the repertoire of substrates that can be degraded and peptides that are generated.

## IV. Assembly of the 20S Core

Assembly of the 20S core is a chaperone-mediated process (reviewed in Ref. 35). The initial step involves the formation of the seven-member ring of α-subunits utilizing dimers of the proteasome-assembling chaperone (PAC1 and PAC2). The PAC1/2 chaperones function as scaffolding and prevent the α-subunits from associating with other nascent α-rings.[36] PAC 3 and PAC 4 also assist with α-ring assembly and connect the end subunits together to form the heptameric ring. PAC3/4 also recruits and correctly positions β2, which is the first β-subunit of the standard proteasome to be incorporated into the nascent complex.[36,37] PAC3/4 then dissociates from the complex and the remaining β-subunits are incorporated in a defined order (β3, β4, β5, β6, β1, and finally β7) with the assembled α-ring to form a half proteasome. Data from chimeric proteins containing different N-terminal propeptides have shown that the mutual interaction of the propeptides of the immature subunits drives the orderly assembly of the β-subunits.[38] In the final step, two half proteasome core particles are combined with the aid of the chaperone proteasome maturation protein (POMP or proteassemblin, the mammalian orthologue to Ump1 in yeast). Formation of the complete 20S core initiates cleavage of the N-terminal propeptide from the three pairs of catalytic β-subunits, which exposes and activates the catalytic threonine residue (Thr1 in the mature 20S). The 20S proteasome then degrades the PAC and POMP chaperones, signaling successful assembly of the mature 20S core.

Although the general mechanism of 20S assembly is similar for all proteasome subtypes, the assembly of i-proteasome is favored over the standard proteasome for several reasons. In cells expressing both i-proteasome and standard proteasome subunits, the LMP2 subunit is the first subunit added, whereas the standard subunit homologue (β1) is incorporated much later in the assembly process. The presence of LMP2 facilitates incorporation of MECL-1.

Data also support the idea that LMP2 is strictly required for MECL-1 incorporation, although this point remains controversial.[38–41] The LMP7 subunit facilitates maturation of the nascent 20S core by assisting in the removal of the propeptides from LMP2 and MECL-1.[39,40] The preferential incorporation of all three i-proteasome subunits, which is known as cooperative assembly, is also aided by the action of POMP. POMP expression is upregulated by IFN-γ, so POMP expression is concomitantly increased in tandem with the inducible subunits. Importantly, POMP binds the propeptide of LMP7 with greater affinity than that of β5, resulting in preferred biogenesis of the i-proteasome over formation of the standard proteasome.[42]

The rapid production of i-proteasome, estimated to be four times faster than the standard proteasome in response to IFN-γ, ensures that the cell can quickly expand that repertoire of peptides to aid in immune defense in a challenged organism.[42] Conversely, the turnover of i-proteasome is also substantially faster compared with the standard proteasome. Pulse-chase experiments in T2 cells expressing either the standard or i-proteasome reported a half-life of 133 versus 27 h, respectively.[42] Taken together, these data suggest that production and degradation of i-proteasome is a highly dynamic and transient process that permits the rapid response to environmental challenges and subsequent return to baseline levels that favor the standard proteasome subtype once the challenge has subsided. A physiologically relevant situation where the rapid adjustment in i-proteasome levels would be critical is in the initial stage of viral infection when cells are exposed to cytokines that are secreted by the innate immune system. As discussed later, i-proteasome plays a key role in regulating the cellular response to an inflammatory challenge.

While it was generally believed that the newly assembling proteasome cores preferentially incorporate all i-proteasome subunits, recent data have challenged this idea. Intermediate proteasomes containing a mixture of standard and inducible proteasome subunits have been reported in multiple cell types and tissues, including skeletal and cardiac muscle, liver, colon, small intestines, and IFN-γ-treated HeLa cells.[15–17,41] In some tissues, intermediate proteasomes make up a significant portion of the proteasome population. Using newly developed antibodies that can distinguish specific subunits complexed in the 20S core, Guillaume and colleagues estimated the proportion of intermediate proteasome in normal human kidney and liver to be 30% and 50%, respectively, with the balance mainly standard proteasome.[41] Intestines contained only intermediate and i-proteasome (containing only the inducible subunits), while heart contained mainly standard proteasome (83%) with the balance as intermediate and i-proteasome. This recent report is consistent with previous suggestions that the proteasome population is very diverse and that the distribution of subtypes is tissue specific.[15–17,41]

Knockout (KO) mice provide additional evidence that assembly of an intermediate proteasome is possible. For example, the content of LMP2 in lymphocytes, cultured retinal pigment epithelial cells, and retina of mice deficient in both LMP7 and MECL-1 is equivalent to that found in wild-type (WT) mice.[30,38] Under conditions of stress, such as aging or exposure of cells to peroxide, WT cells respond by upregulating the inducible subunits.[30] In cells and tissues of KO mice, this response was not replicated; that is, LMP2 expression was not increased in mice double-deficient in LMP7 and MECL-1.[30] These data suggest that formation of the intermediate proteasome occurs under basal conditions. However, the full complement of i-proteasome subunits is required to obtain the maximum induction and incorporation of i-proteasome subunits into the core particle in response to stress.

## V. Enzymatic Activity

As discussed earlier, the three standard catalytic subunits (β1, β2, β5) perform distinct proteolytic activities. Based on the proteolysis of model peptide substrates, the active sites have been classified as caspase-like, trypsin-like, and chymotrypsin-like for cleavage after acidic, basic, and hydrophobic amino acids, respectively. For the i-proteasome, some differences in catalytic activity and peptide generation have been noted (see discussion in section VII.A). Comparing activity of the β2 and β5 standard subunits with the corresponding i-proteasome subunits MECL-1 and LMP7, the specificity is generally the same. However, comparison of cleavage after branched chain and hydrophobic residues for standard and i-proteasome has not been consistent; both increased (reviewed in Ref. 35) and decreased[30] activity have been reported for the i-proteasome. The discrepancy in data is due in part to heterogeneity of proteasome subtypes and cell-specific endogenous regulators in the tissue analyzed, and the inability of model peptide substrates to distinguish between proteasome subtypes. Cleavage after acidic residues, which is accomplished by the β1 standard subunit, is nearly abolished in the i-proteasome. Instead, there is a shift to chymotrypsin-like activity for LMP2, which promotes the generation of MHC class I-compatible peptides containing hydrophobic C-terminal anchors.

While the mechanism of hydrolysis (involving the active-site Thr1, Asp17, and Lys33) is the same for each subunit, the specificity of cleavage for each active site is determined by the amino acids that make up the S1 binding pocket (residues 20, 31, 35, 45, 49, 53, 115, 116),[43,44] which is where protein substrates bind prior to cleavage (Fig. 2). Alignment of sequences comparing the standard with their i-proteasome subunit correlate generally shows high conservation of the amino acids that make up the binding pocket, except for LMP2 and β5t.

A

```
    1                20              31  35          45   49   53
STG-TTIMAVQFDGGVVLGADSRTTTGSYIANRVTDKLTPIHDRIFCCRSGSAADTQAVAD β1
HTG-TTIMAVEFDGGVVMGSDSRVSAGEAVVNRVFDKLSPLHERIYCALSGSAADAQAVAD LMP2

KTG-TTIAGVVYKDGIVLGADTRATEGMVVADKNCSKIHFISPNIYCCGAGTAADTDMTTQ β2
KTG-TTIAGLVFQDGVILGADTRATNDSVVADKSCEKIHFIAPKIYCCGAGVAADAEMTTR MECL

LHG-TTTLAFKFRHGVIVAADSRATAGAYIASQTVKKVIEINPYLLGTMAGGAADCSFWER β5
AHG-TTTLAFKFQHGVIAAVDSRASAGSYISALRVNKVIEINPYLLGTMSGCAADCQYWER LMP7
AHG-TTTLAFRFRHGVIAAADTRSSCGSYVACPASCKVIPVHQHLLGTTSGTSADCATWYR β5t
```

B

```
         *     *
109 HLYSIYPHGSTDKLPY 124 β2
109 QLYGVHPHGSYSRLPF 124 MECL
           *
109 GLYYVDSEGNRISGAT 124 β5
109 GLYYVDEHGTRLSGNM 124 LMP7
```

FIG. 2. Sequence alignment of standard and immunoproteasome subunits. (A) Sequence from humans includes the last three residues of the propeptide, followed by residues 1–57 of the mature protein. The dash represents the site of autocatalytic cleavage, which is performed by the active-site Thr1. The conserved Gly adjacent to Thr1 is essential for efficient autocatalytic processing of the propeptide. Conserved residues essential for catalytic activity in the mature protein include Thr1, Asp17, and Lys33 (dark gray, white letters). The S1 binding pocket that imparts specificity of cleavage includes residues 20, 31, 35, 45, 49, and 53 (light gray). Note that in LMP7, the murine sequence substitutes a Met in place of Val31 in humans. (B) Sequence of the β2 and MECL subunits highlighting four amino acids that contribute to the S1 binding pocket of the adjacent β1 or LMP2 subunit. Residues 114, 116, 118, and 120 are highlighted in gray. Stars indicate nonconservative substitutions between the β2 and MECL subunits. The sequence of the β5 and LMP7 subunits highlights two amino acid substitutions at positions 115 and 116, which are critical for substrate binding.

LMP2 contains two prominent substitutions compared with residues in β1; the β1 Thr21 is replaced by Val and the Arg45 is replaced by Leu. These substitutions minimize the size of the S1 pocket of LMP2 and change the overall charge of the local environment from positive to neutral. These changes in primary sequence and charge state of the LMP2 binding pocket could explain the drastic reduction in the caspase-like activity of the i-proteasome. One other noteworthy change occurs in positions 115 and 116 in β5 and LMP7; substitution of Ser115 for a Glu and substitution of Glu116 for His in β5 and LMP7, respectively, would alter the size and polarity of the binding pocket. These structural changes could alter the substrate preference of each subunit and, consequently, result in the production of different peptides.

For the thymus-specific β5 subunit, β5t, sequence comparison with the two other β5 family members reveals five nonconservative substitutions for residues in the binding pocket (Fig. 2). While the majority of the amino acids in the S1 binding pocket of β5 and LMP7 are hydrophobic, hydrophilic residues populate the binding pocket of β5t.[11] This difference in amino acid composition selectively reduces the chymotrypsin-like activity by 60–70%, but has no effect on the caspase-like or trypsin-like activities in the thymoproteasome.[11]

Amino acids in subunits adjacent to the catalytic subunits also contribute to the S1-binding pocket and, thus, influence catalytic activity. Residues 114, 116, 118, and 120 in β3 and β6, which are subunits constitutively present in all 20S core particles, contribute to the S1-pocket of subunits β2 and MECL and subunits β5 and LMP7, respectively[43] (Fig. 2). These same residues in β2 and MECL contribute to the S1-pocket of β1 and LMP2. Comparison of residues in β2 and MECL reveals nonconservative substitutions at positions 114 (Tyr and His) and 120 (Asp and Ser). These differences in S1-pocket residues could have dramatic effects on the activity of the adjacent catalytic subunits (β1 or LMP2). This point is particularly relevant for the activity of the intermediate proteasomes, which contain a mixture of both standard and i-proteasome catalytic subunits.

Since i-proteasome, standard, and intermediate proteasomes produce some peptides that are unique to each subtype,[15,17] the repertoire of peptides generated within a cell is determined by the proteasome population that is present at any given time. Proteasome composition is quite dynamic, and changes dramatically due to challenges, such as cytokine or oxidative stress, in the cell environment. The diverse spectrum of peptides produced by different proteasome populations may be biologically important not only in producing different antigenic peptides, but also for generating biologically active peptides that regulate critical processes, such as cell signaling.[45]

## VI. Immunoproteasome Knockout Mice

To clarify the physiological role of the i-proteasome subunits, mice deficient in one or more i-proteasome subunits have been generated through the targeted disruption of the gene. These KO mice were developed by immunologists to investigate the putative role of specific i-proteasome subunits in generating immunogenic peptides for antigen presentation. The LMP2$^{-/-}$ and LMP7$^{-/-}$ strains were developed in 1994 by two different laboratories.[46,47] Development of the MECL-1$^{-/-}$ and the thymus-specific proteasome (β5t) strains occurred a decade later.[11,48]

For each single KO mouse, the gene disruption was accomplished using a deletion-type targeting vector containing a neomycin resistance cassette, which eliminates a portion of the gene upon homologous recombination in embryonic stem cells. Germline transmission of the targeted allele was observed in all KO mice. To generate the LMP2 KO mouse, an 800 base pair region spanning parts of exon 2 and intron 2 in the lmp2 gene was deleted.[46] Even though the LMP2 and TAP-1 genes are in close proximity in the WT mouse, this genetic manipulation did not disrupt the TAP-1 gene in the mutant mouse. In the LMP7-deficient mouse, the targeting vector deleted exons 1–5, which corresponds to the first 247 of the 276 total amino acids in the protein.[47] To produce the MECL-1 KO mouse, the targeting vector eliminated exons 5–7 and also disrupted the splicing of exon 8.[48] KO of the thymus-specific proteasome (β5t) gene was accomplished by disrupting the β5t coding sequence with a vector containing the neomycin-resistant cassette and the cDNA encoding the Venus protein.[11] The Venus protein in the mutant mouse was under the control of the β5t promoter, so expression of the Venus protein provided a means for identifying the β5t-expressing cells within the thymus. All strains of KO mice were viable, appeared healthy, and had no obvious gross physical abnormalities.

Double KO mice have also been developed by crossing MECL-1$^{-/-}$ with either LMP7$^{-/-}$ or LMP2$^{-/-}$ to produce mice with two i-proteasome subunits eliminated. Simultaneous elimination of both LMP7 and LMP2 genes by simply crossing single KO mice is highly unlikely due to the close proximity of these genes within the MHC class II locus. To overcome the limitations imposed by the gene structure, a new sequential deletion strategy was employed to generate mice double-deficient in both LMP7 and LMP2.[49] This was accomplished by first introducing a construct to eliminate exon 1 in the *PSMB8* gene; exons 2–5 were deleted by homologous recombination. After the neomycin resistance gene was removed by FLP recombinase activity, a second construct (phosphatase *loxP–neo–loxP*) was fused in frame to the start codon of the *PSMB9* gene. This removed exon 1 by homologous recombination. Elimination of all i-proteasome subunits to generate a triple KO mouse was accomplished by crossing the double-deficient mice (LMP2$^{-/-}$/LMP7$^{-/-}$) with the MECL-1 KO mice.

Another approach to produce mice devoid of i-proteasome activity has been to combine the LMP2$^{-/-}$/MECL-1$^{-/-}$ double KO with use of the LMP7-specific inhibitor, ONX 0914 (also known as PR-957). This approach produces a catalytically inactive i-proteasome.[50] With the development of additional i-proteasome subunit-specific inhibitors, this strategy could be taken to investigate how enzymatic inactivity, rather than complete absence of specific subunits, affects cellular response to stimuli. The use of i-proteasome subunit-specific inhibitors helps eliminate some of the drawbacks of KO mice, which include the potential for compensatory mechanisms that would complicate interpretation of the response. Another limitation of KO mice is that genetic ablation of one or more i-proteasome subunits could alter the structure of the 20S core and affect activity or binding of regulatory molecules. The observation that KO mice do not replicate the phenotype of human diseases caused by mutations in specific i-proteasome subunits supports the idea that the presence of an inactive (mutant) protein may influence results in a way that is not consistent with the complete elimination of the protein. For example, LMP7 KO mice do not spontaneously develop the disease symptoms of the *PSMB8*-associated autoinflammatory syndromes,[51–54] which will be discussed later.

In summary, i-proteasome KO mice have been extensively used by immunologists to investigate the putative role of specific i-proteasome subunits in immune function, emphasizing the generation of peptide substrates for loading into MHC class I. While the initial functional and phenotypic analysis of KO mice has reported surprisingly modest changes in immunologic-based measures, recent studies with KO mice have revealed novel roles for the i-proteasome that are unrelated to antigen presentation.

## VII. Immunoproteasome Function

The structural and sequence similarities between the standard and i-proteasome cores strongly suggested that the generation of antigenic peptides for MHC class I presentation would be a shared function. A schematic representation of a minimal antigen-processing pathway for MHC class I-restricted epitopes derived from cytosolic proteins is shown in Fig. 3A.[55] A representative protein with an epitope (round symbol) is ubiquitinated, targeting the protein to an i-proteasome 20S core. After proteolysis and ejection from the core, peptides are further cut by a battery of aminopeptidases, and arrive at TAP-1 and TAP-2, for transport into the ER. In the ER, the N terminus of peptides may be further clipped by ERAP1 or 2, in preparation for loading into the MHC class I complex. The MHC class I complex forms around nascent class I $\alpha$ chains chaperoned by calnexin (CNX), tapasin (TPN), and calreticulin (CRT), which also serve to promote proper folding of the MHC class I $\alpha$ chain and $\beta$2m chain, and enhance receptiveness of the groove for peptides. TPN promotes the association of the complex with TAP-1/2. Peptides displaying complementarity to the binding site in the class I heavy chain occupy the groove, which leads to dissociation of the heavy chain/peptide/$\beta$2m complex from the chaperones, and subsequent transport to the plasma membrane (PM). Peptides derived from membrane-associated proteins, and from proteins destined for secretion, are also generated in the ER, providing additional sequences for occupancy of the MHC class I groove. Novel peptides derived by splicing noncontiguous sequences have also been shown to be produced by the standard and i-proteasomes (Fig. 3B). These peptides can enter the MHC class I antigen presentation pathway, and have been shown to be rare

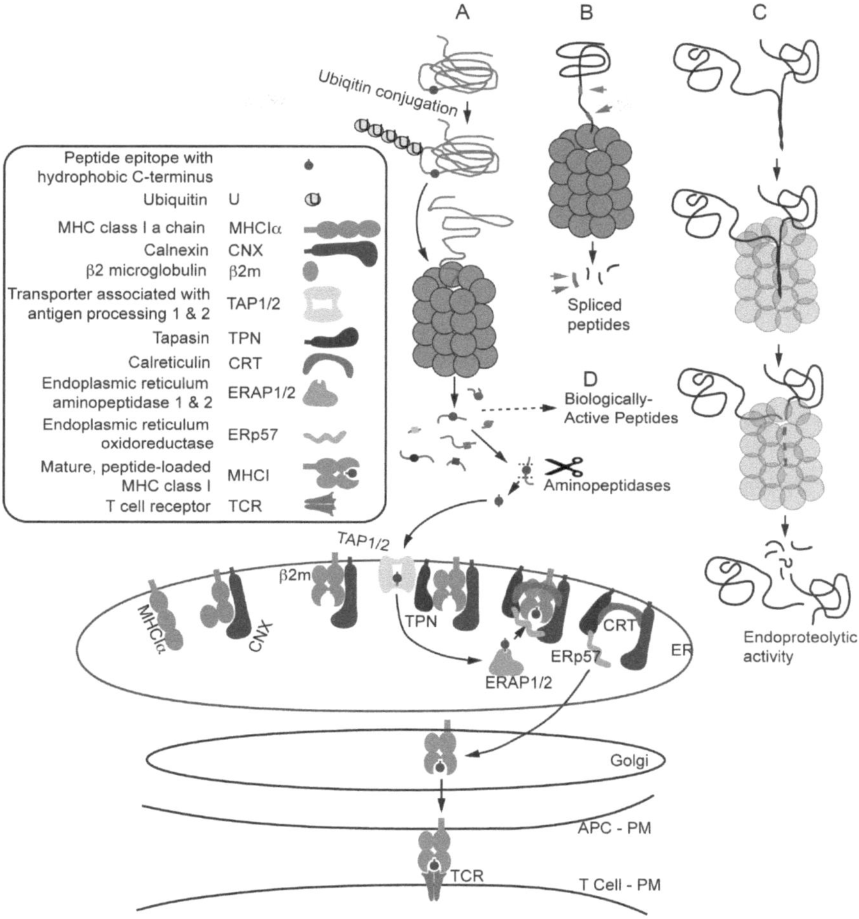


FIG. 3. I-proteasome-dependent pathways of protein degradation. I-proteasome-dependent proteolysis in an antigen-presenting cell (APC) generates peptides and polypeptides for MHC class I-restricted antigen presentation, and for functions other than antigen presentation. (A) The antigen-processing pathway for degradation of a cytosolic protein leading to loading of peptides into MHC class I and antigen presentation to the T cell receptor (TCR). (B) Degradation of a protein by i-proteasome resulting in the generation of a spliced peptide candidate for antigen presentation. (C) I-proteasome-mediated endoproteolytic activity generating large, potentially biologically active polypeptides from a precursor protein. Small peptides may also be produced and would be available for antigen presentation. (D) Some of the small peptides produced by i-proteasome activity may have biological activity. PM, plasma membrane; ER, endoplasmic reticulum. (See Color Insert.)

targets of CD8 T cells.[56] Since the only spliced peptides characterized to date were discovered as a result of MHC class I occupancy and antigen presentation leading to CD8 T cell responses, the frequency and significance of splicing reactions unrelated to antigen processing is unknown.

More recent evidence has expanded i-proteasome's role to include regulation of key processes in maintaining cellular homeostasis and in responding to stress. The i-proteasome achieves these diverse functions through the selective degradation of protein substrates (Fig. 3C). For example, partial degradation of proteins, known as endoproteolysis, is one well-known mechanism for releasing active transcription factors of the NF-κB family.[57] Degradation of a protein can stop signal transduction by uncoupling the pathway. Alternatively, degradation of an inhibitory protein may promote the activity of a signaling pathway. Peptides generated by either partial or complete degradation of proteins can be loaded into MHC class I molecules for recognition by CD8 T cells as part of immune surveillance. Degradation of proteins by i-proteasome could also produce biologically active peptides that regulate key cellular processes (Fig. 3D).[45] The following section will discuss the diverse roles of i-proteasome in antigen processing and other nonimmune functions that have been recently discovered.

## A. Antigen Processing and the Immune Response

T lymphocyte recognition of peptides generated from self and foreign proteins is essential for maintenance of the host organism in its niche. A major class of T lymphocytes, the CD8 T cells, recognizes peptides in the groove of MHC class I molecules. The peptides that are loaded into this groove are substantially provided by the proteolytic activity of proteasomes.[58] The proteasomes access both endogenous and exogenous proteins through direct and cross-presentation pathways, respectively, in antigen-presenting cells of the immune system. Other MHC class I-positive cells process endogenous proteins through the direct pathway.[59,60] The multiple proteasome subtypes efficiently perform the first step of antigen processing that provides the peptides of eight or nine residues ultimately required to fit into the groove of nascent MHC class I molecules in the ER.[60] The cleft in MHC class I molecules includes accommodation for hydrophobic C-terminal anchor residues, which are generated mainly by the chymotrypsin-like activity of the β5/LMP7 subunits. In the i-proteasome, the shift in proteolytic activity in LMP2 to chymotrypsin-like activity promotes the generation of peptides with a hydrophobic C-terminal anchor, providing more, and different, peptides for occupancy of the peptide-binding cleft in MHC class I molecules.[61–63] Extra residues on the N terminus are removed by proteases downstream of the proteasome (review Ref. 44). While there are a number of cytosolic proteases distinct from i-proteasome that also generate peptides, the i-proteasome is a major provider of peptides with a hydrophobic C terminus.

The location of the LMP2 and LMP7 i-proteasome subunits in the MHC region,[4–6] and their production in response to proinflammatory cytokines, suggested their role in antigen processing, and led to the production of KO mice lacking one or both of these subunits to investigate subunit-specific differences. Although the KO mice lacked an obvious phenotype, more detailed analyses revealed differences. Mice deficient in LMP7 exhibited a modest reduction in surface expression of MHC class I, and resulted in mice with a reduced response to the male HY antigen.[47] Deficiency in LMP2 was found to result in a reduced level of CD8 T cells and depressed the response to a nucleoprotein epitope of influenza A.[46] Mice deficient in MECL-1 had a slightly reduced number of CD8 T cells, and a decreased response to two class I-restricted epitopes of lymphocytic choriomeningitis virus.[48] The reduced response to the GP276–286 epitope was attributed to decreased precursor frequency of the specific T cells, rather than impaired production of the GP276–286 epitope. In summary, use of these KO mice, lacking specific subunits of the i-proteasomes, have demonstrated differences in the CD8 T cell responses, confirming their roles in antigen-specific interactions with microbial challenges. The combined results of a growing body of evidence show that the repertoires of peptides generated by i-proteasomes and standard proteasomes for both selection of the CD8 T cell repertoire and presentation by antigen-presenting cells in the periphery have substantial overlap, as well as distinct differences.[46,48,64–67]

van Helden and colleagues[68] reported that MECL-1 and LMP7 deficiency resulted in reduced MHC class I expression, which made those cells susceptible to natural killer cell-mediated killing in mice whose immune system had been activated by a viral infection. The data are compelling, and several reports point to i-proteasome deficiency leading to the lack of suitable peptides for class I occupancy.[47,69] The correlation between MECL-1 and LMP7 deficiency and reduced surface expression of MHC class I is based on two observations: that expression of MHC class I is dependent on occupancy of the cleft in the MHC $\alpha$ chain,[70,71] and that chymotrypsin-like activity of the immune subunit provides a larger pool of precursors. While these hypotheses are sound, direct evidence for a shortage of i-proteasome-generated peptides being the rate-limiting step for maturation of MHC class I molecules and their expression on the cell surface is lacking, as total proteasome content is relatively constant, and standard proteasome would continue to provide peptide precursors, even if the repertoire was altered. Conversely, evidence continues to appear for other roles played by i-proteasome subunits. One of the most significant of these is the linkage of LMP2 to NF-κB.[72] LPS stimulation of LMP2-deficient cells was less able to activate NF-κB due to reduced IκB degradation. As a transcription factor at the center of an important hub in innate and adaptive immune responses, NF-κB is positioned to have substantial influence on immunity, including the regulation of transcription of MHC class I.[73]

Although differences in antigen processing were found in the KO mice, there was no evidence for a global deficit in immune function; i-proteasome KO mice were almost without phenotype with respect to their overall health. The coordinated expression of the i-proteasome subunits following stimulation by TNFα and IFN-γ has led to speculation that loss or inhibition of all three subunits might yield a more definitive outcome. Mice deficient in two subunits, LMP2 and MECL-1, were examined in the absence and presence of ONX0914, an LMP7 inhibitor. Quantitative differences in the cytokine responses of splenocytes treated with polyclonal activators were found if the ONX0914 inhibitor was included.[50] Comparison of responses to several epitopes of lymphocytic choriomeningitis virus by lymphocytes recovered from infected mice showed epitope-specific differences, but no widespread qualitative loss of responses.[50]

The absence of global effects on CD8 T cell responses in mice deficient in one or more i-proteasome subunits contrasts with several reports in which responses to specific epitopes are substantially altered, and can lead to differences in resistance to specific infectious diseases. Two examples clearly illustrate this point. Infection by the intracellular parasite *Listeria monocytogenes* strongly upregulates expression of i-proteasome subunits, including expression in liver cells.[74] Despite the presence of cytotoxic T lymphocytes specific for the protective LLO296–304 epitope, clearance of the bacteria was substantially compromised in the liver of LMP7-deficient mice, but similar to WT in the spleen.[75] Clearance of another parasite, *Toxoplasma gondii*, was also severely compromised in LMP7-deficient mice, and to a lesser degree in LMP2-deficient mice.[76]

Another type of proteasome was recently found to be specifically expressed in the cortical epithelial cells of the thymus (cortical thymic epithelial cell, cTEC).[11] Like i-proteasomes, it contained the LMP2 and MECL-1 subunits, but contained a variant of β5, the β5t subunit, which substituted for the LMP7 subunit. The role of cTECs in positive selection of CD8 T cells and the reduced production of CD8 T cells in β5t-deficient mice suggests a role for this subunit in generating the peptide repertoire available for MHC class I occupancy in CD8 T cell maturation (review Ref. 77). The function of immuno- and thymoproteasomes in generating the universe of CD8 T cell specificities for peptides, and their reciprocal role in generating the peptides that will occupy MHC class I on the cells of the body, that is, identifying them as potential targets for cytotoxic T lymphocytes, is increasingly established as a critical underpinning shaping self-recognition by CD8 T cells.

To date, the single and double i-proteasome-deficient mice have been examined for differences in antigen presentation, yielding modest differences overall. On the possibility that overlap in the cleavage specificities compensated for loss of subunits, triple i-proteasome-deficient (LMP2, LMP7, MECL-1) mice were made.[49] The triple-deficient mice were viable, fertile, and immunocompetent. A decrease in CD8 T cell numbers was found, as well as differences in antigen presentation. These were especially apparent when responses to immunodominant peptides by the corresponding T cell receptor (TCR) transgenic mice were examined. Mass spectrometry analysis of peptides collected from class I molecules expressed on the surface of the respective splenocytes showed that approximately one-third of the peptides were unique to the deficient mice, one-third each were associated with the WT MHC class I, and MHC class I occupied the overlap of peptides produced by standard and i-proteasomes. Identification of the peptides revealed that an average of approximately one peptide was found per protein of origin. Since pathogens usually contain a number of proteins, these results reveal the basis for the limited effect of i-proteasome deficiencies on CD8 T cell immune responses, despite the occasional substantial difference.

A number of studies have concentrated on the reactivity of TCR transgenic CD8 T cells to assess the role of i-proteasomes on the peptides that are presented in MHC class I. In instances in which the TCR in these transgenic mice is specific for a peptide that is differentially produced by i-proteasomes versus standard proteasomes, substantial differences in the T cells can be found by this strategy.[78] Other studies have looked at the outcome of microbial infections, especially those in which a limited number of epitopes play a large role in the outcome of the immune response. Proteomic analysis of peptides eluted from MHC class I molecules (peptidomics) provides a powerful tool for assessing the diversity and identity of peptides in the class I cleft. Using WT and i-proteasome LMP2- and LMP7-deficient double KO mice, de Verteuil and colleagues isolated bone marrow dendritic cells that were induced to become mature and activated.[79] Peptides were stripped from cell-surface MHC class I of dendritic cells and thymocytes, and assessed by mass spectrometry. Peptide diversity from WT dendritic cells was greater than that from WT thymocytes, and the proteins represented in MHC class I from each cell type differed, in part, with respect to cell functions. I-proteasome-deficient dendritic cells had a less diverse population of peptides in the cleft. In both cases, most proteins yielded only a single peptide sequence capable of loading into MHC class I, and there was substantial overlap between the dendritic cells and thymocytes. The presence or absence of i-proteasomes also altered mRNA levels of clusters of genes associated with dendritic cell function by unknown mechanisms that do not appear to be associated with peptide loading of MHC class I. Together, the data both reinforce the role of i-proteasome in generating peptides for class I, and also argue for functions distinct from peptide generation.

In addition to the induction of i-proteasome subunits by proinflammatory cytokines, adapter molecules are also induced, including PA28.[8,80] PA28 alters the activity of both i-proteasome and standard proteasome, and leads to differences in the peptide output, without altering the active sites of the catalytic subunits. Effects on the production and availability of specific epitopes have been found. Among the examples are the $TRP2_{288}$ and $TRP2_{360}$ epitopes of a melanoma antigen.[81,82] Flanking sequences adjacent to the epitope have been found to be required for the PA28 effect, but are not yet resolved at the molecular level.[83] Although widely thought to support the i-proteasome function of generating epitopes for MHC class I, this function is not well supported by data. Other than rare examples, PA28 has not been shown to possess activities that significantly alter the production and repertoire of peptides for antigen presentation, and invites speculation that it alters immunoproteasome activity for another unknown purpose.

It is clear that standard, immuno- and thymoproteasomes, as well as hybrids containing a mixture of the catalytic subunits, provide peptides that go on to occupy the cleft in MHC class I molecules. The overlapping, but distinct, sets of peptides provide an enormous repertoire that enables the protective functions of adaptive immunity. However, analysis strategies affect the outcome of the investigations. For example, if the peptides produced by i-proteasome or standard proteasome or mixed proteasomes are assessed based on isolating them from cell-surface MHC class I molecules, the repertoire of peptides will necessarily reflect the selection pressure of the requirement for the peptides to form stable, mature MHC class I molecules. Due to this selection pressure, such an analysis provides limited information about the total repertoire of proteasome-produced peptides, and limits the extrapolation of the results to the total repertoire of proteins that may be substrates for degradation via proteasomes. The role of i-proteasome in providing a set of peptides for MHC class I occupancy that is distinct, though overlapping, from that generated by standard proteasome is clear. It is also inescapable that i-proteasome, like the standard proteasome, performs more functions than provision of peptides for CD8 T cell recognition of MHC class I complexes. Expression of i-proteasome in cells that have little, if any, expression of conventional MHC class I molecules, neurons, supports the search for functions that are distinct from outcomes due to differences in antigen processing. I-proteasome expansion of the peptide repertoire may be but one of its activities. The new challenge is to explain how the moderate differences in the activities of standard proteasome or i-proteasomes or mixed proteasomes can account for the diversity of differences that are increasingly being found, and which are not easily attributed to peptides in MHC class I. The following section reviews the accruing evidence that suggests new roles for i-proteasomes that appear unrelated to generating peptides for MHC class I occupancy.

## B. Nonimmune Roles for I-proteasome

Existing evidence that i-proteasome performs functions that are unrelated to antigen processing includes its expression in uninjured, immune privileged tissues, such as the retina[30,84–87] and brain.[85,88,89] Localization of i-proteasome to synapses and Purkinje cells in the brain,[88,89] and the photoreceptors and neurons in the outer plexiform layer in the retina,[85] implies a role in normal, neuronal function. Evidence supporting this idea includes the significantly reduced retinal function, as assessed by electroretinography, measured in both single- (LMP2) and double-deficient (LMP7/MECL-1) i-proteasome KO mice.[86] It is important to note that neurons express no, or at best minimal, levels of the MHC class I molecule that provides antigen presentation for CD8 T lymphocytes.

In addition to helping regulate cellular homeostasis, i-proteasome may also take part in the cellular response to stress and injury. This idea was initially based on the observation that i-proteasome expression was substantially upregulated following stress or injury in a number of cells that do not typically present antigen. For example, i-proteasome is significantly upregulated in the central nervous system in response to acute injury to the retina[85] and brain,[85,90] in diseased retina[84] and brain,[88,89] and aging.[30,89,91] It is also found and in cultured cells exposed to oxidative stress.[30,92] Low levels of i-proteasome subunits that are expressed in both cardiac and skeletal muscle are significantly upregulated with injury due to viral infection,[93] ischemia reperfusion,[94] and muscle atrophy due to aging[95,96] or diabetes.[97] These initial studies provided the rationale for more in-depth investigations using KO mice.

Mice deficient in i-proteasome have been invaluable in providing keen insight into functions that go beyond i-proteasome's contribution to antigen processing and provide compelling evidence for its role in multiple cellular processes. One idea is that i-proteasome helps to protect against oxidative damage, in part, due to its enhanced ability to degraded oxidized protein substrates.[97–99] Increased levels of oxidized proteins in the brain and liver of LMP2 KO mice[100] and the greater sensitivity to an oxidative challenge exhibited by retinal pigment epithelial cells from mice double-deficient in LMP7 and MECL-1 subunits[85] support a role for protecting against oxidative damage.

I-proteasome may also participate in regulating tumor development. Uterine leiomyosarcoma is a tumor not known to be associated with human papilloma virus. This cancer has a 5-year survival rate of 35%, so there is an urgent need to understand the factors that influence tumor development. LMP2-deficient mice develop this tumor starting at 6 months of age, reaching a 40% incidence at 14 months.[101] Importantly, LMP2 expression is absent in leiomyosarcoma tumors in human patients, but present in the benign uterine tumor leiomyoma. To identify potential defects in LMP2-deficient cells, primary cultures of tumor cells and splenocytes from LMP2-deficient mice were investigated. The significant findings were that LMP2-deficient cells exhibited reduced expression of IRF-1, and its expression was not upregulated by IFN-$\gamma$. IRF-1 regulates the activity of proteins involved in cell-cycle progression and, hence, cell transformation and proliferation. Additional testing will be required to determine if this initial association between LMP2 and IRF-1 is the critical factor in the development of this devastating cancer.

The link between i-proteasome with lipid metabolism and diabetes is supported by the data from humans with mutations in the *PSMB8* gene that encodes LMP7 and data from KO mice. Lipodystrophy is a prominent clinical symptom of several *PSMB8*-associated syndromes that have been recently described in humans.[51–53] These patients have low levels of mature i-proteasome containing LMP7 due to a defect in i-proteasome assembly. To investigate the potential mechanism for lipodystrophy, Kitamura and colleagues used human and murine preadipocyte cell lines, which both contain high levels of LMP7, and tested the effect of knocking down LMP7 expression using siRNA against the *PSMB8* gene.[53] Downregulation of the *PSMB8* mRNA levels inhibited adipocyte differentiation, suggesting that LMP7 is required for adipocyte maturation. Additionally, subcutaneous injection of *PSMB8* siRNA into BALB/c mice resulted in atrophic adipose tissue (smaller total size and number of adipocytes) compared with control mice. Taken together, these data suggest that LMP7 is critically involved in adipocyte maturation and may help explain the progressive lipodystrophy exhibited by patients with *PSMB8*-associated syndromes.

Heart disease, a well-known complication of diabetes, leads to progressive loss of cardiac function, that is, decreased contractility and cardiac muscle mass.[102–104] Using the streptozotocin (STZ) model of diabetes, Zu and colleagues showed that when STZ-treated mice developed diabetes, LMP7 levels increased in cardiac tissue, but LMP2 and MECL-1 decreased.[105] Exposure of isolated hearts to high glucose replicated these changes in i-proteasome expression, suggesting that the high glucose associated with diabetes could alter proteasome composition in cardiac muscle. Of note, altered proteasome composition was also observed in diabetic skeletal muscle.[97] To determine if the loss in LMP2 expression had physiological consequences for the heart, cardiac mass and function were measured in LMP2 KO mice. The cardiomyopathy observed in the LMP2 KO mouse was similar to the cardiac phenotype in WT diabetic heart. These results suggest that altered proteasome composition, which can occur due to metabolic changes associated with diabetes, can have detrimental effects on cardiac muscle mass and function.

Insulin-dependent diabetes mellitus and diabetes insipidus are both autoimmune diseases with well-known etiologies. Diabetes mellitus is caused by an autoimmune attack on the insulin-producing cells in the pancreas, and diabetes insipidus is caused by a reduction in the antidiuretic vasopressin due to the autoimmune killing of vasopressin-producing cells in the hypothalamus. CD8 T cells may be involved in the targeted cell killing of the insulin-producing beta cells in the pancreas that is the pathologic event in type 1 diabetes.[106,107] To investigate the role of i-proteasome in progression of autoimmune disease, WT and i-proteasome KO mice lacking LMP7 and MECL-1 subunits were irradiated, and then given bone marrow transfer.[108] Following bone marrow reconstitution from either WT or i-proteasome KO mice, i-proteasome-deficient mice exhibited symptoms of multiorgan autoimmune disease, including inflamed skin, diabetes insipidus, and a latent form of diabetes mellitus. The authors also supply supporting evidence in humans from their disease-association analysis using the Type 1 Diabetes Genetics Consortium Major Histocompatibility Complex fine-mapping data set. The analysis identified two single-nucleotide polymorphisms (SNPs) in the *PSMB8* gene encoding LMP7 that were genetic risk factors for development of type 1 diabetes. These data support a role for i-proteasome in protecting from the onset and progression of autoimmune diseases, such as diabetes.

I-proteasome appears to also limit inflammatory damage, potentially by ridding the cell of damaged (i.e., misfolded, oxidized) proteins and/or by regulating the profiles of cytokines produced in response to an inflammatory challenge. Data from Seifert and colleagues suggest that the i-proteasome is more efficient at degrading the "defective ribosomal products" (DRIPS), which are misfolded and/or oxidized proteins that are produced during protein synthesis.[99] Cytokine-induced inflammation significantly upregulates protein synthesis and production of DRIPS, which, if not cleared from the cell, can produce toxic protein aggregates. In i-proteasome-deficient cells, challenges with either lipopolysaccharide- or IFN-$\gamma$-induced aggresome-like structures and increased levels of oxidized and ubiquitinated proteins. In contrast, aggresomes and oxidized proteins were much reduced in cells containing the full complement of i-proteasome subunits. *In vivo* experiments support the *in vitro* observations. In an experimental model of lipopolysaccharide-induced experimental autoimmune encephalitis, LMP7 KO mice exhibited a more severe clinical score, and higher levels of oxidized proteins and protein aggregates in the inflamed liver and brain.[99]

Other studies also support the idea that i-proteasome preserves protein homeostasis and viability in cells under an inflammatory challenge. In a murine model of acute enterovirus myocarditis, which is one of the most common viral diseases among young people, mice deficient in LMP7 exhibited severe myocardial destruction and large inflammatory foci in cardiac cells.[93] Additionally, cardiomyocytes from the LMP7 KO mice accumulated substantially more oxidized and polyubiquitinated proteins compared with WT cells following exposure to IFN-$\gamma$.

In addition to enhanced proteolysis of damaged proteins, i-proteasome could also protect cells by regulating the profile of cytokines produced in response to an inflammatory challenge. Several cytokines that are secreted in innate and adaptive immune responses (i.e., IL-6, TNFα, IFN-γ, etc.) can amplify the inflammation in surrounding tissues and cause collateral damage. Endogenous regulatory mechanisms dampen the cytokine response to achieve a balance of destructive inflammation and effective control of infections. In cells from patients with *PSMB8*-associated mutations in LMP7, basal production of IL-6 was significantly higher compared to cells containing nonmutant LMP7.[52,53] Stimulation with TNFα enhanced production of IL-6 in cultured fibroblasts from patients having Nakajo–Nashimura syndrome (NNS) who have lower levels of LMP7.[52] In thioglycollate-elicited macrophages, exposure to lipopolysaccharide elicited lower IL-6 production in macrophages from LMP7/MECL-1 double-deficient i-proteasome mice.[109] These examples show that the i-proteasome content can affect cytokine production in response to an inflammatory challenge.

Examples provided to this point have shown that i-proteasome deficiency is detrimental to the course of the disease, that is, experimental autoimmune encephalitis, acute enterovirus myocarditis, and *PSMB8*-associated syndromes. There are, however, several examples where i-proteasome deficiency or inactivation can be protective. Schmidt and colleagues used the dextran sodium sulfate-induced colitis model to examine the influence of LMP7 deficiency on intestinal inflammation.[110] Several findings showed that LMP7 KO mice exhibited less inflammation in the tissue, produced lower levels of proinflammatory cytokines, and recruited fewer Gr-1-positive inflammatory cells and fewer cytokine-producing CD4 T cells. The positive effect of i-proteasome deficiency was replicated in a second study of experimentally induced colitis for mice with individual i-proteasome subunits removed (LMP2, LMP7, and MECL-1).[111] A significant reduction in inflammatory markers was also found when the LMP7-specific inhibitor, PR-957, was used. This inhibitor was also successful in blocking cytokine production and reducing the inflammatory symptoms of experimental arthritis.[112]

Whether i-proteasome deficiency exerts a protective effect or exacerbates disease symptoms may depend on the underlying pathogenesis and/or tissue affected by the disease. What may be especially relevant to disease outcomes is how i-proteasome regulates specific cellular signaling pathways. There is a substantial body of literature that shows how i-proteasome can directly regulate several key signaling pathways, including NF-κB. The NF-κB pathway is one of the major mechanisms for responding to a broad range of stressors, such as cytokines, viral and bacterial products, toxic chemicals, UV light, and oxidative damage.[113] NF-κB activation elicits rapid induction of early-response genes that help protect the cell from damage. Production of several proinflammatory cytokines is also regulated by NF-κB, so this pathway is particularly important for cells under an inflammatory challenge. Aberrant NF-κB regulation can lead to pathologies such as toxic shock, and neurodegenerative and inflammatory diseases.[114]

The NF-κB family consists of multiple proteins, including kinases, inhibitors (IκB), and transcription factors (p65, Rel-B, c-rel, p105/p50, p100/p52). Transcription factors associate to form homo- or heterodimers and are sequestered in the cytosol by inhibitory proteins (IκB) until upstream signaling events cause phosphorylation of select NF-κB family members (Fig. 4). Phosphorylation triggers ubiquitination, which signals selective proteolysis by the proteasome and allows for the subsequent translocation of the NF-κB dimer into the nucleus. Two separate pathways for proteasome-dependent activation include

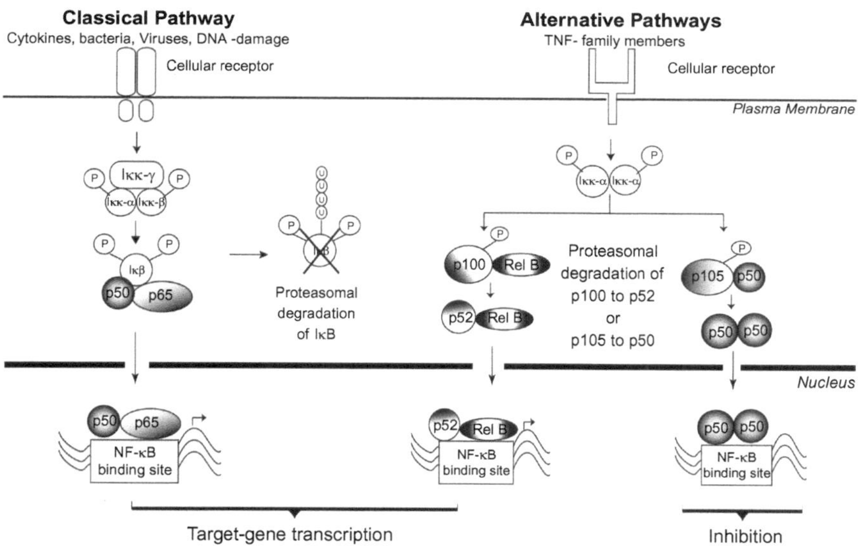


FIG. 4. Proteasome-dependent activation of NF-κB. Receptor-mediated pathways of NF-κB activation include the classical pathway, which involves the binding of a ligand (i.e., cytokines, virus, or bacteria) to a receptor (i.e., IFN-γ or toll-like receptors), and subsequent activation of intracellular kinases that phosphorylate the inhibitory protein, IκB. Phosphorylation is the signal for targeted ubiquitination of IκB, followed by degradation by the proteasome. The release of the inhibitory protein from the transcription factor dimer exposes the nuclear localization signal, which facilitates the movement of the NF-κB dimer into the nucleus. The alternative pathway uses signals from ligand-bound TNF receptors, which trigger the phosphorylation of p100 or p105 by intracellular kinases and the subsequent endoproteolytic degradation of p100 or p105 to form the active transcription factor p52 or p50. Translocation and localization of heterodimers (i.e., p50/p65 or p52/relB) to the NF-κB binding sites initiates transcription of the NF-κB-responsive gene. In contrast, binding of the p50 or p52 homodimers inhibits transcription of NF-κB responsive genes.

(1) the *classical pathway*, where the IκB inhibitory protein is degraded, thereby releasing the NF-κB dimer and allowing it to enter the nucleus, and (2) the *alternative pathway*, where proteolytic processing of the p100 or p105 precursor proteins removes the inhibitory portion of the protein via endoproteolysis, thereby generating the activated transcription factor p52 or p50.[115] Defects in proteolysis of IκB and processing of p100/p105 have been reported in i-proteasome-deficient cells.

Inhibition and termination of the NF-κB signal is also regulated by proteasome. Dimers of p50 or p52, which are produced by proteasome-dependent processing of p105 and p100, are transcriptionally inactive and can prevent active transcription by blocking the NF-κB promoter region. Proteasome is also involved in terminating the signal, by degrading transcription factors (i.e., p65) that are docked on the promoter region.[115]

Evidence from i-proteasome-deficient mice has provided compelling support for the role of i-proteasome in NF-κB activation. In cell lines lacking LMP2, defects in the proteolytic processing of NF-κB precursors (p100/p105) to the active transcription factors (p52/p50), and decreased degradation of the NF-κB inhibitory protein IκBα, were observed.[116–118] While findings reported from these initial studies were controversial,[119,120] more recent studies have provided substantial evidence for i-proteasome regulation of NF-κB. For example, enhanced processing of NF-κB precursors p105 and degradation of IκBα were reported for i-proteasome isolated from the inflamed intestine of patients with Crohn's disease.[121] In TNFα-stimulated embryonic fibroblasts from LMP7 KO mice, reduced nuclear translocation of p65 suggested inhibited NF-κB activation.[110] Findings consistent with these were found in studies of B cells from LMP2-deficient mice; delayed and less complete degradation of IκBα following lipopolysaccharide stimulation was observed.[72]

Recent evidence for regulation by i-proteasomes on events upstream of NF-κB was found using lipopolysaccharide to stimulate activated macrophages collected from i-proteasome-deficient mice.[109] Lipopolysaccharide can induce production of the cytokine TNFα using the MyD88/TRAF6 pathway, while production of nitric oxide is accomplished via the TRIF/TRAF3 pathway. Both pathways ultimately intersect with and activate NF-κB. Stimulation by lipopolysaccharide alone resulted in a severalfold reduction in nitric oxide production for LMP2/LMP7 double KO macrophages but had no effect on TNFα production, suggesting defects in signaling through the TRIF/TRAF3 pathway. Western blots of key signaling molecules in this pathway also confirmed that LMP2/LMP7 double KO macrophages were unable to effectively signal via TRIF/TRAM.

I-proteasome may also regulate the signaling molecule phosphatase and tensin homologue deleted on chromosome 10 (PTEN). PTEN is a phosphatase that functions as a regulator of cell growth by inhibiting Akt signaling. Recent studies have shown that PTEN is an important determinant of cardiac muscle size.[94,105] Notably, these studies also show that PTEN content in cardiac muscle is regulated by the presence of LMP2. The supporting data show that PTEN KO mice have larger hearts compared with WT mice, whereas hearts from LMP2 KO mice, which contain high PTEN, were significantly smaller and exhibited impaired contractility. Treatment of LMP2 KO hearts with a PTEN inhibitor improved cardiac function. PTEN is also involved in the cellular changes associated with cardioprotection following ischemic preconditioning. In WT mice, ischemic preconditioning led to lower PTEN content and activation of Akt. In LMP2 KO mice, PTEN content and Akt were unchanged, and functional measures of cardiac output showed no cardioprotection after ischemic preconditioning.

In summary, there is substantial data supporting a role for i-proteasome-mediated regulation of signaling pathways, including NF-κB and PTEN. Regulation may occur through degradation of proteins in the signaling cascade or through degradation of regulatory molecules that either inhibit or activate the signal. With most biological systems, there is considerable endogenous redundancy that serves to prevent catastrophic outcomes. This strategy is likely to apply to the functions provided by the i-proteasome, especially since i-proteasome deletion is not lethal. Its regulation may be subtle, perhaps by fine tuning the rate of signaling or downstream events, such as cytokine production. It is also highly likely that i-proteasome's involvement also depends on the cell type and stimulation. We really are only beginning to understand some of the nuances in regulation by the i-proteasome.

## VIII. Mutations and Linkage to Human Disease

The list of human diseases that have been linked to deregulation of the i-proteasome has grown exponentially over the past decade. However, the data supporting this link are often indirect. For example, increased i-proteasome expression, altered proteasome activity, and/or accumulation of ubiquitinated protein in the diseased tissue have been reported in multiple diseases, including several neurodegenerative diseases of the brain and retina.[84,88,89] These diseases share oxidation and/or inflammation as part of the disease mechanism and, thus, changes in proteasome expression and activity are likely a consequence of the altered cellular condition rather than part of the primary disease mechanism. Experiments in cultured cells support this idea. Exposure of cells to sublethal levels of oxidative stress,[30,92] inflammatory cytokines,[32,33] or cellular expression of aggregate-prone proteins,[32,122] which is an integral part of the mechanism of Huntington disease, all induce expression of i-proteasome.

Therefore, one needs to be cautious in assuming that the correlation between altered i-proteasome expression or function and disease has mechanistic implications.

While the increased expression of i-proteasome in diseased tissue provides limited mechanistic information, it has opened the door to novel therapeutic options that use inhibitors (i.e., PR-957, PR-924, IPSI-001) that selectively target the i-proteasome catalytic subunits.[123,124] This approach is designed to either specifically kill cells expressing high levels of i-proteasome, as is the goal with cancer, or inhibit signaling pathways, such as the proinflammatory NF-κB pathway in autoimmune disease. Since i-proteasome inhibitors should not affect the activity of the standard proteasome, which is constitutively present in all cells, i-proteasome inhibitors should be less toxic and have reduced off-target side effects compared with other broad-spectrum proteasome inhibitors, such as bortezomib and carfilozomib. Currently, preclinical studies with i-proteasome inhibitors are under way in the treatment of rheumatoid arthritis, inflammatory bowel disease, and cancer.[111,112,125] Initial results have been promising. As a caveat, continued discovery of i-proteasome's ever-expanding role in maintaining cellular homeostasis and in responding to stress suggests that evidence for unexpected side effects in off-target sites need to be vigilantly monitored.

Recent reports of the association between specific diseases and point mutations in i-proteasome subunits provide direct evidence that a loss in i-proteasome function can lead to disease. To date, discrete mutations in the LMP7 protein have defined a spectrum of *PSMB8*-associated autoinflammatory syndromes, which includes an inflammatory response in the absence of infection. These syndromes are distinguished from autoimmune disease because they primarily result from deregulation of the innate immune system rather than adaptive immunity. Notably, there is considerable overlap in symptoms between the different *PSMB8*-associated syndromes potentially because the primary cause is the altered function of the LMP7 protein. In some instances, the detailed structural analysis of the mutant protein has provided significant insight into the molecular details responsible for the development of the disease. Discovery of these *PSMB8*-associated syndromes highlights the profound effect i-proteasome dysfunction has on inflammation and function of specific organs.

One of the first described *PSMB8*-associated syndromes was an autosomal-recessive disease known as joint contractures, muscle atrophy, microcytic anemia, and panniculitis-induced lipodystrophy syndrome (JMP).[126] Patients with JMP exhibit mild metabolic disturbances (i.e., hypertriglycerides, low high-density lipoprotein cholesterol), elevated liver enzymes, muscle atrophy, hypergammaglobulinemia, and joint contractures mainly in the hands and feet. Genome-wide homozygosity mapping identified a missense mutation (c.224C>T) causing substitution at residue 75 (T75M) that resulted in loss of function for the LMP7 i-proteasome catalytic subunit due to a disruption in the tertiary structure.[51]

Another i-proteasome-associated inflammatory disorder is chronic atypical neutrophilic dermatosis with lipodystrophy and elevated temperature (CANDLE) syndrome. The onset of CANDLE disease symptoms occurs within the first year of life and is characterized by recurrent fever, skin lesions, progressive lipodystrophy, and delayed development. The genetic cause was identified as two different mutations in the *PSMB8* gene.[54] In a subpopulation of patients, the missense mutation (c.224C>T) causing a methionine substitution at residue 75 (T75M) in the LMP7 protein was discovered. While there is some overlap in symptoms in patients with JMP and CANDLE, joint contracture and muscle atrophy is not present in CANDLE syndrome, which is distinguished by unique features, such as immature neutrophils in cellular infiltrates. The reason for the discrepancy in symptoms between patients harboring the same mutation is still unclear. A second mutation (c.405C>A) identified in CANDLE patients caused a substitution of a highly conserved cysteine at residue 135 to a stop codon (p.C135X), which resulted in truncation of 141 amino acids from the C-terminal end of the protein.

Patients with NNS exhibit periodic fever, skin rash, lipomuscular atrophy, and joint contractures due to a mutation (G201V) in LMP7.[52] This mutation causes a conformational change that alters the catalytic binding pocket around residues Thr73 and Lys105, resulting in significantly decreased chymotrypsin-like activity. The loss in activity prevents the autolytic cleavage between Gly72 and Thr73, which is required for the LMP7 protein to be incorporated into the 20S core particle. Consequently, the assembly defect caused by the mutation-induced change in protein structure results in lower total content of the 20S and 26S proteasomes in cells normally expressing high levels of i-proteasome. This decrease in i-proteasome content was suggested to be responsible for the increased levels of oxidized and ubiquitinated proteins and activation of the p38 pathway in cells (monocytes, fibroblasts) from patients with NNS.

Autoinflammation and lipodystrophy are characteristics associated with Japanese autoinflamatory syndrome with lipodystrophy, which shares many of the characteristics exhibited by NNS patients but is due to a unique homozygous missense mutation (G197V) in the *PSMB8* gene.[53] The G197V mutation caused a reduction in cellular levels of the LMP7 protein, which consequently resulted in the accumulation of immature i-proteasome cores due to the loss of LMP7 processing. Disturbed adipocyte maturation is another key feature of this disease and highlights the widespread effect that i-proteasome defects can have on cells outside of the immune system.

One of the clinical features shared by several *PSMB8*-associated syndromes was deregulation of lipid processing, suggesting a potential role for i-proteasome in the metabolism and, more specifically, insulin regulation. In support of this idea, two SNPs in the *PSMB8* gene (rs3763365, located 1048 base pairs downstream, and rs9276810 in intron 3) have been identified as an independent genetic risk factor for the development of type 1 diabetes.[108]

SNPs in the *PSMB9* gene that encodes the LMP2 protein have also been associated with disease susceptibility. The SNP in exon 3 is a nonconservative base pair change that encodes two different amino acids (arginine (R) or histidine (H)) at position 60 in the LMP2 protein. Patients with the *PSMB9* SNP rs17587 in exon 3, which results in an R60H substitution in LMP2, are at risk for developing ankylosing spondylitis, a chronic inflammatory joint disease that predominantly affects the spine.[127] Genotype analysis showed an increased frequency and younger age at disease onset for patients with the R/R genotype.[128] It was postulated that the excess bone growth associated with ankylosing spondylitis was likely due to altered NF-κB regulation by the mutant LMP2 protein.[127] The NF-κB pathway affects osteoclastic activity by modulating the Wnt/β-catenin pathway, which is overactive in this disease due to low levels of the Wnt/β-catenin inhibitor, Dicckopf-1. This also raises the possibility that Dicckopf-1 could be a substrate for i-proteasome degradation. The R60H polymorphism in LMP2 has also demonstrated a strong association with psoriasis, another autoimmune disease.[129]

Specific LMP2 codon 60 polymorphisms can also confer resistance to disease. Genotype analysis of a large Italian population showed that individuals with the LMP2 H/H genotype had a reduced risk for developing multiple sclerosis, potentially due to the decreased production of immunodominant peptides from myelin basic protein.[130] To experimentally test this idea, *in vitro* digestion of a 28-mer peptide from myelin basic protein ($MBP_{102-129}$) was performed using purified 20S i-proteasome containing LMP2 with either histidine or arginine in codon 60. Histidine-containing i-proteasomes demonstrated reduced degradation of the 28-mer peptide to the major immunodominant ($MBP_{111-119}$) epitope. These results provide a potential mechanism that could explain the reduced risk for multiple sclerosis in individuals with the H/H genotype.

The strong association of the LMP2 codon 60 polymorphism with disease susceptibility has prompted a more in-depth examination of how this polymorphism can affect subunit structure and function.[131] Structural considerations were studied using an in silico model that was based on the crystal structure of i-proteasome described by Unno and colleagues[132] and taking into account the biochemical differences conferred by the two amino acids. In the mature protein, which has undergone proteolytic cleavage of the N terminus, the amino acid in codon 60 corresponds to residue 40. The crystal structure places this residue on the external surface of the core, close to the interface of the α- and β-rings but not in direct contact with either the α-subunits or the catalytic active site. The biochemical differences conferred by substitution of Arg for His could be quite substantial due to the increased size, which could introduce steric hindrance, and altered charge state of the amino acid from neutral (His) to positive (Arg). Based on these structural and biochemical considerations, three hypotheses were proposed to explain how the polymorphisms could alter i-proteasome function: (i) the residue at amino acid 40 could be the binding site for regulatory molecules (protein and nonprotein) whose affinity is dependent on the amino acid that is present; (ii) the difference in charge state could alter the electrical field at the catalytic site and affect activity; and (iii) the charge change could also affect the stability of the 20S core. The net charge of the α-subunits is −12, and with arginine in residue 40, the net charge of the β-subunits is +12. The resultant electrical interaction between these opposing charge states likely contributes to the affinity between the two rings and aids in the assembly of the core particle. The presence of the neutral-charged His may have a negative effect on this process. The positive charge conferred by Arg at residue 40 also helps stabilize the structure via hydrogen bonds between water molecules and may also affect the positioning of the protein loops that form the channels through which substrates access the catalytic active sites in the core. On the other hand, the reduced hydrogen bonding and consequent greater structural flexibility conferred by the presence of the neutral His may help open the channel and permit greater access to the catalytic active site. These results suggest that the structural and biochemical changes caused by substitution of Arg or His at residue 40 in the LMP2 protein, which occurs as an SNP in the *PSMB9* gene, can have a significant impact on i-proteasome function and may explain the disease susceptibility of specific genotypes.

Without a doubt, the number of human diseases directly linked to defects in i-proteasome will continue to grow. That information will greatly aid our understanding of i-proteasome's ever-expanding role in cellular processes that are unrelated to the generation of peptides for antigen presentation.

## References

1. Ciehanover A, Hod Y, Hershko A. A heat-stable polypeptide component of an ATP-dependent proteolytic system from reticulocytes. *Biochem Biophys Res Commun* 1978;**81**:1100–5.
2. Etlinger JD, Goldberg AL. A soluble ATP-dependent proteolytic system responsible for the degradation of abnormal proteins in reticulocytes. *Proc Natl Acad Sci USA* 1977;**74**:54–8.
3. Brown MG, Driscoll J, Monaco JJ. Structural and serological similarity of MHC-linked LMP and proteasome (multicatalytic proteinase) complexes. *Nature* 1991;**353**:355–7.

4. Glynne R, Powis SH, Beck S, Kelly A, Kerr LA, Trowsdale J. A proteasome-related gene between the two ABC transporter loci in the class II region of the human MHC. *Nature* 1991;**353**:357–60.
5. Kelly A, Powis SH, Glynne R, Radley E, Beck S, Trowsdale J. Second proteasome-related gene in the human MHC class II region. *Nature* 1991;**353**:667–8.
6. Ortiz-Navarrete V, Seelig A, Gernold M, Frentzel S, Kloetzel PM, Hammerling GJ. Subunit of the '20S' proteasome (multicatalytic proteinase) encoded by the major histocompatibility complex. *Nature* 1991;**353**:662–4.
7. Aki M, Shimbara N, Takashina M, Akiyama K, Kagawa S, Tamura T, et al. Interferon-gamma induces different subunit organizations and functional diversity of proteasomes. *J Biochem* 1994;**115**:257–69.
8. Groettrup M, Kraft R, Kostka S, Standera S, Stohwasser R, Kloetzel PM. A third interferon-gamma-induced subunit exchange in the 20S proteasome. *Eur J Immunol* 1996;**26**:863–9.
9. Hayashi M, Ishibashi T, Tanaka K, Kasahara M. The mouse genes encoding the third pair of beta-type proteasome subunits regulated reciprocally by IFN-gamma: structural comparison, chromosomal localization, and analysis of the promoter. *J Immunol* 1997;**159**:2760–70.
10. Nandi D, Jiang H, Monaco JJ. Identification of MECL-1 (LMP-10) as the third IFN-gamma-inducible proteasome subunit. *J Immunol* 1996;**156**:2361–4.
11. Murata S, Sasaki K, Kishimoto T, Niwa S, Hayashi H, Takahama Y, et al. Regulation of $CD8^+$ T cell development by thymus-specific proteasomes. *Science* 2007;**316**:1349–53.
12. Baumeister W, Walz J, Zuhl F, Seemuller E. The proteasome: paradigm of a self-compartmentalizing protease. *Cell* 1998;**92**:367–80.
13. Davies KJ. Degradation of oxidized proteins by the 20S proteasome. *Biochimie* 2001;**83**:301–10.
14. Reinheckel T, Sitte N, Ullrich O, Kuckelkorn U, Davies KJ, Grune T. Comparative resistance of the 20S and 26S proteasome to oxidative stress. *Biochem J* 1998;**335**(Pt 3):637–42.
15. Dahlmann B, Ruppert T, Kuehn L, Merforth S, Kloetzel PM. Different proteasome subtypes in a single tissue exhibit different enzymatic properties. *J Mol Biol* 2000;**303**:643–53.
16. Klare N, Seeger M, Janek K, Jungblut PR, Dahlmann B. Intermediate-type 20 S proteasomes in HeLa cells: "asymmetric" subunit composition, diversity and adaptation. *J Mol Biol* 2007;**373**:1–10.
17. Drews O, Wildgruber R, Zong C, Sukop U, Nissum M, Weber G, et al. Mammalian proteasome subpopulations with distinct molecular compositions and proteolytic activities. *Mol Cell Proteomics* 2007;**6**:2021–31.
18. Groettrup M, Kirk CJ, Basler M. Proteasomes in immune cells: more than peptide producers? *Nat Rev Immunol* 2010;**10**:73–8.
19. Monaco JJ, McDevitt HO. H-2-linked low-molecular weight polypeptide antigens assemble into an unusual macromolecular complex. *Nature* 1984;**309**:797–9.
20. Monaco JJ, McDevitt HO. The LMP antigens: a stable MHC-controlled multisubunit protein complex. *Hum Immunol* 1986;**15**:416–26.
21. Zhou P, Zanelli E, Smart M, David C. Genomic organization and tissue expression of mouse proteasome gene Lmp-2. *Genomics* 1993;**16**:664–8.
22. Fruh K, Yang Y, Arnold D, Chambers J, Wu L, Waters JB, et al. Alternative exon usage and processing of the major histocompatibility complex-encoded proteasome subunits. *J Biol Chem* 1992;**267**:22131–40.
23. Wright KL, White LC, Kelly A, Beck S, Trowsdale J, Ting JP. Coordinate regulation of the human TAP1 and LMP2 genes from a shared bidirectional promoter. *J Exp Med* 1995;**181**:1459–71.

24. Chatterjee-Kishore M, Wright KL, Ting JP, Stark GR. How Stat1 mediates constitutive gene expression: a complex of unphosphorylated Stat1 and IRF1 supports transcription of the LMP2 gene. *EMBO J* 2000;**19**:4111–22.
25. Yang Y, Waters JB, Fruh K, Peterson PA. Proteasomes are regulated by interferon gamma: implications for antigen processing. *Proc Natl Acad Sci USA* 1992;**89**:4928–32.
26. James AB, Conway AM, Morris BJ. Regulation of the neuronal proteasome by Zif268 (Egr1). *J Neurosci* 2006;**26**:1624–34.
27. Zanelli E, Zhou P, Cao H, Smart MK, David CS. Genomic organization and tissue expression of the mouse proteasome gene Lmp-7. *Immunogenetics* 1993;**38**:400–7.
28. Hisamatsu H, Shimbara N, Saito Y, Kristensen P, Hendil KB, Fujiwara T, et al. Newly identified pair of proteasomal subunits regulated reciprocally by interferon gamma. *J Exp Med* 1996;**183**:1807–16.
29. Cruz M, Elenich LA, Smolarek TA, Menon AG, Monaco JJ. DNA sequence, chromosomal localization, and tissue expression of the mouse proteasome subunit lmp10 (Psmb10) gene. *Genomics* 1997;**45**:618–22.
30. Hussong SA, Kapphahn RJ, Phillips SL, Maldonado M, Ferrington DA. Immunoproteasome deficiency alters retinal proteasome's response to stress. *J Neurochem* 2010;**113**:1481–90.
31. Bose S, Brooks P, Mason GG, Rivett AJ. gamma-Interferon decreases the level of 26 S proteasomes and changes the pattern of phosphorylation. *Biochem J* 2001;**353**:291–7.
32. Diaz-Hernandez M, Martin-Aparicio E, Avila J, Hernandez F, Lucas JJ. Enhanced induction of the immunoproteasome by interferon gamma in neurons expressing mutant Huntingtin. *Neurotox Res* 2004;**6**:463–8.
33. Gregerson DS, Lew KL, McPherson SW, Heuss ND, Ferrington DA. RPE cells resist bystander killing by CTLs, but are highly susceptible to antigen-dependent CTL killing. *Invest Ophthalmol Vis Sci* 2006;**47**:5385–94.
34. Kotamraju S, Matalon S, Matsunaga T, Shang T, Hickman-Davis JM, Kalyanaraman B. Upregulation of immunoproteasomes by nitric oxide: potential antioxidative mechanism in endothelial cells. *Free Radic Biol Med* 2006;**40**:1034–44.
35. Marques AJ, Palanimurugan R, Matias AC, Ramos PC, Dohmen RJ. Catalytic mechanism and assembly of the proteasome. *Chem Rev* 2009;**109**:1509–36.
36. Hirano Y, Hendil KB, Yashiroda H, Iemura S, Nagane R, Hioki Y, et al. A heterodimeric complex that promotes the assembly of mammalian 20S proteasomes. *Nature* 2005;**437**:1381–5.
37. Yashiroda H, Mizushima T, Okamoto K, Kameyama T, Hayashi H, Kishimoto T, et al. Crystal structure of a chaperone complex that contributes to the assembly of yeast 20S proteasomes. *Nat Struct Mol Biol* 2008;**15**:228–36.
38. De M, Jayarapu K, Elenich L, Monaco JJ, Colbert RA, Griffin TA. Beta 2 subunit propeptides influence cooperative proteasome assembly. *J Biol Chem* 2003;**278**:6153–9.
39. Groettrup M, Standera S, Stohwasser R, Kloetzel PM. The subunits MECL-1 and LMP2 are mutually required for incorporation into the 20S proteasome. *Proc Natl Acad Sci USA* 1997;**94**:8970–5.
40. Griffin TA, Nandi D, Cruz M, Fehling HJ, Kaer LV, Monaco JJ, et al. Immunoproteasome assembly: cooperative incorporation of interferon gamma (IFN-gamma)-inducible subunits. *J Exp Med* 1998;**187**:97–104.
41. Guillaume B, Chapiro J, Stroobant V, Colau D, Van Holle B, Parvizi G, et al. Two abundant proteasome subtypes that uniquely process some antigens presented by HLA class I molecules. *Proc Natl Acad Sci USA* 2010;**107**:18599–604.
42. Heink S, Ludwig D, Kloetzel PM, Kruger E. IFN-gamma-induced immune adaptation of the proteasome system is an accelerated and transient response. *Proc Natl Acad Sci USA* 2005;**102**:9241–6.

43. Groll M, Ditzel L, Lowe J, Stock D, Bochtler M, Bartunik HD, et al. Structure of 20S proteasome from yeast at 2.4 A resolution. *Nature* 1997;**386**:463–71.
44. Borissenko L, Groll M. Diversity of proteasomal missions: fine tuning of the immune response. *Biol Chem* 2007;**388**:947–55.
45. Yewdell JW. Immunoproteasomes: regulating the regulator. *Proc Natl Acad Sci USA* 2005;**102**:9089–90.
46. Van Kaer L, Ashton-Rickardt PG, Eichelberger M, Gaczynska M, Nagashima K, Rock KL, et al. Altered peptidase and viral-specific T cell response in LMP2 mutant mice. *Immunity* 1994;**1**:533–41.
47. Fehling HJ, Swat W, Laplace C, Kuhn R, Rajewsky K, Muller U, et al. MHC class I expression in mice lacking the proteasome subunit LMP-7. *Science* 1994;**265**:1234–7.
48. Basler M, Moebius J, Elenich L, Groettrup M, Monaco JJ. An altered T cell repertoire in MECL-1-deficient mice. *J Immunol* 2006;**176**:6665–72.
49. Kincaid EZ, Che JW, York I, Escobar H, Reyes-Vargas E, Delgado JC, et al. Mice completely lacking immunoproteasomes show major changes in antigen presentation. *Nat Immunol* 2011;**13**:129–35.
50. Basler M, Beck U, Kirk CJ, Groettrup M. The antiviral immune response in mice devoid of immunoproteasome activity. *J Immunol* 2011;**187**:5548–57.
51. Agarwal AK, Xing C, DeMartino GN, Mizrachi D, Hernandez MD, Sousa AB, et al. PSMB8 encoding the beta5i proteasome subunit is mutated in joint contractures, muscle atrophy, microcytic anemia, and panniculitis-induced lipodystrophy syndrome. *Am J Hum Genet* 2010;**87**:866–72.
52. Arima K, Kinoshita A, Mishima H, Kanazawa N, Kaneko T, Mizushima T, et al. Proteasome assembly defect due to a proteasome subunit beta type 8 (PSMB8) mutation causes the autoinflammatory disorder, Nakajo-Nishimura syndrome. *Proc Natl Acad Sci USA* 2011;**108**:14914–9.
53. Kitamura A, Maekawa Y, Uehara H, Izumi K, Kawachi I, Nishizawa M, et al. A mutation in the immunoproteasome subunit PSMB8 causes autoinflammation and lipodystrophy in humans. *J Clin Invest* 2011;**121**:4150–60.
54. Liu Y, Ramot Y, Torrelo A, Paller AS, Si N, Babay S, et al. Mutations in PSMB8 cause CANDLE syndrome with evidence of genetic and phenotypic heterogeneity. *Arthritis Rheum* 2011;**64**:895–907.
55. Neefjes J, Jongsma ML, Paul P, Bakke O. Towards a systems understanding of MHC class I and MHC class II antigen presentation. *Nat Rev Immunol* 2011;**11**:823–36.
56. Dalet A, Stroobant V, Vigneron N, Van den Eynde BJ. Differences in the production of spliced antigenic peptides by the standard proteasome and the immunoproteasome. *Eur J Immunol* 2011;**41**:39–46.
57. Rape M, Jentsch S. Taking a bite: proteasomal protein processing. *Nat Cell Biol* 2002;**4**: E113–6.
58. York IA, Goldberg AL, Mo XY, Rock KL. Proteolysis and class I major histocompatibility complex antigen presentation. *Immunol Rev* 1999;**172**:49–66.
59. Brossart P, Bevan MJ. Presentation of exogenous protein antigens on major histocompatibility complex class I molecules by dendritic cells: pathway of presentation and regulation by cytokines. *Blood* 1997;**90**:1594–9.
60. Rock KL, York IA, Saric T, Goldberg AL. Protein degradation and the generation of MHC class I-presented peptides. *Adv Immunol* 2002;**80**:1–70.
61. Cardozo C, Kohanski RA. Altered properties of the branched chain amino acid-preferring activity contribute to increased cleavages after branched chain residues by the "immunoproteasome" *J Biol Chem* 1998;**273**:16764–70.

62. Chapiro J, Claverol S, Piette F, Ma W, Stroobant V, Guillaume B, et al. Destructive cleavage of antigenic peptides either by the immunoproteasome or by the standard proteasome results in differential antigen presentation. *J Immunol* 2006;**176**:1053–61.
63. Lei B, Abdul Hameed MD, Hamza A, Wehenkel M, Muzyka JL, Yao XJ, et al. Molecular basis of the selectivity of the immunoproteasome catalytic subunit LMP2-specific inhibitor revealed by molecular modeling and dynamics simulations. *J Phys Chem B* 2010; **114**:12333–9.
64. Sijts AJ, Ruppert T, Rehermann B, Schmidt M, Koszinowski U, Kloetzel PM. Efficient generation of a hepatitis B virus cytotoxic T lymphocyte epitope requires the structural features of immunoproteasomes. *J Exp Med* 2000;**191**:503–14.
65. van Hall T, Sijts A, Camps M, Offringa R, Melief C, Kloetzel PM, et al. Differential influence on cytotoxic T lymphocyte epitope presentation by controlled expression of either proteasome immunosubunits or PA28. *J Exp Med* 2000;**192**:483–94.
66. Morel S, Levy F, Burlet-Schiltz O, Brasseur F, Probst-Kepper M, Peitrequin AL, et al. Processing of some antigens by the standard proteasome but not by the immunoproteasome results in poor presentation by dendritic cells. *Immunity* 2000;**12**:107–17.
67. Chen W, Norbury CC, Cho Y, Yewdell JW, Bennink JR. Immunoproteasomes shape immunodominance hierarchies of antiviral CD8(+) T cells at the levels of T cell repertoire and presentation of viral antigens. *J Exp Med* 2001;**193**:1319–26.
68. van Helden MJ, de Graaf N, Bekker CP, Boog CJ, Zaiss DM, Sijts AJ. Immunoproteasome-deficiency has no effects on NK cell education, but confers lymphocytes into targets for NK cells in infected wild-type mice. *PLoS One* 2011;**6**:e23769.
69. de Graaf N, van Helden MJ, Textoris-Taube K, Chiba T, Topham DJ, Kloetzel PM, et al. PA28 and the proteasome immunosubunits play a central and independent role in the production of MHC class I-binding peptides in vivo. *Eur J Immunol* 2011;**41**:926–35.
70. Cerundolo V, Elliott T, Elvin J, Bastin J, Rammensee HG, Townsend A. The binding affinity and dissociation rates of peptides for class I major histocompatibility complex molecules. *Eur J Immunol* 1991;**21**:2069–75.
71. Townsend A, Ohlen C, Foster L, Bastin J, Ljunggren HG, Karre K. A mutant cell in which association of class I heavy and light chains is induced by viral peptides. *Cold Spring Harb Symp Quant Biol* 1989;**54**(Pt 1):299–308.
72. Hensley SE, Zanker D, Dolan BP, David A, Hickman HD, Embry AC, et al. Unexpected role for the immunoproteasome subunit LMP2 in antiviral humoral and innate immune responses. *J Immunol* 2010;**184**:4115–22.
73. Girdlestone J, Isamat M, Gewert D, Milstein C. Transcriptional regulation of HLA-A and -B: differential binding of members of the Rel and IRF families of transcription factors. *Proc Natl Acad Sci USA* 1993;**90**:11568–72.
74. Khan S, van den Broek M, Schwarz K, de Giuli R, Diener PA, Groettrup M. Immunoproteasomes largely replace constitutive proteasomes during an antiviral and antibacterial immune response in the liver. *J Immunol* 2001;**167**:6859–68.
75. Strehl B, Joeris T, Rieger M, Visekruna A, Textoris-Taube K, Kaufmann SH, et al. Immunoproteasomes are essential for clearance of Listeria monocytogenes in nonlymphoid tissues but not for induction of bacteria-specific $CD8^+$ T cells. *J Immunol* 2006; **177**:6238–44.
76. Tu L, Moriya C, Imai T, Ishida H, Tetsutani K, Duan X, et al. Critical role for the immunoproteasome subunit LMP7 in the resistance of mice to Toxoplasma gondii infection. *Eur J Immunol* 2009;**39**:3385–94.
77. Murata S, Takahama Y, Tanaka K. Thymoproteasome: probable role in generating positively selecting peptides. *Curr Opin Immunol* 2008;**20**:192–6.

78. Osterloh P, Linkemann K, Tenzer S, Rammensee HG, Radsak MP, Busch DH, et al. Proteasomes shape the repertoire of T cells participating in antigen-specific immune responses. *Proc Natl Acad Sci USA* 2006;**103**:5042–7.
79. de Verteuil D, Muratore-Schroeder TL, Granados DP, Fortier MH, Hardy MP, Bramoulle A, et al. Deletion of immunoproteasome subunits imprints on the transcriptome and has a broad impact on peptides presented by major histocompatibility complex I molecules. *Mol Cell Proteomics* 2010;**9**:2034–47.
80. Dick TP, Ruppert T, Groettrup M, Kloetzel PM, Kuehn L, Koszinowski UH, et al. Coordinated dual cleavages induced by the proteasome regulator PA28 lead to dominant MHC ligands. *Cell* 1996;**86**:253–62.
81. Sijts A, Sun Y, Janek K, Kral S, Paschen A, Schadendorf D, et al. The role of the proteasome activator PA28 in MHC class I antigen processing. *Mol Immunol* 2002;**39**:165–9.
82. Sun Y, Sijts AJ, Song M, Janek K, Nussbaum AK, Kral S, et al. Expression of the proteasome activator PA28 rescues the presentation of a cytotoxic T lymphocyte epitope on melanoma cells. *Cancer Res* 2002;**62**:2875–82.
83. Textoris-Taube K, Henklein P, Pollmann S, Bergann T, Weisshoff H, Seifert U, et al. The N-terminal flanking region of the TRP2360-368 melanoma antigen determines proteasome activator PA28 requirement for epitope liberation. *J Biol Chem* 2007;**282**:12749–54.
84. Ethen CM, Hussong SA, Reilly C, Feng X, Olsen TW, Ferrington DA. Transformation of the proteasome with age-related macular degeneration. *FEBS Lett* 2007;**581**:885–90.
85. Ferrington DA, Hussong SA, Roehrich H, Kapphahn RJ, Kavanaugh SM, Heuss ND, et al. Immunoproteasome responds to injury in the retina and brain. *J Neurochem* 2008; **106**:158–69.
86. Hussong SA, Roehrich H, Kapphahn RJ, Maldonado M, Pardue MT, Ferrington DA. A novel role for the immunoproteasome in retinal function. *Invest Ophthalmol Vis Sci* 2011;**52**:714–23.
87. Kapphahn RJ, Bigelow EJ, Ferrington DA. Age-dependent inhibition of proteasome chymotrypsin-like activity in the retina. *Exp Eye Res* 2007;**84**:646–54.
88. Diaz-Hernandez M, Hernandez F, Martin-Aparicio E, Gomez-Ramos P, Moran MA, Castano JG, et al. Neuronal induction of the immunoproteasome in Huntington's disease. *J Neurosci* 2003;**23**:11653–61.
89. Mishto M, Bellavista E, Santoro A, Stolzing A, Ligorio C, Nacmias B, et al. Immunoproteasome and LMP2 polymorphism in aged and Alzheimer's disease brains. *Neurobiol Aging* 2006;**27**:54–66.
90. Yao X, Liu J, McCabe JT. Alterations of cerebral cortex and hippocampal proteasome subunit expression and function in a traumatic brain injury rat model. *J Neurochem* 2008;**104**:353–63.
91. Gavilan MP, Castano A, Torres M, Portavella M, Caballero C, Jimenez S, et al. Age-related increase in the immunoproteasome content in rat hippocampus: molecular and functional aspects. *J Neurochem* 2009;**108**:260–72.
92. Ding Q, Reinacker K, Dimayuga E, Nukala V, Drake J, Butterfield DA, et al. Role of the proteasome in protein oxidation and neural viability following low-level oxidative stress. *FEBS Lett* 2003;**546**:228–32.
93. Opitz E, Koch A, Klingel K, Schmidt F, Prokop S, Rahnefeld A, et al. Impairment of immunoproteasome function by beta5i/LMP7 subunit deficiency results in severe enterovirus myocarditis. *PLoS Pathog* 2011;**7**:e1002233.
94. Cai ZP, Shen Z, Van Kaer L, Becker LC. Ischemic preconditioning-induced cardioprotection is lost in mice with immunoproteasome subunit low molecular mass polypeptide-2 deficiency. *FASEB J* 2008;**22**:4248–57.
95. Ferrington DA, Husom AD, Thompson LV. Altered proteasome structure, function, and oxidation in aged muscle. *FASEB J* 2005;**19**:644–6.

96. Husom AD, Peters EA, Kolling EA, Fugere NA, Thompson LV, Ferrington DA. Altered proteasome function and subunit composition in aged muscle. *Arch Biochem Biophys* 2004;**421**:67–76.
97. Merforth S, Kuehn L, Osmers A, Dahlmann B. Alteration of 20S proteasome-subtypes and proteasome activator PA28 in skeletal muscle of rat after induction of diabetes mellitus. *Int J Biochem Cell Biol* 2003;**35**:740–8.
98. Teoh CY, Davies KJ. Potential roles of protein oxidation and the immunoproteasome in MHC class I antigen presentation: the 'PrOxI' hypothesis. *Arch Biochem Biophys* 2004; **423**:88–96.
99. Seifert U, Bialy LP, Ebstein F, Bech-Otschir D, Voigt A, Schroter F, et al. Immunoproteasomes preserve protein homeostasis upon interferon-induced oxidative stress. *Cell* 2010;**142**:613–24.
100. Ding Q, Martin S, Dimayuga E, Bruce-Keller AJ, Keller JN. LMP2 knock-out mice have reduced proteasome activities and increased levels of oxidatively damaged proteins. *Antioxid Redox Signal* 2006;**8**:130–5.
101. Hayashi T, Horiuchi A, Sano K, Hiraoka N, Kanai Y, Shiozawa T, et al. Molecular approach to uterine leiomyosarcoma: LMP2-deficient mice as an animal model of spontaneous uterine leiomyosarcoma. *Sarcoma* 2011;**2011**:476498.
102. Buchanan J, Mazumder PK, Hu P, Chakrabarti G, Roberts MW, Yun UJ, et al. Reduced cardiac efficiency and altered substrate metabolism precedes the onset of hyperglycemia and contractile dysfunction in two mouse models of insulin resistance and obesity. *Endocrinology* 2005;**146**:5341–9.
103. Aneja A, Tang WH, Bansilal S, Garcia MJ, Farkouh ME. Diabetic cardiomyopathy: insights into pathogenesis, diagnostic challenges, and therapeutic options. *Am J Med* 2008;**121**:748–57.
104. Hu J, Klein JD, Du J, Wang XH. Cardiac muscle protein catabolism in diabetes mellitus: activation of the ubiquitin-proteasome system by insulin deficiency. *Endocrinology* 2008;**149**:5384–90.
105. Zu L, Bedja D, Fox-Talbot K, Gabrielson KL, Van Kaer L, Becker LC, et al. Evidence for a role of immunoproteasomes in regulating cardiac muscle mass in diabetic mice. *J Mol Cell Cardiol* 2010;**49**:5–15.
106. Mallone R, Martinuzzi E, Blancou P, Novelli G, Afonso G, Dolz M, et al. $CD8^+$ T-cell responses identify beta-cell autoimmunity in human type 1 diabetes. *Diabetes* 2007;**56**:613–21.
107. Skowera A, Ellis RJ, Varela-Calvino R, Arif S, Huang GC, Van-Krinks C, et al. CTLs are targeted to kill beta cells in patients with type 1 diabetes through recognition of a glucose-regulated preproinsulin epitope. *J Clin Invest* 2008;**118**:3390–402.
108. Zaiss DM, Bekker CP, Grone A, Lie BA, Sijts AJ. Proteasome immunosubunits protect against the development of CD8 T cell-mediated autoimmune diseases. *J Immunol* 2011;**187**:2302–9.
109. Reis J, Hassan F, Guan XQ, Shen J, Monaco JJ, Papasian CJ, et al. The immunoproteasomes regulate LPS-induced TRIF/TRAM signaling pathway in murine macrophages. *Cell Biochem Biophys* 2011;**60**:119–26.
110. Schmidt N, Gonzalez E, Visekruna A, Kuhl AA, Loddenkemper C, Mollenkopf H, et al. Targeting the proteasome: partial inhibition of the proteasome by bortezomib or deletion of the immunosubunit LMP7 attenuates experimental colitis. *Gut* 2010;**59**:896–906.
111. Basler M, Dajee M, Moll C, Groettrup M, Kirk CJ. Prevention of experimental colitis by a selective inhibitor of the immunoproteasome. *J Immunol* 2010;**185**:634–41.
112. Muchamuel T, Basler M, Aujay MA, Suzuki E, Kalim KW, Lauer C, et al. A selective inhibitor of the immunoproteasome subunit LMP7 blocks cytokine production and attenuates progression of experimental arthritis. *Nat Med* 2009;**15**:781–7.
113. Chen F, Demers LM, Shi X. Upstream signal transduction of NF-kappaB activation. *Curr Drug Targets Inflamm Allergy* 2002;**1**:137–49.

114. Malek R, Borowicz KK, Jargiello M, Czuczwar SJ. Role of nuclear factor kappaB in the central nervous system. *Pharmacol Rep* 2007;**59**:25–33.
115. Hayden MS, Ghosh S. Shared principles in NF-kappaB signaling. *Cell* 2008;**132**:344–62.
116. Hayashi T, Faustman D. NOD mice are defective in proteasome production and activation of NF-kappaB. *Mol Cell Biol* 1999;**19**:8646–59.
117. Hayashi T, Faustman D. Essential role of human leukocyte antigen-encoded proteasome subunits in NF-kappaB activation and prevention of tumor necrosis factor-alpha-induced apoptosis. *J Biol Chem* 2000;**275**:5238–47.
118. Hayashi T, Faustman DL. Selected contribution: association of gender-related LMP2 inactivation with autoimmune pathogenesis. *J Appl Physiol* 2001;**91**:2804–15.
119. Kessler BM, Lennon-Dumenil AM, Shinohara ML, Lipes MA, Ploegh HL. LMP2 expression and proteasome activity in NOD mice. *Nat Med* 2000;**6**:1064 [Author reply 1065–1066].
120. Runnels HA, Watkins WA, Monaco JJ. LMP2 expression and proteasome activity in NOD mice. *Nat Med* 2000;**6**:1064–5 [Author reply 1065–1066].
121. Visekruna A, Joeris T, Seidel D, Kroesen A, Loddenkemper C, Zeitz M, et al. Proteasome-mediated degradation of IkappaBalpha and processing of p105 in Crohn disease and ulcerative colitis. *J Clin Invest* 2006;**116**:3195–203.
122. Ding Q, Lewis JJ, Strum KM, Dimayuga E, Bruce-Keller AJ, Dunn JC, et al. Polyglutamine expansion, protein aggregation, proteasome activity, and neural survival. *J Biol Chem* 2002;**277**:13935–42.
123. Lee W, Kim KB. The immunoproteasome: an emerging therapeutic target. *Curr Top Med Chem* 2011;**11**:2923–30.
124. Kuhn DJ, Hunsucker SA, Chen Q, Voorhees PM, Orlowski M, Orlowski RZ. Targeted inhibition of the immunoproteasome is a potent strategy against models of multiple myeloma that overcomes resistance to conventional drugs and nonspecific proteasome inhibitors. *Blood* 2009;**113**:4667–76.
125. Singh AV, Bandi M, Aujay MA, Kirk CJ, Hark DE, Raje N, et al. PR-924, a selective inhibitor of the immunoproteasome subunit LMP-7, blocks multiple myeloma cell growth both in vitro and in vivo. *Br J Haematol* 2011;**152**:155–63.
126. Garg A, Hernandez MD, Sousa AB, Subramanyam L, Martinez de Villarreal L, dos Santos HG, et al. An autosomal recessive syndrome of joint contractures, muscular atrophy, microcytic anemia, and panniculitis-associated lipodystrophy. *J Clin Endocrinol Metab* 2010;**95**:E58–63.
127. Haroon N, Maksymowych W, Rahman P, Tsui F, O'Shea F, Inman R. Radiographic severity in ankylosing spondylitis is associated with polymorphism in large multifunctional peptidase 2 (LMP2) in the SPARCC cohort. *Arthritis Rheum* 2011;.
128. Vargas-Alarcon G, Gamboa R, Zuniga J, Fragoso JM, Hernandez-Pacheco G, Londono J, et al. Association study of LMP gene polymorphisms in Mexican patients with spondyloarthritis. *Hum Immunol* 2004;**65**:1437–42.
129. Kramer U, Illig T, Grune T, Krutmann J, Esser C. Strong associations of psoriasis with antigen processing LMP and transport genes TAP differ by gender and phenotype. *Genes Immun* 2007;**8**:513–7.
130. Mishto M, Bellavista E, Ligorio C, Textoris-Taube K, Santoro A, Giordano M, et al. Immunoproteasome LMP2 60HH variant alters MBP epitope generation and reduces the risk to develop multiple sclerosis in Italian female population. *PLoS One* 2010;**5**:e9287.
131. Mishto M, Santoro A, Bellavista E, Sessions R, Textoris-Taube K, Dal Piaz F, et al. A structural model of 20S immunoproteasomes: effect of LMP2 codon 60 polymorphism on expression, activity, intracellular localisation and insight into the regulatory mechanisms. *Biol Chem* 2006;**387**:417–29.
132. Unno M, Mizushima T, Morimoto Y, Tomisugi Y, Tanaka K, Yasuoka N, et al. The structure of the mammalian 20S proteasome at 2.75 A resolution. *Structure* 2002;**10**:609–18.

# Interactions of the Proteasomal System with Chaperones: Protein Triage and Protein Quality control

Marc Kästle and
Tilman Grune

*Institute of Nutrition, Friedrich Schiller University, Jena, Germany*

I. Introduction ........ 114
II. Cytosolic Chaperones: The Hsp Family ........ 115
A. High-Molecular-Weight Hsps ........ 115
B. Small Hsps: Hsp27 ........ 119
C. Regulation of Hsps ........ 121
III. Refolding or Degradation? ........ 123
A. BAG-1 and CHIP Inhibition of Chaperones and Promotion of Proteasomal Degradation ........ 123
B. Hip and Hop: Stabilizing the Refolding Process ........ 126
C. The Co-chaperone-Independent Decision of Folding or Degradation by sHsps ........ 126
IV. Heat Shock Response After Proteasome Inhibition ........ 127
A. Regulation of Classical Hsps After Proteasome Inhibition ........ 127
B. Regulation of Heme Oxygenase-1 After Proteasome Inhibition ........ 129
V. Inhibition of Hsps: Influence on Proteasomal Degradation ........ 131
VI. Hsps and the Proteasomal System During Oxidative Stress and the Aging Process ........ 133
VII. Neurodegenerative Diseases: Connection to the Hsp Family and the Proteasome ........ 136
A. Alzheimer Disease ........ 137
B. Parkinson Disease ........ 138
VIII. Early-Stage Protein Quality Control and ERAD ........ 139
A. Yeast: ERAD-L ........ 144
B. Yeast: ERAD-M/C ........ 146
C. The Mammalian ERAD Process ........ 147
D. The ERAD Pathway of Non-glycosylated Proteins ........ 147
E. Proteasome Inhibition and Its Influence on ER Stress and the Unfolded Protein Response ........ 148
IX. Conclusion ........ 150
References ........ 150

Unfolded, misfolded, or modified proteins are able to induce proteotoxic cell stress. To prevent proteotoxic stress, it is crucial to have a functional protein quality control system, especially in the cytosol and in the endoplasmic reticulum where proteins are newly synthesized. The leading actors in this protein

Progress in Molecular Biology
and Translational Science, Vol. 109
DOI: 10.1016/B978-0-12-397863-9.00004-3

quality control system are the ubiquitin–proteasomal system and the huge family of heat shock proteins and chaperones. Both systems interact with each other, influencing the decision of whether a protein becomes (re)folded or degraded. Especially upon cellular stress, such as heat shock or oxidative stress, heat shock proteins are drastically upregulated, supporting, and regulating proteasomal degradation of defect proteins. Failure of one of the systems can be compensated partially by the upregulation of the other. Nevertheless, prolonged failure of the proteasome or chaperones results in protein aggregation and cellular dysfunction.

**Abbreviations:** 17-AAG, 7-*N*-allylamino-17-demethoxygeldanamycin; 2-PES, 2-phenylethynesulfonamide; ATF, activating transcription factor; BAG, bcl-2-associated athanogene; EGFR, epithelial growth factor receptor; eIF2α, eukaryotic initiation factor 2α; ER, endoplasmic reticulum; GSH, glutathione (reduced); HNE, 4-hydroxy-2-nonenal; Hsp, heat shock protein; IRE1a, inositol-requiting enzyme 1a; Keap-1, Kelch-like ECH-associated protein 1; macrophage inflammatory protein1β; NRF-2, nuclear factor erythroid 2-related factor 2; PD, Parkinson diseases; PERK, RNA-dependent protein kinase-like ER-kinase; ROS, reactive oxygen species; TNF-α, tumor necrosis factor-α; UPS, ubiquitin–proteasome system; VCP, valosin-containing protein;

## I. Introduction

During life, the cell is exposed to a multitude of stress situations such as heat, oxidative stress, inflammation, irradiation, heavy metals, or other toxic components. Besides the DNA (and maybe fatty acids with respect to oxidative stress), proteins are the most vulnerable molecules in the cellular system. For example, a small increase in temperature can cause protein unfolding, unspecific aggregation, and thus an imbalance of the protein homeostasis.[1] Therefore, a multitude of studies in a variety of cells and organisms were undertaken to determine the response of the cell to a number of stress stimuli. As a result, it was shown that a significant induction of more than 100 genes takes place, which can be grouped into seven different classes.[1] The first and the second classes comprise the family of heat shock proteins (Hsps) and the components of the proteolytic system.[1] Enzymes repairing and maintaining nucleic acid modifications belong to the third class of stress response proteins.[2] The fourth group involves metabolic enzymes, which stabilize and regulate energy supply of the cell.[3] The fifth class includes regulatory proteins, such as transcription factors or kinases.[4] The sixth protein class stabilizes the cytoskeleton during stress[5] and the last group comprises proteins that are involved in transport, detoxification, and membrane stability.[6] Especially molecular chaperones and the proteasomal system play a crucial role in sustaining protein homeostasis during cellular stress.

The proteasomal system and molecular chaperones are closely connected and the interaction of these two systems is well regulated. Binding of different co-chaperones defines whether a protein becomes stabilized and refolded or the substrate becomes ubiquitinated and degraded by the proteasome. If a protein defect and refolding fails, chaperones and co-chaperones assist and facilitate the ubiquitination process and, therefore, proteasomal degradation. Failure, mutation, or inhibition of one of these systems has automatic effects on the other members of the protein quality control system—effects that are on the one hand responsible for some diseases such as neurodegenerative diseases, cataract formation, or myopathies and on the other, "useful" mechanisms in cancer treatment.

Moreover, interaction of chaperones and the proteasome is not only limited to cytosolic proteins and chaperones, but also rather important for the folding machinery of the endoplasmic reticulum (ER). In the ER, it has also to be decided whether a newly synthesized protein is functional and correctly folded or not, in which case the protein passes through the ER-associated degradation (ERAD) process. The latter is a complex process where several chaperones interact with the proteasome.

For a better understanding of the narrow network of chaperones, co-chaperones, and the proteasome, it is important to have a more detailed look at the characteristics of the single chaperones and co-chaperones.

## II. Cytosolic Chaperones: The Hsp Family

Cytosolic, molecular chaperones comprise the huge family of Hsps and several co-chaperones that are required for the functionality of Hsps. Hsps are a highly conserved protein family that can be subdivided into two major fractions. The traditional classification distinguishes between the high-molecular-weight Hsps and the small Hsps (sHsps).[7] Both fractions have more or less different functions and control mechanisms, but all of them exhibit one or more heat shock elements (HSEs) in their promotor region, which is necessary for them to be classified as a member of the Hsp family.

### A. High-Molecular-Weight Hsps

The high-molecular-weight Hsps involve Hsp100, Hsp90, Hsp70, and Hsp60/chaperonins.[1,5] During physiological conditions, high-molecular-weight chaperones are needed constantly for assistance in protein (re)folding and protein stabilization in the cytosolic compartment to avoid protein aggregation. In situations of cellular stress, such as heat shock, oxidative stress, UV radiation, or metal intoxications, Hsps are heavily upregulated, trying to keep up cell and protein homeostasis. To discharge this obligation, high-molecular-weight Hsps need energy in the form of ATP. In contrast to many other enzymes, chaperones have to act at stoichiometric concentrations to maintain protein (re) folding. Thereby, chaperones have short-lived, low-affinity interactions with folded proteins, ensuring that Hsps do not interfere with substrates and affect their normal functionality—a fact that makes molecular chaperones perfectly suitable candidates to maintain a functional dynamic cell environment.[8] Many of the Hsps interact with other Hsps or influence one another's regulation, but some of them also have several other functions, which makes it necessary for us to have a closer look at the single subgroups of the Hsp family.

### 1. Hsp100

Proteins of the Hsp100 family can be found in bacteria, yeast, plants, and mammals.[9] The literature refers to Hsp100 using a variety of different nomenclatures depending on the species: HspH1, Hsp101, Hsp105, or Hsp104 (eukaryotes) and ClpA, ClpB, ClpC, ClpD, ClpE, ClpX, or ClpY (bacteria). All of them obtain a typical N-terminal domain for substrate binding, two AAA+ nucleotide-binding domains, a wing domain, and a small C-terminal domain. Interestingly, the AAA+ domain can also be found in the 19S regulator of the 26S proteasome.[9,10] Six of these subunits build the functional hexameric Hsp100 cylinder.[9,10] The main function of Hsp100 is to solubilize proteins that have become aggregated after cellular stress. Therefore, Hsp100 is strongly induced after drastic cellular stress. Furthermore, Hsp104 or Hsp101 can cooperate with Hsp70 to dissolve aggregated proteins.[11] It was suggested that loops in the central hexamer of Hsp100-rings translocate bound proteins by ATP hydrolysis axially down the channel, exerting a mechanical pulling force. Because the protein segments have to pass a narrow channel with an approximate diameter of 12 Å, this pulling action is associated with an unfolding process.[12]

### 2. Hsp90

Under normal physiological conditions, Hsp90 is already present at very high concentrations in the cytosol and can be further upregulated in stress situations.[13] High basal levels of Hsp90 allow augmented binding occurrence and short-lived interactions with client proteins. The Hsp90 chaperone is a flexible dimer, whereas every monomer consists of an N-terminal domain, which is connected via a long linker sequence with an M-domain and ends up with a C-terminal domain.[14] ATP binding and hydrolysis are conducted in a very slow fashion by the N-terminal domain, resulting in conformational changes of the chaperone, which are required for an activating state of the bound substrate. The slow hydrolysis of ATP is therefore the rate-limiting step of Hsp90 action.[14]

Unlike the other Hsps, Hsp90 is, in the first place, not responsible for binding unfolded proteins and preventing their aggregation, but rather binds to a special, defined set of proteins, called clients, in the cytosol and the nucleus.

The wide field of client proteins comprises protein kinases, nuclear hormone receptors, cell surface receptors, or transcription factors in the cytosol.[8] Thus, Hsp90 acts on these prefolded proteins to assist in the opening and stabilization of ligand-binding clefts. A prominent example for this mechanism is the binding of steroid hormone receptors in the cytosol by Hsp90.[15] At the beginning of Hsp research, it was suggested that Hsp90 interacts only with cytosolic proteins. Later on, it was established that Hsp90 also plays a role in RNA transcription, RNA processing, DNA replication, DNA recombination, DNA repair, and DNA metabolism.[8]

The activity and function of Hsp90 is significantly influenced by several co-chaperones, which display a special binding preference for distinct Hsp90 conformations. At the C terminus of the chaperone, Hsp90 possesses EEVD motifs, which function as binding sites for small tetratricopeptide repeat (TPR) domains. Importantly, the TPR domain-containing co-chaperones are Hop/Sti1 (Hsp organizing proteins), CHIP (ubiquitin ligase), and Cyp40 (protein maturation). Hop/Sti1 maintains the linkage of Hsp90 and Hsp70/Hsp40, which is needed, for example, for steroid hormone receptor stabilization, protein kinase binding, or the transfer of a client protein from Hsp90 to Hsp70.[7,14] The second listed co-chaperone, called carboxyl terminus of Hsp70 interacting protein (CHIP), functions as a ubiquitin E3 ligase of proteins that are bound on Hsp90/Hsp70 and aim for protein degradation. The detailed function of CHIP is discussed later in this chapter.

It has already been mentioned that Hsp90 is not responsible in the first place for refolding or maintenance of misfolded proteins; instead, it is responsible for keeping (ligand free) proteins in a functional folded state. This is all the more true for a situation of cellular stress. The upregulated co-chaperone Sti1 functions as a noncompetitive inhibitor of the Hsp90 ATPase,[16] keeping Hsp90 in a conformation that could facilitate its interaction with its substrate but inhibit further molecular changes in it, leading to a more basic holding function and preventing aggregation of the bound protein.[1,17]

### 3. Hsp70

The family of Hsp70s is a group of molecular chaperones with the most variable functions in the cell.[11] These tasks comprise, on the one hand, housekeeping functions, involving folding of newly synthesized proteins, stabilization and control of signal transduction proteins, and membrane translocation of secretory proteins and, on the other hand, quality control functions, such as proofreading of proteins and repair of misfolded, damaged, or aggregated proteins.[18–21] To cope with all these different obligations, the 70-kDa chaperones are expressed as a constitutive form (Hsc70; gene: *HSPA8*), maintaining the normal cell function, and as a stress-inducible form (Hsp70/72; gene: *HSPA1A*, *HSPA1B*, and *HSPA1L*), which is mainly responsible for the prevention of protein damage or protein aggregation as well as the reestablishment of functional proteins. Although there are different expression types of these chaperones with different functions, all of them possess a similar structure: an N-terminal ATPase domain and a C-terminal substrate-binding domain (SBD), which can be subdivided into a β-sandwich domain and an α-helical subdomain.[21] The ATPase domain indicates the ATP dependency of Hsp70 for its chaperone and folding activity. The ATP-bound form of Hsp70 shows a low affinity to the bound substrate and is, therefore, responsible for a rapid binding, processing, and release of the substrate, while the adenosine diphosphate (ADP)-bound form possesses a much higher affinity to the bound substrate, and thus, binding and release of proteins expires in a very slow way.[22] In the ADP-bound state, one of the α-helices of the SBD forms salt bridges and hydrogen bonds to the outer loops, thereby closing the substrate-binding pocket like a lid.[11] This transition from the ATP to the ADP state is markedly facilitated by the *J*-domain protein Hsp40 (prokaryotes: DnaJ). Further on, it has been shown that Hsp40 is also necessary to keep up the substrate specificity of Hsp70. The SBD of Hsp70 is specially meant for the recognition and binding of hydrophobic patches of the substrate,[11,21] Therefore, Hsp40 scans hydrophobic protein surfaces, recognizing aromatic and aliphatic side chains and marks them as substrates for Hsp70.[23]

Like Hsp90, Hsp70 is also equipped with an EEVD domain at the C terminus, which allows the binding of TPR domain-containing co-chaperones, such as Hop/Sti1 and CHIP,[24] as well as the interaction with Hsp90. Two other competing proteins that also act as co-chaperones for Hsp70 interact with the N-terminal ATPase domain. These two proteins are Hip (human homolog: p48) and BAG-1 (bcl-2-associated athanogene-1). It is supposed that Hip/p48 binds to the ATPase domain of Hsp70, stabilizes the ADP state of the chaperone, and thus promotes the folding activity of Hsp70 by preventing an early substrate release.[22] This can be of importance for the efficient interaction of Hsp70 with other chaperones such as Hsp90.

Hip/p48 binding to the N terminus is in competition with BAG-1. BAG-1 belongs to the BAG-protein family (BAG-1,-2,-3,-4,-5,-6). All of them obtain at least one BAG domain that, among others, is responsible for Hsp70 binding. In contrast to HIP/p48, BAG-1 has exactly the contrary effect on Hsp70 chaperones. This means that interaction of BAG-1 with the amino-terminal ATPase domain of Hsp70 results in a much faster release of ADP and a quicker return of Hsp70 to a new ATP-bound state.[22]

Interaction of these co-chaperones with each other and Hsp70 has a pronounced influence on the decision of whether a protein becomes refolded by the molecular chaperone system or is transferred to the proteasomal system where it is going to be degraded. The detailed mechanisms for this decision are discussed below.

4. Hsp40

The protein family of Hsp40s is an intermediate between molecular chaperones and small chaperones. Although Hsp40 binds diverse unfolded client proteins, it is more defined as an Hsp70-associated co-chaperone.[25] Hsp40s are homologous to the *Escherichia coli* co-chaperone DnaJ.[25] As indicated in the name of DnaJ, the conserved *J*-domain is the most important domain of Hsp40. Moreover, Hsp40s are classified into three types, containing different protein domains. Type 1 Hsp40s possess a zinc-finger-like domain, a glycine- and phenylalanine-rich (G/P) domain, and the *J*-domain. Type 2 Hsp40s also have a G/P domain and a *J*-domain, whereas type 3 Hsp40s contain only a *J*-domain.[25] Certainly the *J*-domain is adequate to assist Hsp70 during protein binding.[26] For the main part, Hsp40 is responsible for recognizing and binding unfolded proteins and transferring them to Hsp70. Therefore, Hsp40 binds with its *J*-domain to Hsp70 when Hsp70 is in its ATP-bound state. Delivery of the substrate from Hsp40 to Hsp70 has not been absolutely clarified, but it is suggested that the association of Hsp40 to Hsp70 results in an allosteric shift in the Hsp40 protein structure, reducing its binding affinity to the substrate. Moreover, binding of the *J*-domain of Hsp70 takes the substrate very near to the Hsp70 SBD. The following ATP hydrolysis enhances Hsp70 substrate affinity, which exceeds the forces binding Hsp40 to the substrate, resulting in the release of Hsp40 from the substrate and from Hsp70.[25] The other existing domains of Hsp40 can also be involved in the protein transfer to Hsp70. Type 1 Hsp40s, for example, use their zinc-finger-like region for an exact positioning of the substrate during their transfer to Hsp70.[27] Besides the Hsp70 substrate delivery, Hsp40s fulfill some "optional functions." ERdj5, an ER-located type 3 Hsp40, which plays an important role in ERAD, contains three cystein-rich thioredoxin domains that are involved in substrate recognition and retranslocation into the cytosol. ERdj5 cleaves disulfide bonds of nonnative proteins, targeted for ERAD, and facilitates therewith the retranslocation process.[25] In summary, Hsp40 is a crucial co-chaperone, functioning as a linker between the substrate and Hsp70. Hsp40 recognizes suitable Hsp70 substrates, delivers them to Hsp70, and assists during the substrate transfer to Hsp70.

## B. Small Hsps: Hsp27

As sHsps are members of the Hsp superfamily, they have some differences compared with their larger relatives: They (i) are ATP independent, (ii) possess much smaller molecular masses (12–42 kDa), (iii) are able to form large oligomers, and (iv) contain a conserved α-crystallin domain.[28,29] Despite the common α-crystallin domain, sHsps possess a less conserved structure than larger Hsps and find their expression in the numerous diverse sHsps across several different organisms. In humans, ten representatives of the sHsp family were identified.[28]

All of them show an N-terminal region followed by a representative α-crystallin domain and a C-terminal domain. The α-crystallin domain is admittedly the signature motif of all sHsps, but except for a few amino acids, the sequence of the α-crystallin domain is highly variable.[28] The α-crystallin domain is mainly organized in β-sheet sandwiches, which are responsible for dimer formation of several sHsps. This dimer formation is the basic structure of many sHsps, but to mediate their function as chaperones, sHsps have to form larger oligomers.[29] The size of these oligomers varies between different species and the different types of sHsps. For example, yeast Hsp26 assembles oligomers of up to 24 subunits, whereas α-crystallins of the human lens can form polydisperse oligomers with up to 50 subunits per complex.[28] The mechanism by which oligomerization of sHsps is accomplished is still under debate and also underlies the overabundance of different sHsps. It seems evident that the α-crystallin structure is essential for oligomerization, but the roles of the N-terminal and C-terminal domains are very diverse. For example, the hydrophobic N-terminal domain of wheat Hsp16.9 is incorporated in oligomeric assembly,[30] whereas shortened yeast Hsp26 (removal of the first 30 N-terminal residues) can also build large oligomers, avoiding the aggregation of substrate proteins.[28] In contrast, the C terminus of mammalian sHsps, such as Hsp25, possesses highly flexible polar extensions. The high polarity of these residues mediates the solubility of hydrophobic sHsps and stabilizes the interaction between sHsps and the bound substrate.[31] In the case of Hsp27 and α-crystallins, oligomerization seems to depend on the phosphorylation status of some serine (Ser) residues. Phosphorylation of these Ser residues is catalyzed by the two downstream p38 kinases MAPKAP 2/3.[32] If the Ser residues (Hsp27: $Ser^{15}$, $Ser^{78}$, and $Ser^{82}$; α-crystallin: $Ser^{19}$, $Ser^{45}$, and $Ser^{59}$) are dephosphorylated, Hsp27 and α-crystallins are able to build large oligomers, enabling the chaperone function. In the phosphorylated state, the sHsps remain in their dimerized form.

Oligomerization of sHsps is a dynamic process with a continuous exchange of subunits, which allows sHsps to shift between active and inactive states. In general, activation of sHsp complexes takes place upon the introduction of stress such as heat stress. The main function of activated sHsps is the prevention of protein denaturation and irreversible aggregation. After heat shock, approximately 33% of cytosolic proteins are retained in a soluble state with the help of sHsps.[33] Target proteins of sHsps after stress are diverse and widespread. Immunoprecipitation experiments and MALDI-TOF mass spectrometry experiments have elucidated that sHsps bind to DNA-binding proteins, actin and intermediate filaments, proteoglycans, heme oxygenase, ubiquitin ligases, and proteasomal subunits.[34] The question of substrate specificity for sHsps is still unacknowledged, but observations in *Saccharomyces cerevisiae* reveal that Hsp26 and Hsp42 have overlapping, but not identical, substrate patterns, suggesting some selectivity.[33]

The interaction and cooperation with other molecular chaperones, such as Hsp70, Hsp40, and Hsp100, facilitates a possible refolding process of the bound substrate and the release from the sHsp complex (Table I).

## C. Regulation of Hsps

The proteins of the Hsp family have at least two things in common: they are heat and stress inducible and obtain one or more HSEs, composed of a minimum of three inverted repeats of the consensus sequence nGAAn in the proximal promotor region. Ligands of the HSEs are the heat shock factors (HSFs). Currently, there are four known HSFs (HSF-1 to -4) in vertebrates, whereas the function of HSF-3 and HSF-4 is not yet fully understood.[35]

The best examined HSF is HSF-1. In most tissues, HSF-1 is constitutively expressed and localized in the nucleus and in the cytosol. Upon heat shock and activation of HSF-1, the export out of the nucleus into the cytosol is inhibited, resulting in an increased accumulation of active HSF-1 in the nucleus. In unstressed cells, the majority of HSF-1 exists in an inert monomeric form, which is bound to Hsp90.[36] During cellular stress, there is a rapid reduction of the HSF-1/Hsp90 interaction, inducing the conversion of HSF-1 from an inactive monomer to an active homotrimer. Trimerized HSF-1 binds to the nGAAn sequence of HSE and starts the transcription of the heat shock responsive genes.[35] In response to distinct stress and developmental stimuli, it is also possible that there is a hetero-trimerization of HSF-1 and HSF-2, facilitating transcriptional activation.[37]

The fact that HSF-1 plays an essential role in the cellular stress response raises a very interesting question: What are the sensor mechanisms activating the stress response via HSF-1? It has to be mentioned beforehand that the understanding of how the proteotoxic stress is sensed and converted by HSF-1 is far from complete. There are different theories about HSF-1 activation.

One of them is the theory of RNA-mediated activation of HSF-1. In 2006, Shamovsky *et al.* described the involvement of a ribonucleoprotein complex in HSF-1-linked stress sensing. This complex consists of the translation elongation factor eEF1A and the constitutively expressed noncoding RNA called heat shock RNA-1 (HSR-1).[38] The addition of HSR-1 and eEF1A *in vitro* results in an increased assembly of trimerized HSF-1, whereas an RNAi-mediated down-regulation of HSR-1 is accompanied with a higher susceptibility to heat stress.[38]

Furthermore, there are some *Caenorhabditis elegans* and *in vitro* experiments that propose the occurrence of additional intrinsic and neuronal control mechanisms to maintain HSF-1 trimerization. In the latter experiments, which were conducted in *C. elegans*, it was shown that there are thermosensory neurons that guide the temperature-dependent behavior and response of the

TABLE I

Overview of the Most Important Chaperones

| Heat shock protein | Importaomnt domains | Function |
|---|---|---|
| Hsp100 | – N-terminal substrate-binding domain[9,10]<br>– Two AAA$^{+}$ nucleotide-binding domain[9,10]<br>– Small C-terminal domain[9,10] | – Dilution and disaggregation of aggregated proteins[9,10]<br>– Strong induction after cellular stress[9,10] |
| Hsp90 | – N-terminal ATP-binding domain[7,14]<br>– C-terminal domain with EEVD motifs to bind TPR-domain-containing proteins[7,14]<br>– M-domain: interaction with N-terminal ATPase domain and co-chaperones[7,14] | – Binding and stabilization of client proteins (e.g., steroid hormone receptor)[8]<br>– Involvement in RNA transcription/processing[8]<br>– Involvement in DNA recombination, repair, and metabolism[8]<br>– Protein folding[15] |
| Hsp70 | – N-terminal ATP-binding domain[21]<br>– C-terminal substrate-binding domain[21]<br>– EEVD motifs at the C terminus to bind TPR-domain-containing proteins[21] | – Protein folding/proofreading/stabilization[18]<br>– Control of signal transduction proteins[20]<br>– Membrane translocation of secretory proteins[19]<br>– Involved in CHIP-dependent protein ubiquitination and proteasomal degradation[21] |
| Hsp40 | – J-domain for interaction and binding to Hsp70[26]<br>– Glycine–phenylalanine-rich domain[25]<br>– Zinc-finger-like domain facilitates protein transfer to Hsp70[25,27] | – Recognition of unfolded proteins and suitable Hsp70 substrates[25]<br>– Transfer of substrates to Hsp70[27]<br>– Involvement in the ERAD retranslocation process[25] |
| Small Hsps—Hsp27 | – N-terminal domain is often involved in chaperone oligomerization[28,29]<br>– α-crystallin domain functions as a binding domain for chaperone dimerization[28,29]<br>– C-terminal domain: high polarity stabilizes and mediates solubility of substrates[31] | – Stress-inducible oligomerization of small Hsps[28,33,34]<br>– Prevention of protein denaturation and irreversible aggregation[28,33,34]<br>– Interaction/cooperation with high-molecular-weight chaperones[28,33,34] |

nematodes in a non-cell-autonomous manner.[39] Mutant *C. elegans*, lacking these neurons, exhibited a drastic decrease in their heat shock response and were less viable compared with wild-type *C. elegans.*[39]

Due to the very fast heat shock response by HSF-1 upon heat or toxic stress, it is also supposed that HSF-1 has an intrinsic mechanism to sense stress. In fact, some experiments with isolated HSF-1 (produced by bacterial expression or *in vitro* translation) were reported that show an autonomous, intrinsic trimerization of HSF-1 after treatment with hydrogen peroxide, heat stress, low pH, or increased calcium concentrations.[35,40,41] It is suggested that some cystein and aromatic amino acids in the DNA-binding domain of HSF-1 play a crucial role as "built-in sensors" in intrinsic HSF-1 activation.[35]

A better understood activation mechanism of HSF-1 is the already mentioned interaction of HSF-1 with Hsp90. Reasons for dissociation of Hsp90 from HSF-1 upon stress have been investigated after proteasome inhibition, and the detailed mechanisms are discussed in a later chapter that deals with proteasome inhibition and the heat shock response.

## III. Refolding or Degradation?

If a nonfunctional and unfolded protein is bound by a molecular chaperone, such as Hsp70 and Hsp90, the decision has to be made as to whether it should/can be repaired or, in case it is not repairable, be degraded by the proteasomal system. The switch between refolding and degradation is mainly mediated by some co-chaperones, influencing the binding affinity of Hsp70 and Hsp90. The leading actors in this mechanism have already been introduced: CHIP, BAG-1, Hip/p48, and Hop/Sti1.

### A. BAG-1 and CHIP Inhibition of Chaperones and Promotion of Proteasomal Degradation

CHIP and BAG-1 are co-chaperones that induce the termination of the refolding process, the release of the substrate, ubiquitination of the protein, and transport to the proteasome.[7,42] Interaction of BAG-1 with the amino-terminal ATPase domain of Hsp70 results in a fast release of ADP, which in turn decreases the affinity of Hsp70 to its bound substrate[43] (Fig. 1). In addition, it has been shown that BAG-1 reduced Hsp70-mediated protein folding *in vitro.*[44] The second important co-chaperone that is necessary for proteasomal degradation of Hsp-bound substrates is CHIP. As already mentioned, CHIP is a ubiquitin ligase with the ability to interact with Hsp70 and Hsp90. Therefore, CHIP is equipped with a TPR domain at its amino terminus. On the C terminus of CHIP, there is a "U-box" domain that interacts with the

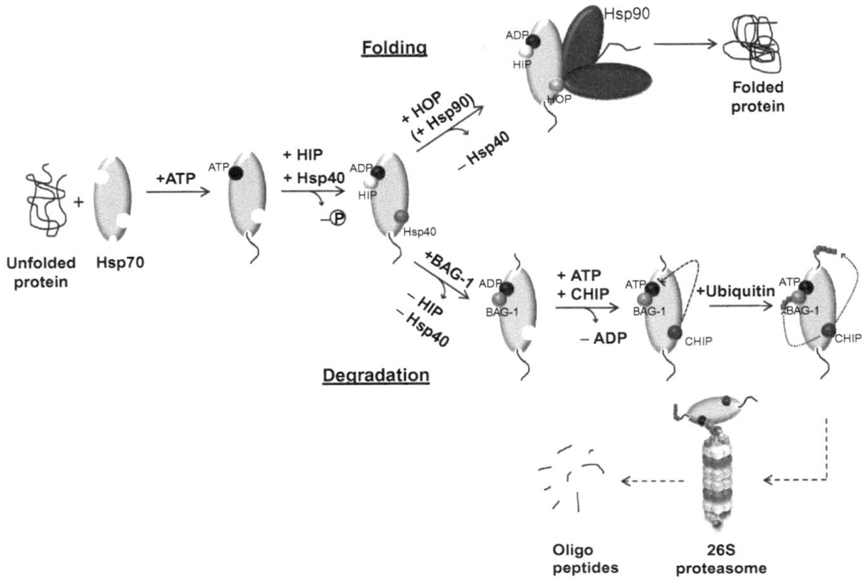


FIG. 1. Involvement of chaperones and their cofactors in protein folding or degradation. The chaperone function of Hsp70 depends on the presence of ATP. After ATP and the substrate have bound to Hsp70, HIP binds to the N-terminal ATPase domain and Hsp40 binds to the C terminus, resulting in a fast hydrolysis of ATP to ADP and a stronger affinity to ADP by the chaperone. At this step, Hsp70 also has a strong binding affinity to the substrate. *Folding*: Allowing proper folding of the protein, HOP replaces Hsp40, resulting in a stabilization of the ADP state. In addition, HOP also functions as a linker protein between Hsp70 and Hsp90, which facilitates the proper folding of the protein. *Degradation*: In the state where Hsp40 and HIP have bound to Hsp70, BAG-1 replaces HIP at the ATPase-binding domain. Following the dissociation of Hsp40 and HIP, ADP is exchanged by a new ATP and the TPR-containing-protein CHIP binds to the C-terminal EEVD domain. CHIP inhibits the hydrolysis of ATP and ubiquitinates to the bound substrate, as well as Hsp70 and BAG-1. K-11-linked ubiquitination of Hsp70 and BAG-1 functions as a binding signal to the 26S proteasome, where the ubiquitinated substrate becomes degraded.

UBCH5 family of E2 ubiquitin-conjugating enzymes.[45] The U-box domain functions similar to the RING-finger domains of some E3 ubiquitin ligases and ubiquitinate Hsp substrates to target them for the proteasome.[42] In contrast to RING-finger domains, U-box domains are not structured by metal-chelating residues, but rather by intramolecular interactions. CHIP and especially the "U-box" domain are very highly conserved amino sequences among human, mouse, and Drosophila.[42] Several results support the important role of CHIP during protein folding or degradation decisions, due to the fact that CHIP is particularly highly expressed in metabolic active tissues such as the heart, muscle, or brain.[46] On the other hand, CHIP can also be identified in all other tissues and most cell culture systems.[46]

Binding of CHIP to the TPR acceptor site of Hsp70 results, on the one hand, in a retardation of the protein-refolding process and, on the other hand, in a transformation of the Hsp70 chaperone to a "chaperone-dependent ubiquitin ligase"[42] (Fig. 1). Like BAG-1, CHIP also attenuates the protein-refolding process by Hsp70. But in contrast to BAG-1, CHIP promotes the substrate release not by a faster release of ADP, but rather by inhibition of ATP hydrolysis to ADP, which keeps Hsp70 in the low substrate-affinity state[42,45] (Fig. 1). It is supposed that this "slowing down" mechanism assists the loading of misfolded proteins from chaperones into the ubiquitin–proteasome machinery.

To check whether CHIP is a ubiquitin ligase, it should be possible to reconstitute its ubiquitination activity *in vitro* in the presence of CHIP, E1, E2, and ubiquitin. Indeed, Jiang *et al.* and Murata *et al.* could prove *in vitro* that CHIP is a chaperone-dependent ubiquitin E3 ligase.[45,47] But it is suggested that CHIP is not able to detect misfolded proteins on its own. Therefore, CHIP is dependent on the substrate-binding function of the associated chaperones. This is also equivalent to the RING-finger E3 ligases, which also need the so-called F-box-containing proteins as substrate recognition modules.[42,48]

Due to its TPR domain, Hsp90 is also able to associate with CHIP.[49] Binding of CHIP to Hsp90 results in a remodeling of Hsp90 into a folding-incompetent structure. The consequences of CHIP–Hsp90 binding have been studied in the context of glucocorticoid receptor (GR) binding by Hsp90. To fulfill its normal function, GR is complexed with Hsp90 and some other co-chaperones in the cytosol. Binding of CHIP to Hsp90 leads to (i) decreased substrate (GR) binding, (ii) decreased stability of the GR–Hsp90 complex, (iii) ubiquitination of GR, and (iv) proteasomal degradation of GR.[49] These effects were U-box and TPR dependent, suggesting that GR was ubiquitinated by CHIP.

In addition, BAG-1 and CHIP are also involved in the guidance of Hsp70–substrate complexes to the 26S proteasome. Therefore, BAG-1 and CHIP build a heterotrimeric complex with Hsp70, whereas this complex formation is promoted by the direct interaction of BAG-1 and CHIP.[50] Within this complex, CHIP ubiquitinates not only the misfolded proteins, but also BAG-1 and the chaperone itself. Certainly, the ubiquitin chains detected on BAG-1 and Hsp70 were short, noncanonical lysine (K)-11-linked, and had no signal for proteasomal degradation.[42,51] Moreover, it can be assumed that this K-11-linked oligoubiquitin chain mediates BAG-1 binding to the proteasome in a degradation-independent manner and is, therefore, an integral part of the sorting process[51] (Fig. 1). Additionally, BAG-1 obtains a ubiquitin-like domain that is also necessary for proteasome association.[43] Deletion fragments of the BAG-1 molecule lacking the ubiquitin-like domain can bind to Hsp70 and have the ability to regulate its function, but they cannot bind to the proteasome.[52] A possible acceptor site of the proteasome, which interacts with ubiquitinated BAG-1 and CHIP, could be the S5a/Rpn10 subunit of the 19S proteasome regulator. Evidence for these suggestions is in the form of the two different ubiquitin-binding domains of S5a on the one hand, and on the other, the findings of co-immunoprecipitated CHIP with the S5a subunit.[42,53]

In summary, CHIP and BAG-1 are the leading cofactors mediating the switch from protein refolding to proteasomal degradation. Therefore, CHIP is needed for (i) retarding protein refolding, (ii) substrate labeling for proteasomal degradation by K-63-linked polyubiquitination, and (iii) K-11-linked ubiquitination of Hsp70 and BAG-1 for transport and association with the 26S proteasome. BAG-1 is responsible for the association with the proteasome by its ubiquitin-like domain and the fast release of chaperone-bound substrates by interaction with the chaperone ATPase domain.

## B. Hip and Hop: Stabilizing the Refolding Process

Hip and Hop are two co-chaperones that act completely contrary to BAG-1 and CHIP by promoting the protein (re)folding process of Hsp70 and Hsp90.[54] Indeed, Hip (which also possesses a TPR domain) competes directly with BAG-1 for its Hsp70-binding site.[22] Hip and Hop act as key factors in the distinct defined folding procedure of Hsp70 and Hsp90. After the first part of Hsp70-dependent protein folding, when Hsp40 had bound to the chaperone and catalyzed the hydrolysis from ATP to ADP, the 42-kDa protein Hip binds to the ATPase domain of Hsp70 and prevents the release of ADP. Thus, Hsp70 remains in a stabilized position, ensuring an accurate protein folding by the chaperone[54] (Fig. 1). In many cases, it is also necessary to build a complex between Hsp70 and Hsp90 for accurate and final protein (re)folding or receptor activation.[54] Therefore, Hop binds to the Hsp70–Hip–substrate complex and prepares the Hsp70 complex to cooperate with Hsp90. Because Hop possesses two binding sites, one for Hsp70 and one for Hsp90, it functions as the link between these two chaperones.[54] Further, Hop induces the ADP/ATP exchange of Hsp70, resulting in the release of the bound substrate to the cytosol or to the connected Hsp90, which determines the preparations of the substrate protein[54] (Fig. 1).

## C. The Co-chaperone-Independent Decision of Folding or Degradation by sHsps

It has been mentioned that the phosphorylation status of Hsp27 and α-crystallins is important for their ability to form oligomers. In addition, phosphorylation also influences the decision of whether Hsp27 promotes ubiquitination and degradation by the proteasome or the refolding process of an unfolded protein and the inhibition of the proteasome. Hsp27 has the ability to interact directly with the 19S regulator of the 26S proteasome and was found to be associated with long-chain polyubiquitin molecules.[55–57] During stress conditions, Hsp27 favors the ubiquitination of proteins such as $p27^{Kip1}$ or

I-κBα and is thereby involved in the regulation of apoptosis or the cell cycle. I-κBα is the inhibitor of the anti-apoptotic transcription factor NF-κB. Liberation and, thus, the anti-apoptotic activity of NF-κB are regulated by the ubiquitin-dependent proteasomal degradation of I-κBα. It was demonstrated that Hsp27 stimulates the ubiquitination and degradation of I-κBα.[57] If cell stress provokes cell arrest in the $G_0/G_1$ phase, Hsp27 is able to accelerate the passage of cells into the S phase by promoting the ubiquitination of the cyclin-dependent kinase inhibitor $p27^{KIP1}$.[56] The function of Hsp27 in the ubiquitination process is quite unknown. However, there is the suggestion of the function as an E4 enzyme, acting as a scaffold during substrate ubiquitination or as a transporter, which delivers polyubiquitin chains to the substrate.[7]

Contrarily, it was also observed that Hsp27 can prevent ubiquitination of some substrates. This may be mediated by sumoylation of these substrates. SUMO (small ubiquitin-like modifier) is transferred as ubiquitin to lysine residues of the substrates. Therefore, sumoylation stays in direct competition to ubiquitination and proteasomal degradation. It was demonstrated that Hsp27 is involved in protein sumoylation. Hsp27 binds to Ubc9 (an E2 SUMO-conjugating enzyme) and acts as an E3-SUMO ligase.[58,59] Regulation of the participation of Hsp27s in ubiquitination or sumoylation is likely attributable to the oligomerization/phosphorylation status of Hsp27. Lanneau *et al.* proposed a model where dephosphorylated Hsp27 large oligomers promote sumoylation of proteins, preventing ubiquitination and proteasomal degradation, and phosphorylated Hsp27 dimers facilitate the ubiquitination-dependent proteasomal degradation of proteins.[7]

## IV. Heat Shock Response After Proteasome Inhibition

### A. Regulation of Classical Hsps After Proteasome Inhibition

Proteasome inhibition in several different cell lines, tissues, and species is always accompanied with the upregulation of numerous members of the Hsp family and some co-chaperones.[60–64] Especially in cancer therapies, where bortezomib is used as a proteasome inhibitor, Hsp response decreases the efficiency of antitumor medication, thereby the upregulation of Hsps (mainly Hsp27 and Hsp70) and some co-chaperones leading to decreased apoptosis by interfering with cytochrome c release and apoptosome formation,[64] stabilization of undegraded proteins, and prohibition of protein aggregation.

Additionally, it is evident that the rise of Hsps after proteasome inhibition is mainly mediated by activation of HSF-1 and its translocation into the nucleus. As has been already explained in nonstressed cells, HSF-1 monomers are bound to Hsp90, which suppresses the trimerization and phosphorylation of HSF-1 monomers. Upon stress, Hsp90 dissociates from HSF-1, allowing the hyperphosphorylation of HSF-1 at distinct serine residues and the formation of the DNA-binding HSF-1-trimers.[35,62] This raises the following question: What is the reason for Hsp90 dissociation from HSF-1? First, it had been supposed that the rising quantity of nonnative proteins, which emerge during cellular stress, competes with HSF-1 for the binding sites of Hsp90, resulting in the release of HSF-1.[36] However, in 2007, Boyault *et al.* demonstrated that this mechanism is oversimplifying the problem.[62] For the first time, the involvement of the histone deacetylase-6 (HDAC-6) and p97/VCP (valosin-containing protein) in the activation of HSF-1 upon proteasome inhibition and accumulation of ubiquitinated proteins was demonstrated. Already in 2003 and 2005, it had been investigated that, first, HDAC-6 mediates the clearance of cytotoxic misfolded proteins by autophagy and, second, that HDAC-6 deacetylates Hsp90, modulating its chaperone functions.[65–67] Furthermore, the additional identification of a complex between HDAC-6 and the chaperone-like AAA+ ATPase p97/VCP, which is essential for the degradation of misfolded proteins by the proteasome,[68] indicated the role of HDAC-6 in Hsp90-dependent HSF-1 activation upon the accumulation of nonnative proteins.

In regard to the results of Boyault *et al.*, HDAC-6, p97/VCP, and Hsp90 remain in an inactive complex in the absence of cellular stress. Thereby, p97/VCP undertakes the task as the essential connector for the HDAC-6–Hsp90 complex. During cellular stress, such as proteasome inhibition, HDAC-6 detects the rising amount of nonnative, ubiquitinated proteins through its mono- and polyubiquitin-binding sites (Fig. 2). Following this, the complex of HDAC-6 and p97/VCP disaggregates, which is attended by the dissociation of Hsp90 and HSF-1. Sensing of ubiquitinated proteins by HDAC-6 concurrently triggers three individual stress response mechanisms: (i) release of HDAC-6, which mediates degradation of aggregated proteins by autophagy[65]; (ii) release of p97/VCP, which guides ubiquitinated proteins to the proteasome and itself functions as a chaperone[69]; and finally (iii) release of HSF-1, which induces the expression and accumulation of major Hsps (Fig. 2).

The final step of Hsp90 dissociation from HSF-1 is not yet fully understood. Experiments of Boyault *et al.* with mice, expressing an HDAC-6 protein with two mutated catalytic deacetylase domains, showed that deacetylation of Hsp90 plays no clear role in HSF-1 activation. Furthermore, they suggested that free p97/VCP uses its segregase activity to dissolve the Hsp90 complex, leading to HSF-1 release. Alternatively, they supposed an indirect or direct activation of Hsp90s ATPase activity by p97/VCP.[62] However, some authors have shown a clear regulation of Hsp90 and its chaperone function by acetylation and deacetylation.[67,70] Therefore, the aggregation of ubiquitinated proteins functions as a strong regulator of HDAC-6–p97/VCP-mediated activation of Hsp stress response upon proteasome inhibition.

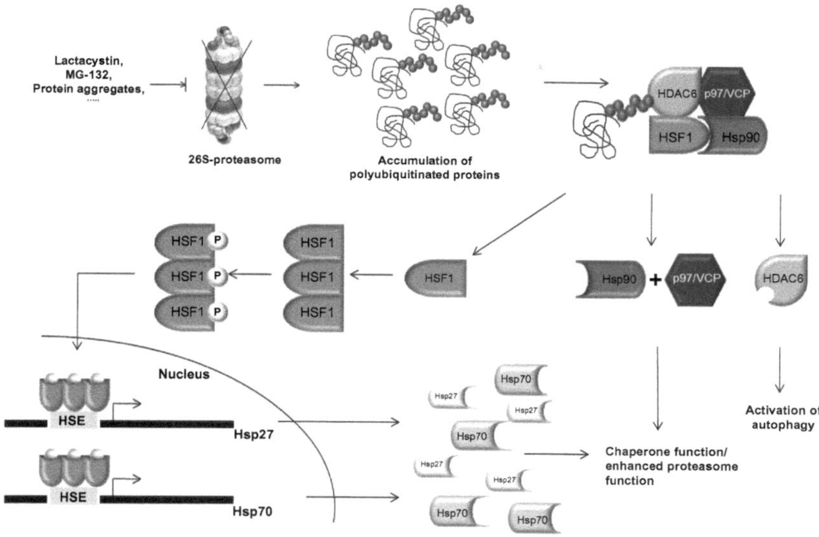


FIG. 2. The role of HDAC-6 and p97/VCP in cytosolic protein quality control during proteasomal inhibition. Inhibition of the proteasome leads to an accumulation of undegraded, ubiquitinated proteins. The accumulation of ubiquitinated proteins is detected by a ubiquitin-binding domain of HDAC-6. HDAC-6 is a component of a cytosolic complex consisting of HDAC-6, p97/VCP, Hsp90, and HSF-1. Upon the detection of ubiquitinated proteins by HDAC-6, the complex disintegrates, liberating the single components of the complex. HDAC-6 is able to promote an autophagic degradation of accumulated proteins. p97/VCP and Hsp90 act as chaperones and enhance the proteasomal function. Free HSF-1 trimerizes, becomes phosphorylated in the cytosol, and migrates into the nucleus. In the nucleus, HSF-1 binds to HSE complexes of target genes, promoting the transcription of Hsp27 and Hsp70.

## B. Regulation of Heme Oxygenase-1 After Proteasome Inhibition

Besides the traditional Hsps, proteasome inhibition also influences the expression and activity of another atypical Hsp: heme oxygenase-1 (HO-1), also known as Hsp32. Although HO-1 lacks conventional Hsp functions, such as protein stabilization or refolding of denatured proteins, it belongs to the Hsp family. This is due to the high stress inducibility of HO-1 and the occurrence of several HSE sequences in its promotor region.[71–73] In human cells, HO-1 expression is encouraged by oxidative stress, UV irradiation, and heavy metals,[74] but not by heat shock.[71,72] HO-1 degrades free heme into the three components, carbon monoxide (CO), free iron ($Fe^{2+}$), and biliverdin. All these three components contribute to the antioxidative, anti-inflammatory, and anti-apoptotic effects of HO-1.[75] $Fe^{2+}$, per se a highly oxidative reagent by inducing the Fenton reaction,[76] induces the expression of the iron-sequestering ferritin and activates the $Fe^{2+}$ transporter Fe-ATPase, resulting in a decrease of free intracellular iron.[75] Biliverdin and its decomposition product bilirubin are both strong antioxidants that scavenge peroxyl radicals and decrease membrane peroxidation.[77] The anti-inflammatory effects of HO-1 are mainly mediated by the release of CO. CO acts similarly to nitrogen monoxide (NO) as a second messenger, by inducing the soluble guanylyl cyclase or by activating the MAP kinase kinase (MKK/p38) pathway.[75,78] Therefore, CO attenuates the expression of the pro-inflammatory cytokines TNF-α (tumor necrosis factor-α), IL-1β, and MIP-1β and promotes the upregulation of anti-inflammatory cytokines such as IL-10.[78] It was also shown that induction of HO-1 and the accompanied increase in CO could suppress TNF-α-mediated apoptotic cell death in endothelial cells after actinomycine D treatment.[79] In addition, during antitumor therapies, it was observed that elevated HO-1/CO expressions lower apoptotic tumor cell death by inhibited p53 expression and blockage of mitochondrial cytochrome c release.[80]

Due to all these cell-protective effects of HO-1, it is not surprising that there is also a great upregulation of HO-1 during or after proteasome inhibition. The whole mechanism of HO-1 upregulation upon proteasome inhibition is not fully elucidated. But it seems very likely that NRF-2 (nuclear factor erythroid 2-related factor 2), HSF-1, and p38/MAPK activation are involved in HO-1 upregulation in the case of proteasome inhibition. Inhibition of the proteasome with lactacystin or MG-132 treatment elevated HO-1 levels in several different cell types and tissues.[81–83] Coincidently, phosphorylated p38 and NRF-2 are also upregulated after treatment with proteasome inhibitors.[83,84] In unstressed cells, NRF-2 is bound to Keap-1 (Kelch-like ECH-associated protein 1), inhibiting its translocation into the nucleus and facilitating the ubiquitin-dependent degradation of NRF-2 by the proteasome.[85] In stressed cells, the connection between Keap-1 and NRF-2 is disrupted, allowing NRF-2 to translocate in the nucleus and bind to EpRE(ARE)/StRE domains of target genes, such as hmox-1 (human HO-1 gene), which contains four EpRE elements in its promotor region.[73] It is suggested that dissociation of NRF-2 from Keap-1 can be mediated by various mechanisms. Besides direct modification of distinct cysteins of Keap-1–NRF-2 by electrophiles, NRF-2 is liberated after phosphorylation by protein kinase C (PKC), phosphoinositol-3-kinase (PI3K), or MAPK.[85] In response to proteasome inhibition, the MAPK/p38 pathway seems to be responsible for NRF-2 activation and, thus, elevation of HO-1 expression.[61,81,84] Due to the fact that NRF-2 homeostasis is normally kept up by the ubiquitin–proteasome system (UPS), inhibition of the proteasome leads per se to an elevation of nondegraded (ubiquitinated) NRF-2.

## V. Inhibition of Hsps: Influence on Proteasomal Degradation

As has been described, the living cell is able to cope with some disturbance of the protein quality control mechanism. In case of proteasome inhibition, there is an adaptation by the upregulation of the chaperone system.

Due to the promising anticancer effects of some Hsp inhibitors, a counter regulation of the proteasome might also be interesting, which may attenuate the anticancer treatment by keeping up the proteasomal clearance of damaged proteins. Substances such as herbimycin, geldanamycin, and the "new generation" geldanamycin derivates 17-dimethylaminoethylamino-17-demethoxygeldanamycin (17-DMAG) and 7-*N*-Allylamino-17-demethoxygeldanamycin (17-AAG) are potent Hsp90 inhibitors in anticancer therapies.[54,86,87] However, in many studies, Hsp90 inhibition was accompanied with increased proteasomal degradation.[15] At the beginning, it was suggested that the accompanying promotion of proteasomal degradation was induced by the Hsp90-inhibition-induced HSF-1-dependent stress response, but when it was shown that there is still a proteasomal degradation of protein aggregates in HSF-1$^{-/-}$ cells after Hsp90 inhibition, this hypothesis was disapproved.[15] Another possible explanation for increased ubiquitin-dependent proteasomal degradation after Hsp90 inhibition would be this: Under normal cellular conditions, without the presence of Hsp90 inhibitors, Hsp90 stabilizes and prevents client proteins from ubiquitination and proteasomal degradation.[15,88–90] Thus, inhibition or perturbation of Hsp90 by geldanamycin and its other benzoquinone ansamycin derivates impairs the stabilization of client proteins, resulting in ubiquitination and proteasomal degradation of these proteins.[91,92] This has been proven for some proteins, such as the cystic fibrosis transmembrane conductance regulator.[88] Additionally, in the 1990s, it was shown that Hsp90 also has the ability to work as a direct inhibitor of the proteasomal Z-Leu-Leu-Leu-MCA-degrading activity.[89] Similar findings have been made with isolated Hsp90 and α-crystallins of isolated bovine lenses. Hsp90 and α-crystallin inhibited the proteasome depending on the concentration of the proteasomal substrate.[90] However, induction of proteasomal degradation of Hsp90 client/substrate proteins during Hsp90 inhibition in anticancer treatment does not attenuate the effect of the inhibitors, but is rather an integral part of its success. Because Hsp90 client proteins comprise steroid hormone receptors, transcription factors, and kinases, proteasomal degradation of these client proteins during Hsp90 inhibition leads to enormous disturbances in many signal transduction pathways essential for tumor progression and survival.[92]

In most cancer cells, inhibition of Hsp90 is accompanied with an adapted elevation of Hsp70 expression, while CHIP expression remains largely unaffected.[93] Therefore, it can be suggested that the enhanced degradation of

Hsp90 client proteins during Hsp90 inhibition is mediated or initiated by the Hsp70–CHIP system. Further support for this theory comes from the fact that the suppression of CHIP results in decreased ability to degrade Hsp90 client proteins and overexpression of CHIP leads to an augmented degradation of these client proteins. These experiments were done with the leucine-rich repeat kinase-2 (LRRK2) protein, which is the most common reason for late-onset Parkinson disease (PD).[94] As expected, inhibition of Hsp90 with 17-AAG also reduced the amount of LRRK2.

However, Hsp70 induction after Hsp90 inhibition is an additional adaption mechanism to cope with all "free" Hsp90 client proteins, which could reduce the anticancer effect of Hsp90 inhibitors and confer resistance to the apoptotic effect of Hsp90 inhibitors.[95,96]

Interestingly, not all Hsp90 client proteins are degraded by the proteasome after Hsp90 inhibition. There is also the possibility of lysosomal degradation by autophagy. IκB kinase (IKK) is one example for a proteasome-independent, autophagy-mediated degradation of a client protein after Hsp90 inhibition.[97] Cells treated with geldanamycin and coincident Atg5 (a crucial component of the autophagy pathway) knockout reduced the Hsp90-inhibition-dependent degradation of IKK, while geldanamycin combined with proteasome inhibition or ubiquitination inhibition had no effect on IKK degradation.[97]

The observed upregulation of Hsp70 during Hsp90 or proteasome inhibition and the investigation of further functions of the Hsp70 molecule make it especially important to develop potent inhibitors of Hsp70. Hsp70 inhibition is mediated by specific RNAi constructions or a variety of pharmacological substances, such as KNK-437,[98] 2-phenylethynesulfonamide (2-PES),[93] or methylene blue.[99] In contrast to the inhibition of Hsp90, silencing or impairment of Hsp70 is accompanied with a decline in proteasomal degradation, resulting in the accumulation of aggregated, insoluble proteins.[93] Due to its key position during the decision of protein refolding or protein degradation, it is not surprising that inhibition of Hsp70 also affects proteasomal degradation. Impairment of Hsp70 reduces, on the one hand, the interaction with the chaperone-dependent E3 ubiquitin ligase CHIP, and on the other hand, the interaction with Hsp90 to mediate folding or stabilization of Hsp90 client proteins.[93,99] As already explained, Hsp90 client proteins become stabilized by Hsp70, ubiquitinated by CHIP, and degraded by the proteasome in case of Hsp90 impairment. If Hsp70 is inhibited, both possibilities are abrogated: Hsp70 cannot support Hsp90 dealing with client proteins and Hsp70 cannot sustain the CHIP-mediated ubiquitination of these client proteins, resulting in the accumulation of (client)-protein aggregates.[93,99] In 2011, this was proven by Leu *et al.* with the Hsp70 inhibitor 2-PES. A detergent-soluble and a detergent-insoluble cell fraction after the application of 2-PES or 17-AAG were isolated. Samples treated with 2-PES exhibit a huge accumulation of Hsp90 client proteins such as HER2/ErB2 and

EGER. Additionally, treatment with 20 μM 2-PES reduced proteasomal activity nearly as much as treatment with the specific proteasome inhibitor bortezomib.[93] Treatment with the Hsp90 inhibitor 17-AAG had no remarkable effect on proteasome activity. Furthermore, samples treated with 17-AAG exhibited undetectable or really low quantities of epithelial growth factor receptor (EGFR) or HER2/ErbB2 in the detergent-insoluble fraction.[93] The latter fact can be traced to a functional Hsp70–CHIP unit.

Summarizing this, it is clearly demonstrated that inhibition of one part of the protein quality control system affects directly or indirectly the other parts of this system. Inhibition of Hsp90 enhances ubiquitination of its client proteins by Hsp70 and CHIP, leading to an augmented degradation by the 26S proteasome. The other way round, inhibition of Hsp70 impairs Hsp90 and the CHIP-dependent proteasomal degradation, resulting in the accumulation of Hsp70 and Hsp90 substrates. In regard to anticancer therapies, both Hsps seem to be potent aims for pharmacological intervention. By inhibition of Hsp90, most client proteins become degraded by the proteasome, which means a drastic incision into the cellular metabolism. The inhibition of Hsp70 leads contrarily to the accumulation of (toxic) protein aggregates, impairing the protein quality control system and thus the whole-cell metabolism of cancer cells. To improve the cell killing effect of Hsp inhibitors, simultaneous inhibition of Hsp90 and Hsp70 can reduce the Hsp90-inhibitor-dependent induction of Hsp70 and promote apoptosis.[95,100]

## VI. Hsps and the Proteasomal System During Oxidative Stress and the Aging Process

Oxidative stress is defined as the imbalance of reactive oxygen species (ROS) and the antioxidative defense mechanisms, with an overweight for the amount of ROS.[101] An effective defense mechanism should therefore include the following characteristics: It should (i) prevent the formation of oxidants, or catch and quench oxidants; (ii) detect and repair intracellular changes/oxidative damages; (iii) have the ability to protect the cell against oxidative damages, such as protein misfolding and aggregation; and (iv) protect the cell from accompanying downstream events, such as apoptosis and inflammatory reactions.[102] Besides point one, Hsps and the proteasomal system fulfill all of these requirements. Especially Hsp32, better known as heme oxygenase, senses intracellular redox changes. This is mediated by several highly reactive cystein residues which interact with zinc ions, resulting in the release of zinc ions and the activation of chaperone function. Similar findings have already been made for Hsp70.[102,103]

Vulnerable targets of oxidative modifications during oxidative stress are the DNA, lipids, and, in particular, proteins. The different oxidative modifications, such as protein carbonylation or disruption of disulfide bonds, hamper proteins from preserving their native conformation. The results are, depending on the degree of the oxidative burden, a huge number of misfolded and unfolded proteins tending to aggregate.[104] To prevent and diminish protein misfolding or aggregation, Hsps bind these modified proteins trying to refold or to stabilize the protein structure and the proteasome degrades unfolded, oxidized proteins which have not been refolded.

As already mentioned, there is a direct connection between molecular chaperones and the proteasome, especially in the early response to oxidative stress. In a paper recently published by our group, it is demonstrated that there is a switch between the 20S and the 26S proteasomes in response to oxidative stress early after $H_2O_2$ exposure. In the first 0.5–1 h after oxidative stress, there is a drastic decrease of the 26S proteasomal form, whereas the amount of 20S proteasome remains largely unchanged.[105] The 19S complex did lose contact with the 20S proteasome; however, it did not fall apart into subunits, nor was it proteolytically degraded. Due to the fact that the reconstitution of the 26S proteasome after 3 h of $H_2O_2$ exposure was cycloheximide-independent, it was suggested that the 19S regulator was detached and reattached to the 20S proteasome. Additionally, it was suggested that the stabilization of the 19S regulator during detachment is chaperone mediated. Indeed, impairment of Hsp70 by RNAi or the specific Hsp70 inhibitor KNK abolished 19S reattachment after oxidative stress.[105] For this reason, it was proposed that during the early stages of oxidative stress, Hsp70 binds to the 19S regulator of the proteasome, resulting in the dissociation and coincident stabilization of the regulator. It is supposed that this rapid release of the 20S proteasome core enhances the cellular capacity to degrade oxidized proteins and to cope with the oxidative stress.[105] Further support for this theory comes from yeast studies. Upon oxidative stress, the proteasome-interacting protein Ecm29 binds to the 19S regulator, enhancing and facilitating proteasomal degradation of oxidative modified proteins.[106]

Generation of cellular oxidative stress is highly variable and affects the proteasome and its catalytical activities in different ways. For example, it was shown that metal-catalyzed oxidation impairs the proteasomal trypsin- and peptidyl–glutamyl-like activity of isolated proteasome in bovine lens cells.[107] Contrarily, the lipid peroxidation product 4-hydroxy-2-nonenal (HNE) affected only the trypsin-like activity, whereas the peptidyl–glutamyl- and the chymotrypsin-like activity remained unaffected.[107] Interestingly, it is reported that proteasomal activity declines in aged proteasomes, whereas the concentration of HNE increases during the aging process.[108] Further, in several studies it was recognized that Hsp90 was easily copurified with the proteasome.[90,107] Admittedly, interaction of Hsp90 with the proteasome attenuated the Z-Leu-Leu-Leu-7-amido-4-methylcoumarin-degrading activity but did not affect the other hydrolase activities. Due to Hsp90 being upregulated upon oxidative stress, it was obviously suggested that the chaperone may have protective effects on the proteasome. Indeed, it could be shown that increased levels of Hsp90 protected the trypsin-like activity of the proteasome in the presence of HNE or during metal-catalyzed oxidation.[107] Unfortunately, the authors left the question of "how" Hsp90 protected the trypsin-like activity of the proteasome unanswered. Naturally, it can be supposed that stabilization and refolding of the responsible proteasomal subunits by Hsp90 upon oxidative stress maintain the trypsin-like activity, but then the question arises as to why Hsp90 had no effect on the functional decline of the peptidyl–glutamyl activity upon metal-catalyzed oxidation.

Nevertheless, it seems evident that the proteasomal system interacts with several chaperones (especially Hsp70 and Hsp90) during or upon oxidative stress, whether to immediately enhance the capability of the proteasome to degrade oxidative modified proteins in a ubiquitin-independent manner by Hsp70 and protect the proteasome from oxidative damages by Hsp90 or to support the ubiquitin-dependent degradation in a later response to oxidative stress by the CHIP–Hsp70–Hsp90 machinery.

It has been mentioned that the highly reactive lipid oxidation product HNE is increased in aged organisms. But this is not the only highly reactive oxidation product that is augmented during the aging process. All in all, it can be summarized that the pro-oxidative status, accompanied with oxidative damaged/modified proteins, lipids, and DNA, increases during the aging process. This is further supported by the "Free Radical Theory of Ageing" from D. Harman, which is still one of the most conclusive theories describing the aging process and the associated physiological changes and diseases.[109] Furthermore, there are also some changes of the proteasomal and chaperon systems during aging.

Because this book includes a whole chapter, written by Bertrand Friguet, about the changes of the proteasomal system during the aging process only a brief mention is made at this point about what happens with the chaperone system and its interactions with the proteasome in the aging process. There are a great variety of publications describing the functional decline of several chaperones in aged cells or aged organisms.[110–112] While constitutive levels of chaperones seem to be increased in aged cells, the induction upon stress and the functionality of chaperones are drastically decreased. This was tested, for example, in young and adult rats; Hsp70 response after heat shock was much higher in young rats[113] or in cell culture systems (lung fibroblasts[114] and retinal cells[115]), and Hsp70 induction of younger cells exceeded the induction in older cells. Equal findings were also made for Hsp90 or α-crystallins.[110] The enhanced levels of molecular chaperones in older cells may be evoked by the accumulation of misfolded/unfolded proteins in aged organisms.[112] The malfunction of the chaperone system in aged cells upon stress is probably traced to the reduced DNA-binding activity of HSF during aging.[116] Another reason for the functional decline of the chaperone function during aging are insults by oxidative modifications to the Hsp protein structure. Indeed, chaperones seem to be prone to oxidative damages. Binding of a plentitude of (oxidized) substrates requires an extensive binding surface. Therefore, Soti *et al.* suggested that these exposed areas make Hsps susceptible targets for oxidative, proteotoxic damages, and that chaperones function as "suicide proteins" during the aging process, sacrificing themselves instead of normal proteins.[111] This indicates that even if there are "enough" chaperone molecules, they might be either oxidative modified and nonfunctional or already bound to misfolded, oxidative modified proteins, leading to an imbalance between free available and already bound chaperones in aged organisms. This theory suits very well the symptoms of many age-related diseases, such as Alzheimer disease (AD), PD, Huntington disease, or cataracts, where huge deposited protein aggregates include molecular chaperones.[110,112] Aggregation of oxidized and undegradable proteins and the accompanying indirect functional decline of Hsps are partly due to the shrinking proteolytic activity of the proteasome during aging. In turn, functional decline of the proteasome is also negatively influenced by shrinking chaperone capacity. It has been mentioned earlier that Hsp90 protects the proteasome from age-related, oxidative-dependent decline. However, association of Hsp90 and the proteasome decreases with advancing age.[107,117] Moreover, it has to be assumed that functional decline of Hsp70 and Hsp90 affects the CHIP/ubiquitin-dependent degradation of proteins. Thus, it is not surprising that most age-related diseases, which are related to the decreased proteasome activity, are also mentioned when there is talk of diseases caused by the functional decline of Hsps. This accounts for many neurodegenerative diseases such as Alzheimer, Parkinson, or Huntington,[118–124] impairments of the aging eye such as cataract formation,[125,126] and myopathies of aged muscle fibers.[110,127]

## VII. Neurodegenerative Diseases: Connection to the Hsp Family and the Proteasome

Neurodegenerative diseases comprise diseases in different regions of the brain that are connected to denaturation, functional impairment, deposition, and aggregation of specific proteins or protein fragments, resulting in the formation of several inclusion bodies and the death of selective neuronal cell populations.[123,128] Genetic defects, life style, and the aging process are reasons for the formation of neurodegenerative diseases. AD and PD are the two most often occurring neurodegenerative diseases—especially after people have reached the age of 65 years.[124]

Corresponding with many other neurodegenerative diseases, PD and AD are clearly associated with an advancing impairment of the proteasomal system.[123] As already mentioned, inhibition of the proteasome, provoked by pharmacological inhibitors or aggregated proteins, implies a compensating induction of Hsp expression. In neurodegenerative diseases, it has been shown that there is an upregulation of Hsp90, Hsp70, Hsp27, and α-crystallins.[128–130] Upregulation of all these Hsps supports the maintenance of protein homeostasis and the dealing with aggregated proteins. For example, Hsp70 helps to recruit misfolded proteins for the two E3 ubiquitin ligases CHIP and Parkin. Especially in the juvenile form of PD, Hsp70 and CHIP could support ubiquitination of mutated Parkin. The endothelin receptor-like receptor (Pael-R) is a highly expressed protein in the brain and is normally ubiquitinated by Parkin and degraded in the ER. Dysfunction of Parkin results in Pael-R deposition and aggregation. Upregulation of Hsp70 could support the Parkin-dependent ubiquitination of Pael-R by binding to Pael-R and provision to CHIP. CHIP dissociates Pael-R from Hsp70 and stimulates its ubiquitination by Parkin.[131] Furthermore, it has been observed that Hsp70 reduced α-synuclein-mediated toxicity and neuron loss in PD, but did not prevent the aggregation of these proteins. It is therefore suggested that Hsp70 does not prevent the aggregation of α-synucleins but rather attenuates their toxicity by avoiding toxic conformational changes in the process of aggregation.[128]

## A. Alzheimer Disease

Classification ofAD progression can be divided into 2–3 states. In the beginning, AD is characterized by common "age-related" symptoms such as difficulty to remember recent events or impaired learning ability. If AD is more progressed, patients suffer from confusion aggressions, mood swings, long-term memory loss, and language breakdown. The third step comprises the loss of crucial body functions and finally death.[132] Three proteins are closely related with AD: the amyloid precursor protein (APP), the APP cleavage product Aβ, and tau.[133–135] APP is a neuronal transmembrane protein functioning as a membrane receptor. Its cleavage product Aβ is a 40- to 42-amino acid long peptide, tending to form polymeric β-sheet structures. Accumulation of this polymeric Aβ structures results in the formation of amyloid plaques.[133,134] It is a matter of debate which progression state of amyloid plaque formation has the most influence on AD progression and proteasomal inhibition. But it is suggested that oligomeric forms of Aβ have the most toxic potential.[136,137] The second major pathological component is the hyperphosphorylation and aggregation of tau proteins, forming the large neurofibrillary tangels,[135] which are clearly associated with neuron loss and cognitive deficits.[133] The majority of available data in regard to chaperones and AD focused on the aggregation of the tau protein. Affected regions from AD brain tissue have been positively stained for several Hsps, especially α-crystallins,[133] Hsp27,[138] Hsp70,[139] ubiquitin, and E3 ubiquitin ligases such as CHIP[140] or Parkin.[138] The role of the chaperone system and the link to the proteasome in AD is unclear. Data from Dou *et al.* indicates an inverse correlation of Hsp70/90 and aggregated tau protein in various cellular models.[141] However, the reduction of aggregated tau protein was not a result of diminished total tau levels, but rather a shift from insoluble, aggregated tau to soluble, microtubule-bound tau protein—an effect that was abrogated when Hsps were silenced by RNAi.[141] Another study from Petrucelli *et al.*[140] demonstrated the interaction of Hsp70 and CHIP with tau proteins. It was shown that CHIP-mediated ubiquitination of tau contributes to an increase of ubiquitinated tau aggregates. Further, inhibition of the proteasome augmented this CHIP-mediated tau aggregation *in vivo.*[140] Interestingly, CHIP-dependent ubiquitination was not only via K-48 ubiquitin linkages, which trigger the degradation by the proteasome, but also via K-63 ubiquitin linkages, suggesting a possible degradation by autophagy.[135,140] Although Hsp70 is correlated with decreased aggregated tau proteins, it facilitates the direct association of CHIP with tau and, therefore, tau ubiquitination.[133]

Furthermore, it is reported that treatment with Hsp90 ATPase inhibitors facilitates the CHIP-mediated degradation of tau in a ubiquitin–proteasome-dependent manner. The suggested reasons for facilitated degradation of tau are multifold. Akt, an oncogenic master kinase, is elevated in AD and responsible for tau phosphorylation at S262/S356 via the microtubules' affinity-regulating kinase 2 (Par1/MARK2).[133,142] It was shown that S262/S356-phosphorylated tau protein could not be degraded by the Hsp90/CHIP complex. Furthermore, Akt is also a client protein of the Hsp90/CHIP complex. In case of Hsp90-ATPase-inhibition by further co-chaperones, the complex switched to degradation and diminished the amount of Akt. Consequently, there was a reduction in phosphorylated tau, making it a good substrate for CHIP-dependent ubiquitination and proteasomal degradation.[133,142]

## B. Parkinson Disease

PD is characterized by dramatic movement disorders, evoked by a progressive degeneration of dopaminergic neurons in the substantia nigra pars compacta.[129] Neuronal cell death is mediated by three different mechanisms, including the (i) aggregation of α-synucleins, (ii) mutation of the E3 ubiquitin ligase Parkin, and (iii) mutation of the ubiquitin carboxyl-terminal hydrolase L1 (UHC-L1).[129] Deposited and aggregated α-synucleins are the basis of Lewy body formation; they impair the proteasome-mediated protein degeneration and hamper the ability of cells to cope with age-related stress.

Impairment of Parkin is a drastic incision in the ubiquitination reaction and, thus, not a neglectable disruption of the ubiquitin–proteasomal system. Moreover, Parkin also promotes a proteasome-independent but calpain-dependent cleavage of α-synucleins and diminishes thereby the neurotoxicity of α-synucleins.[143] Mutation and alteration of the UCH-L1 attenuates the recycling of monoubiquitin molecules, so inhibition of the enzyme results in failure to provide sufficient amounts of ubiquitin to target abnormal proteins, which tend to accumulate.[129]

Analysis of deposited α-synucleins and Lewy bodies reveals that these aggregates are highly ubiquitinated, and are also associated with Parkin, Hsp70, CHIP, and BAG-5.[124] While Parkin mediates a calpain-dependent cleavage of α-synucleins, CHIP is another E3 ubiquitin ligase that contributes to α-synuclein degradation in a ubiquitin–proteasomal and/or autophagic way.[144] But if Parkin and CHIP are associated with α-synucleins and have the ability to cleave or tag them for the ubiquitin-dependent proteasomal degradation, why is the body of α-synuclein aggregates still growing and why are dopaminergic neurons still dying during PD? Certainly, one reason is the ongoing inhibition/impairment of the proteasome during the disease process. Another reason is the preceding mutation of Parkin and UCH-L1.[145] An additional cause is connected to BAG-5, which is also associated to α-synuclein. BAG-5, like BAG-1, belongs to the BAG family and is also a co-chaperone of Hsp70. However, the effect of BAG-5 on Hsp70 and the coupled E3 ligases is absolutely contrary to the co-chaperone functions of BAG-1. BAG-5 binds to Hsp70, CHIP, or Parkin and inhibits, on the one hand, Hsp70-mediated refolding of misfolded proteins and, on the other hand, the ubiquitination of substrate proteins by CHIP and Parkin.[124,146] Interestingly, BAG-5-mediated inhibition of Parkin and CHIP enhance also their sequestration to Lewy body aggregates.[124,146] The influence of BAG-5 on Hsp70 and the ubiquitin ligase stands is related directly to the cell death of dopaminergic neurons. Overexpression of BAG-5 in dopaminergic neurons leads to enhanced cell death, whereas downregulation of BAG-5 is accompanied by lower progression of PD.[124,146] It is suggested that binding of co-chaperones belonging to the BAG family is a fine-tuned regulation of the "protein-handling pathways".[124] In the case of some diseases such as PD, this system may go out of control.

## VIII. Early-Stage Protein Quality Control and ERAD

About 7500 proteins are folded in the ER to reach their functional three-dimensional structure.[147] Interestingly, about 30–80% of the newly synthesized proteins are misfolded.[148] Due to the disposition of misfolded proteins to accumulate and aggregate, one of the major tasks of the protein quality control system, including the proteasome and the family of chaperones, is to recognize, refold, or degrade unfolded and misfolded proteins. The protein quality control protects the cell from toxic, unfunctional proteins, keeps up protein homeostasis, and recycles the amino acids for the synthesis of new proteins. Because the UPS is not located in the ER, unfolded proteins have to be recognized and transported out of the ER to the cytosolic UPS. Stabilization and transport of ER-located proteins is mediated by ER-specific chaperones. Moreover, ER-specific chaperones are kept bound to their substrate, ensuring that the substrates remain soluble. This is an important fact, because insoluble proteins are improper substrates for translocation into and out of the ER lumen. The main components of this system are BiP/Grp78 (the ER homologue to Hsp70),[149] its co-chaperone Hsp40/DnaJ,[150] calreticulin, and calnexin[151] (Table II).

In eukaryotes, many proteins are directly translocated from the ribosome into the ER via a special translocation channel in the ER membrane. Recognition of the protein–ribosome couple by the translocation channel needs an additional signal recognition particle (SRP). Interaction of the SRP with the channel allows for binding of the protein–ribosome–SRP complex to the channel and translocation of the substrate.

Proteins, translocated posttranslationally, obtain only a small hydrophobic signal sequence, which is not recognized by SRP. Thus, this translocation mechanism into the ER is different compared to the ribosome-dependent one. First, it is necessary that the proteins remain unfolded after the release of the ribosome into the cytosol (in general, these proteins are stabilized by cytosolic chaperones during their presence in the cytosol). In contrast to ribosomal translocation, posttranslational translocation needs some other supporting proteins. Therefore, the translocation channel builds a complex with Sec62/Sec63. Translocation starts with binding of the substrate to the new translocation complex. Within the channel, the substrate slides in either direction by Brownian molecular motion. Because the substrate is captured by several BiP/Grp78 proteins in the ER lumen, the backward movement is prevented, resulting in a net forward movement through the channel. During the translocation process, all cytosolic chaperones are removed from the nascent protein.[152] Entering the ER lumen, the polypeptide becomes N-glycated by the oligosaccharyl transferase, transferring the oligosaccharide complex from dolichol to an asparagine residue of the polypeptide.[153] In the next step, two of three N-terminal glucose residues of the protein sugar residues are cut off by glucosidase (Gls) 1 and 2, which makes the monoglucosylated protein a suitable substrate for calnexin and calreticulin[147] (Fig. 3). To bind the monoglucosylated protein, calreticulin builds a complex with the oxidoreductase ERp57/GRp58. Calreticulin escorts the substrate and ERp57/GRp58 to the transmembrane chaperone calnexin. Binding to the complex of ERp57/GRp58, calreticulin enters the protein and into the calnexin/calreticulin binding cycle, providing time for the

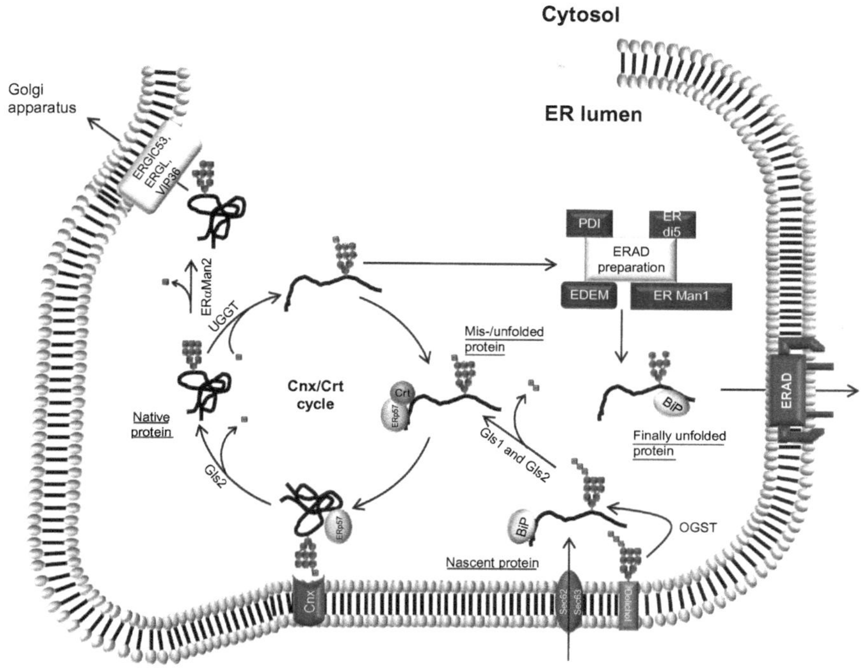


FIG. 3. Protein quality control system in the ER. Nascent proteins are translocated into the ER by a Sec62/Sec63 channel and BiP. The oligosaccharyl transferase (OGST) transfers the oligosaccharides from dolichol to the nascent protein. The following deglycosylation of the nascent protein by glucosidases 1 and 2 (Gls1 and Gls2) enables the binding to the chaperones Calreticulin (Crt) and ERp57. Crt transfers the substrate with ERp57 to the transmembrane chaperone Calnexin (Cnx). Removal of the last glucose molecule results in the release of the native protein from Cnx. If protein folding by Crt/Cnx is successful, the protein becomes demannosylated by ERαMan2 and retranslocates to the Golgi apparatus or the cytosol. If protein folding is unsuccessful, the protein becomes reglycosylated by UDP-glucose:glycoprotein glucosyltransferase (UGGT) and passes the Crt/Cnx cycle once again. If the protein is not foldable, it becomes demannosylated by ER Man1 and ER-degradation-enhancing α-mannosidase-like proteins (EDEMs). Extensive demannosylated proteins are bound by BiP, escorting the protein to the ERAD machinery (modified from Ref. 147).

protein to reach its final, functional structure. When the remaining N-terminal glucose molecule is detached by Gls 2, the native protein resolves the attachment to calnexin and ERp57/GRp58.[151] If protein folding is successful, the ER-α mannosidase 2 (ER-α Man2) trims the native protein by one mannose molecule. The correctly folded protein is then translocated out of the ER by lectins for further processing in the Golgi apparatus. The most prominent lectins in this process are ERGIC53, ERGL, and VIP36[154] (Fig. 3).

If protein folding is improper or incomplete, the protein becomes "reglycosylated" at the end of mannose branch *A* by the UDP-glucose, glycoprotein glucosyltransferase (UGGT), and is redirected to the calreticulin/calnexin cycle. Repeated cycling in the calreticulin/calnexin cycle gives the protein the chance to be correctly folded (Fig. 3). Reentering this cycle is also termed "retention" or "first phase of ER retention".[155] Impairment of any of the mentioned chaperones connected to the ER-dependent protein folding disrupts the whole protein-folding mechanism in the ER and is, therefore, often lethal.[147]

As long as an improperly folded protein continues to become reglycosylated, it passes the calreticulin/calnexin cycle. The longer such a defective protein traverses this cycle, the more probable is the chance that the protein is deposited and builds aggregates with other proteins, interfering with other newly synthesized proteins.[155] Therefore, it is crucial to mark these proteins for ERAD. In contrast to the cytosol, where proteins become ubiquitinated, the proteins in the ER become demannosylated. Extensive demannosylation of all three sugar branches (including mannose A) of an ER protein is, on the one hand, a direct signal for the disposal of folding-incompetent polypeptides and, on the other hand, a convenient way to prevent further reglycosylation on mannose branch *A* and reentry into the calnexin cycle.[155] The ER-Man-1 is one candidate, trimming mannose molecules from proteins. However, *in vitro* and *in vivo* experiments show that ER-Man-1 is too inefficient, suggesting the involvement of other demannosylating proteins[155–157] (Fig. 3).

The other group of proteins that demannosylate proteins to terminate the calnexin/calreticulin cycle are the ER-degradation-enhancing-α-mannosidase-like proteins (EDEMs). The family of EDEMs consists of EDEM-1, EDEM-2, and EDEM-3. EDEMs belong also to the group of lectins and accelerate protein disposal during ER stress conditions.[155,158] Like ER-Man-1, EDEMs have the ability to progressively demannosylate folding-defective peptides and mark them for ERAD. Inhibition of mannose removal slows down the release of folding-reluctant proteins from the calreticulin/calnexin cycle and reducing significantly their degradation by ERAD[157,159] (Fig. 3).

In addition, especially EDEM-1 seems to play a bigger role in ER-dependent protein quality control. It was shown that EDEM-1 lacking its glycanase activity still acts as a molecular chaperone and facilitates the degradation of misfolded proteins. The rate-limiting step in the removal of "folding-defective glycoproteins" is the formation of disulfide-bound glycoprotein aggregates after the liberation from calnexin.[155] EDEM-1 is able to discriminate between folding-capable and folding-defective proteins and inhibits the formation of disulfide-bound aggregates upon release from calnexin.[155,160,161]

The ER lumen has an oxidative environment, facilitating the arrangement of disulfide bonds, which stabilizes the tertiary structure of newly synthesized/folded proteins.[162] Further on, these disulfide bonds are formed and isomerized by the protein disulfide isomerase (PDI). PDI is an ER-located sulfhydryl oxidase, mediating the formation of disulfide bonds of polypeptides in the folding process to their native structures. It also acts in concert with several ER chaperones such as BiP, Grp94, ERp57, and ERp72.[163] The redox ability of PDI is applied by CXXC motifs in the protein.[164] Due to the fact that there are oxidized disulfide bonds as well in misfolded and useless proteins, demannosylation is not the only process that is necessary for ERAD preparation. Chaperones such as BiP, PDI, and ERdj5 are also needed for the ERAD process.[163,165–168] The ERAD-targeting process is a good example for the collaboration of several chaperones and lectins. While BiP and Erp72 stabilize the protein and keep misfolded proteins in a reversible aggregated state, PDI and ERdj5 resolve existing disulfide bonds of misfolded proteins and target them directly for retranslocation.[165,167,168] Erdj5 is a mammalian disulfide reductase containing a *J*-domain that is able to interact with BiP and four thioredoxin-like domains with CXXC motifs.[164,168] Erdj5 resolves disulfide bonds, prevents, therefore, the formation of dimers, and keeps the proteins, with the help of BiP, in a single, nonaggregated state. Artificial modifications of ERdj5 such as inactivation of the CXXC motifs or a general knockdown of ERdj5 result in a disruption of the ERAD process. Interestingly, inhibition of EDEM demannosidase activity by kifunensine also enables ERdj5 to accelerate the ERAD process, indicating a clear interaction or at least a dependency of ERdj5, EDEM, and BiP.[164,168] So far, ERdj5 has been the first and only disulfide reductase reported for the mammalian ER. In yeast, Pdi1p, a homologue to PDI, seems to have a function similar to ERdj5 function in mammalian cells. Pdi1p also interacts with the yeast EDEM homologue and accelerates the ERAD process[169] (Table II).

Typical substrates for PDI-mediated retranslocation are BACE457 (a pancreatic isoform of human beta-secretase[160]) and choleratoxin.[165,167] However, PDI-mediated retranslocation seems to be dependent on the redox state of the chaperone. Thus, it is suggested that only reduced PDI is able to bind the cholera toxin, whereas the oxidative constitution of PDI is a more open conformation and fails to bind/unfold choleratoxin.[165]

The retranslocation process of misfolded proteins from the ER into the cytosol is still under discussion and till today not completely elucidated. Therefore, several different models of ERAD translocation following ubiquitination are suggested. Additionally, in the literature, yeast and mammalian ERAD components are often mixed or commuted in one model. To prevent such confusion, we first describe the suggested retranslocation models for yeast, followed by the current knowledge about the mammalian ERAD process for glycosylated proteins.

In yeast, ERAD translocation is divided into three different ERAD pathways: ERAD-L, ERAD-M, and ERAD-C. Protein translocation and degradation of the ER *l*umen pass through the ERAD-L pathway. Misfolded ER-membrane

TABLE II

OVERVIEW OF THE MOST IMPORTANT ER CHAPERONES

| Chaperone/lectin | Function |
|---|---|
| BiP/Grp78 | – Stabilization of unfolded proteins[151,152,164]<br>– Involvement in protein translocation into the ER[152]<br>– Supply of proteins to the ERAD machinery[151,164] |
| Calreticulin | – Stabilization of unfolded proteins in the calreticulin/calnexin cycle[147,151]<br>– Transport of unfolded monoglucosylated proteins to calnexin[147,151] |
| Calnexin | – Binding to unfolded proteins delivered from the calreticulin complex[147,151]<br>– Folding of unfolded, monoglucosylated proteins[147,151] |
| PDI | – Is not an "ordinary" chaperone<br>– Oxidation/reduction of consisting disulfide bonds[163]<br>– Central part in protein preparation for ERAD[164] |
| Yos9p/OS9/XTP3-B | – Binding/stabilization of unfolded proteins<br>– Transport to the ERAD machinery[151,164] |

proteins are degraded by ERAD-M and ERAD-C—ERAD-M for proteins with a *m*embrane-spanning domain and ERAD-C for ER-membrane proteins with a *c*ytosolic domain. All of these pathways are based on multi-protein complexes containing or interacting with several E2-ubiquitin-conjugating enzymes, E3 ubiquitin ligases, and the 26S proteasome.

## A. Yeast: ERAD-L

The ERAD-L pathway is based on the complex of the ubiquitin ligase Hrd1p and the ER-membrane protein Hrd3p, which contains a TPR motif. Hrd1p is a sixfold-spanning transmembrane protein containing a RING-finger domain on the cytosolic side. To ubiquitinate the misfolded proteins, oligomerization of two Hrd1ps and interaction with some E2-ubiquitin-conjugating enzymes is necessary.[164,170,171] Oligomerization of the complex is mediated by the scaffolding protein Usa1p. Usa1p possesses the two important cytosolic domains H and U. The H domain binds one Hrd1p protein at its C terminus and the U domain interacts with the U domain of another Usa1p that is in turn bound to a second Hrd1p protein. Furthermore, Usa1p is responsible for the recruitment of Der1p (the yeast homologue in the mammalian Derlin-1). The role of Der1p is not completely clear. It seems that Der1p is involved, together with Hrd3p, in substrate delivery to the ubiquitin ligase Hrd1p.[170] Der1p is also discussed to be a major component of the retranslocation channel, due to a large amount of transmembrane segments.[164,170,171] Another possible candidate, forming the retranslocation channel, is Sec61. Sec61 is normally involved in translocation into the ER, but some studies also suggest a participation in protein transport out of the ER lumen. Within this model, there is a requirement of PDI to reduce and unfold proteins during the preparation for the retranslocation process.[172,173] It is also suggested that protein retranslocation out of the ER lumen into the cytosol is mediated by Hdr1p on its own[170] (Fig. 4A).

Proteins destined for degradation are recognized and recruited to the Hdr1p complex by the lectin Yos9p, the chaperone Kar2p (yeast homologue of BiP), and Der1p. All of these proteins stay in interaction with Hrd3p, which

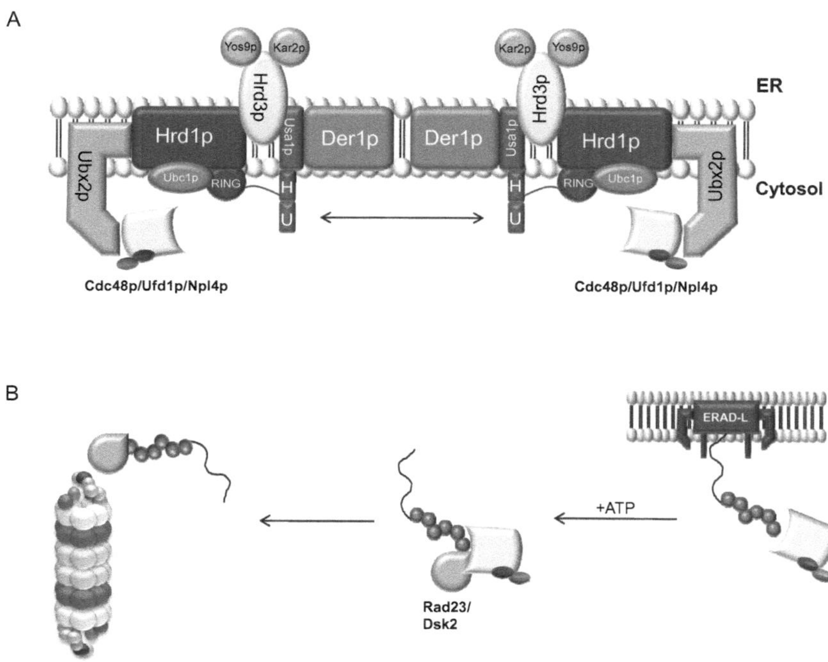


FIG. 4. Retranslocation of misfolded proteins from the ER to the proteasome. In (A) the ERAD-L complex of yeast is schematically depicted. Yos9p and Kar2p carry misfolded proteins to the complex and transfer it to the Hrd3p/Hrd1p/Der1p complex. Hrd3p and Der1p provide the substrate to the E3 ubiquitin ligase Hrd1p. Hrd1p ubiquitinates the substrate via its RING domain in assistance with one of the E2 ubiquitin-conjugating enzymes (e.g., Ubc1p). Together with Der1p, Hrd1p translocates the ubiquitinated protein through the ER membrane (modified from Refs. 164,170). (B) The AAA+ ATPase complex Cdc48p/Ufd1p/NpI4p binds to the ubiquitinated substrate and mediates its transport out of the ER by hydrolysis of ATP. In the cytosol, Cdc48p/Ufd1p/NpI4p transfers the ubiquitinated substrate to the UBA/UBL-domain-containing proteins Rad32/Dsk2, which deliver the ERAD substrate to the proteasome.

passes substrates on to Hdr1p. To enable ubiquitination, Hdr1p is connected with the E2-ubiquitin-conjugating enzymes Ubc7p/Cue1lp or Ubc1p.[164,170,171] Withdrawal of polyubiquitinated proteins of the Hdr1p complex and transport/delivery to the 26S proteasome is accomplished by the AAA+ ATPase Cdc48p and its two cofactors Ufd1p and Npl4p. The linkage of the Cdc48p complex to the cytosolic side of the ER membrane occurs through several different interactions with Ubx2p. Ubx2p spans the membrane twice and possesses an additional ubiquitin-binding (Uba) domain. The exact mechanism of how Ubx2p is recruited to the Hrd1p complex is unknown, but it is supposed that there is an interaction between the Uba domain of Ubx2p and some polyubiquitinated proteins.[174] Cdc48p is a barrel-like, homohexameric protein complex, comprising two ATPase domains (D1 and D2) and one N domain. The N domain connects Cdc48p with the cofactors Ufd1p and Npl4p. Both cofactors are responsible for the binding of polyubiquitinated proteins sticking out of the Hdr1p complex. By hydrolyzing ATP, the Cdc48p/Ufd1p/Npl4p complex changes its conformation, resulting in a pulling force that delivers the captured polyubiquitinated protein through the ER channel into the cytosol.[175] Depending on other interacting cofactors, Cdc48p can elongate or trim the polyubiquitin chain, leading to proteasome-dependent degradation, or rescue the protein from proteolysis.[176] Within the Cdc48p/Ufd1p/Npl4p complex, polyubiquitinated proteins are transferred to the ubiquitin-fold (UBL)/ubiquitin-associated (UBA)-domain-containing proteins Rad23 and Dsk2. Rad23 and Dsk2 function as shuttle factors, delivering (ERAD-) ubiquitinated proteins to the 26S proteasome. The link between these two UBL–UBA proteins and Cdc48p is the bound cofactor Ufd2, which elongates the ubiquitination chain of target proteins for optimized interaction with Rad32 and Dsk2[175] (Fig. 4B).

## B. Yeast: ERAD-M/C

The ERAD-M-pathway for membrane proteins requires ERAD-L Hrd1p and Hrd3p for polyubiquitination and retranslocation, but according to current knowledge, ERAD-M is Usa1p and Der1p independent. Yos9p might also be needed for protein targeting to Hrd3p/Hrd1p.[171] Equal to ERAD-L, the polyubiquitinated, misfolded protein is pulled out and transferred to the proteasome by the Cdc48p/Ufd1p/Npl4p machinery. Because Der1p and Usa1p are not involved, the ERAD-M retranslocation process uses a smaller channel that may consist of one Hrd1p or the misfolded membrane protein is extracted directly into the cytosol after polyubiquitination.[171]

In contrast to ERAD-L and ERAD-M, ERAD-C does not use Hdr1p and Hdr3p for polyubiquitination and retranslocation. Instead, ERAD-C requires the ubiquitin ligase Doa10p, which has also been found in polyubiquitinates of nuclear and cytosolic proteins, indicating a lesser specificity for ERAD than Hdr1p.[174,177] In addition to the ERAD-L and -M process, polyubiquitinated proteins move to the cytosol by Cdc48p.

## C. The Mammalian ERAD Process

Because ERAD is a highly conserved pathway, the mammalian ERAD process shows many similarities to the yeast process. Interestingly, the mammalian retranslocation process includes only one retranslocation complex, which is related to the ERAD-L pathway.

In mammalian cells, the Cdc48p homologue p97 (also called VCP), which is also responsible for protein locomotion through the retranslocation channel and guidance of the ubiquitinated protein to the proteasome, is connected to the ERAD complex via the Der1p homologue Derlin1/2/3 and, instead of Ubx2p, via the protein VIMP.[178,179] The ubiquitin–ligase complex is also similar to the ERAD-L proteins, consisting of HRD1 (Hrd1p homologue) and SEL1L (Hrd3p homologue).[164] Protein transport and guidance inside the ER to the HRD1–SEL1l complex is also mediated by some chaperones: OS-9 and XTP3-B (both are orthologous proteins of Yos9p), BiP, and Grp94.[180,181] To reduce existing disulfide bonds of proteins that are designated for retranslocation, the mammalian ERAD complex is also associated with ERdj5. In addition, in 2009, a new member of the mammalian ERAD protein complex was described: ERFAD. ERFAD contains binding motifs for FAD and NAPDH and is associated with OS-9, SEL1L, and ERdj5. Thus, one suggested function of ERFAD may be the reduction of oxidized ERdj5, keeping Erdj5 in an active state to crack further disulfide bonds.[182]

## D. The ERAD Pathway of Non-glycosylated Proteins

It has been described above that degradation of glycosylated ER proteins needs the involvement of lectins and lectin-like proteins. Thus, the ERAD process/complex for non-glycosylated proteins is lectin independent. Without lectin proteins or lectin-associated proteins, the ERAD membrane complex is minimized to three components: the ubiquitin–ligase and channel protein HRD1, the substrate-binding protein Derlin-1, and another protein called HERP. Additionally, substrate transfer to the ERAD complex is only conducted by BiP, which forms a complex with Derlin-1 and HERP, providing the misfolded protein to HRD1.[164] It is suggested that Erdj5 could accelerate the degradation process of non-glycosylated proteins, too.

## E. Proteasome Inhibition and Its Influence on ER Stress and the Unfolded Protein Response

In the preceding paragraphs, it has been explained that ERAD-L/-C/-M in the mammalian ERAD process is responsible for the retranslocation and degradation of defective and unfolded ER-located proteins. However, there are some ER quality control (ERQC) substrates (CPY*, PrA*) that are targeted for the proteasome by a HRD/DER-independent mechanism or were also degraded when the "normal" HRD/DER-dependent mechanism is saturated.[183,184] Thus, it was suggested that under conditions where ordinary substrates accumulate and overcharge the normal ERAD pathways, there must be an inducible pathway, disburdening the HRD/DER pathway. In studies where PrA* was overexpressed, an induction of an HRD-independent pathway (HIP) was described.[184] The clear mechanism of HIP retranslocation and transport to the 26S proteasome has to be elucidated, but it is evident that HIP utilizes, instead of HRD1, the E3 ubiquitin ligase Rsp5p and the associated E2-ubiquitin-conjugating enzymes Ubc4p and Ubc5p.[184] But before the unfolded protein becomes a substrate of the HIP machinery, it has first to be transported to the Golgi apparatus. This was shown for the HIP substrate CPY*. CPY*is transported to COPII post-ER vesicles containing the *cis*-Golgi cargo receptor Erv29p.[184] Some other important ERQC components, such as UGGT, have also been found in transport vesicles to the *cis*-Golgi, suggesting a further processing of some HIP substrates in the *cis*-Golgi.[185] Once the protein has entered the *cis*-Golgi compartment, it becomes modified by α1,6-mannose addition. Following the mannose addition, the protein becomes ubiquitinated by the Rsp5p–Ubc4p–Ubc5p complex. However, it is unclear where ubiquitination of the substrate takes place. There are two suggestions: The first one postulates a retransport via COPI vesicles into the ER, where the protein is ubiquitinated and channeled out of the ER lumen into the cytosol via a complex of Sec61p and Rsp5p–Ubc4p–Ubc5p; in the cytosol, the ubiquitinated protein can be degraded by the proteasome.[184] The second suggests the possibility of a direct export of the protein from the *cis*-Golgi into the cytosol, where the protein becomes ubiquitinated by the HIP complex and degraded by the proteasome.[184]

So, there are at least two ERAD pathways protecting the ER for protein aggregation and they are able, therefore, to diminish ER stress. However, both pathways are proteasome dependent. Therefore, proteasome inhibition can disable both ERAD pathways and cause proteotoxic ER stress by the presence of accumulating unfolded protein, which results in the unfolded protein response (UPR).[186–188] Prolonged presence of unfolded proteins and aggregates is accompanied by the generation of ROS.[186] UPR-dependent generation of ROS is due to ER- and mitochondrial-located mechanisms. In the ER, ROS accumulation is closely connected to three enzymes: Pdi1p, Ero1p, and GSH (glutathione (reduced)). Pdi1p and Ero1p are two proteins involved in the formation of disulfide bonds, transferring electrons of substrates to molecular oxygen.[189] By-products of the reaction is an increase in ROS by augmented Pdi1p and Ero1p activity during the UPR. Additionally, glutathione also plays a role in concert with Pdi1p, Ero1p, and disulfide bonds. GSH can directly reduce unstable disulfide bonds and reduce oxidized Pdi1p/Ero1p. Thus, the coincidence of a high level of newly generated ROS and a depletion of GSH leads to a synergistic augmentation of total ROS in the ER.[186] The depletion of GSH also affects the mitochondrial respiratory chain, leading to further accumulation of ROS outside the ER. It is also suggested that the high amount of ROS generated in the ER signals causes mitochondrial dysfunction.[186]

To eliminate the growing mass of unfolded proteins and prevent the associated synthesis of ROS, the UPR includes the upregulation/liberation of some specific transcription factors. Activation of the UPR and these transcription factors starts with the ER chaperone BiP. In unstressed cells, BiP is bound to ER-lumenal domains of ATF (activating transcription factor) 6 and the transmembrane kinases PERK (RNA-dependent protein kinase-like ER-kinase) and IRE1a (inositol-requiting enzyme 1a).[187] Thereby, BiP keeps the kinases in a monomeric state and prevents their activation. In case of protein accumulation, the chaperone capacity of BiP is exceeded, resulting in the liberation of PERK, IRE1a, and ATF6. ATF6 binds to ER stress response elements in the promotor region of UPR target genes, directly enhancing the UPR gene response.[188] Dimerization of free PERK entails the phosphorylation of the protein translation factor eIF2$\alpha$ (eukaryotic initiation factor 2$\alpha$), inhibiting the recycling of eIF2$\alpha$ to its GTP-bound form and therefore to a general inhibition of protein translation. Moreover, the phosphorylation of eIF2$\alpha$ leads to an upregulation of the crucial UPR transcription factor ATF4.[187,188] Most data of proteasome-inhibition-dependent induction of ATF4 come from cancer studies where tumors are treated with the proteasome inhibitor bortezomib. Like ATF6, ATF4 also induces the synthesis of UPR-related genes. In addition, ATF 4 is also able to build dimers with NRF-2 and enhance the translation of antioxidative genes, such as HO-1, NAD(P)H:quinine oxidoreductase1, and glutathione S transferase, reinforcing the antioxidative capacity of the UPR.[190]

If proteasomal inhibition persists and the accumulation of unfolded proteins and ROS exceeds the capacity of the UPR-related defense efforts, the UPR program switches from defense and repair to programmed cell death. Again the responsible mechanisms are manifold. On the one hand, mitochondria plays an important role; due to high concentrations of ROS, there is a release of cytochrome c and the cleavage of caspases.[187] On the other hand, ATF4 induces the expression of the two leucine zipper transcription factors CHOP/GADD153, which are associated with Bcl-2 downregulation, disturbing the redox state.[191,192] The generation of ER stress is one major reason for the bortezomib-induced cell death in cancer therapy. In spite of this, bortezomib susceptibility varies between different tumor types and in accordance with bortezomib concentration. Possibly, this is due to the differences in UPR response.[188]

## IX. Conclusion

Protein quality control and preservation of protein homeostasis is an integral part of cellular metabolism, ensuring cell functionality and viability. The proteasome and the huge family of chaperones are two essential systems that help cope with this challenge. In general, these two systems decide whether it is reasonable to degrade an unfolded/unfunctional protein or to stabilize it by trying to refold the protein. Criteria for this decision are largely unknown and have to be a part of further investigations, but they are likely related to the kind and intensity of the stress stimuli. Thus, heat shock, for example, evokes a response of chaperones other than oxidative stress or proteasomal inhibition. However, it is evident that the ubiquitin–proteasomal system is closely connected with many chaperones. Molecular chaperones can promote the ubiquitination process and transfer the substrates to the proteasome or they divest the proteasome of proteins by repairing them. In fact, Hsp27 is able to prevent ubiquitination of substrates by preceding sumoylation. Because of this close connection, it is not surprising that failure of one system affects the other. This is expressed in the upregulation of chaperones after proteasomal inhibition and the partly enhanced proteolytic capacity after chaperone inhibition or dysfunction. Knowledge of these compensating mechanisms explains the limited effects of Hsp or proteasome inhibitors in cancer therapy and promotes the combined application of Hsp90 and Hsp70 and proteasome inhibitors. Further understanding of the fate of cellular proteins will improve not only cancer therapy, but also the treatment of age-related diseases such as neurodegeneration, diabetes, cataract formation, and myopathies.

### References


1. Richter K, Haslbeck M, Buchner J. The heat shock response: life on the verge of death. *Mol Cell* 2010;**40**:253–66.
2. Jantschitsch C, Trautinger F. Heat shock and UV-B-induced DNA damage and mutagenesis in skin. *Photochem Photobiol Sci* 2003;**2**:899–903.
3. Malmendal A, Overgaard J, Bundy JG, Sorensen JG, Nielsen NC, Loeschcke V, et al. Metabolomic profiling of heat stress: hardening and recovery of homeostasis in Drosophila. *Am J Physiol Regul Integr Comp Physiol* 2006;**291**:R205–12.

4. Al RA, Alix JH. Ribosome biogenesis is temperature-dependent and delayed in Escherichia coli lacking the chaperones DnaK or DnaJ. *Mol Microbiol* 2009;**71**:748–62.
5. Toivola DM, Strnad P, Habtezion A, Omary MB. Intermediate filaments take the heat as stress proteins. *Trends Cell Biol* 2010;**20**:79–91.
6. Welker S, Rudolph B, Frenzel E, Hagn F, Liebisch G, Schmitz G, et al. Hsp12 is an intrinsically unstructured stress protein that folds upon membrane association and modulates membrane function. *Mol Cell* 2010;**39**:507–20.
7. Lanneau D, Wettstein G, Bonniaud P, Garrido C. Heat shock proteins: cell protection through protein triage. *ScientificWorldJournal* 2010;**10**:1543–52.
8. Dezwaan DC, Freeman BC. HSP90: the Rosetta stone for cellular protein dynamics? *Cell Cycle* 2008;**7**:1006–12.
9. Barends TR, Werbeck ND, Reinstein J. Disaggregases in 4 dimensions. *Curr Opin Struct Biol* 2010;**20**:46–53.
10. Doyle SM, Wickner S. Hsp104 and ClpB: protein disaggregating machines. *Trends Biochem Sci* 2009;**34**:40–8.
11. Mayer MP. Gymnastics of molecular chaperones. *Mol Cell* 2010;**39**:321–31.
12. Bukau B, Weissman J, Horwich A. Molecular chaperones and protein quality control. *Cell* 2006;**125**:443–51.
13. Welch WJ, Feramisco JR. Purification of the major mammalian heat shock proteins. *J Biol Chem* 1982;**257**:14949–59.
14. Wandinger SK, Richter K, Buchner J. The Hsp90 chaperone machinery. *J Biol Chem* 2008;**283**:18473–7.
15. Pratt WB, Morishima Y, Peng HM, Osawa Y. Proposal for a role of the Hsp90/Hsp70-based chaperone machinery in making triage decisions when proteins undergo oxidative and toxic damage. *Exp Biol Med (Maywood)* 2010;**235**:278–89.
16. Richter K, Muschler P, Hainzl O, Reinstein J, Buchner J. Sti1 is a non-competitive inhibitor of the Hsp90 ATPase. Binding prevents the N-terminal dimerization reaction during the atpase cycle. *J Biol Chem* 2003;**278**:10328–33.
17. Hessling M, Richter K, Buchner J. Dissection of the ATP-induced conformational cycle of the molecular chaperone Hsp90. *Nat Struct Mol Biol* 2009;**16**:287–93.
18. Bukau B, Deuerling E, Pfund C, Craig EA. Getting newly synthesized proteins into shape. *Cell* 2000;**101**:119–22.
19. Neupert W, Brunner M. The protein import motor of mitochondria. *Nat Rev Mol Cell Biol* 2002;**3**:555–65.
20. Pratt WB, Toft DO. Regulation of signaling protein function and trafficking by the hsp90/hsp70-based chaperone machinery. *Exp Biol Med (Maywood)* 2003;**228**:111–33.
21. Mayer MP, Bukau B. Hsp70 chaperones: cellular functions and molecular mechanism. *Cell Mol Life Sci* 2005;**62**:670–84.
22. Hohfeld J, Jentsch S. GrpE-like regulation of the hsc70 chaperone by the anti-apoptotic protein BAG-1. *EMBO J* 1997;**16**:6209–16.
23. Rudiger S, Schneider-Mergener J, Bukau B. Its substrate specificity characterizes the DnaJ co-chaperone as a scanning factor for the DnaK chaperone. *EMBO J* 2001;**20**:1042–50.
24. Odunuga OO, Hornby JA, Bies C, Zimmermann R, Pugh DJ, Blatch GL. Tetratricopeptide repeat motif-mediated Hsc70-mSTI1 interaction. Molecular characterization of the critical contacts for successful binding and specificity. *J Biol Chem* 2003;**278**:6896–904.
25. Summers DW, Douglas PM, Ramos CH, Cyr DM. Polypeptide transfer from Hsp40 to Hsp70 molecular chaperones. *Trends Biochem Sci* 2009;**34**:230–3.
26. Sahi C, Craig EA. Network of general and specialty J protein chaperones of the yeast cytosol. *Proc Natl Acad Sci USA* 2007;**104**:7163–8.

27. Ramos CH, Oliveira CL, Fan CY, Torriani IL, Cyr DM. Conserved central domains control the quaternary structure of type I and type II Hsp40 molecular chaperones. *J Mol Biol* 2008;**383**:155–66.
28. Haslbeck M, Franzmann T, Weinfurtner D, Buchner J. Some like it hot: the structure and function of small heat-shock proteins. *Nat Struct Mol Biol* 2005;**12**:842–6.
29. Sun Y, MacRae TH. Small heat shock proteins: molecular structure and chaperone function. *Cell Mol Life Sci* 2005;**62**:2460–76.
30. van Montfort RL, Basha E, Friedrich KL, Slingsby C, Vierling E. Crystal structure and assembly of a eukaryotic small heat shock protein. *Nat Struct Biol* 2001;**8**:1025–30.
31. Lindner RA, Carver JA, Ehrnsperger M, Buchner J, Esposito G, Behlke J, et al. Mouse Hsp25, a small shock protein. The role of its C-terminal extension in oligomerization and chaperone action. *Eur J Biochem* 2000;**267**:1923–32.
32. Kostenko S, Moens U. Heat shock protein 27 phosphorylation: kinases, phosphatases, functions and pathology. *Cell Mol Life Sci* 2009;**66**:3289–307.
33. Haslbeck M, Braun N, Stromer T, Richter B, Model N, Weinkauf S, et al. Hsp42 is the general small heat shock protein in the cytosol of Saccharomyces cerevisiae. *EMBO J* 2004;**23**:638–49.
34. Basha E, Lee GJ, Breci LA, Hausrath AC, Buan NR, Giese KC, et al. The identity of proteins associated with a small heat shock protein during heat stress in vivo indicates that these chaperones protect a wide range of cellular functions. *J Biol Chem* 2004;**279**:7566–75.
35. Anckar J, Sistonen L. Regulation of HSF1 function in the heat stress response: implications in aging and disease. *Annu Rev Biochem* 2011;**80**:1089–115.
36. Zou J, Guo Y, Guettouche T, Smith DF, Voellmy R. Repression of heat shock transcription factor HSF1 activation by HSP90 (HSP90 complex) that forms a stress-sensitive complex with HSF1. *Cell* 1998;**94**:471–80.
37. Sandqvist A, Bjork JK, Akerfelt M, Chitikova Z, Grichine A, Vourc'h C, et al. Heterotrimerization of heat-shock factors 1 and 2 provides a transcriptional switch in response to distinct stimuli. *Mol Biol Cell* 2009;**20**:1340–7.
38. Shamovsky I, Ivannikov M, Kandel ES, Gershon D, Nudler E. RNA-mediated response to heat shock in mammalian cells. *Nature* 2006;**440**:556–60.
39. Prahlad V, Cornelius T, Morimoto RI. Regulation of the cellular heat shock response in Caenorhabditis elegans by thermosensory neurons. *Science* 2008;**320**:811–4.
40. Mosser DD, Kotzbauer PT, Sarge KD, Morimoto RI. In vitro activation of heat shock transcription factor DNA-binding by calcium and biochemical conditions that affect protein conformation. *Proc Natl Acad Sci USA* 1990;**87**:3748–52.
41. Ahn SG, Thiele DJ. Redox regulation of mammalian heat shock factor 1 is essential for Hsp gene activation and protection from stress. *Genes Dev* 2003;**17**:516–28.
42. McDonough H, Patterson C. CHIP: a link between the chaperone and proteasome systems. *Cell Stress Chaperones* 2003;**8**:303–8.
43. Luders J, Demand J, Hohfeld J. The ubiquitin-related BAG-1 provides a link between the molecular chaperones Hsc70/Hsp70 and the proteasome. *J Biol Chem* 2000;**275**:4613–7.
44. Takayama S, Bimston DN, Matsuzawa S, Freeman BC, Aime-Sempe C, Xie Z, et al. BAG-1 modulates the chaperone activity of Hsp70/Hsc70. *EMBO J* 1997;**16**:4887–96.
45. Jiang J, Ballinger CA, Wu Y, Dai Q, Cyr DM, Hohfeld J, et al. CHIP is a U-box-dependent E3 ubiquitin ligase: identification of Hsc70 as a target for ubiquitylation. *J Biol Chem* 2001; **276**:42938–44.
46. Ballinger CA, Connell P, Wu Y, Hu Z, Thompson LJ, Yin LY, et al. Identification of CHIP, a novel tetratricopeptide repeat-containing protein that interacts with heat shock proteins and negatively regulates chaperone functions. *Mol Cell Biol* 1999;**19**:4535–45.
47. Murata S, Minami Y, Minami M, Chiba T, Tanaka K. CHIP is a chaperone-dependent E3 ligase that ubiquitylates unfolded protein. *EMBO Rep* 2001;**2**:1133–8.

48. Jackson PK, Eldridge AG, Freed E, Furstenthal L, Hsu JY, Kaiser BK, et al. The lore of the RINGs: substrate recognition and catalysis by ubiquitin ligases. *Trends Cell Biol* 2000; **10**:429–39.
49. Connell P, Ballinger CA, Jiang J, Wu Y, Thompson LJ, Hohfeld J, et al. The co-chaperone CHIP regulates protein triage decisions mediated by heat-shock proteins. *Nat Cell Biol* 2001;**3**:93–6.
50. Demand J, Alberti S, Patterson C, Hohfeld J. Cooperation of a ubiquitin domain protein and an E3 ubiquitin ligase during chaperone/proteasome coupling. *Curr Biol* 2001;**11**:1569–77.
51. Alberti S, Demand J, Esser C, Emmerich N, Schild H, Hohfeld J. Ubiquitylation of BAG-1 suggests a novel regulatory mechanism during the sorting of chaperone substrates to the proteasome. *J Biol Chem* 2002;**277**:45920–7.
52. Luders J, Demand J, Papp O, Hohfeld J. Distinct isoforms of the cofactor BAG-1 differentially affect Hsc70 chaperone function. *J Biol Chem* 2000;**275**:14817–23.
53. Young P, Deveraux Q, Beal RE, Pickart CM, Rechsteiner M. Characterization of two polyubiquitin binding sites in the 26 S protease subunit 5a. *J Biol Chem* 1998;**273**:5461–7.
54. Frydman J, Hohfeld J. Chaperones get in touch: the Hip-Hop connection. *Trends Biochem Sci* 1997;**22**:87–92.
55. Garrido C, Brunet M, Didelot C, Zermati Y, Schmitt E, Kroemer G. Heat shock proteins 27 and 70: anti-apoptotic proteins with tumorigenic properties. *Cell Cycle* 2006;**5**:2592–601.
56. Parcellier A, Brunet M, Schmitt E, Col E, Didelot C, Hammann A, et al. HSP27 favors ubiquitination and proteasomal degradation of p27Kip1 and helps S-phase re-entry in stressed cells. *FASEB J* 2006;**20**:1179–81.
57. Parcellier A, Schmitt E, Gurbuxani S, Seigneurin-Berny D, Pance A, Chantome A, et al. HSP27 is a ubiquitin-binding protein involved in I-kappaBalpha proteasomal degradation. *Mol Cell Biol* 2003;**23**:5790–802.
58. Brunet SM, De TA, Hammann A, Joly AL, Bossis G, Fourmaux E, et al. Heat shock protein 27 is involved in SUMO-2/3 modification of heat shock factor 1 and thereby modulates the transcription factor activity. *Oncogene* 2009;**28**:3332–44.
59. Joanisse DR, Inaguma Y, Tanguay RM. Cloning and developmental expression of a nuclear ubiquitin-conjugating enzyme (DmUbc9) that interacts with small heat shock proteins in Drosophila melanogaster. *Biochem Biophys Res Commun* 1998;**244**:102–9.
60. Kawazoe Y, Nakai A, Tanabe M, Nagata K. Proteasome inhibition leads to the activation of all members of the heat-shock-factor family. *Eur J Biochem* 1998;**255**:356–62.
61. Shi YY, Small GW, Orlowski RZ. Proteasome inhibitors induce a p38 mitogen-activated protein kinase (MAPK)-dependent anti-apoptotic program involving MAPK phosphatase-1 and Akt in models of breast cancer. *Breast Cancer Res Treat* 2006;**100**:33–47.
62. Boyault C, Zhang Y, Fritah S, Caron C, Gilquin B, Kwon SH, et al. HDAC6 controls major cell response pathways to cytotoxic accumulation of protein aggregates. *Genes Dev* 2007;**21**: 2172–81.
63. Young JT, Heikkila JJ. Proteasome inhibition induces hsp30 and hsp70 gene expression as well as the acquisition of thermotolerance in Xenopus laevis A6 cells. *Cell Stress Chaperones* 2010;**15**:323–34.
64. Du ZX, Zhang HY, Meng X, Gao YY, Zou RL, Liu BQ, et al. Proteasome inhibitor MG132 induces BAG3 expression through activation of heat shock factor 1. *J Cell Physiol* 2009;**218**: 631–7.
65. Kawaguchi Y, Kovacs JJ, McLaurin A, Vance JM, Ito A, Yao TP. The deacetylase HDAC6 regulates aggresome formation and cell viability in response to misfolded protein stress. *Cell* 2003;**115**:727–38.

66. Murphy PJ, Morishima Y, Kovacs JJ, Yao TP, Pratt WB. Regulation of the dynamics of hsp90 action on the glucocorticoid receptor by acetylation/deacetylation of the chaperone. *J Biol Chem* 2005;**280**:33792–9.
67. Kovacs JJ, Murphy PJ, Gaillard S, Zhao X, Wu JT, Nicchitta CV, et al. HDAC6 regulates Hsp90 acetylation and chaperone-dependent activation of glucocorticoid receptor. *Mol Cell* 2005;**18**:601–7.
68. Seigneurin-Berny D, Verdel A, Curtet S, Lemercier C, Garin J, Rousseaux S, et al. Identification of components of the murine histone deacetylase 6 complex: link between acetylation and ubiquitination signaling pathways. *Mol Cell Biol* 2001;**21**:8035–44.
69. Rumpf S, Jentsch S. Functional division of substrate processing cofactors of the ubiquitin-selective Cdc48 chaperone. *Mol Cell* 2006;**21**:261–9.
70. Kekatpure VD, Dannenberg AJ, Subbaramaiah K. HDAC6 modulates Hsp90 chaperone activity and regulates activation of aryl hydrocarbon receptor signaling. *J Biol Chem* 2009;**284**:7436–45.
71. Okinaga S, Takahashi K, Takeda K, Yoshizawa M, Fujita H, Sasaki H, et al. Regulation of human heme oxygenase-1 gene expression under thermal stress. *Blood* 1996;**87**:5074–84.
72. Alam J, Cook JL. How many transcription factors does it take to turn on the heme oxygenase-1 gene? *Am J Respir Cell Mol Biol* 2007;**36**:166–74.
73. Ryter SW, Alam J, Choi AM. Heme oxygenase-1/carbon monoxide: from basic science to therapeutic applications. *Physiol Rev* 2006;**86**:583–650.
74. Morse D, Choi AM. Heme oxygenase-1: the "emerging molecule" has arrived. *Am J Respir Cell Mol Biol* 2002;**27**:8–16.
75. Jozkowicz A, Was H, Dulak J. Heme oxygenase-1 in tumors: is it a false friend? *Antioxid Redox Signal* 2007;**9**:2099–117.
76. Halliwell BB, Gutteridge JMC. *Free radicals in biology and medicine*. Oxford: Oxford University Press; 2007.
77. Baranano DE, Snyder SH. Neural roles for heme oxygenase: contrasts to nitric oxide synthase. *Proc Natl Acad Sci USA* 2001;**98**:10996–1002.
78. Otterbein LE, Bach FH, Alam J, Soares M, Tao LH, Wysk M, et al. Carbon monoxide has anti-inflammatory effects involving the mitogen-activated protein kinase pathway. *Nat Med* 2000;**6**:422–8.
79. Brouard S, Otterbein LE, Anrather J, Tobiasch E, Bach FH, Choi AM, et al. Carbon monoxide generated by heme oxygenase 1 suppresses endothelial cell apoptosis. *J Exp Med* 2000;**192**:1015–26.
80. Fang J, Akaike T, Maeda H. Antiapoptotic role of heme oxygenase (HO) and the potential of HO as a target in anticancer treatment. *Apoptosis* 2004;**9**:27–35.
81. Wu WT, Chi KH, Ho FM, Tsao WC, Lin WW. Proteasome inhibitors up-regulate haem oxygenase-1 gene expression: requirement of p38 MAPK (mitogen-activated protein kinase) activation but not of NF-kappaB (nuclear factor kappaB) inhibition. *Biochem J* 2004;**379**:587–93.
82. Yamamoto N, Izumi Y, Matsuo T, Wakita S, Kume T, Takada-Takatori Y, et al. Elevation of heme oxygenase-1 by proteasome inhibition affords dopaminergic neuroprotection. *J Neurosci Res* 2010;**88**:1934–42.
83. Dreger H, Westphal K, Weller A, Baumann G, Stangl V, Meiners S, et al. Nrf2-dependent upregulation of antioxidative enzymes: a novel pathway for proteasome inhibitor-mediated cardioprotection. *Cardiovasc Res* 2009;**83**:354–61.
84. Du ZX, Yan Y, Zhang HY, Liu BQ, Gao YY, Niu XF, et al. Proteasome inhibition induces a p38 MAPK pathway-dependent antiapoptotic program via Nrf2 in thyroid cancer cells. *J Clin Endocrinol Metab* 2011;**96**:E763–71.
85. Itoh K, Tong KI, Yamamoto M. Molecular mechanism activating Nrf2-Keap1 pathway in regulation of adaptive response to electrophiles. *Free Radic Biol Med* 2004;**36**:1208–13.

86. Koll TT, Feis SS, Wright MH, Teniola MM, Richardson MM, Robles AI, et al. HSP90 inhibitor, DMAG, synergizes with radiation of lung cancer cells by interfering with base excision and ATM-mediated DNA repair. *Mol Cancer Ther* 2008;**7**:1985–92.
87. Gimenez OA, Montalar SJ. Heat shock proteins as targets in oncology. *Clin Transl Oncol* 2010;**12**:166–73.
88. Loo MA, Jensen TJ, Cui L, Hou Y, Chang XB, Riordan JR. Perturbation of Hsp90 interaction with nascent CFTR prevents its maturation and accelerates its degradation by the proteasome. *EMBO J* 1998;**17**:6879–87.
89. Tsubuki S, Saito Y, Kawashima S. Purification and characterization of an endogenous inhibitor specific to the Z-Leu-Leu-Leu-MCA degrading activity in proteasome and its identification as heat-shock protein 90. *FEBS Lett* 1994;**344**:229–33.
90. Wagner BJ, Margolis JW. Age-dependent association of isolated bovine lens multicatalytic proteinase complex (proteasome) with heat-shock protein 90, an endogenous inhibitor. *Arch Biochem Biophys* 1995;**323**:455–62.
91. Blagg BS, Kerr TD. Hsp90 inhibitors: small molecules that transform the Hsp90 protein folding machinery into a catalyst for protein degradation. *Med Res Rev* 2006;**26**:310–38.
92. Zhang H, Burrows F. Targeting multiple signal transduction pathways through inhibition of Hsp90. *J Mol Med (Berl)* 2004;**82**:488–99.
93. Leu JI, Pimkina J, Pandey P, Murphy ME, George DL. HSP70 inhibition by the small-molecule 2-phenylethynesulfonamide impairs protein clearance pathways in tumor cells. *Mol Cancer Res* 2011;**9**:936–47.
94. Ko HS, Bailey R, Smith WW, Liu Z, Shin JH, Lee YI, et al. CHIP regulates leucine-rich repeat kinase-2 ubiquitination, degradation, and toxicity. *Proc Natl Acad Sci USA* 2009;**106**:2897–902.
95. Guo F, Rocha K, Bali P, Pranpat M, Fiskus W, Boyapalle S, et al. Abrogation of heat shock protein 70 induction as a strategy to increase antileukemia activity of heat shock protein 90 inhibitor 17-allylamino-demethoxy geldanamycin. *Cancer Res* 2005;**65**:10536–44.
96. Ghoshal S, Rao I, Earp JC, Jusko WJ, Wetzler M. Down-regulation of heat shock protein 70 improves arsenic trioxide and 17-DMAG effects on constitutive signal transducer and activator of transcription 3 activity. *Cancer Chemother Pharmacol* 2010;**66**:681–9.
97. Qing G, Yan P, Xiao G. Hsp90 inhibition results in autophagy-mediated proteasome-independent degradation of IkappaB kinase (IKK). *Cell Res* 2006;**16**:895–901.
98. Yokota S, Kitahara M, Nagata K. Benzylidene lactam compound, KNK437, a novel inhibitor of acquisition of thermotolerance and heat shock protein induction in human colon carcinoma cells. *Cancer Res* 2000;**60**:2942–8.
99. Wang AM, Morishima Y, Clapp KM, Peng HM, Pratt WB, Gestwicki JE, et al. Inhibition of hsp70 by methylene blue affects signaling protein function and ubiquitination and modulates polyglutamine protein degradation. *J Biol Chem* 2010;**285**:15714–23.
100. Davenport EL, Zeisig A, Aronson LI, Moore HE, Hockley S, Gonzalez D, et al. Targeting heat shock protein 72 enhances Hsp90 inhibitor-induced apoptosis in myeloma. *Leukemia* 2010;**24**:1804–7.
101. Sies H. Oxidative stress: oxidants and antioxidants. *Exp Physiol* 1997;**82**:291–5.
102. Kalmar B, Greensmith L. Induction of heat shock proteins for protection against oxidative stress. *Adv Drug Deliv Rev* 2009;**61**:310–8.
103. Jakob U, Muse W, Eser M, Bardwell JC. Chaperone activity with a redox switch. *Cell* 1999;**96**:341–52.
104. Kastle M, Grune T. Protein oxidative modification in the aging organism and the role of the ubiquitin proteasomal system. *Curr Pharm Design* 2011;**17**:4007–22.

105. Grune T, Catalgol B, Licht A, Ermak G, Pickering AM, Ngo JK, et al. HSP70 mediates dissociation and reassociation of the 26S proteasome during adaptation to oxidative stress. *Free Radic Biol Med* 2011;**51**:1355–64.
106. Wang X, Yen J, Kaiser P, Huang L. Regulation of the 26S proteasome complex during oxidative stress. *Sci Signal* 2010;**3**:ra88.
107. Conconi M, Friguet B. Proteasome inactivation upon aging and on oxidation-effect of HSP 90. *Mol Biol Rep* 1997;**24**:45–50.
108. Bulteau AL, Szweda LI, Friguet B. Age-dependent declines in proteasome activity in the heart. *Arch Biochem Biophys* 2002;**397**:298–304.
109. Harman D. Aging: a theory based on free radical and radiation chemistry. *J Gerontol* 1956;**11**:298–300.
110. Macario AJ, Conway de ME. Sick chaperones, cellular stress, and disease. *N Engl J Med* 2005;**353**:1489–501.
111. Soti C, Csermely P. Aging and molecular chaperones. *Exp Gerontol* 2003;**38**:1037–40.
112. Soti C, Csermely P. Molecular chaperones and the aging process. *Biogerontology* 2000; **1**:225–33.
113. Wu B, Gu MJ, Heydari AR, Richardson A. The effect of age on the synthesis of two heat shock proteins in the hsp70 family. *J Gerontol* 1993;**48**:B50–6.
114. Gutsmann-Conrad A, Heydari AR, You S, Richardson A. The expression of heat shock protein 70 decreases with cellular senescence in vitro and in cells derived from young and old human subjects. *Exp Cell Res* 1998;**241**:404–13.
115. Bernstein SL, Liu AM, Hansen BC, Somiari RI. Heat shock cognate-70 gene expression declines during normal aging of the primate retina. *Invest Ophthalmol Vis Sci* 2000;**41**: 2857–62.
116. Heydari AR, You S, Takahashi R, Gutsmann-Conrad A, Sarge KD, Richardson A. Age-related alterations in the activation of heat shock transcription factor 1 in rat hepatocytes. *Exp Cell Res* 2000;**256**:83–93.
117. Conconi M, Szweda LI, Levine RL, Stadtman ER, Friguet B. Age-related decline of rat liver multicatalytic proteinase activity and protection from oxidative inactivation by heat-shock protein 90. *Arch Biochem Biophys* 1996;**331**:232–40.
118. Widmer R, Ziaja I, Grune T. Protein oxidation and degradation during aging: role in skin aging and neurodegeneration. *Free Radic Res* 2006;**40**:1259–68.
119. Gregori L, Hainfeld JF, Simon MN, Goldgaber D. Binding of amyloid beta protein to the 20 S proteasome. *J Biol Chem* 1997;**272**:58–62.
120. Yoo BC, Kim SH, Cairns N, Fountoulakis M, Lubec G. Deranged expression of molecular chaperones in brains of patients with Alzheimer's disease. *Biochem Biophys Res Commun* 2001;**280**:249–58.
121. Valera AG, az-Hernandez M, Hernandez F, Ortega Z, Lucas JJ. The ubiquitin-proteasome system in Huntington's disease. *Neuroscientist* 2005;**11**:583–94.
122. Zabel C, Chamrad DC, Priller J, Woodman B, Meyer HE, Bates GP, et al. Alterations in the mouse and human proteome caused by Huntington's disease. *Mol Cell Proteomics* 2002;**1**: 366–75.
123. Grimm S, Hoehn A, Davies KJ, Grune T. Protein oxidative modifications in the ageing brain: consequence for the onset of neurodegenerative disease. *Free Radic Res* 2011;**45**:73–88.
124. Kalia LV, Kalia SK, Chau H, Lozano AM, Hyman BT, McLean PJ. Ubiquitinylation of alpha-synuclein by carboxyl terminus Hsp70-interacting protein (CHIP) is regulated by Bcl-2-associated athanogene 5 (BAG5). *PLoS One* 2011;**6**:e14695.
125. Bloemendal H, de JW, Jaenicke R, Lubsen NH, Slingsby C, Tardieu A. Ageing and vision: structure, stability and function of lens crystallins. *Prog Biophys Mol Biol* 2004;**86**:407–85.

126. Wride MA, Geatrell J, Guggenheim JA. Proteases in eye development and disease. *Birth Defects Res C Embryo Today* 2006;**78**:90–105.
127. Husom AD, Peters EA, Kolling EA, Fugere NA, Thompson LV, Ferrington DA. Altered proteasome function and subunit composition in aged muscle. *Arch Biochem Biophys* 2004; **421**:67–76.
128. Meriin AB, Sherman MY. Role of molecular chaperones in neurodegenerative disorders. *Int J Hyperthermia* 2005;**21**:403–19.
129. Luo GR, Chen S, Le WD. Are heat shock proteins therapeutic target for Parkinson's disease? *Int J Biol Sci* 2007;**3**:20–6.
130. Goldbaum O, Richter-Landsberg C. Proteolytic stress causes heat shock protein induction, tau ubiquitination, and the recruitment of ubiquitin to tau-positive aggregates in oligodendrocytes in culture. *J Neurosci* 2004;**24**:5748–57.
131. Imai Y, Soda M, Hatakeyama S, Akagi T, Hashikawa T, Nakayama KI, et al. CHIP is associated with Parkin, a gene responsible for familial Parkinson's disease, and enhances its ubiquitin ligase activity. *Mol Cell* 2002;**10**:55–67.
132. Waldemar G, Dubois B, Emre M, Georges J, McKeith IG, Rossor M, et al. Recommendations for the diagnosis and management of Alzheimer's disease and other disorders associated with dementia: EFNS guideline. *Eur J Neurol* 2007;**14**:e1–e26.
133. Koren III J, Jinwal UK, Lee DC, Jones JR, Shults CL, Johnson AG, et al. Chaperone signalling complexes in Alzheimer's disease. *J Cell Mol Med* 2009;**13**:619–30.
134. Nunan J, Small DH. Regulation of APP cleavage by alpha-, beta- and gamma-secretases. *FEBS Lett* 2000;**483**:6–10.
135. Lehman NL. The ubiquitin proteasome system in neuropathology. *Acta Neuropathol* 2009;**118**:329–47.
136. Shankar GM, Bloodgood BL, Townsend M, Walsh DM, Selkoe DJ, Sabatini BL. Natural oligomers of the Alzheimer amyloid-beta protein induce reversible synapse loss by modulating an NMDA-type glutamate receptor-dependent signaling pathway. *J Neurosci* 2007;**27**: 2866–75.
137. Cleary JP, Walsh DM, Hofmeister JJ, Shankar GM, Kuskowski MA, Selkoe DJ, et al. Natural oligomers of the amyloid-beta protein specifically disrupt cognitive function. *Nat Neurosci* 2005;**8**:79–84.
138. Nemes Z, Devreese B, Steinert PM, Van BJ, Fesus L. Cross-linking of ubiquitin, HSP27, parkin, and alpha-synuclein by gamma-glutamyl-epsilon-lysine bonds in Alzheimer's neurofibrillary tangles. *FASEB J* 2004;**18**:1135–7.
139. Perez N, Sugar J, Charya S, Johnson G, Merril C, Bierer L, et al. Increased synthesis and accumulation of heat shock 70 proteins in Alzheimer's disease. *Brain Res Mol Brain Res* 1991;**11**:249–54.
140. Petrucelli L, Dickson D, Kehoe K, Taylor J, Snyder H, Grover A, et al. CHIP and Hsp70 regulate tau ubiquitination, degradation and aggregation. *Hum Mol Genet* 2004;**13**:703–14.
141. Dou F, Netzer WJ, Tanemura K, Li F, Hartl FU, Takashima A, et al. Chaperones increase association of tau protein with microtubules. *Proc Natl Acad Sci USA* 2003;**100**:721–6.
142. Dickey CA, Koren J, Zhang YJ, Xu YF, Jinwal UK, Birnbaum MJ, et al. Akt and CHIP coregulate tau degradation through coordinated interactions. *Proc Natl Acad Sci USA* 2008; **105**:3622–7.
143. Kim SJ, Sung JY, Um JW, Hattori N, Mizuno Y, Tanaka K, et al. Parkin cleaves intracellular alpha-synuclein inclusions via the activation of calpain. *J Biol Chem* 2003;**278**:41890–9.
144. Shin Y, Klucken J, Patterson C, Hyman BT, McLean PJ. The co-chaperone carboxyl terminus of Hsp70-interacting protein (CHIP) mediates alpha-synuclein degradation decisions between proteasomal and lysosomal pathways. *J Biol Chem* 2005;**280**:23727–34.

145. Berke SJ, Paulson HL. Protein aggregation and the ubiquitin proteasome pathway: gaining the UPPer hand on neurodegeneration. *Curr Opin Genet Dev* 2003;**13**:253–61.
146. Kalia SK, Lee S, Smith PD, Liu L, Crocker SJ, Thorarinsdottir TE, et al. BAG5 inhibits parkin and enhances dopaminergic neuron degeneration. *Neuron* 2004;**44**:931–45.
147. Jung T, Catalgol B, Grune T. The proteasomal system. *Mol Aspects Med* 2009;**30**:191–296.
148. Rivett AJ, Hearn AR. Proteasome function in antigen presentation: immunoproteasome complexes, peptide production, and interactions with viral proteins. *Curr Protein Pept Sci* 2004;**5**:153–61.
149. Awad W, Estrada I, Shen Y, Hendershot LM. BiP mutants that are unable to interact with endoplasmic reticulum DnaJ proteins provide insights into interdomain interactions in BiP. *Proc Natl Acad Sci USA* 2008;**105**:1164–9.
150. Nishikawa SI, Fewell SW, Kato Y, Brodsky JL, Endo T. Molecular chaperones in the yeast endoplasmic reticulum maintain the solubility of proteins for retrotranslocation and degradation. *J Cell Biol* 2001;**153**:1061–70.
151. Roth J, Yam GH, Fan J, Hirano K, Gaplovska-Kysela K, Le FV, et al. Protein quality control: the who's who, the where's and therapeutic escapes. *Histochem Cell Biol* 2008;**129**:163–77.
152. Rapoport TA. Protein translocation across the eukaryotic endoplasmic reticulum and bacterial plasma membranes. *Nature* 2007;**450**:663–9.
153. Alberts B, Johnson A, Walter P, Lewis J, Raff M, Roberts K. *Molecular biology of the cell.* Wiley-VCH; 1995.
154. Hauri HP, Nufer O, Breuza L, Tekaya HB, Liang L. Lectins and protein traffic early in the secretory pathway. *Biochem Soc Symp* 2002;**69**:73–82.
155. Olivari S, Cali T, Salo KE, Paganetti P, Ruddock LW, Molinari M. EDEM1 regulates ER-associated degradation by accelerating de-mannosylation of folding-defective polypeptides and by inhibiting their covalent aggregation. *Biochem Biophys Res Commun* 2006;**349**:1278–84.
156. Herscovics A, Romero PA, Tremblay LO. The specificity of the yeast and human class I ER alpha 1,2-mannosidases involved in ER quality control is not as strict previously reported. *Glycobiology* 2002;**12**:14G–NaN.
157. Wu Y, Swulius MT, Moremen KW, Sifers RN. Elucidation of the molecular logic by which misfolded alpha 1-antitrypsin is preferentially selected for degradation. *Proc Natl Acad Sci USA* 2003;**100**:8229–34.
158. Olivari S, Galli C, Alanen H, Ruddock L, Molinari M. A novel stress-induced EDEM variant regulating endoplasmic reticulum-associated glycoprotein degradation. *J Biol Chem* 2005; **280**:2424–8.
159. Liu Y, Choudhury P, Cabral CM, Sifers RN. Oligosaccharide modification in the early secretory pathway directs the selection of a misfolded glycoprotein for degradation by the proteasome. *J Biol Chem* 1999;**274**:5861–7.
160. Molinari M, Galli C, Piccaluga V, Pieren M, Paganetti P. Sequential assistance of molecular chaperones and transient formation of covalent complexes during protein degradation from the ER. *J Cell Biol* 2002;**158**:247–57.
161. Molinari M, Calanca V, Galli C, Lucca P, Paganetti P. Role of EDEM in the release of misfolded glycoproteins from the calnexin cycle. *Science* 2003;**299**:1397–400.
162. Hwang C, Sinskey AJ, Lodish HF. Oxidized redox state of glutathione in the endoplasmic reticulum. *Science* 1992;**257**:1496–502.
163. Nuss JE, Choksi KB, Deford JH, Papaconstantinou J. Decreased enzyme activities of chaperones PDI and BiP in aged mouse livers. *Biochem Biophys Res Commun* 2008;**365**:355–61.
164. Hoseki J, Ushioda R, Nagata K. Mechanism and components of endoplasmic reticulum-associated degradation. *J Biochem* 2010;**147**:19–25.
165. Appenzeller-Herzog C, Ellgaard L. The human PDI family: versatility packed into a single fold. *Biochim Biophys Acta* 2008;**1783**:535–48.

166. Kimura T, Hosoda Y, Sato Y, Kitamura Y, Ikeda T, Horibe T, et al. Interactions among yeast protein-disulfide isomerase proteins and endoplasmic reticulum chaperone proteins influence their activities. *J Biol Chem* 2005;**280**:31438–41.
167. Kleizen B, Braakman I. Protein folding and quality control in the endoplasmic reticulum. *Curr Opin Cell Biol* 2004;**16**:343–9.
168. Ushioda R, Hoseki J, Araki K, Jansen G, Thomas DY, Nagata K. ERdj5 is required as a disulfide reductase for degradation of misfolded proteins in the ER. *Science* 2008;**321**: 569–72.
169. Sakoh-Nakatogawa M, Nishikawa S, Endo T. Roles of protein-disulfide isomerase-mediated disulfide bond formation of yeast Mnl1p in endoplasmic reticulum-associated degradation. *J Biol Chem* 2009;**284**:11815–25.
170. Carvalho P, Stanley AM, Rapoport TA. Retrotranslocation of a misfolded luminal ER protein by the ubiquitin-ligase Hrd1p. *Cell* 2010;**143**:579–91.
171. Carvalho P, Goder V, Rapoport TA. Distinct ubiquitin-ligase complexes define convergent pathways for the degradation of ER proteins. *Cell* 2006;**126**:361–73.
172. Willer M, Forte GM, Stirling CJ. Sec61p is required for ERAD-L: genetic dissection of the translocation and ERAD-L functions of Sec61P using novel derivatives of CPY. *J Biol Chem* 2008;**283**:33883–8.
173. Tsai B, Rodighiero C, Lencer WI, Rapoport TA. Protein disulfide isomerase acts as a redox-dependent chaperone to unfold cholera toxin. *Cell* 2001;**104**:937–48.
174. Neuber O, Jarosch E, Volkwein C, Walter J, Sommer T. Ubx2 links the Cdc48 complex to ER-associated protein degradation. *Nat Cell Biol* 2005;**7**:993–8.
175. Raasi S, Wolf DH. Ubiquitin receptors and ERAD: a network of pathways to the proteasome. *Semin Cell Dev Biol* 2007;**18**:780–91.
176. Schrader EK, Harstad KG, Matouschek A. Targeting proteins for degradation. *Nat Chem Biol* 2009;**5**:815–22.
177. Ravid T, Kreft SG, Hochstrasser M. Membrane and soluble substrates of the Doa10 ubiquitin ligase are degraded by distinct pathways. *EMBO J* 2006;**25**:533–43.
178. Ye Y, Shibata Y, Kikkert M, van VS, Wiertz E, Rapoport TA. Recruitment of the p97 ATPase and ubiquitin ligases to the site of retrotranslocation at the endoplasmic reticulum membrane. *Proc Natl Acad Sci USA* 2005;**102**:14132–8.
179. Sun F, Zhang R, Gong X, Geng X, Drain PF, Frizzell RA. Derlin-1 promotes the efficient degradation of the cystic fibrosis transmembrane conductance regulator (CFTR) and CFTR folding mutants. *J Biol Chem* 2006;**281**:36856–63.
180. Christianson JC, Shaler TA, Tyler RE, Kopito RR. OS-9 and GRP94 deliver mutant alpha1-antitrypsin to the Hrd1-SEL1L ubiquitin ligase complex for ERAD. *Nat Cell Biol* 2008;**10**:272–82.
181. Hosokawa N, Wada I, Nagasawa K, Moriyama T, Okawa K, Nagata K. Human XTP3-B forms an endoplasmic reticulum quality control scaffold with the HRD1-SEL1L ubiquitin ligase complex and BiP. *J Biol Chem* 2008;**283**:20914–24.
182. Riemer J, Appenzeller-Herzog C, Johansson L, Bodenmiller B, Hartmann-Petersen R, Ellgaard L. A luminal flavoprotein in endoplasmic reticulum-associated degradation. *Proc Natl Acad Sci USA* 2009;**106**:14831–6.
183. Hill K, Cooper AA. Degradation of unassembled Vph1p reveals novel aspects of the yeast ER quality control system. *EMBO J* 2000;**19**:550–61.
184. Haynes CM, Caldwell S, Cooper AA. An HRD/DER-independent ER quality control mechanism involves Rsp5p-dependent ubiquitination and ER-Golgi transport. *J Cell Biol* 2002;**158**:91–101.

185. Zuber C, Fan JY, Guhl B, Parodi A, Fessler JH, Parker C, et al. Immunolocalization of UDP-glucose:glycoprotein glucosyltransferase indicates involvement of pre-Golgi intermediates in protein quality control. *Proc Natl Acad Sci USA* 2001;**98**:10710–5.
186. Haynes CM, Titus EA, Cooper AA. Degradation of misfolded proteins prevents ER-derived oxidative stress and cell death. *Mol Cell* 2004;**15**:767–76.
187. Rzymski T, Milani M, Singleton DC, Harris AL. Role of ATF4 in regulation of autophagy and resistance to drugs and hypoxia. *Cell Cycle* 2009;**8**:3838–47.
188. Obeng EA, Carlson LM, Gutman DM, Harrington Jr. WJ, Lee KP, Boise LH. Proteasome inhibitors induce a terminal unfolded protein response in multiple myeloma cells. *Blood* 2006;**107**:4907–16.
189. Tu BP, Weissman JS. Oxidative protein folding in eukaryotes: mechanisms and consequences. *J Cell Biol* 2004;**164**:341–6.
190. Cullinan SB, Diehl JA. PERK-dependent activation of Nrf2 contributes to redox homeostasis and cell survival following endoplasmic reticulum stress. *J Biol Chem* 2004;**279**:20108–17.
191. Zinszner H, Kuroda M, Wang X, Batchvarova N, Lightfoot RT, Remotti H, et al. CHOP is implicated in programmed cell death in response to impaired function of the endoplasmic reticulum. *Genes Dev* 1998;**12**:982–95.
192. McCullough KD, Martindale JL, Klotz LO, Aw TY, Holbrook NJ. Gadd153 sensitizes cells to endoplasmic reticulum stress by down-regulating Bcl2 and perturbing the cellular redox state. *Mol Cell Biol* 2001;**21**:1249–59.

# Proteasome Inhibitors

Boris Cvek

*Department of Cell Biology & Genetics, Palacky University, Olomouc, Czech Republic*

I. Short History of Bortezomib ........ 161
II. Bortezomib in Multiple Myeloma ........ 163
III. Bortezomib in Mantle Cell Lymphoma ........ 166
IV. Bortezomib in Other Cancers ........ 167
V. Bortezomib in Diseases Different from Cancer ........ 170
VI. The Adverse Effects of Bortezomib ........ 171
VII. Second-Generation of 20S Proteasome Inhibitors ........ 173
A. Marizomib ........ 200
B. Carfilzomib ........ 202
C. CEP-18770 and MLN9708/MLN2238 ........ 204
VIII. Early Drugs as Proteasome Inhibitors ........ 205
IX. Immunoproteasome Inhibitors ........ 207
X. Beyond the 20S Proteasome ........ 208
References ........ 209

In May 2003, the US Food and Drug Administration (FDA) granted accelerated approval for the use of the first-in-class proteasome inhibitor bortezomib as a third-line therapy in multiple myeloma, and the European Union followed suit a year later. Bortezomib has subsequently been approved for multiple myeloma as a second-line treatment on its own and as a first-line therapy in combination with an alkylating agent and a corticosteroid. Furthermore, bortezomib has also been approved as a second-line therapy for mantle cell lymphoma. In this chapter, the focus is on the current clinical research on bortezomib, its adverse effects, and the resistance of multiple myeloma patients to bortezomib-based therapy. The various applications of bortezomib in different diseases and recent advances in the development of a new generation of inhibitors that target the proteasome or other parts of the ubiquitin–proteasome system are also reviewed.

## I. Short History of Bortezomib

Bortezomib (Velcade®) is the first and still the only proteasome inhibitor that has been U.S. Food and Drug Administration (FDA)-approved as a first-line treatment for a cancer, that is, for multiple myeloma, and it has been in use since May 2003.[1] Currently, there are about 200 active clinical trials of bortezomib for various kinds of cancer listed in ClinicalTrials.gov. The basic principles of the ubiquitin–proteasome system and proteasome inhibition are

Progress in Molecular Biology
and Translational Science, Vol. 109
DOI: 10.1016/B978-0-12-397863-9.00005-5


1877-1173/12 $35.00

reviewed elsewhere in this book. Furthermore, a comprehensive and current summary of the chemical principles of proteasome inhibition is available, for instance, in a useful work published in Chemical Reviews.[2] The complex and not yet fully understood mechanism of the action of bortezomib in cancer cells has recently been reviewed in Current Pharmaceutical Design.[3]

The first mention of bortezomib, formerly known as PS-341, LDP-341, and MLM341,[4] was reported in December 1998; it described how the drug was able to attenuate inflammation in rats through proteasome and nuclear factor-κB (NF-κB) inhibition.[5] In a follow-up study published in April 1999, proteasome inhibitors (including bortezomib) were still used only as a tool to study the NF-κB pathway.[6] The focus first shifted in an article[7] authored by Julian Adams at ProScript Inc., a biotechnology company started by Harvard professors A. Goldberg, T. Maniatis, M. Rosenblatt, and K. Rock in 1994 under the name Myogenics (therefore, bortezomib was originally named MG-341 (Myogenics) or PS-341 (ProSscript)).

As referred to in Adams *et al.*,[7] one of the first proteasome inhibitors[8] was MG-132, a peptide aldehyde based on calpain inhibitor I.[9] However, MG-132 is not a sufficiently selective inhibitor of the proteasome, as it also inhibits cathepsin B and calpains.[10,11] The (20S) proteasome as a threonine protease[12] is very efficiently targeted by dipeptidyl boronic acids.[13–16] Consequently, the average concentration of bortezomib[a] required to inhibit cell growth by 50% across the entire National Cancer Institute (NCI) cell panel (60 cell lines derived from multiple human tumors) was 7 nM.[17] The NCI algorithm COMPARE showed that bortezomib had a unique toxicity "fingerprint" among the 60,000 compounds tested.[18] Such results strongly encouraged further testing in animals, which showed bortezomib to be a potent cancer suppressor in several murine tumors and human xenografts.[19] Moreover, no adverse effects of intravenously (i.v.) administered bortezomib were noted by Adams *et al.*,[7] although the proteasome was significantly inhibited in the tumor as well as in murine white blood cells, colon, liver, muscle, and prostate. Intravenous dosing with radiolabeled [$^{14}$C]bortezomib produced the highest radioactivity levels in the adrenals, kidney cortex, liver, prostate, and spleen 10 min after drug administration. In contrast, levels of the drug in brain, spinal cord, eye, and testes tissues were below the limits of quantitation. The majority (66%) of the radiolabeled drug was excreted into the bile, and the remainder was excreted in the urine. However, neither fluid exhibited proteasome inhibitory activity, suggesting that only metabolized bortezomib was excreted.

A full toxicological evaluation of bortezomib in both rodents and primates revealed the drug's side effect profile.[7] In primates, bortezomib caused anorexia, emesis, and diarrhea, all in a dose-dependent manner. Among 40 other examined

[a][(1*R*)-3-methyl-1-({(2*S*)-3-phenyl-2-[(pyrazin-2-ylcarbonyl)amino]propanoyl}amino)butyl] boronic acid.

tissues, modest effects of bortezomib were seen in the thymus and the spleen. The drug was therefore approved for phase I clinical trials in human patients with cancer.[19] During the trials, the patients experienced low-grade fever and/or fatigue after several cycles of bortezomib at doses of about 1.0 mg $m^{-2}$. Thrombocytopenia was also observed but was not dose-limiting, and some patients experienced low-grade diarrhea. Peripheral neuropathy was the most serious adverse effect of bortezomib, and it has remained so throughout the drug's history. The phase I results showed that four malignancies seemed to be the most susceptible to proteasome inhibition: multiple myeloma,[20] androgen-independent non-small-cell lung cancer,[21] prostate cancer,[22] and melanoma.[23]

A pivotal phase II trial of bortezomib was performed in patients with relapsed and/or refractory multiple myeloma whose cancer had progressed after a median of six previous therapeutic regimens.[24] The incidence of grade 4 adverse events was low, and most could be managed with standard approaches. Among all 202 patients involved in the trial, the median time to progression was 7 months with bortezomib treatment, as compared with 3 months for the last treatment received before entering the trial. The positive therapeutic response to bortezomib was not influenced by the type of multiple myeloma or by the type or number of previous therapies. In May 2003, the FDA granted accelerated approval for the use of bortezomib as a third-line therapy in multiple myeloma, and the European Union followed suit a year later.[25–27] Bortezomib has subsequently been approved for multiple myeloma as a second-line treatment on its own[28] and as a first-line therapy in combination with an alkylating agent and a corticosteroid.[29] Furthermore, bortezomib has also been approved as a second-line therapy for mantle cell lymphoma.[30]

In the following sections, the focus is on the current clinical research on bortezomib, its adverse effects, and the resistance of multiple myeloma patients to bortezomib-based therapy. The various applications of bortezomib in different diseases and recent advances in the development of a new generation of inhibitors that target the proteasome or other parts of the ubiquitin–proteasome system are also reviewed.

## II. Bortezomib in Multiple Myeloma

The activity of bortezomib as a therapy for multiple myeloma has recently been reviewed from a clinical point of view.[31,32] However, many new findings in the increasingly complex field of bortezomib clinical applications have further illuminated the promises and limits of the drug. To begin, the basic facts are summed up.

The anticancer effects of bortezomib on multiple myeloma were confirmed in the above-mentioned phase II trial[24] that showed a 27% overall response rate (partial response and complete remission) in heavily pretreated patients who received an i.v. bolus of 1.3 mg $m^{-2}$ on days 1, 4, 8, and 11 of every 21-day cycle. Further follow-up determined that the median duration of response was 12.7 months, the median time to progression was 7 months, and the median overall survival was 17.0 months.[33] A subsequent phase III randomized trial comparing dexamethasone with bortezomib showed that the latter induced a better overall response rate, better response quality, and longer median time to progression and overall survival.[34–37]

To overcome resistance and induce the chemosensitization of multiple myeloma, bortezomib has been combined with the following:

- melphalan, showing encouraging activity with controllable toxicity[38];
- samarium lexidronam, as a well-tolerated regimen that produced responses in 21% of patients[39];
- vorinostat, leading to an overall response rate of 42% and low toxicity[40];
- pegylated liposomal doxorubicin, leading to demonstrated safety and promising antitumor activity.[41]

In a randomized clinical trial comparing bortezomib with bortezomib and pegylated liposomal doxorubicin in relapsed and/or refractory multiple myeloma, the combination induced a superior time to progression than bortezomib alone. The combination therapy also improved the duration of response and progression-free survival.[42,43] Although the results of bortezomib/dexamethasone combination therapy have been encouraging,[44–47] another induction regimen combining bortezomib with melphalan and prednisone (M/P) showed an impressive overall response rate and durability, even in patients with high-risk cytogenetic markers.[48] This led to an international randomized study of M/P versus bortezomib/M/P, which showed a superior overall response rate and a better response quality of bortezomib/M/P.[49] The percentages of patients with a partial response or better were 71% in the bortezomib group and 35% in the control group. Complete response rates were 30% and 4%, respectively. The median duration of the response was 19.9 months in the bortezomib group and 13.1 months in the control group. As a result, bortezomib/M/P became a standard of care for initial therapy in multiple myeloma patients who were unable to tolerate, or presented with contraindications to, thalidomide, as well as patients who would probably not be transplant candidates.[29] Furthermore, bortezomib and low-dose intravenous melphalan combination therapy has been reported to be a safe and highly effective regimen for patients with relapsed multiple myeloma.[50]

Further effective regimens for the treatment of multiple myeloma tested in phase II studies include bortezomib with thalidomide and dexamethasone,[51–53] bortezomib with ascorbic acid and melphalan,[54] and bortezomib with perifosine and dexamethasone.[55]

The principal biotransformation pathway of bortezomib is oxidative deboronation, most notably to a pair of diastereomeric carbinolamide metabolites. Further, metabolism of the leucine and phenylalanine moieties produces tertiary hydroxylated metabolites and a metabolite that is hydroxylated at the benzylic position, respectively. The carbinolamides may also be converted to the corresponding amide and carboxylic acid. Several P450 isoforms contribute to the metabolism of bortezomib, including CYP3A4, CYP2C19, CYP1A2, CYP2D6, and CYP2C9.[56] The oxidase activity of P450 most probably represents a deboronation mechanism.[57] Confirming this model, in patients with various cancers, the concomitant administration of bortezomib and the CYP3A inhibitor ketoconazole resulted in a mean increase of 35% in bortezomib exposure.[58] No major P450-mediated clinical drug–drug interactions are anticipated for bortezomib or its major metabolites.[59] No impact on the pharmacokinetics, pharmacodynamics, or safety profile of bortezomib was seen upon the coadministration of omeprazole, a CYP2C19 inhibitor, in patients with advanced tumors.[60]

The efflux of bortezomib from cancer cells could be responsible for secondary or acquired resistance to the drug, as *P*-glycoprotein blockade enhanced bortezomib sensitivity in models of leukemia[61] and Ewing's sarcoma.[62] Individual patients can achieve excellent responses to retreatment with bortezomib,[63] but in larger studies, the ability of bortezomib to reinduce response in patients with previously sensitive disease was halved.[64,65] Bortezomib-adapted cells show increased transcription rates, activities, and polypeptide levels of not only the bortezomib-sensitive β5, but also the β2 proteasome subunit; these cells also consistently retained elevated levels of active β1/β5-type proteasome subunits in the presence of therapeutic levels of bortezomib.[66] Mutation and overexpression of the proteasome subunit β5 have been demonstrated to be an important mechanisms of bortezomib resistance.[67–70]

These changes in proteasome expression and subunit composition were accompanied by an increase in proteasomal peptidase activities and could not be inhibited by high doses of bortezomib.[71] Moreover, it was found that patient diet can alter or even inhibit the anticancer effects of bortezomib, as was reported for dietary flavonoids.[72] One recent study suggests that other proteasome inhibitors may act through a noncompetitive mechanism to overcome bortezomib resistance.[73]

Markers for bortezomib response in multiple myeloma patients are largely unknown. However, initial attempts to elucidate the bortezomib response markers in multiple myeloma patients have been published. The plasma levels of two potential efficacy response markers, apolipoprotein C-I and apolipoprotein C-I′, have been shown to be significantly more abundant in nonresponsive patients than in responders at 24 h postdose.[74] Another study has demonstrated that patients who expressed cyclin D1 were more likely to achieve a response. In contrast, patients who expressed p16 and cytoplasmic p53 and displayed particularly intense Bcl-2 staining had a worse response. Patients who expressed fibroblast growth receptor 3 (FGFR3), a marker of poor prognosis, responded equally well and had similar outcomes after bortezomib treatment compared with FGFR3-negative patients.[75]

## III. Bortezomib in Mantle Cell Lymphoma

In a phase II clinical trial that led to the approval of bortezomib as a second-line treatment for mantle cell lymphoma,[30] the response rate was 33% (8% complete response) and the median duration of the response was 9.2 months.[76] In a follow-up study, the median duration of the response was the same, and the median time to progression and median overall survival in responding patients were 12.4 months and 35.4 months, respectively.[77] Another clinical trial of bortezomib in mantle cell lymphoma compared the responses in relapsed and refractory patients.[78] The response rates in relapsed and refractory patients were 50% and 43%, respectively. Progression-free survival was 5.6 months in relapsed patients and 3.8 months in refractory patients. According to another phase II clinical study, the response rate was 46.4% with a median response duration of 10 months.[79]

Bortezomib also has significant activity in mantle cell lymphoma and follicular lymphoma in combination with rituximab[80]: the 2-year progression-free survival rate was 24% in all patients and 60% in responding patients. However, the high incidence of grade 3 neurotoxicity was a limiting factor of the combination of bortezomib and rituximab. Bortezomib and rituximab were also used in combination with bendamustine in patients with indolent and mantle cell lymphoma. Among these patients, 83% achieved an objective response, and the 2-year progression-free survival was 47% in all patients with no grade 3 adverse effects.[81] According to a phase II clinical trial published by the GOELAMS group, bortezomib was proposed as a first-line therapy for elderly mantle cell lymphoma patients in combination with rituximab, doxorubicin, dexamethasone, and chlorambucil.[82]

Elevated NF-κB subunit p65, cyclin-dependent kinase inhibitor p27, and low proteasome subunit alpha type-5 were identified as biomarkers of bortezomib activity in patients with mantle cell lymphoma.[83] However, resistance to bortezomib in mantle cell lymphoma was associated with plasmacytic differentiation.[84]

## IV. Bortezomib in Other Cancers

According to further recent clinical studies (Table I), bortezomib seems to be active as a single agent in phase II clinical trials. These trials include the following:

- a study of 12 patients with cutaneous T-cell lymphoma,[85] with an overall response rate of 67%, including two (17%) complete remissions and six (50%) partial remissions;
- a study of 16 patients with mucosa-associated lymphoid tissue lymphoma,[86] with an overall response rate of 80%, including 7 patients with complete remission (43%), 6 with partial response (37%), and 3 with stable disease;
- a study of 22 patients with B-cell chronic lymphocytic leukemia[87]; although none of the 19 patients who could be evaluated for response achieved complete remission or partial response, signs of biologic activity based on disease site responses (e.g., reduction in lymphocytosis, splenomegaly, and lymphadenopathy) were observed.

TABLE I

CLINICAL TRIALS OF BORTEZOMIB AS A SINGLE-AGENT IN VARIOUS CANCERS

| Cancer | Results |
|---|---|
| Multiple myeloma | FDA-approved[29] |
| Mantle cell lymphoma | FDA-approved[30] |
| Cutaneous T-cell lymphoma | Significant activity in a phase II trial[85] |
| MALT lymphoma | Significant activity in a phase II trial[86] |
| Waldenström's macroglobulinemia | Significant activity in two phase II trials[90,91] |
| B-cell chronic lymphocytic leukemia | Biologic activity in a phase II trial[87] |
| AIDS-related plasmablastic lymphoma | Two case reports[95,96] |
| Children's leukemia | Little activity in a phase I trial[97] |
| Iatinib-refractory chronic myeloid leukemia | Minimal efficacy in a phase I trial[98] |
| Platinum-sensitive ovarian, primary peritoneal cancer | Little activity in a phase II trial[99] |
| Relapsed, refractory, or metastatic colorectal cancer | Inactive in two phase II trials[100,101] |
| Recurrent or metastatic soft tissue sarcomas | Little activity in a phase II trial[102] |
| Metastatic malignant melanoma | Inactive in a phase II trial[103] |
| Neuroendocrine tumors | Inactive in a phase II trial[104] |
| Metastatic breast cancer | Inactive in two phase II trials[105,106] |
| Metastatic gastric/gastroesophageal adenocarcinoma | Inactive in a phase II trial[107] |
| Non-small-cell lung cancer (relapsed/refractory) | Modest activity in two phase II trials[108] |
| Non-small-cell lung cancer (chemotherapy-naive) | Inactive in a phase II trial[109] |
| Small-cell lung cancer | Little activity in a phase II trial[110] |

In Waldenström's macroglobulinemia, clinical evidence from a handful of clinical trials suggests that bortezomib could be used as a potent single-agent therapy.[88] Among ten previously treated patients with Waldenström's macroglobulinemia, six achieved a partial response, which occurred at a median of 1 month.[89] A further clinical study (phase II) of bortezomib in 27 patients with Waldenström's macroglobulinemia[90] showed 7 partial responses (26%) and stabilization of the disease in 19 patients (70%). In a multicenter phase II trial of bortezomib in relapsed or refractory Waldenström's macroglobulinemia, 27 patients were enrolled; the overall response rate was 85%, with 10 and 13 patients achieving minor and major responses, respectively.[91]

The results of combining bortezomib and rituximab in a multicenter phase II trial seem to support the use of this treatment for relapsed or refractory follicular or marginal-zone B-cell lymphoma.[92] Bortezomib with rituximab versus rituximab alone was also tested in a phase III clinical trial, but the improvement in progression-free survival conferred by the combination was smaller than expected. However, the combination of bortezomib plus rituximab might still be a useful therapy for some subgroups of patients.[93] When the combination of bortezomib and rituximab was tested with bendamustine in a phase II clinical trial in 63 patients with relapsed or refractory follicular lymphoma, the overall response rate was 88% (including 53% complete response), and the median duration of response was 11.7 months.[94]

Two recently published case reports describe dramatic and early responses to bortezomib alone and in combination with other drugs in patients with AIDS-related plasmablastic lymphoma.[95,96]

In contrast to these promising trials, bortezomib has failed as a single agent in phase I clinical studies in patients with pediatric refractory leukemia[97] or imatinib-refractory chronic myeloid leukemia.[98]

Bortezomib monotherapy has also shown minimal anticancer activity in phase II clinical trials in patients with solid tumors,[99] as shown in Table I. For example, although bortezomib effectively inhibited the 20S proteasome in almost all the 26 patients with recurrent platinum-sensitive ovarian or primary peritoneal cancer in a phase II clinical trial, only a few partial responses were observed, with an objective response rate of 3.8%.[100] In two phase II clinical studies (19 and 102 patients), bortezomib was not effective against advanced or metastatic and refractory or relapsed colorectal cancer.[101,102] Furthermore, bortezomib has failed in several phase II clinical trials:

- in 18 patients with recurrent or metastatic soft tissue sarcomas[103];
- in 27 patients with metastatic melanoma[104];
- in 16 patients with metastatic neuroendocrine tumors[105];

- in 24 patients (two trials of 12 patients each) with metastatic breast cancer[106,107];
- in 16 patients with unresectable or metastatic gastric and gastroesophageal junction adenocarcinoma[108];
- in 55 patients with relapsed or refractory non-small-cell lung cancer[109];
- in 14 chemotherapy-naive patients with non-small-cell lung cancer[110];
- in 155 patients with relapsed or refractory small-cell lung cancer.[111]

The crucial question regarding the failure of bortezomib in many solid tumors may be whether the proteasome plays the same role in the clinical presentation of these tumors as it plays in multiple myeloma patients, in whom increased serum proteasome concentrations are correlated with advanced disease.[112] Furthermore, GRP-78, a chaperone protein involved in the unfolded protein response, seems to be responsible for the bortezomib resistance of cell lines derived from solid tumors rather than multiple myeloma cell lines.[113]

However, combination therapy including bortezomib was reported to be successful in some solid tumors. In phase I studies performed in patients with solid tumors, bortezomib was well tolerated in combination regimens with docetaxel,[114] irinotecan,[115] paclitaxel and carboplatin,[116] and gemcitabine and cisplatin.[117] According to further studies, the combination of bortezomib with gemcitabine/carboplatin[118,119] resulted in a notable survival benefit in patients with advanced non-small-cell lung cancer in phase I and phase II clinical trials (114 patients were involved in the phase II trial, with a response rate of 23% and a median overall survival of 11 months). In two phase I clinical trials, bortezomib and carboplatin were active in primary or resistant ovarian cancer.[120,121] Larger clinical trials may show that bortezomib in combination with other treatments is suitable as a chemotherapeutic for various solid tumors.

According to *in vitro* data, bortezomib can enhance the delivery of cisplatin to intraperitoneal ovarian carcinomas by downregulating human copper transporter 1.[122] In addition (and conversely), other chemotherapeutic agents may even target proposed mechanisms of the primary resistance to bortezomib, for example:

- blocking heat-shock protein-27 restores the apoptotic response to bortezomib in a lymphoma cell line[123];
- the inhibition of eIF2$\alpha$ (eukaryotic translation initiation factor 2$\alpha$) dephosphorylation maximizes bortezomib toxicity in quiescent myeloma cells[124];
- cell lines derived from solid tumors may be sensitized to bortezomib by interferon-$\gamma$-mediated induction of proteasome subunit expression.[125]

## V. Bortezomib in Diseases Different from Cancer

In addition to chemotherapy, there is a growing list of potential applications for proteasome inhibitors in medicine. It is very important to consider the possibility of repurposing existing drugs for new uses and to offer future medicines for debilitating diseases that are refractory to currently available drugs.[126,127]

Bortezomib is active in a lethal murine model of myelofibrosis, the most serious myeloproliferative disorder, and it impairs both myelofibrosis and the osteosclerosis that is induced by high thrombopoietin levels. Bortezomib also dramatically improved the survival of mice with high levels of thrombopoietin (89% vs. 8% survival at week 52).[128]

Another potential application of bortezomib is related to the fact that the proteasome is required for the essential immune functions of activated $CD4^+$ T cells and may thus serve as a molecular target for the suppression of deregulated and unwanted T-cell-mediated immune responses.[129] At clinically achievable concentrations, bortezomib induces selective apoptosis and decreases T helper cell 1 response among alloreactive T lymphocytes but barely affects unstimulated T cells.[130] Furthermore, naturally occurring $CD4^+$/$CD25^+$ regulatory T cells are resistant to the proapoptotic effect of bortezomib. Long-term culture of $CD4^+$ T cells in the presence of bortezomib promotes the emergence of a regulatory T-cell population that significantly inhibits proliferation, IFN-$\gamma$ production, and CD40L expression among stimulated effector T cells.[131] These results establish a basis for the clinical use of bortezomib in the management of graft-versus-host disease. Indeed, according to a study focused on transplant rejection in kidney transplant patients, bortezomib provided a prompt reversal of rejection, marked and prolonged reductions in donor-specific antihuman leukocyte antigen antibody levels, improved renal allograft function, and suppressed recurrent rejection for at least 5 months.[132,133] The ability of bortezomib to prevent transplant rejection was confirmed by another study in a series of renal allograft recipients.[134] However, although this limited experience may seem promising for transplant recipients, other studies may suggest otherwise.[135]

Bortezomib eliminated both short- and long-lived plasma cells by activating the terminal unfolded protein response, and it prolonged survival in two mouse strains with lupus-like disease.[136] Furthermore, myasthenia gravis, a disabling autoimmune disease caused by autoantibodies against the acetylcholine receptor,[137] seems to be sensitive to bortezomib. In an experimental model of autoimmune myasthenia gravis induced in Lewis rats, bortezomib reduced the increase of auto-anti-acetylcholine receptor antibodies, prevented damage to the postsynaptic membrane, improved neuromuscular transition, and reduced symptoms of myasthenia.[138]

The proteasome is also a promising target of anti-HIV therapy,[139–141] and proteasome inhibition could explain the success[142–144] and failure[145,146] of ditiocarb therapy in patients with AIDS 20 years ago.[147]

Because the α′,β′-epoxyketone epoxomicin peptide is active against Plasmodium falciparum regardless of the parasite's grade of chloroquine susceptibility, proteasome inhibition is a highly promising strategy for new antimalarials.[148] Furthermore, the selective targeting of the Mycobacterium tuberculosis proteasome could provide an alternative method for treating drug-resistant clones of this pathogen.[149,150]

## VI. The Adverse Effects of Bortezomib

In comparison to traditional cancer chemotherapy, bortezomib treatment has moderate adverse side effects. During phase I studies (mentioned above), dose-limiting toxicities included sensory neuropathy, diarrhea, fatigue, hypotension, fluid retention, hypokalemia, hyponatremia, malaise, nausea, orthostasis, and thrombocytopenia. In a phase II trial of multiple myeloma patients, adverse events were reported in at least 10% of patients and included anemia, anorexia, constipation, dehydration, diarrhea, dizziness, fatigue, headache, limb pain, nausea, neutropenia, peripheral neuropathy, pyrexia, rash, thrombocytopenia, vomiting, and weakness. Subsequent studies have better characterized the thrombocytopenia[151] and neuropathy,[152] which limit dosing in the clinic.

Proteasome inhibitors can impair protein synthesis and induce cell death in primary rat neurons *in vitro*. According to one study, proteasome inhibition impaired neuronal protein synthesis at concentrations similar to the concentrations necessary to induce subsequent neuron death. The inhibition of protein synthesis was reversible during the first 6 h of treatment, and the neurotoxicity of proteasome inhibition was reversible during the first 12 h of treatment.[153]

In rats, the sciatic nerve examination and morphometric determinations after bortezomib administration demonstrated mild to moderate pathologic changes involving Schwann cells, myelin, and axonal degeneration. Bortezomib-induced changes were also observed in dorsal root ganglia neurons, represented by satellite cell intracytoplasmic vacuolization due to mitochondrial and endoplasmic reticulum damage.[154] Another study performed in rats identified dorsal root ganglia neurons as an important target of bortezomib toxicity. In these cells, the accumulation of ubiquitinated proteins, reduction of transcriptional activity, and nuclear retention of poly(A) RNAs in numerous spherical or ring-shaped dense granules were observed. At the cytoplasmic level, most neurons exhibited chromatolysis, supporting the inhibition of mRNA translation.[155]

Results from a study applying quantitative sensory testing in patients receiving bortezomib for the treatment of either multiple myeloma or non-Hodgkin B-cell lymphoma demonstrated that bortezomib-induced pain is associated with deficits in all three major fiber types: Aβ-, Aδ-, and C-caliber primary afferent fibers.[156] Peripheral nerve damage is one of the most significant toxicities of bortezomib and often requires dose modification. Grade 1 or 2 bortezomib-associated peripheral neuropathy (BAPN) occurs in up to 75% and 33% of patients receiving bortezomib therapy for recurrent or newly diagnosed disease, respectively, whereas grades 3 and 4 neurotoxicity may affect up to 30% and 18% of patients with recurrent disease and newly diagnosed disease, respectively.[157] The risk of BAPN seems to be greater in patients who have peripheral neuropathy or diabetes mellitus at baseline.[158] The incidence, severity, and outcome of BAPN are similar in untreated and pretreated multiple myeloma patients, except for neuropathic pain, which has a lower incidence and shorter duration in untreated patients. Age emerges as the most relevant risk factor for peripheral neuropathy.[159] However, BAPN is manageable and reversible in most patients with relapsed myeloma,[160] and dosage modification using a specific guideline improves peripheral neuropathy management without adversely affecting outcome.[161,162]

Approximately 1 year before bortezomib received official approval in Japan and appeared on the open market, several Japanese patients who used imported bortezomib experienced fatal pulmonary complications.[163] In August and November 2006, American hospitals reported that several African Americans had the same severe pulmonary complications with bortezomib.[164,165] One report showed that the risk of pulmonary complications decreased when bortezomib was given in combination with a steroid.[166] As the frequency and lethality of pulmonary complications decreased in a postmarketing clinical trial, several explanations were proposed[167]: First, combined administration with corticosteroids became popular. Fifty-four percent of patients who received imported bortezomib also received a steroid in combination. This frequency increased to 70% in the postmarketing clinical trial. The steroid has an antitumor effect on multiple myeloma and is used in combination with many other types of treatments. Gotoh *et al.*[166] showed that the risk of pulmonary complications decreased when bortezomib and steroid therapy were administered in combination. As a result of this study, some doctors likely began to use bortezomib in combination with a steroid to prevent pulmonary complications. Second, as reports of pulmonary complications with bortezomib increased, some doctors began to choose patients as candidates for this therapy more carefully. During the clinical trial, Miyakoshi *et al.*[163] found that pulmonary complications occurred in succession with patients in poor general condition who were administered an unapproved drug. After bortezomib became available for sale, Janssen set a proper usage criterion that included lung disease history. This probably contributed widely to the decrease in adverse events in postmarketing clinical trials.

In a phase II clinical study, bortezomib reduced the populations of $CD8^+$ cytotoxic T cells, $CD56^+$ natural killer cells, and peripheral blood lymphocytes, and it was clinically associated with high rates of viral reactivation of varicella zoster.[168] Moreover, varicella zoster virus reactivation was also observed in bortezomib-treated patients with relapsed multiple myeloma in a phase III clinical study.[169]

Recently, a case report of a 50-year-old man with multiple myeloma who presented with severe renal failure and developed a colon mucositis following bortezomib has appeared in the literature. In addition to potential autonomic neuropathy,[170] the alternation of normal microflora[171,172] and the suppression of the humoral immune system,[173] mucositis may also be involved in bortezomib-induced watery diarrhea.[174]

In a 59-year-old man receiving bortezomib as treatment for multiple myeloma, the sudden onset of Sweet syndrome, with neutrophilic infiltration of the dermis, leukocytoclasia, and subepidermal edema, was observed.[175] Another case report of a similar syndrome in a 69-year-old patient with multiple myeloma may represent histiocytoid Sweet syndrome, with mononuclear infiltrate that probably indicates immature neutrophils.[176]

When mice were injected with bortezomib, the numbers of T and B lymphocyte precursors decreased dramatically, while mature lymphocytes were only partially affected. Thymocytes were almost depleted 3 days after a single bortezomib injection, and pro-B and pre-B cells were depleted after only 2 days. Thymocytes and B-cell precursors recovered within 2 weeks.[177]

In multiple myeloma patients, bortezomib can cause conformational modifications in structurally sensitive proteins, such as hepatic serpins. However, while these modifications may not be directly associated with the adverse effects of bortezomib, the combination of bortezomib with other drugs (e.g., thalidomide) might exacerbate these conformational effects, with potential pathological consequences.[178]

In pigs, chronic proteasome inhibition is associated with increased coronary artery oxidative stress and early atherosclerosis,[179] but it improved renal endothelial functional responses in experimental hypercholesterolemia.[180]

Recently, one surprising study has shown that in rats, bortezomib targets serine proteases in cells other than those from the proteasome, especially serine cathepsin G in splenocytes. Thus, the adverse effects of bortezomib may not be caused by proteasomal inhibition; consequently, other proteasome inhibitors currently in clinical development could be safer than bortezomib.[181]

## VII. Second-Generation of 20S Proteasome Inhibitors

The limited efficacy of bortezomib, its adverse effects, and the prevalence of acquired resistance to this drug are major reasons to develop new proteasome inhibitors for clinical use.[182] There are already ongoing clinical trials of

TABLE II

Ongoing Clinical Trials of 20S Proteasome Inhibitors (Except for Bortezomib) as Single Agents

| Inhibitor | Cancer | Identifier |
|---|---|---|
| MLN9708 | Multiple myeloma | NCT00932698 |
| | Nonhematologic malignancies | NCT00830869 |
| | Lymphoma | NCT00893464 |
| Marizomib | Solid malignancies, lymphoma | NCT00396864 |
| | Multiple myeloma | NCT00461045 |
| | Advanced malignancies | NCT00629473 |
| Carfilzomib | Solid tumors, multiple myeloma | NCT00531284 |
| | Multiple myeloma | NCT00721734 |
| CEP 18770 | Multiple myeloma | NCT01023880 |

some of the next-generation 20S proteasome inhibitors, such as MLN9708, marizomib (NPI-0052, salinosporamide A), carfilzomib (PR-171), and CEP 18770 (Table II). Furthermore, the 20S proteasome is currently not the only target within the ubiquitin–proteasome system; one disruptor of E3 ligase activity (MLN4924) is being tested in phase I clinical trials against a variety of malignancies.

Traditionally, the fast-growing family of 20S proteasome inhibitors[183,184] has been comprised of both reversible and irreversible inhibitors that are active against various cancers at least *in vitro* but often also *in vivo*. The mechanism of proteasome inhibition has been well described for these traditional proteasomal inhibitors. The reversible inhibitors include peptide aldehydes[185] and peptide boronates[186] such as CEP 18770,[187] BU-32,[188] and MLN9708.[189] Irreversible inhibitors include peptide epoxyketones, such as carfilzomib,[190] syrbactins, such as syringolin A,[191–194] and β-lactones, such as marizomib.[195–197]

Other 20S proteasome inhibitors with no fully elucidated mechanism of reaction with the 20S proteasome have been reported in many recent studies. These molecules are toxic to many cancer cells *in vitro*, but these promising results should be confronted with skepticism because of the history of bortezomib. Bortezomib is also toxic to many cell lines derived from human solid tumors, but it has failed in many phase II clinical trials in patients with solid tumors (Table I). A list of naturally occurring compounds and synthetic compounds that have been described as proteasome inhibitors (including compounds that do not inhibit 20S proteasome) are shown in Table III. As different 20S proteasome inhibitors presumably have different activities toward the 20S proteasome active sites, and the relative contributions of these sites depend on the protein substrate, these inhibitors might have dissimilar impacts on protein breakdown in various cancer cells.[198,199]

TABLE III

Natural and Synthetic Compounds Reported to be Proteasome Inhibitors

| Name | Structure | Reference |
|---|---|---|
| Marizomib | 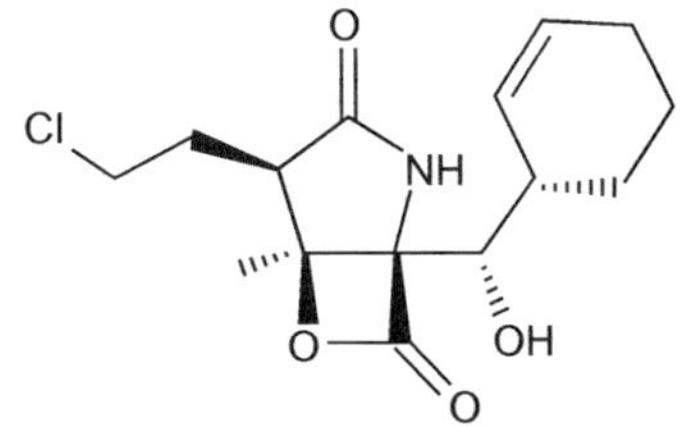 | 218 |
| Bortezomib | 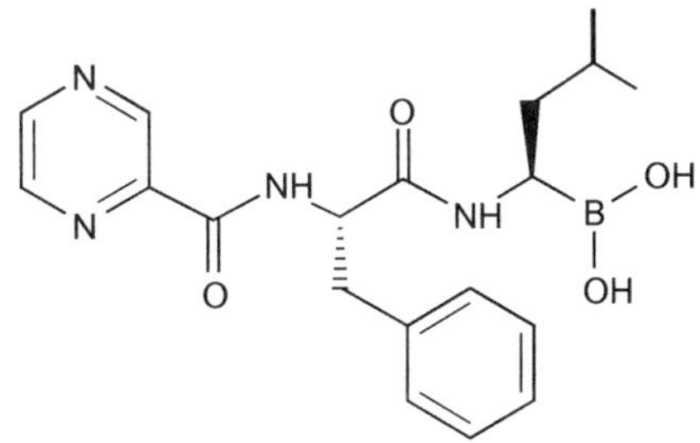 | |
| Carfilzomib | 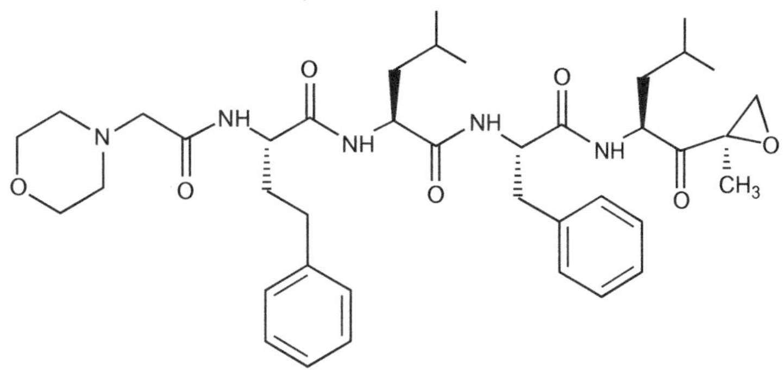 | |

(Continues)

TABLE III (*Continued*)

| Name | Structure | Reference |
| --- | --- | --- |
| CEP-18770 | 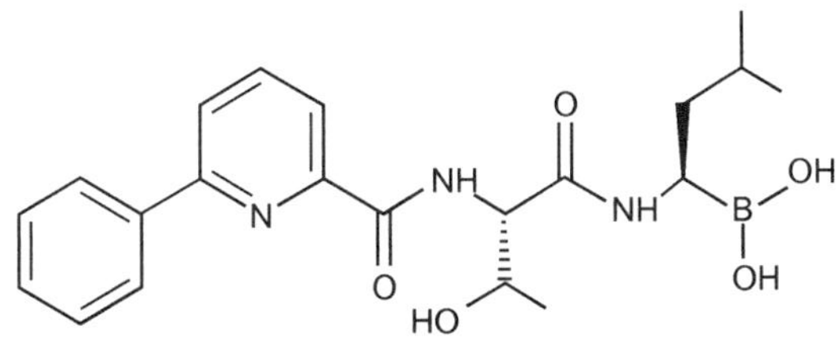 | |
| MLN9708 | 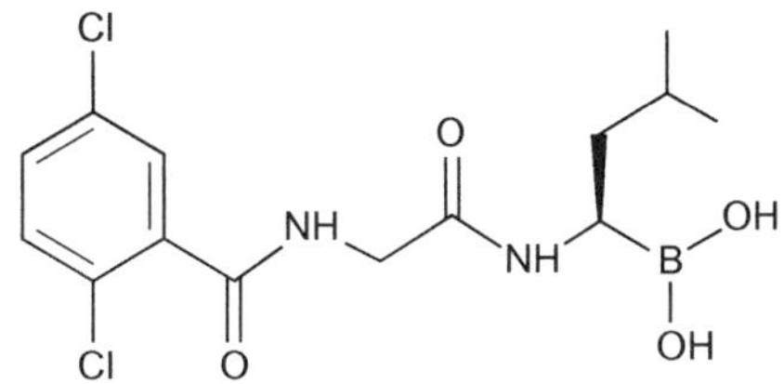 | |
| ONX-0912 | 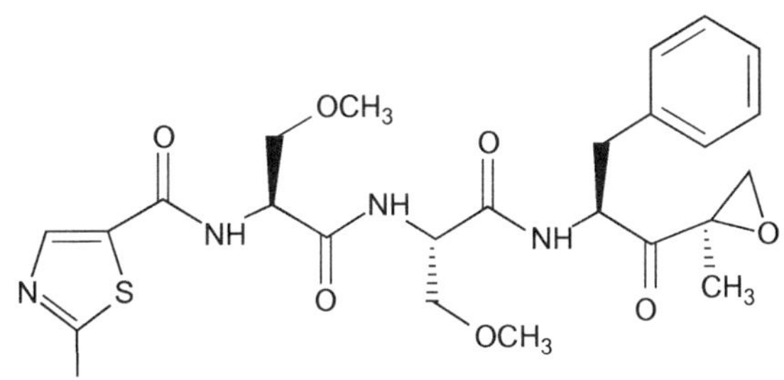 | |

| | | |
|---|---|---|
| PS-519 | 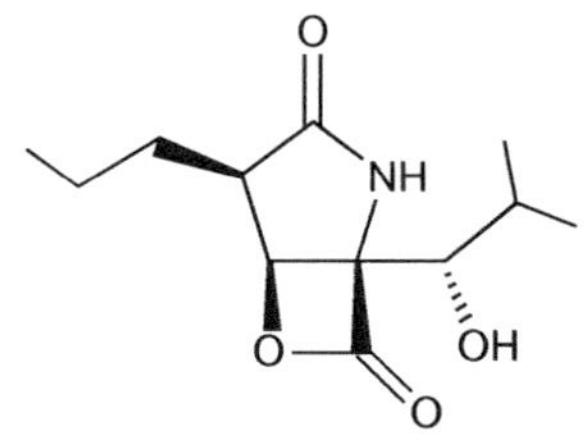 | |
| 5,6,3′,4′-tetrahydroxy-7-methoxyflavone | 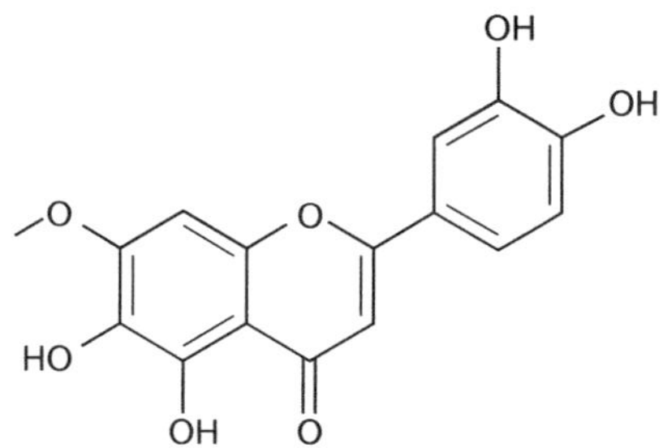 | 205 |
| 5,6,4′-trihydroxy-7,3′-dimethoxyflavone | 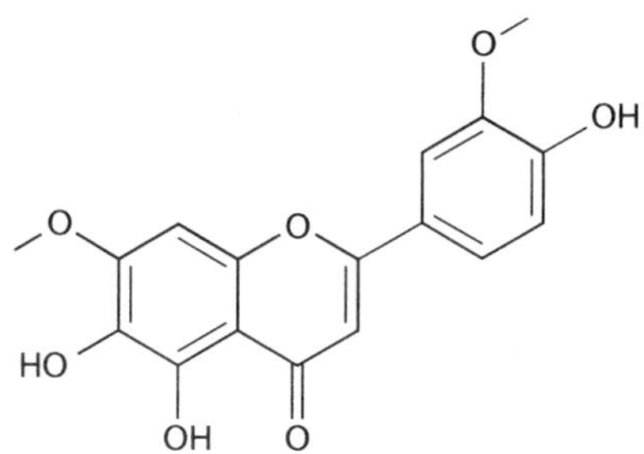 | |

(*Continues*)

TABLE III (*Continued*)

| Name | Structure | Reference |
|---|---|---|
| 6,7,4′-trihydroxyisoflavone | 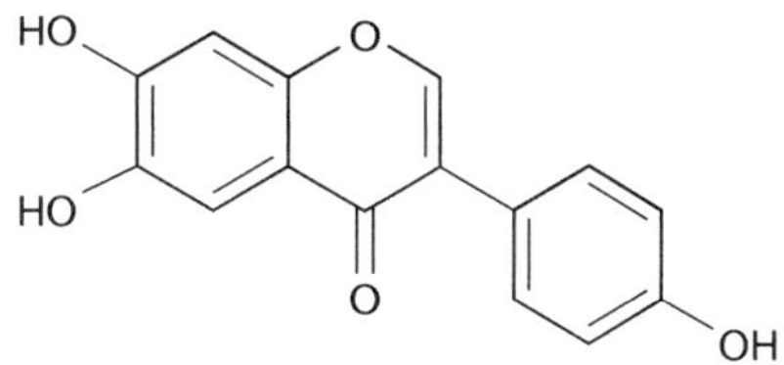 | |
| Apigenin-6-hydroxy | 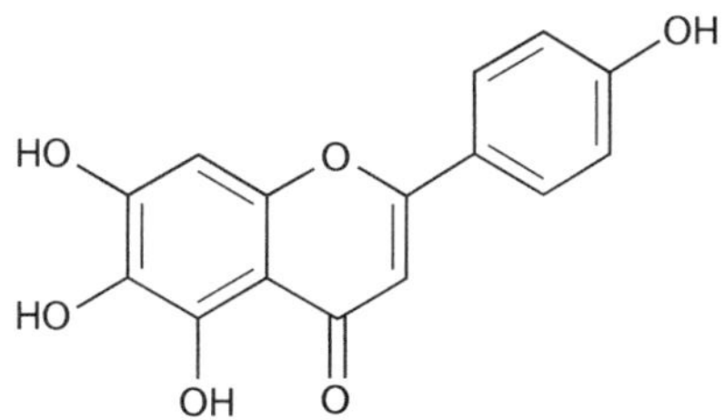 | |

| | |
|---|---|
| Cyanidin | 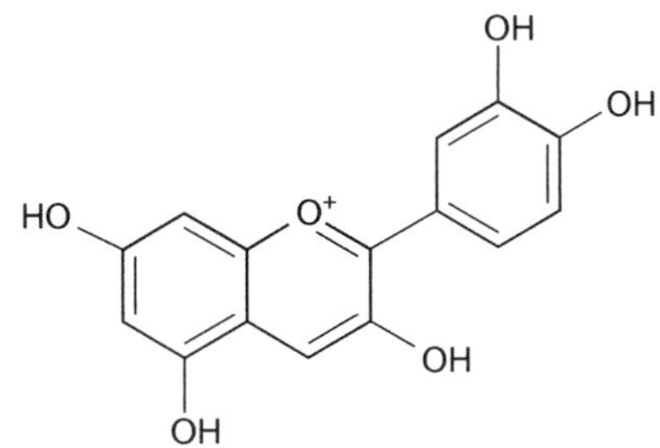 |
| Cyanin | 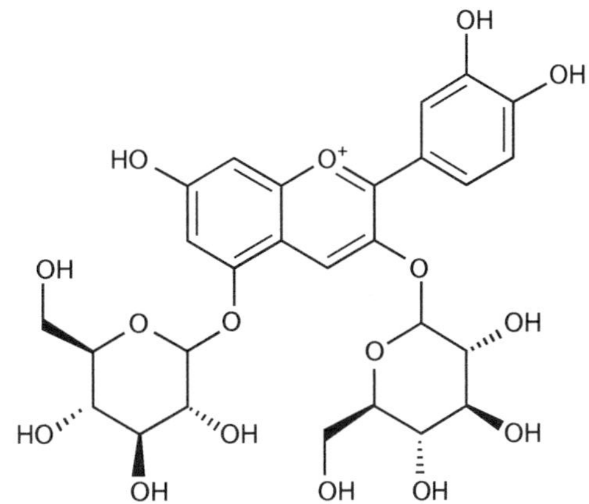 |

(*Continues*)

TABLE III (*Continued*)

| Name | Structure | Reference |
|---|---|---|
| Delphinidin | 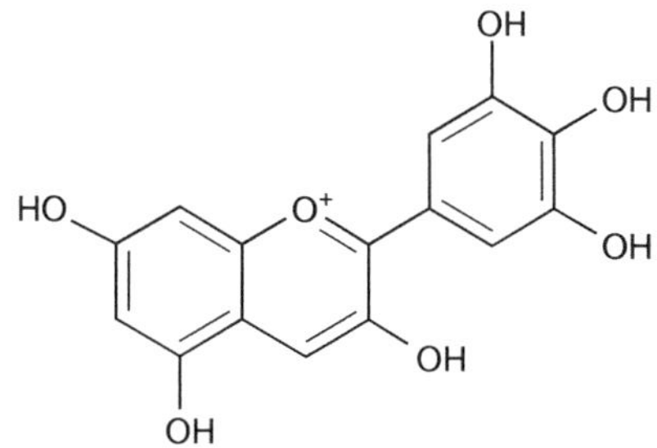 | |
| Glycitein | 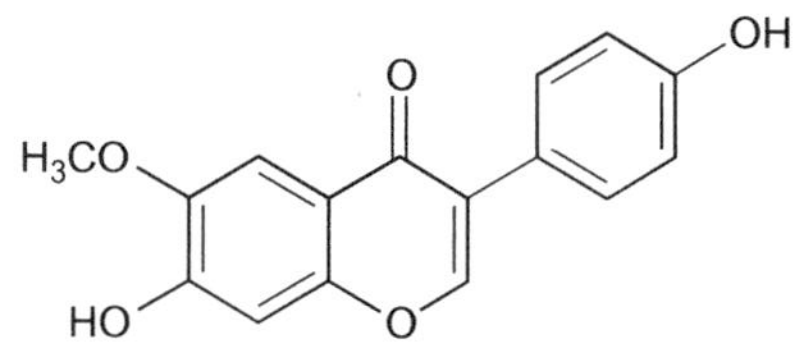 | |
| Ideain | 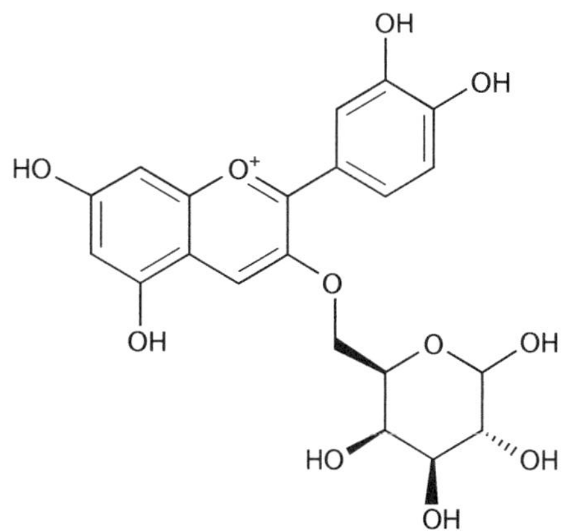 | |

| | |
|---|---|
| Kaempferidinidin | 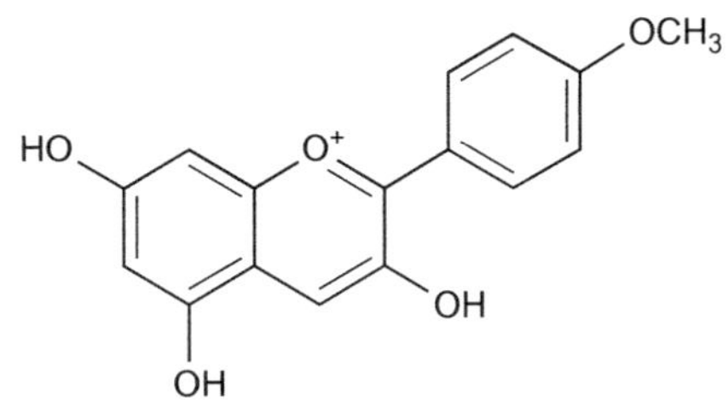 |
| Malvidin | 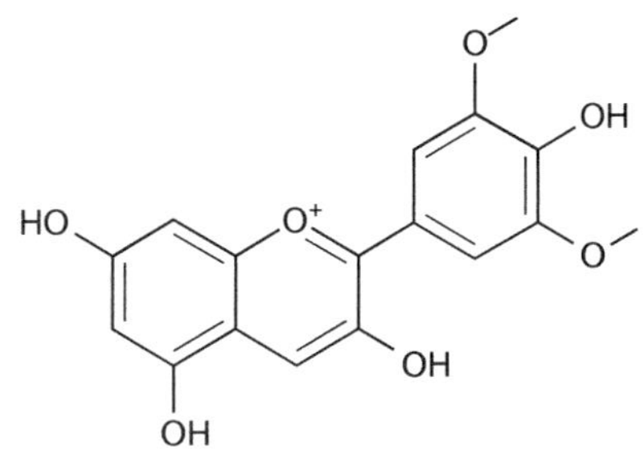 |

(Continues)

TABLE III (*Continued*)

| Name | Structure | Reference |
|---|---|---|
| Malvidin-3-galactoside | 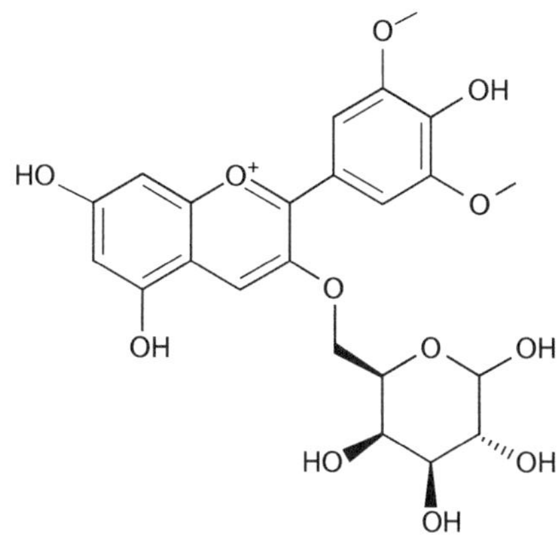 | |
| Malvin | 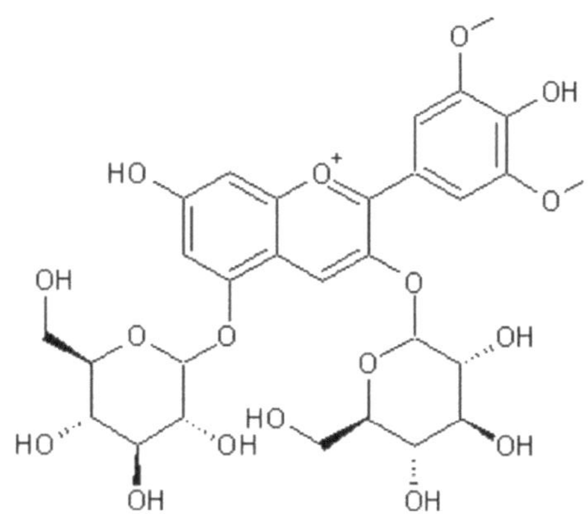 | |

| | |
|---|---|
| Myrtillin | 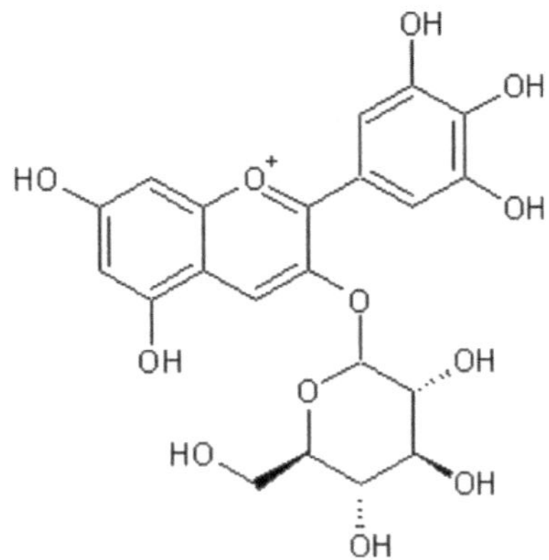 |
| Oenin | 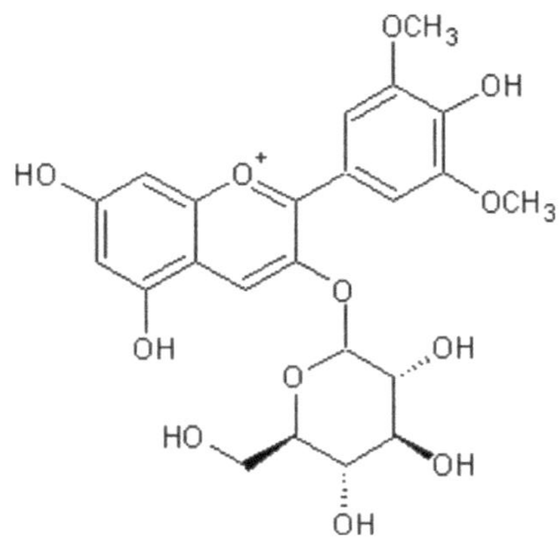 |

(*Continues*)

TABLE III (*Continued*)

| Name | Structure | Reference |
|---|---|---|
| Pelargonidin | 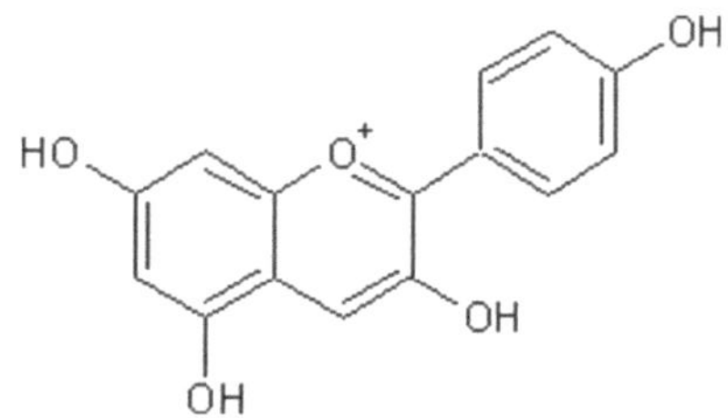 | |
| Pelargonin | 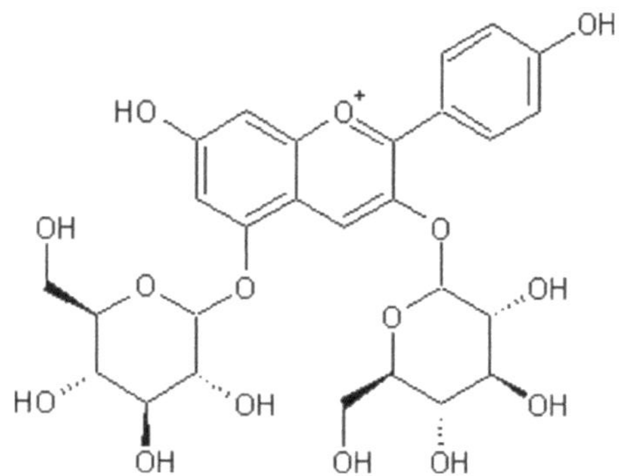 | |
| Peonidin | 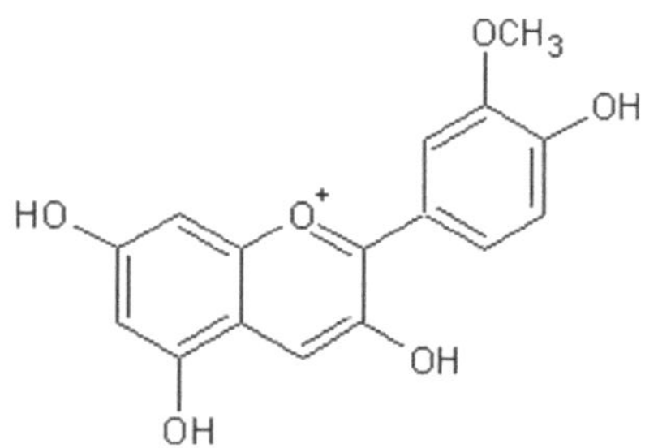 | |

| | |
|---|---|
| Peonidin-3-glucoside | 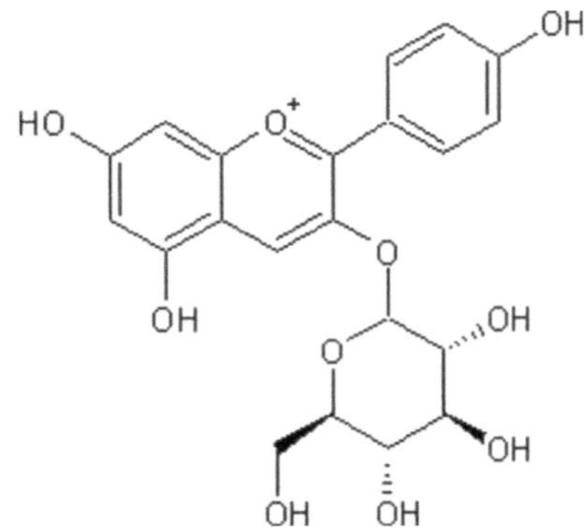 |
| Petunidin | 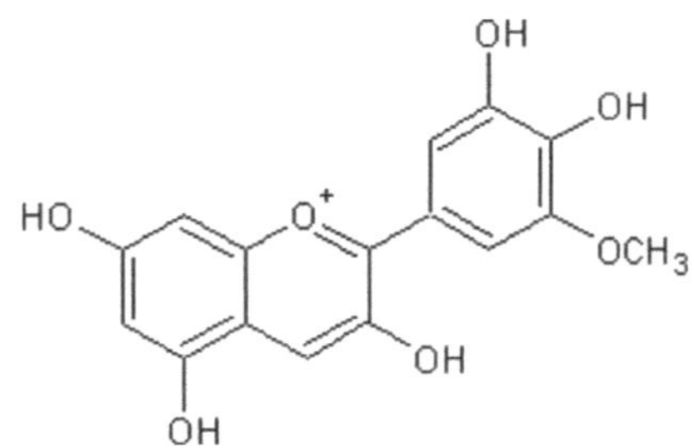 |





*(Continues)*

TABLE III (*Continued*)

| Name | Structure | Reference |
|---|---|---|
| Rutin | 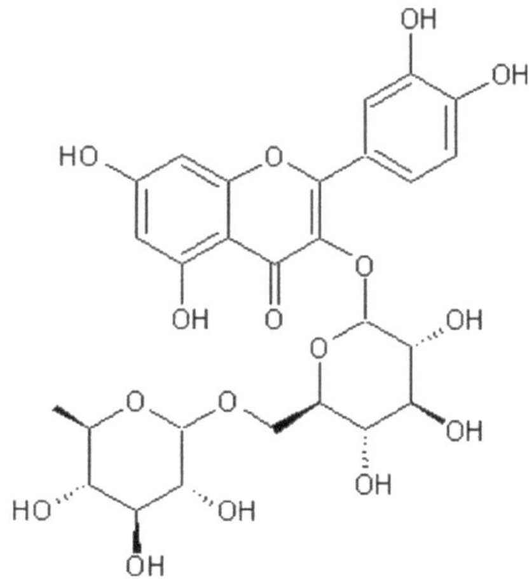 | |
| Sumatranoside | 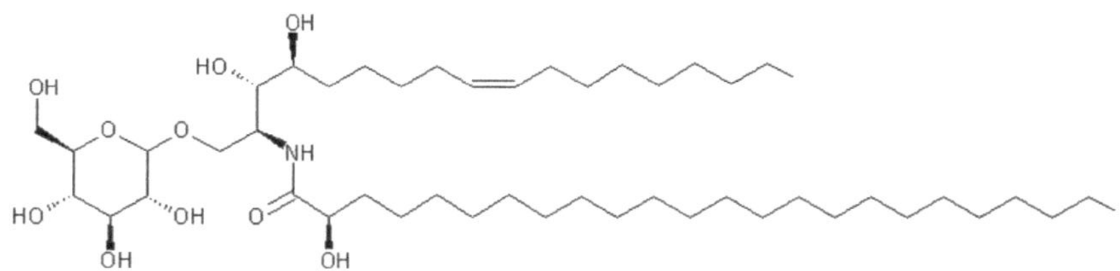 | |

| | |
|---|---|
| Witharfarin | 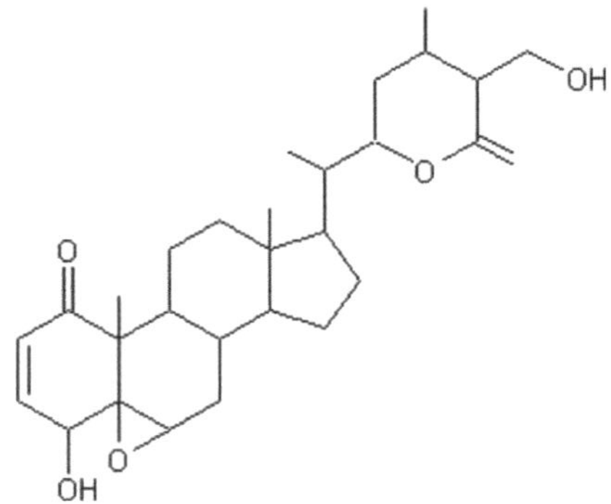 |
| Apigenin | 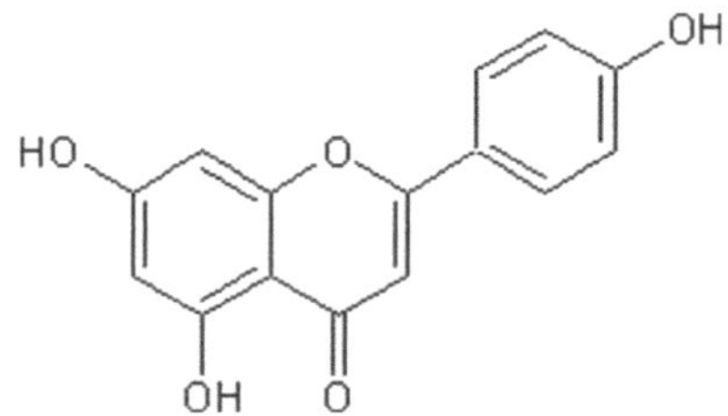 |
| Capsaicin | 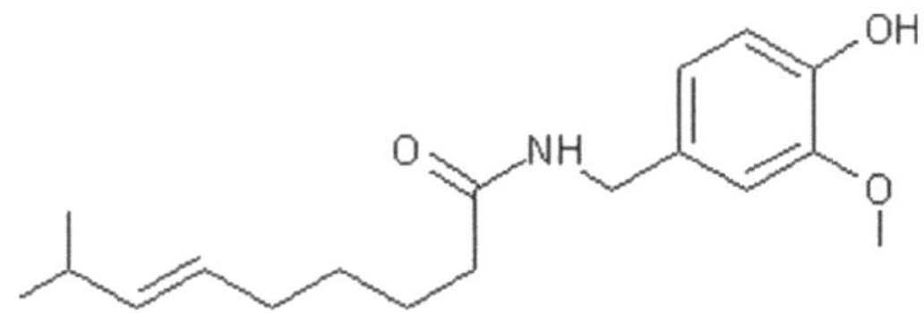 |

(Continues)

TABLE III (*Continued*)

| Name | Structure | Reference |
| --- | --- | --- |
| Celastrol | 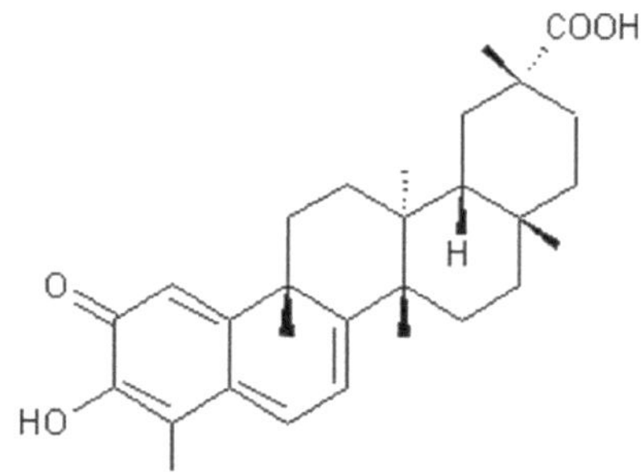 | |
| Chlorogenic_acid | 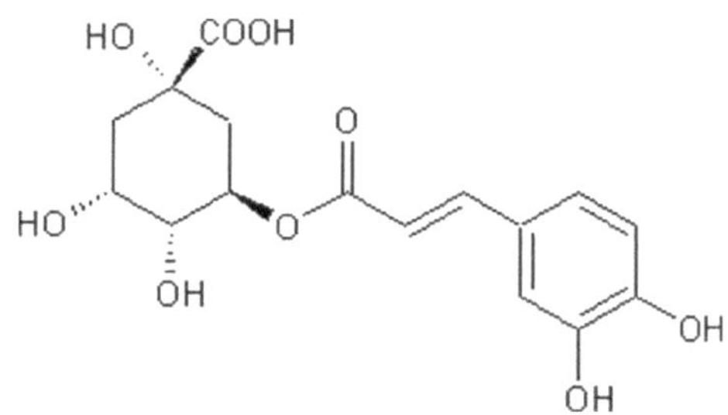 | |
| Chrysin | 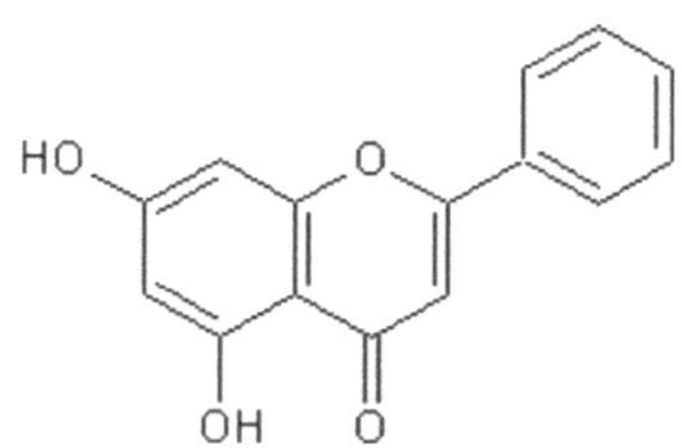 | |

Curcumin

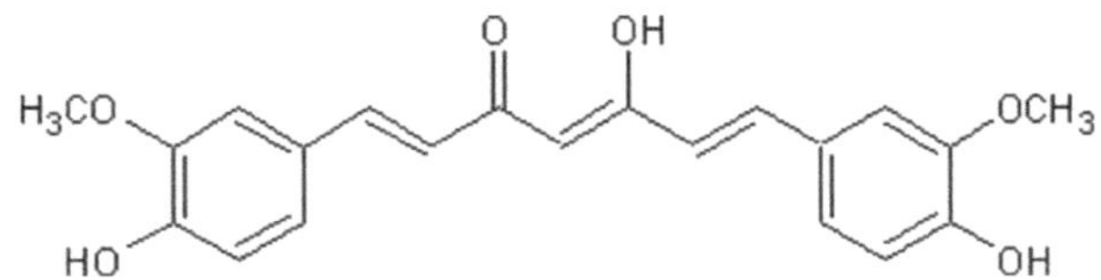


Fisetin

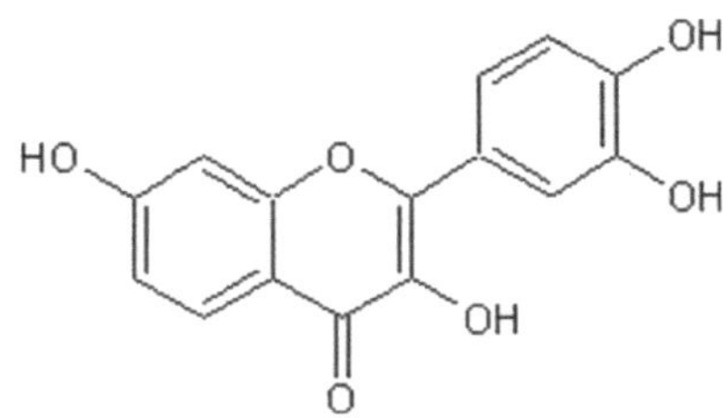


Genistein

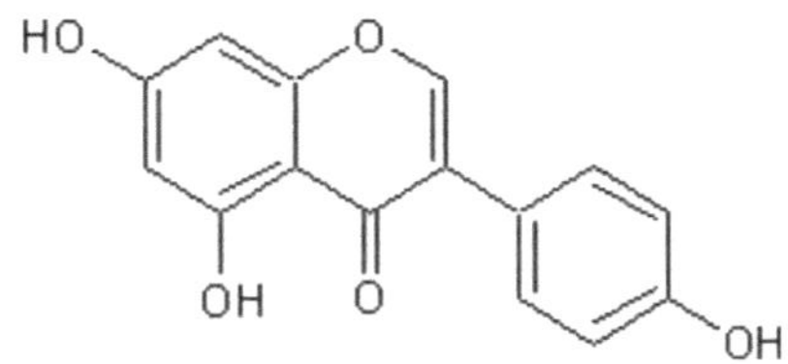


Kaempferol

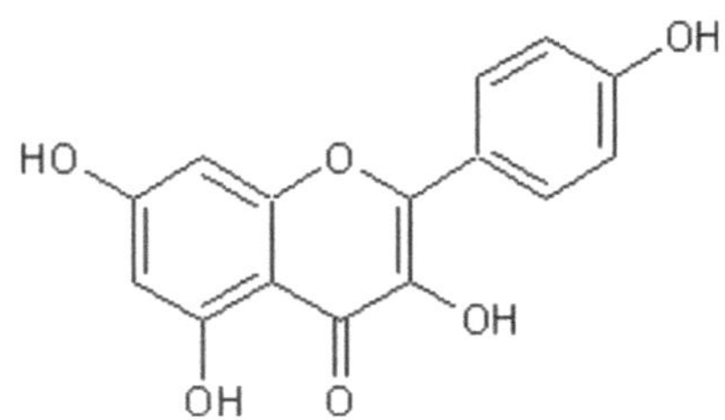


*(Continues)*

TABLE III (*Continued*)

| Name | Structure | Reference |
|---|---|---|
| Luteolin | 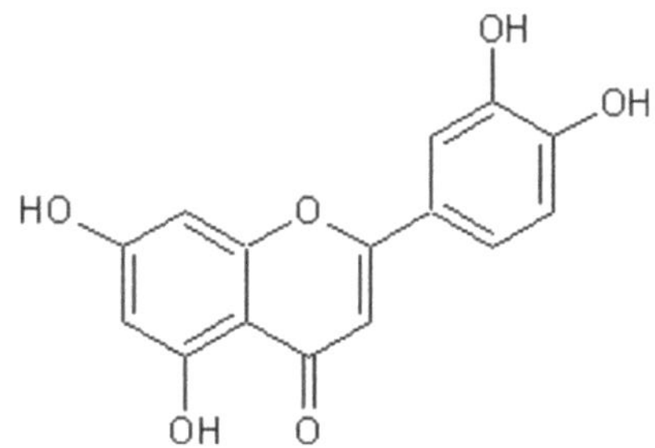 | |
| Myricetin | 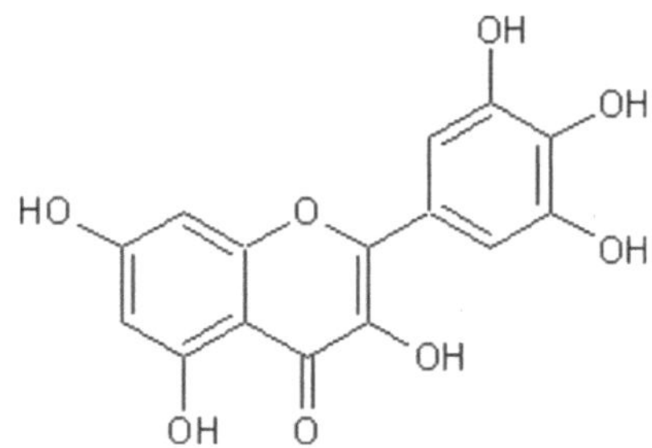 | |
| Quercetin | 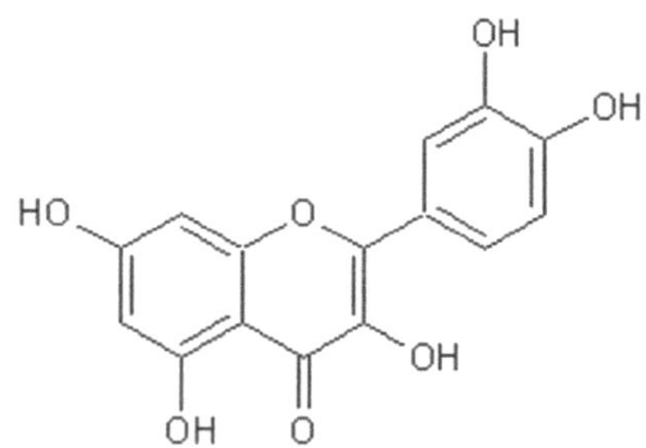 | |

| | |
|---|---|
| Resveratrol | 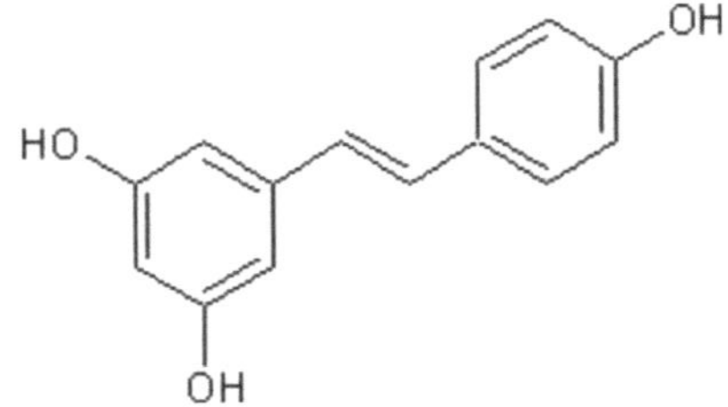 |
| Silibinin | 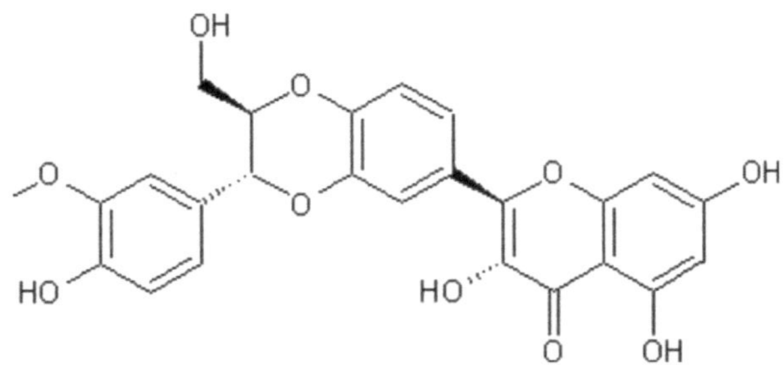 |

(*Continues*)

TABLE III (*Continued*)

| Name | Structure | Reference |
|---|---|---|
| Tannic acid | 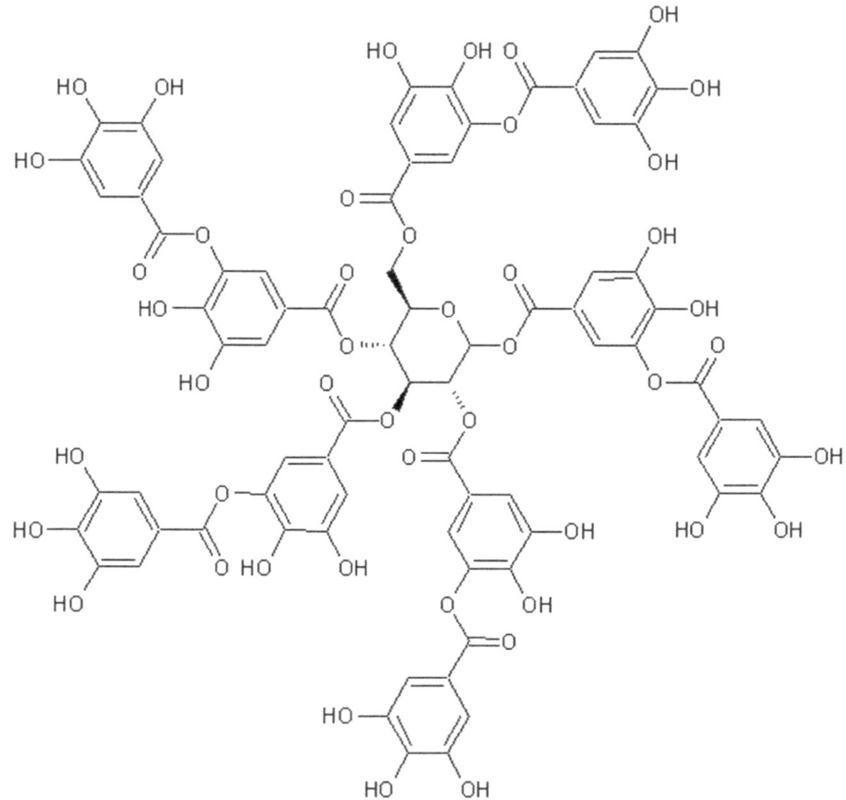 | |

| | |
|---|---|
| Antiprotealide | 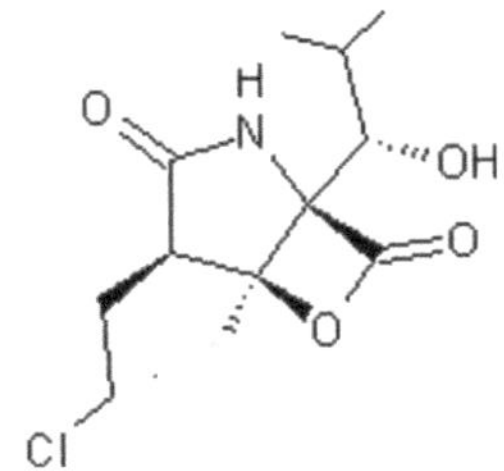 |
| Salinosporamide A | 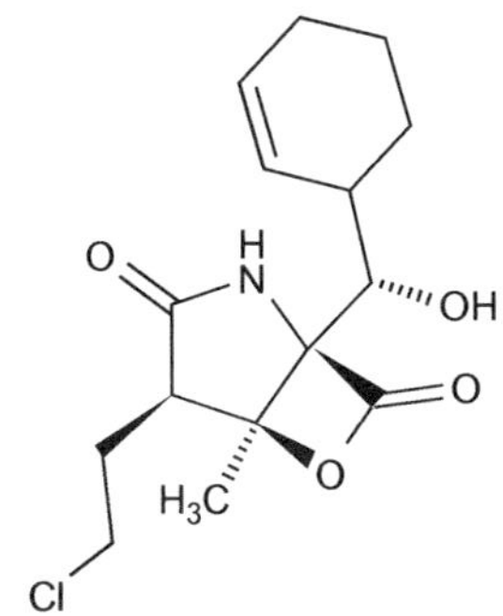 |
| Omuralide | 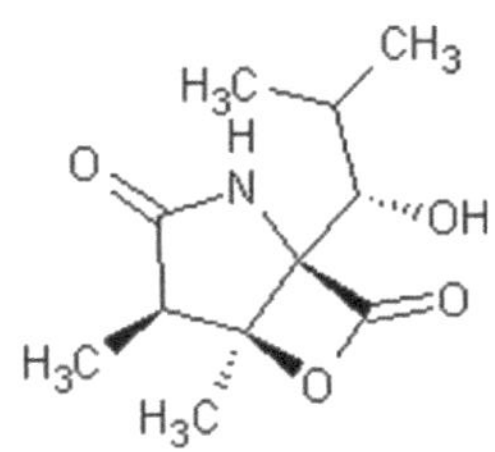 |

(Continues)

TABLE III (*Continued*)

| Name | Structure | Reference |
|---|---|---|
| Au compound 1 | 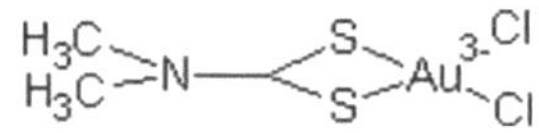 | 201 |
| Au compound 2 | 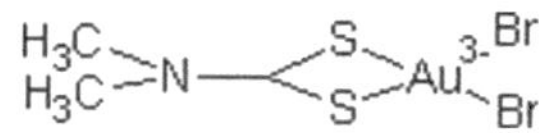 | |
| Au compound 3 | 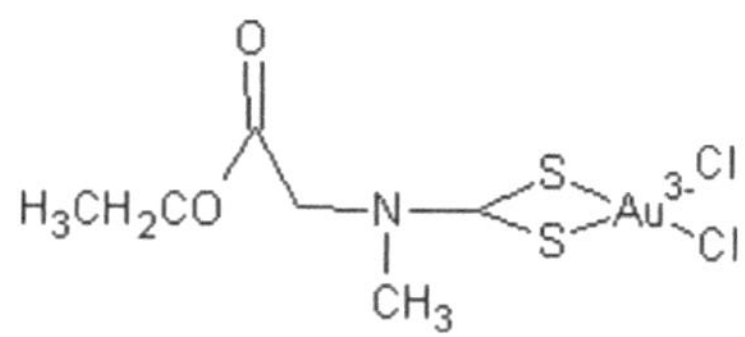 | |
| Au compound 4 | 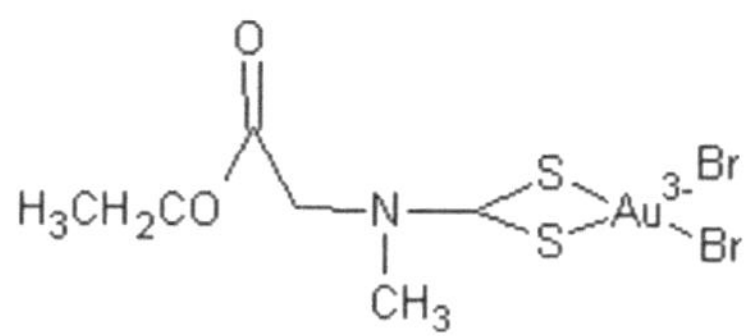 | |
| EGCG | 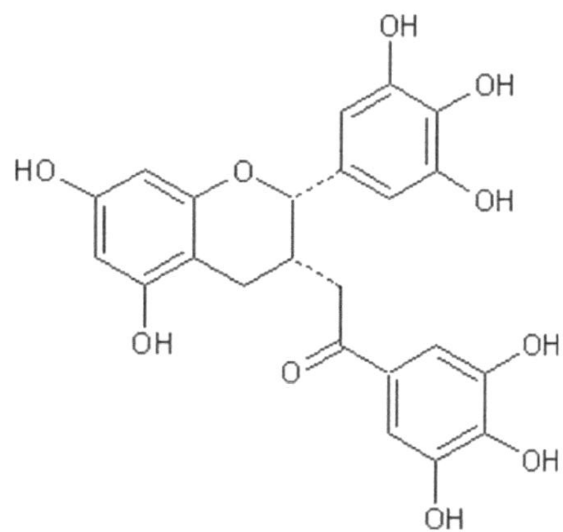 | 204 |

| | | |
|---|---|---|
| Galium complexes | 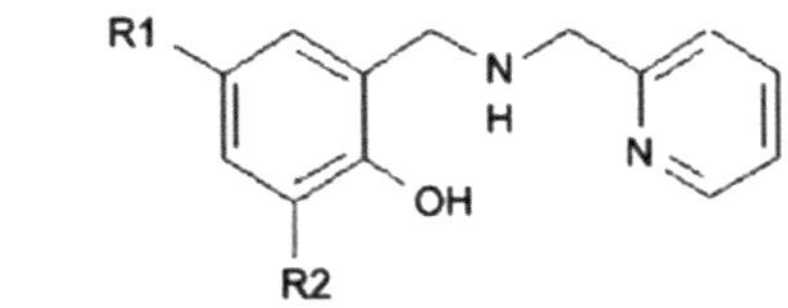<br>***1:*** $[Ga^{III}(L^{OCH3})_2]ClO_4$; (R1 = H, R2 = methoxy)<br>***2:*** $[Ga^{III}(L^{NO2})_2]ClO_4$; (R1 = nitro, R2 = H)<br>***3:*** $[Ga^{III}(L^{Cl2})_2]ClO_4$; (R1 and R2 = chloro)<br>***4:*** $[Ga^{III}(L^{Br2})_2]ClO_4$; (R1 and R2 = bromo)<br>***5:*** $[Ga^{III}(L^{I2})_2]ClO_4$; (R1 and R2 = iodo) | 203 |
| Lactacystin | 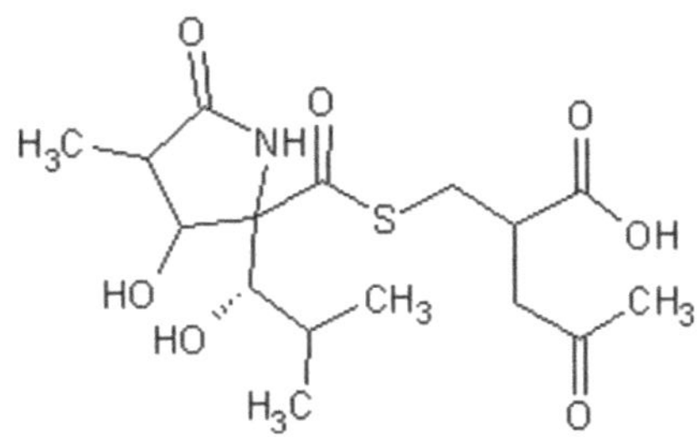 | 223 |
| PI-083 | 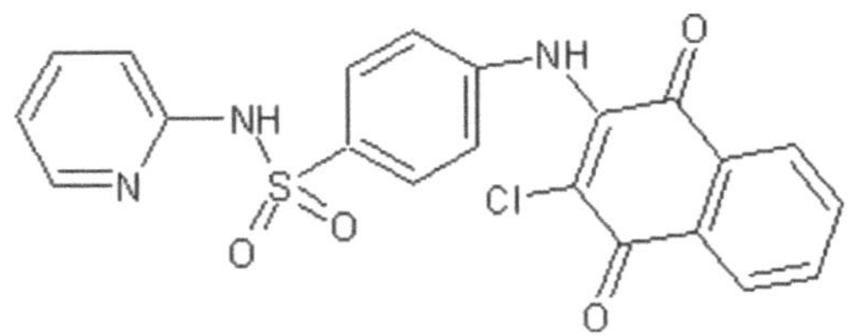 | 213 |

(*Continues*)

TABLE III (*Continued*)

| Name | Structure | Reference |
| --- | --- | --- |
| Pristimerin | 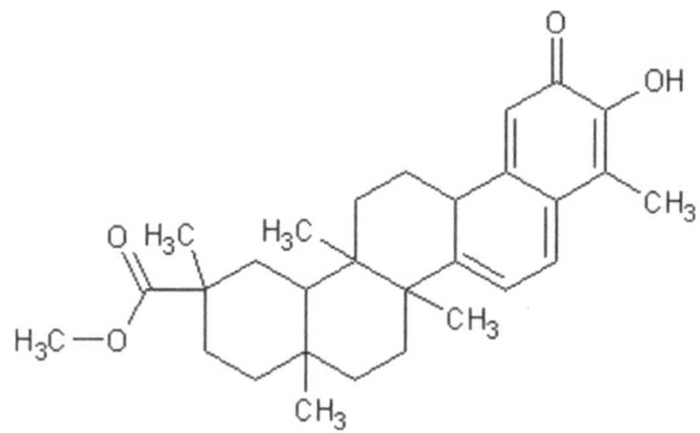 | 208 |
| Syringolin A | 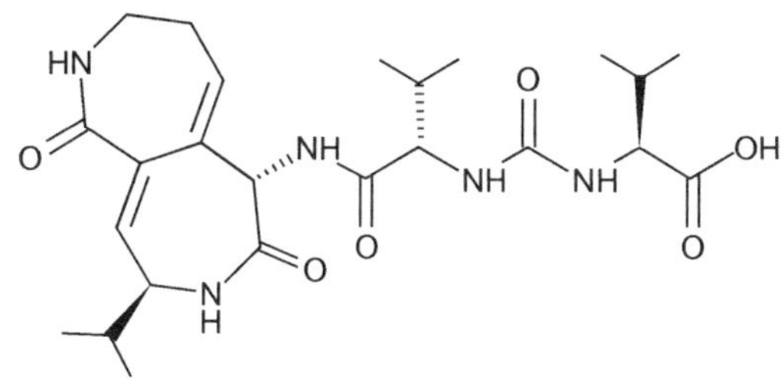 | 193 |
| Syringolin B | 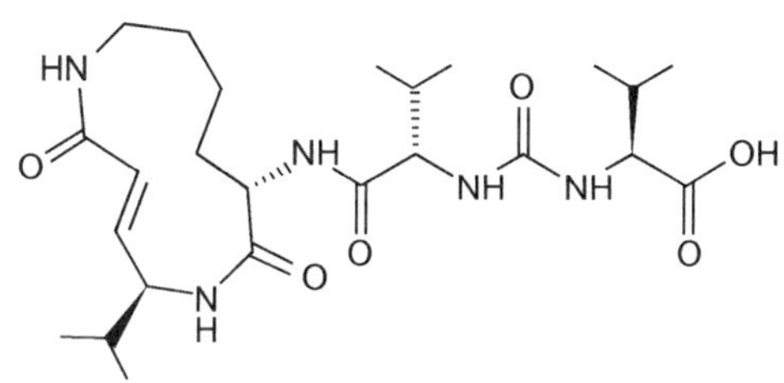 | |

| | | |
|---|---|---|
| CuEt | 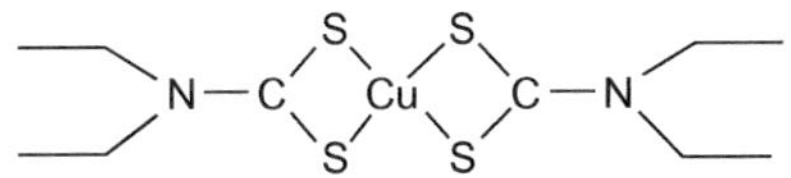 | 261 |
| ZnEt | 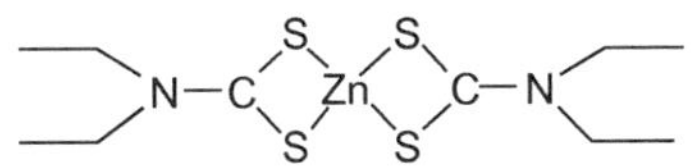 | |
| Glidobactin A | 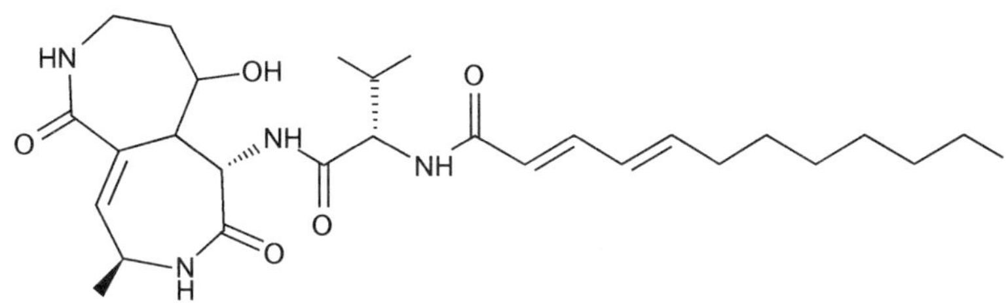 | 193 |

(*Continues*)

TABLE III (*Continued*)

| Name | Structure | Reference |
|---|---|---|
| S 2209 | 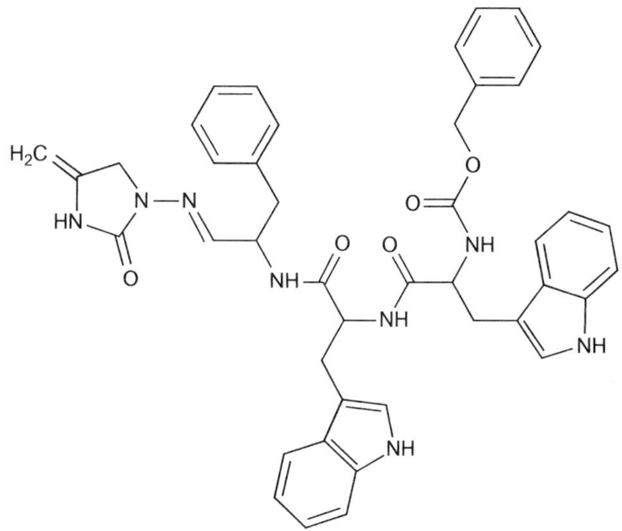 | 211 |
| Clioquinol | 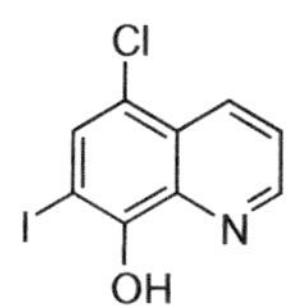 | 266 |

A synthetic gold(III) coordination compound with dithiocarbamate ligands (Au compound 2 in Table III) potently inhibited purified 20S proteasome and also inhibited the 26S proteasome in intact highly metastatic MDA-MB-231 breast cancer cells. The compound was even active *in vivo* in MDA-MB-231 breast tumor-bearing nude mice.[200,201] Other gold(III) and gold(I) dithiocarbamate compounds were also reported to inhibit purified 20S proteasome and 26S proteasome in MDA-MB-231 cells.[202] Another class of metal-based 20S proteasome inhibitors is comprised of gallium complexes in which the gallium ion is bound to an NN′O asymmetrical ligand containing pyridine and substituted phenolate moieties at a 1:2 (metal/ligand) ratio. One of these compounds was able to suppress the growth of prostate cancer xenografts in mice by 66%. This effect was associated with the proteasome inhibition and the induction of apoptosis.[203]

In recent years, many natural compounds have been reported to be 20S proteasome inhibitors, as summarized in Table III, for example, green tea polyphenols,[204] curcumin, apigenin, anthocyanins, hypericin, and resveratrol.[205] The most abundant and biologically active green tea catechin, (-)-epigallocatechin-3-gallate or (-)-EGCG, has been shown to inhibit the 20S proteasome and induce tumor cell death. After the peracetate protections of its reactive hydroxyl groups, the drug was more active against the proteasome and breast cancer in mice with breast cancer xenografts.[206]

The treatment of HCT-116 colon tumor-bearing mice with curcumin resulted in decreased tumor growth associated with proteasome inhibition, proliferation suppression, and induction of apoptosis in tumor tissues. That study showed that proteasome inhibition could be one mechanism underlying the chemopreventive and/or therapeutic roles of curcumin in human colon cancer.[207] More similarly to bortezomib, which is active only in some blood cancers, especially multiple myeloma, the natural triterpene pristimerin was identified as a 20S proteasome inhibitor active against xenografted plasmacytoma tumors in mice and exhibited synergistic cytotoxicity with bortezomib.[208] Celastrol, a triterpene extracted from the root bark of the Chinese medicine "Thunder of God Wine," and withaferin A, a steroidal lactone purified from the medicinal plant "Indian Winter Cherry," were shown to be potent 20S proteasome inhibitors and were active against prostate cancer xenografts in mice.[209,210]

The peptide-semicarbazone S-2209 represents a new class of 20S proteasome inhibitors. Animals treated with S-2209 exhibited greater proteasome inhibition than control animals treated with 0.1 mg $kg^{-1}$ $d^{-1}$ bortezomib. S-2209 is also active in myeloma cells, and the initial *in vitro* studies show a favorable toxicity profile.[211] Another class of 20S proteasome inhibitors is represented by argyrin A, a cyclical peptide derived from the myxobacterium Archangium gephyra. A recent publication showed that argyrin A suppressed growth of colon cancer xenografts in mice in a manner similar to bortezomib.

Functional gene-clustering analyses based on gene ontology terms indicate that while both argyrin A and bortezomib inhibit the 20S proteasome, they cause very divergent perturbations at the cellular level.[212] Another study has shown that a derivative of 2-chloro-1,4-naphthoquinone PI-083 is a 20S proteasome inhibitor that selectively targets the proteasome in cancer cells only. This compound inhibits the 20S proteasome and cell proliferation, and induces apoptosis selectively in cancer cells compared to their normal/immortalized counterparts. In addition, PI-083 required less time than bortezomib to produce antitumor effects in all cancer cell lines tested. Furthermore, in mouse xenografts, this compound, but not bortezomib, suppressed the growth of human breast and lung tumors. Following *in vivo* treatment in mice, the compound inhibited tumor, but not hepatic, 20S proteasome activity, whereas bortezomib inhibited both tumor and liver 20S proteasome activities.[213]

A partially unfolded protein–lipid complex called HAMLET, which is internalized by tumor cells and triggers cell death, bound directly to the 20S proteasome and caused proteasome inhibition. These effects were confirmed in extracts from HAMLET-treated cells, which exhibited signs of fragmentation of the 20S proteasome. This led to speculation that perturbations of the proteasome structure might contribute to the cytotoxic effects of unfolded protein complexes that invade host cells.[214] Consistent with these results, another publication demonstrated that soluble aggregates of toxic proteins may cause prion disease by specifically inhibiting proteasome function.[215,216]

## A. Marizomib

Marizomib (NPI-0052) is an orally active 20S proteasome inhibitor that induces apoptosis in multiple myeloma cells that are resistant to conventional and bortezomib therapies, without affecting normal lymphocyte viability. Marizomib is distinct from bortezomib in its chemical structure, effects on proteasome activities, mechanisms of action, and toxicity profile in normal cells. The pharmacodynamic properties of marizomib are more durable than that of bortezomib, and the recovery of proteasome activity after marizomib treatment is readily observed within 24 h.[217] The irreversible binding properties of marizomib have been correlated with markedly enhanced cytotoxicity in tumor cells, prolonged proteasome inhibition *in vivo*, and sustained inhibition in tumor tissues and packed whole blood associated with reduced tumor growth.[218] In a human MM.1S plasmocytoma xenograft murine model, i.v. marizomib treatment prolongs survival and reduces tumor recurrence. In this model, marizomib leaves the vascular compartment and enters the tumors, where it inhibits 20S proteasome activity in extravascular tumors. MM.1S-bearing immunodeficient mice treated with marizomib exhibit reduced tumor proliferation without marked toxicity and the prolonged inhibition of proteasome activity in tumors, but not in normal tissues.[219]

In combination with bortezomib, marizomib triggers synergistic *in vivo* cytotoxicity in Waldenström's macroglobulinemia, and this combination therapy induces cell death through the Akt pathway.[220] Studies in a xenograft model of multiple myeloma show that a low-dose combination of marizomib and bortezomib is well tolerated and triggers the synergistic inhibition of tumor growth and proteasome activity in tumor cells. Multiple myeloma tumors from marizomib plus bortezomib-treated mice showed growth inhibition, apoptosis, and a decrease in associated angiogenesis.[221] Combination treatment with marizomib and lenalidomide induces synergistic anti-multiple myeloma activity *in vitro* using multiple myeloma cell lines and in cells from multiple myeloma patients. In an animal tumor model of multiple myeloma, a low-dose combination of marizomib and lenalidomide is well tolerated, significantly inhibits tumor growth, and prolongs survival.[222]

Compared with lactacystin, bortezomib, and MG-132, marizomib is the most potent suppressor of NF-κB activation, and it enhances apoptosis, suppresses osteoclastogenesis, and inhibits invasion.[223] In contrast, bortezomib actually activates NF-κB in cancer cells from multiple myeloma patients.[224] In human prostate cancer cell lines, marizomib modifies the NF-κB–Snail–Raf-1 kinase inhibitory pathway and results in the downstream inhibition of antiapoptotic gene products and chemoimmunosensitization.[225] In a Jurkat cell line and primary chronic lymphocytic leukemia cells, marizomib treatment leads to hyperacetylation of histone H3, in contrast to bortezomib. The observed cross-talk between marizomib and histone deacetylase inhibitors, along with caspase-8 activation and oxidative stress, contributes to their synergistic cytotoxic effects in leukemia cells.[226]

Marizomib was shown to be a well-tolerated proteasome inhibitor that, when administered orally, achieves significantly higher and more sustained levels of proteasome inhibition than bortezomib in a colon cancer xenograft. Data show that a single orally administered dose of marizomib leads to the sustained inhibition of the catalytic activity of the 20S proteasome. Significant recovery of the proteasome catalytic activity was observed by day 7. In contrast, a single dose of bortezomib inhibited the catalytic activity of the 20S proteasome but with significant recovery by 24 h. Repeated oral dosing with marizomib at 0.5 mg $kg^{-1}$ results in sustained 20S proteasome inhibition of up to 90–99% in whole blood, with a maximal weight loss of 11.5%. The high level of proteasome inhibition achieved by marizomib is well tolerated and significantly improves the tumoricidal response to multidrug treatment in the colon cancer model.[227]

In contrast, marizomib as a single agent only modestly controlled tumor growth in glioma xenografts, failed to consistently radiosensitize glioma cells *in vitro*, and was not synergistic with radiation *in vivo*.[228] This failure of marizomib can be explained by the observation that reduced proteasome activity is a general feature of glioma cancer-initiating cells.[229]

The clinical development of marizomib began with a phase I dose escalation study in patients with solid tumors or lymphomas.[230,231] Marizomib was administered once weekly, and adverse effects included fatigue, dyspnea, anemia, anorexia, back pain, headache, diarrhea, vomiting, constipation, dizzines, or infusion site pain. Most of the adverse events occurred during the early cycles of therapy. Marizomib did not appear to induce peripheral neuropathy, neutropenia, or thrombocytopenia, in spite of proteasome inhibition levels that exceeded those produced by bortezomib. The anticancer activity of marizomib was observed even in patients with multiple myeloma resistant to bortezomib. Marizomib was also tested in combination with vorinostat in a phase I clinical trial[232] including 22 patients with melanoma, pancreatic carcinoma, or lung cancer. This combination did not lead to increased toxicity. Coadministration did not appear to affect the pharmacokinetics or pharmacodynamics of either drug. Furthermore, 61% of evaluable patients demonstrated stable disease and 39% exhibited decreases in tumor size.

## B. Carfilzomib

Carfilzomib has a high specificity for the proteasome with minimal activity against a wide range of other cellular targets.[233] The potent proteasome inhibition by carfilzomib induces apoptosis that is specific to multiple myeloma cell lines and, in contrast to bortezomib, does not dramatically affect peripheral blood mononuclear cells.[234]

Although carfilzomib (PR-171), bortezomib, and marizomib exhibit comparable effects on the chymotrypsin-like activity of the proteasome, bortezomib and marizomib inhibit the caspase-like and trypsin-like activities, respectively, more potently than does carfilzomib. Despite the irreversible binding of carfilzomib to the 20S proteasome, the recovery of proteasome activity in cultured cells was only moderately slower than that observed with bortezomib. In rats, carfilzomib was rapidly cleared from the plasma compartment following i.v. administration. Noncompartmental analysis of carfilzomib pharmacokinetics revealed an average terminal plasma half-life of about 15 min. Carfilzomib induces a dose-dependent suppression of proteasome activity in all tissues examined except the brain, suggesting that the compound does not readily cross the blood–brain barrier. Tissue recovery from bortezomib-mediated inhibition was comparable to that from carfilzomib but was faster than blood recovery, suggesting that new proteasome synthesis, rather than binding reversibility, plays a dominant role in the recovery of proteasome activity in all tissues other than whole blood. Carfilzomib can be delivered in intensive daily dosing schedules that inhibit proteasome activity by more than 80% in most tissues without excessive toxicity. Furthermore, xenograft studies show that more intensive dosing schedules can yield greater efficacy in both solid and hematologic tumor models.[235]

Carfilzomib also inhibited proliferation and activated apoptosis in patient-derived multiple myeloma cells and in neoplastic cells from patients with other hematologic malignancies. Importantly, carfilzomib had a higher efficacy than bortezomib, and it was active against bortezomib-resistant multiple myeloma cell lines and samples from patients with clinical bortezomib resistance. Carfilzomib also overcame resistance to other conventional agents and acted synergistically with dexamethasone to enhance cell death. In models of multiple myeloma, carfilzomib-induced cell death is associated with c-Jun-N-terminal kinase activation, mitochondrial membrane depolarization, cytochrome c release, and the activation of both intrinsic and extrinsic caspase pathways.[236]

Three phase I clinical trials with carfilzomib were conducted in patients with multiple myeloma, non-Hodgkin lymphoma, Hodgkin disease, and Waldenström's macroglobulinemia who received two or more previous therapy treatments. At the highest dose of carfilzomib, proteasome inhibition in whole blood exceeded 80% 1 h after the first dose. The most common side effects were nausea, diarrhea, and fatigue, at least one of which occurred in 20% of patients, and grade 3 and 4 cytopenias were the principal adverse events. Six of the myeloma patients had stable disease while on the regimen, and one patient who was refractory to bortezomib had a partial response.[237,238] Another phase I clinical study[239] enrolled 29 patients who had relapsed or were refractory to at least two prior therapies. Nonhematologic toxicities included fatigue, nausea, and diarrhea in more than one-third of patients, mostly of grade 1 or 2 in severity. No grade 3 or 4 peripheral neuropathies were reported. One unconfirmed complete response (mantle cell), one partial response (multiple myeloma), and two minimal responses (multiple myeloma and Waldenström's macroglobulinemia) were observed.

Several phase II clinical trials of carfilzomib for relapsed or refractory multiple myeloma have been completed or are ongoing. Part one of an open-label, single-arm clinical study has been completed.[240] This study enrolled 46 patients who had relapsed from more than two prior therapies or immunomodulatory treatment and were refractory to their last treatment. The overall response was 26%, with 5 patients achieving partial response and 16 patients displaying stable disease for at least 6 weeks.

Another phase II trial examined the efficacy of carfilzomib in relapsed myeloma patients with 1–3 prior therapies.[241,242] Patients were divided into two cohorts: bortezomib-naive and bortezomib-exposed. In the bortezomib-naive arm, 51 patients were evaluable, and all were refractory to their most recent treatment including stem cell transplant, lenalidomide, thalidomide, alkylators, and anthracyclines. The overall response rate was 45% and included 1 complete response, 4 very good partial responses, and 18 partial responses. Grade 3 and 4 adverse events were rare and were primarily cytopenic in nature; a few other events, including pneumonia and hyperglycemia, occurred in less than 5% of the cohort. Carfilzomib was safely administered to patients with significant comorbidities without serious complications. The second arm of the phase II study contained patients who had previously received bortezomib therapy. Of this group, 33 patients were evaluable for response, and their previous ineffective therapies were similar to those described above in the bortezomib-naive cohort. The overall response rate was 18%, including 1 complete response and 5 partial responses. Four patients had a minimal response and 13 had stable disease. Grade 3 and 4 adverse events occurred in no more than 10% of patients and included anemia and neutropenia, while grade 1 and 2 events included fatigue, nausea, vomiting, dyspnea, diarrhea, and anemia. These adverse effects did not lead to a dose reduction or discontinuation.

The relationship of cytogenetic abnormalities in patients with relapsed or refractory multiple myeloma to carfilzomib response was tested in another clinical trial.[243] Of the 79 patients in the study, 23 had at least one cytogenetic abnormality; however, the time to progression in these patients was 195 days, which is not significantly different from that of patients without abnormalities. The results of this study were similar to those of bortezomib-treated patients, in which the cytogenetic abnormalities did not significantly affect the efficacy of bortezomib.[244,245] While many side effects, such as nausea, fatigue, and grade 3 and 4 cytopenias, are similar for bortezomib and carfilzomib, one important difference is the lack of peripheral neuropathy experienced by patients treated with carfilzomib.[246] Bortezomib-induced neuropathy seems to be caused by its targeting of proteins other than the 20S proteasome.[181]

## C. CEP-18770 and MLN9708/MLN2238

CEP-18770 induces apoptotic cell death in primary purified CD138-positive explant cultures from untreated and bortezomib-treated multiple myeloma patients. *In vitro*, CEP-18770 has a strong anti-angiogenic activity and potently represses osteoclastogenesis. Intravenous or oral administration of CEP-18770 resulted in a more sustained pharmacodynamic inhibition of proteasome activity in tumors relative to normal tissues, produced complete tumor regression in multiple myeloma xenografts, and improved the overall median survival in a systemic model of the disease.[247,248]

MLN2238, the biologically active form of the novel orally bioavailable inhibitor MLN9708, inhibits growth and induces apoptosis in multiple myeloma cells that are resistant to conventional and bortezomib therapies, without affecting the viability of normal cells. In animal tumor model studies, MLN9708 is well tolerated and inhibits tumor growth with significantly reduced tumor recurrence. A head-to-head comparison of MLN2238 and bortezomib showed a significantly longer survival time in mice treated with MLN2238 than mice receiving bortezomib. Immunostaining of multiple myeloma tumors from MLN2238-treated mice showed growth inhibition, apoptosis, and a decrease in tumor-associated angiogenesis.[249] Both cell line-derived OCI-Ly10 and primary human lymphoma-derived PHTX22L xenograft models of diffuse large B-cell lymphoma were used to evaluate the pharmacodynamics and antitumor effects of MLN2238 and bortezomib. The iMyc(Cα)/Bcl-X(L) genetically engineered mouse model was used to assess the effects of these drugs on de novo plasma cell malignancies and overall survival. The newly developed DP54-Luc-disseminated model of iMyc(Cα)/Bcl-X(L) was used to determine MLN2238 antitumor activity and its effects on osteolytic bone disease. MLN2238 exhibits an improved pharmacodynamic profile and antitumor activity compared with bortezomib in both OCI-Ly10 and PHTX22L models. Although both MLN2238 and bortezomib prolonged overall survival, reduced splenomegaly, and attenuated IgG2a levels in the iMyc(Cα)/Bcl-X(L) GEM model, only MLN2238 alleviated osteolytic bone disease in the DP54-Luc model.[250]

## VIII. Early Drugs as Proteasome Inhibitors

In 1993, Dufour *et al.* published the intriguing results of a clinical trial of a putative adjuvant immunotherapy with ditiocarb (diethyldithiocarbamate) versus a placebo in patients with high-risk breast cancer.[251] Sixty-four patients with nonmetastatic high-risk breast cancer were randomized in a double-blind trial of adjuvant immunotherapy with sodium ditiocarb versus placebo. All patients underwent surgery and then adjuvant chemotherapy with or without ditiocarb. With a median follow-up of 5 years, 6 relapses and 5 deaths were observed in the ditiocarb group, in comparison to 13 relapses and 12 deaths in the control group. At 6 years, overall survival was 81% in the ditiocarb group versus 55% in the control group. Disease-free survival was 76% in the ditiocarb group versus 55% in the control group. These results, however, have been largely ignored because another study published in the same year reported that ditiocarb has no immunomodulatory effect in the human body.[146]

Through the 1990s and 2000s, no one has explained or further investigated these original results. However, some years later, the use of disulfiram (antabuse), which is a prodrug of ditiocarb and has been used in clinics as an aversion therapy for alcoholism,[252] was found to be successful against metastatic melanoma in a patient with stage IV metastatic ocular melanoma when administered with zinc supplementation. The patient remained clinically well and physically active after 53 continuous months of therapy.[253] These findings were the starting point for a new clinical trial at the Huntsman Cancer Institute in Utah (NCT00742911), where disulfiram and copper gluconate are administrated to patients to combat refractory solid tumors involving the liver. Disulfiram has also been used in other clinical trials in combination with other therapeutic agents to target various features of cancer cells.[254]

With this in mind, the results of Dufour *et al.* seem to be straightforward: ditiocarb and disulfiram create a diethylthiocarbamate complex with copper in the body,[255,256] and this complex acts against the proteasome in cancer cells.[257] Disulfiram itself inhibits the proteasome by targeting copper in tumor cells, as has been demonstrated *in vivo*.[258] Other copper coordination compounds are known to be active against the proteasome, although their mechanisms of action on the proteolytic machine are not yet well described; however, they seem to inhibit the 20S proteasome.[259,260]

Bortezomib targets the 20S core particle and, at the same time, is not effective against various solid tumors including melanoma and breast cancer; thus, disulfiram complexes with copper or zinc apparently may inhibit the proteasome in a different way from bortezomib. Indeed, it has been shown that in a breast cancer cell line, disulfiram complexes do not inhibit purified 20S proteasomes, but rather, presumably inhibit the JAMM domain at the proteasome lid in a breast cancer cell line.[261] As a required component of the 26S proteasome that is responsible for deubiquitination of proteins to be degraded in the proteasome,[262] the JAMM domain is an intriguing drug target for cancer therapies.[263]

The concept of repurposing old drugs for new uses seems to be emerging as a major approach in drug discovery and development that will be pursued in the coming decades.[126,264,127] Future experiments and clinical trials will show whether disulfiram will become a pilot case for nonprofit, nonpatentable drugs for use in cancer therapy.[265]

Another example of an old drug that targets the proteasome is clioquinol.[266] From the 1950s to the 1970s, clioquinol was used as a vermicide[267] and in the treatment of gastrointestinal infections such as cholera.[268] Recently, clioquinol was shown to be preferentially cytotoxic to malignant cells compared to normal cells, supporting the novel use of this drug as a cancer therapy. Clioquinol is able to bind metal ions, such as copper or zinc, and consistent with this ability, the supplementation of the cell culture media with zinc or copper enhances the cytotoxicity of the drug.[269] Due to its ability to bind metals, clioquinol has been reported to disrupt lysosome function and target NF-κB and cyclin D1 gene expression causing cell death.[270–272]

When in complex with copper, clioquinol can inhibit the 20S proteasome activity, block proliferation, and induce apoptotic cell death preferentially in breast cancer cells. These effects are less pronounced in premalignant breast cells, and clioquinol is nontoxic to normal/nontransformed breast cells at the concentrations tested. In contrast to clioquinol, the strong copper chelator tetrathiomolybdate or copper alone had no effects on any of the cells. Premalignant or cancerous breast cells that contain copper at concentrations similar to those found in patients undergo proteasome inhibition and apoptosis when treated with clioquinol alone, but not when treated with tetrathiomolybdate.[273]

After binding to copper, clioquinol can inhibit the activity of the 20S proteasome, repress androgen receptor protein expression, and induce apoptotic cell death in human prostate cancer cells. In addition, clioquinol alone exhibits similar effects in prostate cancer cell lines with elevated copper at concentrations similar to those found in patients. Animal studies have shown that clioquinol treatment significantly inhibits the growth of human prostate tumor C4-2B xenografts (by 66%), and this inhibition is associated with *in vivo* proteasome inhibition, androgen receptor protein repression, angiogenesis suppression, and apoptosis induction.[274]

Patients with malignancy have higher levels of copper in their plasma and malignant cells compared to plasma samples and normal cells from healthy volunteers.[275–277] This may contribute to the therapeutic window of clioquinol and its ability to preferentially induce cell death in malignant cells compared to normal cells. Furthermore, clioquinol analogues in which the hydroxyl group is substituted to prevent its binding to copper were able to enter cells, but these analogues were not toxic to breast cancer cells and did not inhibit the proteasome.[278]

## IX. Immunoproteasome Inhibitors

The immunoproteasome, or 20Si, is an inducible variant of the constitutive proteasome in which the active proteolytic subunits have been replaced with immunoproteasome-specific subunits.[3] Bortezomib and, presumably, all 20S proteasome inhibitors target normal and 20Si species to similar extents in myeloma cells. This lack of discrimination is thought to be one cause of the side effects associated with bortezomib and other 20S proteasome inhibitors.[279]

Although 20Si deficiency is an important strategy by which cells derived from solid tumors manipulate immune surveillance,[280,281] and low levels of 20Si subunit expression are associated with shortened survival of cancer patients with solid tumors,[282–285] the expression of one immunoproteasome subunit and leucine aminopeptidase is enhanced in fibrosarcoma lesions compared to normal tissues.[286] Similarly, overexpression of the 20Si has been found in hematologic malignancies, such as lymphomas or multiple myeloma, suggesting that the selective inhibition of 20Si may increase the effectiveness and specificity of proteasome inhibition-associated tumor cell death, alleviating off-target effects.[287]

The development of molecular probes that selectively inhibit the major catalytic subunit of the immunoproteasome was reported in a recent study. Importantly, cancer cells that are rich in this subunit are more sensitive to growth inhibition by immunoproteasome inhibitors, implying that this subunit plays an important role in regulating the cell growth of malignant tumors.[288]

Consequently, the increased expression of the 20Si in multiple myeloma could be an explanation for the observed antimyeloma activity of a 20Si-specific inhibitor. The agent IPSI-001 induces the accumulation of ubiquitin–protein conjugates and proapoptotic proteins, and activates caspase-mediated apoptosis. It also potently inhibits proliferation in samples from myeloma patients and in other hematologic malignancies. Importantly, IPSI-001 is able to overcome conventional and novel drug resistance, including resistance to bortezomib.[289]

However, a selective inhibitor of another immunoproteasome subunit impedes interleukin-23 production by activated monocytes, as well as interferon-$\gamma$ and interleukin-2 production by T cells. In mouse models of rheumatoid arthritis, treatment with this inhibitor reverses signs of disease and reduces cellular infiltration, cytokine production, and autoantibody levels.[290]

## X. Beyond the 20S Proteasome

Within the ubiquitin–proteasome system, there are many potential targets for drug development other than the 20S proteasome.[291–293] The conjugation of proteins with ubiquitin plays numerous regulatory roles through both proteasomal-dependent and proteasomal-independent functions. Changes in ubiquitination are observed in a wide range of pathologic conditions, including numerous malignancies. For this reason, there is great interest in targeting the ubiquitination system in cancer. One recent study published the first inhibitor of an E1 ubiquitin-activating enzyme with promising anticancer activity in cancer cells. This pyrazone-based inhibitor provides proof of principle of the capacity to differentially suppress cancer cells by targeting the ubiquitin pathway.[294]

The first chemical inhibitor of the 19S regulatory particle of the 26S proteasome was also described recently. This compound was identified by screening a library of nucleoside-capped peptoids for binding to the yeast 26S proteasome in a crude extract. The compound indentified to have proteasome-binding activity was resynthesized and shown to block 19S-mediated protein unfolding *in vitro* and proteasome-mediated turnover of p27 in cancer cells.[295] Further study described the identification of the protein target of the inhibitor.[296] However, there is as yet no evidence for the anticancer activity of such compounds.

The driving force of the cell cycle is the activation of cyclin-dependent kinases (CDKs), the activities of which are controlled by the ubiquitin-mediated proteolysis of key regulators, such as cyclins and CDK inhibitors. Two ubiquitin ligases, the SKP1-CUL1-F-box–protein complex and the anaphase-promoting complex/cyclosome, are responsible for the specific ubiquitination of many of these regulators. The deregulation of the proteolytic system might therefore result in uncontrolled proliferation, genomic instability, and cancer. Cumulative clinical evidence shows that alterations in the ubiquitination of cell-cycle regulators participate in the etiology of many human malignancies.[297] The ability to target ubiquitin ligases to selectively block the degradation of certain cellular proteins seems especially attractive. However, no ubiquitin ligase inhibitor has yet entered clinical trials.

The only compound designed to alter ubiquitin ligase function that is undergoing clinical trials is MLN4924, an inhibitor of neural precursor cell-expressed developmentally downregulated protein 8 (NEDD8)-activating enzyme (NAE). NAE is an essential component of the NEDD8 conjugation pathway, which controls the activity of cullin-RING (really interesting new gene) ubiquitin ligases.[298] MLN4924 disrupts cullin-RING ligase-mediated protein turnover, leading to apoptotic death in human tumor cells through a novel mechanism of action, the deregulation of S-phase DNA synthesis. MLN4924 suppressed the growth of human tumor xenografts in mice at compound exposures that were well tolerated.[299]

Other attractive targets within the ubiquitin–proteasome system include the deubiquitinating enzymes.[300,301] The anticancer effect of an isopeptidase inhibitor in comparison to bortezomib has already been demonstrated at the cellular level.[302] More interestingly, a small molecule b-AP15 that abrogates the deubiquitinating activity of the 19S regulatory particle has been described. This molecule inhibits the activity of two 19S regulatory particle-associated deubiquitinases, ubiquitin C-terminal hydrolase 5 and ubiquitin-specific peptidase 14, resulting in the accumulation of polyubiquitin. This induces tumor cell apoptosis that is insensitive to both TP53 status and the overexpression of the apoptosis inhibitor BCL2. Treatment with b-AP15 has also been shown to inhibit tumor progression in four different *in vivo* solid tumor models and inhibit organ infiltration in an acute myeloid leukemia model.[303]

## Acknowledgment

This work was financed by project OP VK CZ.1.07/2.3.00/20.0062 "An inexpensive drug Antabuse as anticancer remedy: mechanism of action and clinical trials" from resources of European Union and the Czech Republic.

## References

1. Orlowski RZ, Kuhn DJ. Proteasome inhibitors in cancer therapy: lessons from the first decade. *Clin Cancer Res* 2008;**14**:1649–57.
2. Borissenko L, Groll M. 20S proteasome and its inhibitors: crystallographic knowledge for drug development. *Chem Rev* 2007;**107**:687–717.
3. Cvek B, Dvorak Z. The ubiquitin-proteasome system (UPS) and the mechanism of action of bortezomib. *Curr Pharm Des* 2011;**17**:1483–99.

4. Adams J, Kauffman M. Development of the proteasome inhihitor Velcade™ (Bortezomib). *Cancer Invest* 2004;**22**:304–11.
5. Palombella VJ, Conner EM, Fuseler JW, Destree A, Davis JM, Laroux FS, et al. Role of the proteasome and NF-κB in streptococcal cell wall-induced polyarthritis. *Proc Natl Acad Sci U S A* 1998;**95**:15671–6.
6. Kalogeris TJ, Laroux FS, Cockrell A, Ichikawa H, Okayama N, Phifer TJ, et al. Effect of selective proteasome inhibitors on TNF-induced activation of primary and transformed endothelial cells. *Am J Physiol* 1999;**276**:C856–64.
7. Adams J, Palombella VJ, Sausville EA, Johnson J, Destree A, Lazarus DD, et al. Proteasome inhibitors: a novel class of potent and effective antitumor agents. *Cancer Res* 1999;**59**:2615–22.
8. Iqbal M, Chatterjee S, Kauer JC, Das M, Messina P, Freed B, et al. Potent inhibitors of proteasome. *J Med Chem* 1995;**38**:2276–7.
9. Tsubuki S, Kawasaki H, Saito Y, Miyashita N, Inomata M, Kawashima S. Purification and characterization of a Z-Leu-Leu-Leu-MCA degrading protease expected to regulate neurite formation: a novel catalytic activity in proteasome. *Biochem Biophys Res Commun* 1993;**196**:1195–201.
10. Adams J, Stein R. Novel inhibitors of the proteasome and their therapeutic use in inflammation. *Annu Rep Med Chem* 1996;**31**:279–88.
11. Adams J, Ma YT, Stein R, Baevsky M, Grenier L, Plamondon L. International patent application WO 96/13266, published May 9, 1996. *Chem Abstr* 1996;**125**:143315.
12. Seemuller E, Lupas A, Stock D, Lowe J, Huber R, Baumeister W. Proteasome from Thermoplasma acidophilum: a threonine protease. *Science* 1995;**268**:579–82.
13. Stein RL, Melandri F, Dick L. Kinetic characterization of the chymotryptic activity of the 20S proteasome. *Biochemistry* 1996;**35**:3899–908.
14. Adams J, Behnke M, Chen S, Cruickshank AA, Dick LR, Grenier L, et al. Potent and selective inhibitors of the proteasome: dipeptidyl boronic acids. *Bioorg Med Chem Lett* 1998;**8**:333–8.
15. Yang W, Gao X, Wang B. Boronic acid compounds as potential pharmaceutical agents. *Med Res Rev* 2003;**23**:346–68.
16. Matteson DS. α-Amido boronic acids: a synthetic challenge and their properties as serine protease inhibitors. *Med Res Rev* 2008;**28**:233–46.
17. Boyd MR, Paull KD. Some practical considerations and applications of the National Cancer Institute in vitro anticancer drug discovery screen. *Drug Dev Res* 1995;**34**:91–109.
18. Weinstein JN, Myers TG, O'Connor PM, Friend SH, Fornace AJ, Kohn KW, et al. An information-intensive approach to the molecular pharmacology of cancer. *Science* 1997;**275**:343–9.
19. Adams J. Development of the proteasome inhibitor PS-341. *Oncologist* 2002;**7**:9–16.
20. Orlowski RZ, Stinchcombe TE, Mitchell BS, Shea TC, Baldwin AS, Stahl S, et al. Phase I trial of the proteasome inhibitor PS-341 in patients with refractory hematologic malignancies. *J Clin Oncol* 2002;**20**:4420–7.
21. Aghajanian C, Soignet S, Dizon DS, Pien CS, Adams J, Elliott PJ, et al. A phase I trial of the novel proteasome inhibitor PS341 in advanced solid tumor malignancies. *Clin Cancer Res* 2002;**8**:2505–11.
22. Papandreou CN, Daliani DD, Nix D, Yang H, Madden T, Wang X, et al. Phase I trial of the proteasome inhibitor bortezomib in patients with advanced solid tumors with observations in androgen-independent prostate cancer. *J Clin Oncol* 2004;**22**:2108–21.
23. Hamilton AL, Eder JP, Pavlick AC, Clark JW, Liebes L, Garcia-Carbonero R, et al. Proteasome inhibition with bortezomib (PS-341): a phase I study with pharmacodynamic end points using a day 1 and day 4 schedule in a 14-day cycle. *J Clin Oncol* 2005;**23**:6107–16.

24. Richardson PG, Barlogie B, Berenson J, Singhal S, Jagannath S, Irwin D, et al. A phase 2 study of bortezomib in relapsed, refractory myeloma. *N Engl J Med* 2003;**348**:2609–17.
25. Kane RC, Bross PF, Farrell AT, Pazdur R. Velcade: U.S. FDA approval for the treatment of multiple myeloma progressing on prior therapy. *Oncologist* 2003;**8**:508–13.
26. Bross PF, Kane R, Farrell AT, Abraham S, Benson K, Brower ME, et al. Approval summary for bortezomib for injection in the treatment of multiple myeloma. *Clin Cancer Res* 2004;**10**:3954–64.
27. Adams J. The development of proteasome inhibitors as anticancer drugs. *Cancer Cell* 2004;**5**:417–21.
28. Kane RC, Farrell AT, Sridhara R, Pazdur R. United States Food and Drug Administration approval summary: bortezomib for the treatment of progressive multiple myeloma after one prior therapy. *Clin Cancer Res* 2006;**12**:2955–60.
29. Doss S, Hay N, Sutcliffe F. NICE guidance on bortezomib and thalidomide for first-line treatment of multiple myeloma. *Lancet Oncol* 2011;**12**:837–8.
30. Kane RC, Dagher R, Farrell A, Ko CW, Sridhara R, Justice R, et al. Bortezomib for the treatment of mantle cell lymphoma. *Clin Cancer Res* 2007;**13**:5291–4.
31. Jung L, Holle L, Dalton WS. Discovery, development, and clinical applications of bortezomib. *Oncology* 2004;**18**:4–13.
32. Schwartz R, Davidson T. Pharmacology, pharmacokinetics, and practical applications of bortezomib. *Oncology* 2004;**18**:14–21.
33. Richardson PG, Barlogie B, Berenson J, Singhal S, Jagannath S, Irwin DH, et al. Extended follow-up of a phase II trial in relapsed, refractory multiple myeloma: final time-to-event results from the SUMMIT trial. *Cancer* 2006;**106**:1316–9.
34. Richardson PG, Sonneveld P, Schuster MW, Irwin D, Stadtmauer EA, Facon T, et al. Assessment of proteasome inhibition for extending remissions (APEX) Investigators. Bortezomib or high-dose dexamethasone for relapsed multiple myeloma. *N Engl J Med* 2005;**352**:2487–98.
35. Richardson PG, Sonneveld P, Schuster M, Irwin D, Stadtmauer E, Facon T, et al. Extended follow-up of a phase 3 trial in relapsed multiple myeloma: final time-to-event results of the APEX trial. *Blood* 2007;**110**:3557–60.
36. Niesvizky R, Richardson PG, Rajkumar SV, Coleman M, Rosinol L, Sonneveld P, et al. The relationship between quality of response and clinical benefit for patients treated on the bortezomib arm of the international, randomized, phase 3 APEX trial in relapsed multiple myeloma. *Br J Haematol* 2008;**143**:46–53.
37. Lee SJ, Richardson PG, Sonneveld P, Schuster MW, Irwin D, San Miguel JF, et al. Bortezomib is associated with better health-related quality of life than high-dose dexamethasone in patients with relapsed multiple myeloma: results from the APEX study. *Br J Haematol* 2008;**143**:511–9.
38. Berenson JR, Yang HH, Sadler K, Jarutirasarn SG, Vescio RA, Mapes R, et al. Phase I/II trial assessing bortezomib and melphalan combination therapy for the treatment of patients with relapsed or refractory multiple myeloma. *J Clin Oncol* 2006;**24**:937–44.
39. Berenson JR, Yellin O, Patel R, Duvivier H, Nassir Y, Mapes R, et al. A phase I study of samarium lexidronam/bortezomib combination therapy for the treatment of relapsed or refractory multiple myeloma. *Clin Cancer Res* 2009;**15**:1069–75.
40. Badros A, Burger AM, Philip S, Niesvizky R, Kolla SS, Goloubeva O, et al. Phase I study of vorinostat in combination with bortezomib for relapsed and refractory multiple myeloma. *Clin Cancer Res* 2009;**15**:5250–7.
41. Orlowski RZ, Voorhees PM, Garcia RA, Hall MD, Kudrik FJ, Allred T, et al. Phase 1 trial of the proteasome inhibitor bortezomib and pegylated liposomal doxorubicin in patients with advanced hematologic malignancies. *Blood* 2005;**105**:3058–65.

42. Biehn SE, Moore DT, Voorhees PM, Garcia RA, Lehman MJ, Dees EC, et al. Extended follow-up of outcome measures in multiple myeloma patients treated on a phase I study with bortezomib and pegylated liposomal doxorubicin. *Ann Hematol* 2007;**86**:211–6.
43. Orlowski RZ, Nagler A, Sonneveld P, Blade J, Hajek R, Spencer A, et al. Randomized phase III study of pegylated liposomal doxorubicin plus bortezomib compared with bortezomib alone in relapsed or refractory multiple myeloma: combination therapy improves time to progression. *J Clin Oncol* 2007;**25**:3892–901.
44. Jagannath S, Richardson PG, Barlogie B, Berenson JR, Singhal S, Irwin D, et al. Bortezomib in combination with dexamethasone for the treatment of patients with relapsed and/or refractory multiple myeloma with less than optimal response to bortezomib alone. *Haematologica* 2006;**91**:929–34.
45. Harousseau JL, Attal M, Leleu X, Troncy J, Pegourie B, Stoppa AM, et al. Bortezomib plus dexamethasone as induction treatment prior to autologous stem cell transplantation in patients with newly diagnosed multiple myeloma: results of an IFM phase II study. *Haematologica* 2006;**91**:1498–505.
46. Mikhael JR, Belch AR, Prince HM, Lucio MN, Maiolino A, Corso A, et al. High response rate to bortezomib with or without dexamethasone in patients with relapsed or refractory multiple myeloma: results of a global phase 3b expanded access program. *Br J Haematol* 2009;**144**:169–75.
47. Berenson JR, Yellin O, Chen CS, Patel R, Bessudo A, Boccia RV, et al. A modified regimen of pegylated liposomal doxorubicin, bortezomib and dexamethasone (DVD) is effective and well tolerated for previously untreated multiple myeloma patients. *Br J Haematol* 2011;**155**:580–7.
48. Mateos MV, Hernandez JM, Hernandez MT, Gutierrez NC, Palomera L, Fuertes M, et al. Bortezomib plus melphalan and prednisone in elderly untreated patients with multiple myeloma: results of a multicenter phase 1/2 study. *Blood* 2006;**108**:2165–72.
49. San Miguel JF, Schlag R, Khuageva NK, Dimopoulos MA, Shpilberg O, Kropff M, et al. Bortezomib plus melphalan and prednisone for initial treatment of multiple myeloma. *N Engl J Med* 2008;**359**:906–17.
50. Popat R, Oakervee H, Williams C, Cook M, Craddock C, Basu S, et al. Bortezomib, low-dose intravenous melphalan, and dexamethasone for patients with relapsed multiple myeloma. *Br J Haematol* 2009;**144**:887–94.
51. Wang M, Giralt S, Delasalle K, Handy B, Alexanian R. Bortezomib in combination with thalidomide-dexamethasone for previously untreated multiple myeloma. *Hematology* 2007;**12**:235–9.
52. Nair B, van Rhee F, Shaughnessy JD, Anaissie E, Szymonifka J, Hoering A, et al. Superior results of Total Therapy 3 (2003-33) in gene expression profiling-defined low-risk multiple myeloma confirmed in subsequent trial 2006-66 with VRD maintenance. *Blood* 2010;**115**:4168–73.
53. van Rhee F, Szymonifka J, Anaissie E, Nair B, Waheed S, Alsayed Y, et al. Total therapy 3 for multiple myeloma: prognostic implications of cumulative dosing and premature discontinuation of VTD maintenance components, bortezomib, thalidomide and dexamethasone, relevant to all phases of therapy. *Blood* 2010;**116**:1220–7.
54. Berenson JR, Yellin O, Woytowitz D, Flam MS, Cartmell A, Patel R, et al. Bortezomib, ascorbic acid and melphalan (BAM) therapy for patients with newly diagnosed multiple myeloma: an effective and well-tolerated frontline regimen. *Eur J Haematol* 2009;**82**:433–9.
55. Richardson PG, Wolf J, Jakubowiak A, Zonder J, Lonial S, Irwin D, et al. Perifosine plus bortezomib and dexamethasone in patients with relapsed/refractory multiple myeloma previously treated with bortezomib: results of a multicenter phase I/II trial. *J Clin Oncol* 2011;**29**(32):4243–9.

56. Pekol T, Daniels JS, Labutti J, Parsons I, Nix D, Baronas E, et al. Human metabolism of the proteasome inhibitor bortezomib: identification of circulating metabolites. *Drug Metab Dispos* 2005;**33**:771–7.
57. Labutti J, Parsons I, Huang R, Miwa G, Gan LS, Daniels JS. Oxidative deboronation of the peptide boronic acid proteasome inhibitor bortezomib: contributions from reactive oxygen species in this novel cytochrome P450 reaction. *Chem Res Toxicol* 2006;**19**:539–46.
58. Venkatakrishnan K, Rader M, Ramanathan RK, Ramalingam S, Chen E, Riordan W, et al. Effect of the CYP3A inhibitor ketoconazole on the pharmacokinetics and pharmacodynamics of bortezomib in patients with advanced solid tumors: a prospective, multicenter, open-label, randomized, two-way crossover drug-drug interaction study. *Clin Ther* 2009;**2**:2444–58.
59. Lu C, Gallegos R, Li P, Xia CQ, Pusalkar S, Uttamsingh V, et al. Investigation of drug-drug interaction potential of bortezomib in vivo in female Sprague-Dawley rats and in vitro in human liver microsomes. *Drug Metab Dispos* 2006;**34**:702–8.
60. Quinn DI, Nemunaitis J, Fuloria J, Britten CD, Gabrail N, Yee L, et al. Effect of the cytochrome P450 2C19 inhibitor omeprazole on the pharmacokinetics and safety profile of bortezomib in patients with advanced solid tumours, non-Hodgkin's lymphoma or multiple myeloma. *Clin Pharmacokinet* 2009;**48**:199–209.
61. Rumpold H, Salvador C, Wolf AM, Tilg H, Gastl G, Wolf D. Knockdown of PgP resensitizes leukemic cells to proteasome inhibitors. *Biochem Biophys Res Commun* 2007;**361**:549–54.
62. Nakamura T, Tanaka K, Matsunobu T, Okada T, Nakatani F, Sakimura R, et al. The mechanism of cross-resistance to proteasome inhibitor bortezomib and overcoming resistance in Ewing's family tumor cells. *Int J Oncol* 2007;**31**:803–11.
63. Bhandari M, Jagannath S. Repeated complete responses with bortezomib in a heavily pretreated primary refractory patient with light chain multiple myeloma. *Clin Lymphoma Myeloma* 2007;**7**:373–5.
64. Conner TM, Doan QD, Walters IB, LeBlanc AL, Beveridge RA. An observational, retrospective analysis of retreatment with bortezomib for multiple myeloma. *Clin Lymphoma Myeloma* 2008;**8**:140–5.
65. Wolf J, Richardson PG, Schuster M, LeBlanc A, Walters IB, Battleman DS. Utility of bortezomib retreatment in relapsed or refractory multiple myeloma patients: a multicenter case series. *Clin Adv Hematol Oncol* 2008;**6**:755–60.
66. Ruckrich T, Kraus M, Gogel J, Beck A, Ovaa H, Verdoes M, et al. Characterization of the ubiquitin-proteasome system in bortezomib-adapted cells. *Leukemia* 2009;**23**:1098–105.
67. Oerlemans R, Franke NE, Assaraf YG, Cloos J, van Zantwijk I, Berkers CR, et al. Molecular basis of bortezomib resistance: proteasome subunit beta5 (PSMB5) gene mutation and overexpression of PSMB5 protein. *Blood* 2008;**112**:2489–99.
68. Lu S, Yang J, Song X, Gong S, Zhou H, Guo L, et al. Point mutation of the proteasome β5 subunit gene is an important mechanism of bortezomib resistance in bortezomib-selected variants of Jurkat T cell lymphoblastic lymphoma/leukemia line. *J Pharmacol Exp Ther* 2008;**326**:423–31.
69. Lu S, Chen Z, Yang J, Chen L, Gong S, Zhou H, et al. Overexpression of the PSMB5 gene contributes to bortezomib resistance in T-lymphoblastic lymphoma/leukemia cells derived from Jurkat line. *Exp Hematol* 2008;**36**:1278–84.
70. Lu S, Yang J, Chen Z, Gong S, Zhou H, Xu X, et al. Different mutants of PSMB5 confer varying bortezomib resistance in T lymphoblastic lymphoma/leukemia cells derived from the Jurkat cell line. *Exp Hematol* 2009;**37**:831–7.
71. Fuchs D, Berges C, Opelz G, Daniel V, Naujokat C. Increased expression and altered subunit composition of proteasomes induced by continuous proteasome inhibition establish apoptosis resistance and hyperproliferation of Burkitt lymphoma cells. *J Cell Biochem* 2008;**103**:270–83.

72. Liu FT, Agrawal SG, Movasaghi Z, Wyatt PB, Rehman IU, Gribben JG, et al. Dietary flavonoids inhibit the anticancer effects of the proteasome inhibitor bortezomib. *Blood* 2008; **112**:3835–46.
73. Li X, Wood TE, Sprangers R, Jansen G, Franke NE, Mao X, et al. Effect of noncompetitive proteasome inhibition on bortezomib resistance. *J Natl Cancer Inst* 2010;**102**:1069–82.
74. Hsieh FY, Tengstrand E, Pekol TM, Guerciolini R, Miwa G. Elucidation of potential bortezomib response markers in mutliple myeloma patients. *J Pharm Biomed Anal* 2009; **49**:115–22.
75. Dawson MA, Opat SS, Taouk Y, Donovan M, Zammit M, Monaghan K, et al. Clinical and immunohistochemical features associated with a response to bortezomib in patients with multiple myeloma. *Clin Cancer Res* 2009;**15**:714–22.
76. Fisher RI, Bernstein SH, Kahl BS, Djulbegovic B, Robertson MJ, de Vos S, et al. Multicenter phase II study of bortezomib in patients with relapsed or refractory mantle cell lymphoma. *J Clin Oncol* 2006;**24**:4867–74.
77. Goy A, Bernstein SH, Kahl BS, Djulbegovic B, Robertson MJ, de Vos S, et al. Bortezomib in patients with relapsed or refractory mantle cell lymphoma: updated time-to-event analyses of the multicenter phase 2 PINNACLE study. *Ann Oncol* 2009;**20**:520–5.
78. O'Connor OA, Moskowitz C, Portlock C, Hamlin P, Straus D, Dumitrescu O, et al. Patients with chemotherapy-refractory mantle cell lymphoma experience high response rates and identical progression-free survivals compared with patients with relapsed disease following treatment with single agent bortezomib: results of a multicentre Phase 2 clinical trial. *Br J Haematol* 2009;**145**:34–9.
79. Belch A, Kouroukis CT, Crump M, Sehn L, Gascoyne RD, Klasa R, et al. A phase II study of bortezomib in mantle cell lymphoma: the National Cancer Institute of Canada Clinical Trials Group trial IND.150. *Ann Oncol* 2007;**18**:116–21.
80. Baiocchi RA, Alinari L, Lustberg ME, Lin TS, Porcu P, Li X, Johnston JS, Byrd JC, Blum KA. Phase 2 trial of rituximab and bortezomib in patients with relapsed or refractory mantle cell and follicular lymphoma. *Cancer* 2010; in press. Epub ahead of publication: http://www.ncbi.nlm.nih.gov/pubmed/21157957.
81. Friedberg JW, Vose JM, Kelly JL, Young F, Bernstein SH, Peterson D, et al. The combination of bendamustine, bortezomib, and rituximab for patients with relapsed/refractory indolent and mantle cell non-Hodgkin lymphoma. *Blood* 2011;**117**:2807–12.
82. Houot R, Le Gouill S, Ojeda Uribe M, Mounier C, Courby S, Dartigeas C, et al. Combination of rituximab, bortezomib, doxorubicin, dexamethasone and chlorambucil (RiPAD+C) as first-line therapy for elderly mantle cell lymphoma patients: results of a phase II trial from the GOELAMS. *Ann Oncol* 2011; in press.
83. Goy A, Bernstein SH, McDonald A, Pickard MD, Shi H, Fleming MD, et al. Potential biomarkers of bortezomib activity in mantle cell lymphoma from the phase 2 PINNACLE trial. *Leuk Lymphoma* 2010;**51**:1269–77.
84. Perez-Galan P, Mora-Jensen H, Weniger MA, Shaffer AL, Rizzatti EG, Chapman CM, et al. Bortezomib resistance in mantle cell lymphoma is associated with plasmacytic differentiation. *Blood* 2011;**117**:542–52.
85. Zinzani PL, Musuraca G, Tani M, Stefoni V, Marchi E, Fina M, et al. Phase II trial of proteasome inhibitor bortezomib in patients with relapsed or refractory cutaneous T-cell lymphoma. *J Clin Oncol* 2007;**25**:4293–7.
86. Troch M, Jonak C, Mullauer L, Puspok A, Formanek M, Hauff W, et al. A phase II study of bortezomib in patients with MALT lymphoma. *Haematologica* 2009;**94**:738–42.
87. Faderl S, Rai K, Gribben J, Byrd JC, Flinn IW, O'Brien S, et al. Phase II study of single-agent bortezomib for the treatment of patients with fludarabine-refractory B-cell chronic lymphocytic leukemia. *Cancer* 2006;**107**:916–24.

88. Chen C, Kouroukis CT, White D, Voralia M, Stadtmauer E, Stewart AK, et al. Bortezomib in relapsed or refractory Waldenstrom's macroglobulinemia. *Clin Lymphoma Myeloma* 2009;**9**:74–6.
89. Dimopoulos MA, Anagnostopoulos A, Kyrtsonis MC, Castritis E, Bitsaktsis A, Pangalis GA. Treatment of relapsed or refractory Waldenstrom's macroglobulinemia with bortezomib. *Haematologica* 2005;**90**:1655–8.
90. Chen CI, Kouroukis CT, White D, Voralia M, Stadtmauer E, Stewart AK, et al. Bortezomib is active in patients with untreated or relapsed Waldenstrom's macroglobulinemia: a phase II study of the National Cancer Institute of Canada Clinical Trials Group. *J Clin Oncol* 2007;**25**:1570–5.
91. Treon SP, Hunter ZR, Matous J, Joyce RM, Mannion B, Advani R, et al. Multicenter clinical trial of bortezomib in relapsed/refractory Waldenstrom's macroglobulinemia: results of WMCTG Trial 03-248. *Clin Cancer Res* 2007;**13**:3320–5.
92. de Vos S, Goy A, Dakhil SR, Saleh MN, McLaughlin P, Belt R, et al. Multicenter randomized phase II study of weekly or twice-weekly bortezomib plus rituximab in patients with relapsed or refractory follicular or marginal-zone B-cell lymphoma. *J Clin Oncol* 2009;**27**:5023–30.
93. Coiffier B, Osmanov EA, Hong X, Scheliga A, Mayer J, Offner F, et al. Bortezomib plus rituximab versus rituximab alone in patients with relapsed, rituximab-naive or rituximab-sensitive, follicular lymphoma: a randomised phase 3 trial. *Lancet Oncol* 2011;**12**:773–84.
94. Fowler N, Kahl BS, Lee P, Matous JV, Cashen AF, Jacobs SA, et al. Bortezomib, bendamustine, and rituximab in patients with relapsed or refractory follicular lymphoma: the phase II VERTICAL study. *J Clin Oncol* 2011;**29**:3389–95.
95. Bose P, Thompson C, Gandhi D, Ghabach B, Ozer H. AIDS-related plasmablastic lymphoma with dramatic, early response to bortezomib. *Eur J Haematol* 2009;**82**:490–2.
96. Bibas M, Grisetti S, Alba L, Picchi G, Del Nonno F, Antinori A. Patient with HIV-associated plasmablastic lymphoma responding to bortezomib alone and in combination with dexamethasone, gemcitabine, oxaliplatin, cytarabine, and pegfilgrastim chemotherapy and lenalidomide alone. *J Clin Oncol* 2010;**28**:704–8.
97. Horton TM, Pati D, Plon SE, Thompson PA, Bomgaars LR, Adamson PC, et al. A phase 1 study of the proteasome inhibitor bortezomib in pediatric patients with refractory leukemia: a Children's Oncology Group study. *Clin Cancer Res* 2007;**13**:1516–22.
98. Santos FP, Kantarjian H, McConkey D, O'Brien S, Faderl S, Borthakur G, et al. Pilot study of bortezomib for patients with imatinib-refractory chronic myeloid leukemia in chronic or accelerated phase. *Clin Lymphoma Myeloma* 2011;**11**:355–60.
99. Caravita T, de Fabritiis P, Palumbo A, Amadori S, Boccadoro M. Bortezomib: efficacy comparisons in solid tumors and hematologic malignancies. *Nat Clin Pract Oncol* 2006;**3**:374–87.
100. Aghajanian C, Blessing JA, Darcy KM, Reid G, Degeest K, Rubin SC, et al. A phase II evaluation of bortezomib in the treatment of recurrent platinum-sensitive ovarian or primary peritoneal cancer: a Gynecologic Oncology Group study. *Gynecol Oncol* 2009;**115**:215–20.
101. Mackay H, Hedley D, Major P, Townsley C, Mackenzie M, Vincent M, et al. A phase II trial with pharmacodynamic endpoints of the proteasome inhibitor bortezomib in patients with metastatic colorectal cancer. *Clin Cancer Res* 2005;**11**:5526–33.
102. Kozuch PS, Rocha-Lima CM, Dragovich T, Hochster H, O'Neil BH, Atiq OT, et al. Bortezomib with or without irinotecan in relapsed or refractory colorectal cancer: results from a randomized phase II study. *J Clin Oncol* 2008;**26**:2320–6.
103. Maki RG, Kraft AS, Scheu K, Yamada J, Wadler S, Antonescu CR, et al. A multicenter Phase II study of bortezomib in recurrent or metastatic sarcomas. *Cancer* 2005;**103**:1431–8.
104. Markovic SN, Geyer SM, Dawkins F, Sharfman W, Albertini M, Maples W, et al. A phase II study of bortezomib in the treatment of metastatic malignant melanoma. *Cancer* 2005;**103**:2584–9.

105. Shah MH, Young D, Kindler HL, Webb I, Kleiber B, Wright J, et al. Phase II study of the proteasome inhibitor bortezomib (PS-341) in patients with metastatic neuroendocrine tumors. *Clin Cancer Res* 2004;**10**:6111–8.
106. Yang CH, Gonzalez-Angulo AM, Reuben JM, Booser DJ, Pusztai L, Krishnamurthy S, et al. Bortezomib (VELCADE) in metastatic breast cancer: pharmacodynamics, biological effects, and prediction of clinical benefits. *Ann Oncol* 2006;**17**:813–7.
107. Engel RH, Brown JA, Von Roenn JH, O'Regan RM, Bergan R, Badve S, et al. A phase II study of single agent bortezomib in patients with metastatic breast cancer: a single institution experience. *Cancer Invest* 2007;**25**:733–7.
108. Shah MA, Power DG, Kindler HL, Holen KD, Kemeny MM, Ilson DH, et al. A multicenter, phase II study of Bortezomib (PS-341) in patients with unresectable or metastatic gastric and gastroesophageal junction adenocarcinoma. *Invest New Drugs* 2010;**29**:1475–81.
109. Fanucchi MP, Fossella FV, Belt R, Natale R, Fidias P, Carbone DP, et al. Randomized phase II study of bortezomib alone and bortezomib in combination with docetaxel in previously treated advanced non-small-cell lung cancer. *J Clin Oncol* 2006;**24**:5025–33.
110. Li T, Ho L, Piperdi B, Elrafei T, Camacho FJ, Rigas JR, et al. Phase II study of the proteasome inhibitor bortezomib (PS-341, Velcade((R))) in chemotherapy-naïve patients with advanced stage non-small cell lung cancer (NSCLC). *Lung Cancer* 2010;**68**:89–93.
111. Lara PN, Chansky K, Davies AM, Franklin WA, Gumerlock PH, Guaglianone PP, et al. Bortezomib (PS-341) in relapsed or refractory extensive stage small cell lung cancer: a Southwest Oncology Group phase II trial (S0327). *J Thorac Oncol* 2006;**1**:996–1001.
112. Jakob C, Egerer K, Liebisch P, Turkmen S, Zavrski I, Kuckelkorn U, et al. Circulating proteasome levels are an independent prognostic factor for survival in multiple myeloma. *Blood* 2007;**109**:2100–5.
113. Kern J, Untergasser G, Zenzmaier C, Sarg B, Gastl G, Gunsilius E, et al. GRP-78 secreted by tumor cells blocks the antiangiogenic activity of bortezomib. *Blood* 2009;**114**:3960–7.
114. Messersmith WA, Baker SD, Lassiter L, Sullivan RA, Dinh K, Almuete VI, et al. Phase I trial of bortezomib in combination with docetaxel in patients with advanced solid tumors. *Clin Cancer Res* 2006;**12**:1270–5.
115. Ryan DP, O'Neil BH, Supko JG, Rocha Lima CM, Dees EC, Appleman LJ, et al. A Phase I study of bortezomib plus irinotecan in patients with advanced solid tumors. *Cancer* 2006;**107**:2688–97.
116. Ma C, Mandrekar SJ, Alberts SR, Croghan GA, Jatoi A, Reid JM, et al. A phase I and pharmacologic study of sequences of the proteasome inhibitor, bortezomib (PS-341, Velcade), in combination with paclitaxel and carboplatin in patients with advanced malignancies. *Cancer Chemother Pharmacol* 2007;**59**:207–15.
117. Voortman J, Smit EF, Honeywell R, Kuenen BC, Peters GJ, van de Velde H, et al. A parallel dose-escalation study of weekly and twice-weekly bortezomib in combination with gemcitabine and cisplatin in the first-line treatment of patients with advanced solid tumors. *Clin Cancer Res* 2007;**13**:3642–51.
118. Davies AM, Ruel C, Lara PN, Lau DH, Gumerlock PH, Bold R, et al. The proteasome inhibitor bortezomib in combination with gemcitabine and carboplatin in advanced non-small cell lung cancer: a California Cancer Consortium Phase I study. *J Thorac Oncol* 2008;**3**:68–74.
119. Davies AM, Chansky K, Lara P, Gumerlock PH, Crowley J, Albain KS, et al. Bortezomib plus gemcitabine/carboplatin as first-line treatment of advanced non-small cell lung cancer: a phase II Southwest Oncology Group Study (S0339). *J Thorac Oncol* 2009;**4**:87–92.
120. Aghajanian C, Dizon DS, Sabbatini P, Raizer JJ, Dupont J, Spriggs DR. Phase I trial of bortezomib and carboplatin in recurrent ovarian or primary peritoneal cancer. *J Clin Oncol* 2005;**23**:5943–9.

121. Ramirez PT, Landen CN, Coleman RL, Milam MR, Levenback C, Johnston TA, et al. Phase I trial of the proteasome inhibitor bortezomib in combination with carboplatin in patients with platinum- and taxane-resistant ovarian cancer. *Gynecol Oncol* 2008;**108**:68–71.
122. Jandial DD, Farshchi-Heydari S, Larson CA, Elliott GI, Wrasidlo WJ, Howell SB. Enhanced delivery of cisplatin to intraperitoneal ovarian carcinomas mediated by the effects of bortezomib on the human copper transporter 1. *Clin Cancer Res* 2009;**15**:553–60.
123. Chauhan D, Li G, Shringarpure R, Podar K, Ohtake Y, Hideshima T, et al. Blockade of Hsp27 overcomes bortezomib/proteasome inhibitor PS-341 resistance in lymphoma cells. *Cancer Res* 2003;**63**:6174–7.
124. Schewe DM, Aguirre-Ghiso JA. Inhibition of eIF2α dephosphorylation maximizes bortezomib efficiency and eliminates quiescent multiple myeloma cells surviving proteasome inhibitor therapy. *Cancer Res* 2009;**69**:1545–52.
125. Busse A, Kraus M, Na IK, Rietz A, Scheibenbogen C, Driessen C, et al. Sensitivity of tumor cells to proteasome inhibitors is associated with expression levels and composition of proteasome subunits. *Cancer* 2008;**112**:659–70.
126. Chong CR, Sullivan DJ. New uses for old drugs. *Nature* 2007;**448**:645–6.
127. Collins FS. Mining for therapeutic gold. *Nat Rev Drug Discov* 2011;**10**:397.
128. Wagner-Ballon O, Pisani DF, Gastinne T, Tulliez M, Chaligne R, Lacout C, et al. Proteasome inhibitor bortezomib impairs both myelofibrosis and osteosclerosis induced by high thrombopoietin levels in mice. *Blood* 2007;**110**:345–53.
129. Berges C, Haberstock H, Fuchs D, Miltz M, Sadeghi M, Opelz G, et al. Proteasome inhibition suppresses essential immune functions of human $CD4^+$ T cells. *Immunology* 2008;**124**: 234–46.
130. Blanco B, Perez-Simon JA, Sanchez-Abarca LI, Carvajal-Vergara X, Mateos J, Vidriales B, et al. Bortezomib induces selective depletion of alloreactive T lymphocytes and decreases the production of Th1 cytokines. *Blood* 2006;**107**:3575–83.
131. Blanco B, Perez-Simon JA, Sanchez-Abarca LI, Caballero-Velazquez T, Gutierrez-Cossio S, Hernandez-Campo P, et al. Treatment with bortezomib of human $CD4^+$ T cells preserves natural regulatory T cells and allows the emergence of a distinct suppressor T-cell population. *Haematologica* 2009;**94**:975–83.
132. Everly MJ, Everly JJ, Susskind B, Brailey P, Arend LJ, Alloway RR, et al. Bortezomib provides effective therapy for antibody- and cell-mediated acute rejection. *Transplantation* 2008;**86**:1754–61.
133. Everly MJ, Everly JJ, Susskind B, Brailey P, Arend LJ, Alloway RR, et al. Proteasome inhibition reduces donor-specific antibody levels. *Transplant Proc* 2009;**41**:105–7.
134. Trivedi HL, Terasaki PI, Feroz A, Everly MJ, Vanikar AV, Shankar V, et al. Abrogation of anti-HLA antibodies via proteasome inhibition. *Transplantation* 2009;**87**:1555–61.
135. Raghavan R, Jeroudi A, Achkar K, Gaber AO, Patel SJ, Abdellatif A. Bortezomib in kidney transplantation. *J Transplant* 2010; 698594.
136. Neubert K, Meister S, Moser K, Weisel F, Maseda D, Amann K, et al. The proteasome inhibitor bortezomib depletes plasma cells and protects mice with lupus-like disease from nephritis. *Nat Med* 2008;**14**:748–55.
137. Atassi MZ, Oshima M, Deitiker P. On the initial trigger of myasthenia gravis and suppression of the disease by antibodies against the MHC peptide region involved in the presentation of a pathogenic T-cell epitope. *Crit Rev Immunol* 2001;**21**:1–27.
138. Gomez AM, Vrolix K, Martinez-Martinez P, Molenaar PC, Phernambucq M, van der Esch E, et al. Proteasome inhibition with bortezomib depletes plasma cells and autoantibodies in experimental autoimmune myasthenia gravis. *J Immunol* 2011;**186**:2503–13.
139. Piccinini M, Rinaudo MT, Chiapello N, Ricotti E, Baldovino S, Mostert M, et al. The human 26S proteasome is a target for antiretroviral agents. *AIDS* 2002;**16**:693–700.

140. Piccinini M, Rinaudo MT, Anselmino A, Buccinna B, Ramondetti C, Dematteis A, et al. The HIV protease inhibitors nelfinavir and saquinavir, but not a variety of HIV reverse transcriptase inhibitors, adversely affect human proteasome function. *Antivir Ther* 2005;**10**:215–23.
141. Lassot I, Latreille D, Rousset E, Sourisseau M, Linares LK, Chable-Bessia C, et al. The proteasome regulates HIV-1 transcription by both proteolytic and nonproteolytic mechanisms. *Mol Cell* 2007;**25**:369–83.
142. Lang JM, Touraine JL, Trepo C, Choutet P, Kirstetter M, Falkenrodt A, et al. Randomised, double-blind, placebo-controlled trial of ditiocarb sodium ('Imuthiol') in human immunodeficiency virus infection. *Lancet* 1988;**332**:702–6.
143. Reisinger EC, Kern P, Ernst M, Bock P, Flad HD, Dietrich M. Inhibition of HIV progression by dithiocarb. German DTC Study Group. *Lancet* 1990;**335**:679–82.
144. Hersh EM, Brewton G, Abrams D, Bartlett J, Galpin J, Gill P, et al. Ditiocarb sodium (diethyldithiocarbamate) therapy in patients with symptomatic HIV infection and AIDS. A randomized, double-blind, placebo-controlled, multicenter study. *JAMA* 1991;**265**:1538–44.
145. The HIV87 Study Group . Multicenter, randomized, placebo-controlled study of ditiocarb (Imuthiol) in human immunodeficiency virus-infected asymptomatic and minimally symptomatic patients. The HIV87 Study Group. *AIDS Res Hum Retroviruses* 1993;**9**:83–9.
146. Vanham GL, Kestens L, Van Hoof J, Penne G, Colebunders R, Goilav C, et al. Immunological parameters during treatment with ditiocarb (Imuthiol). *AIDS* 1993;**7**:525–30.
147. Cvek B. Failure of ditiocarb (diethyldithiocarbamate) therapy: was diet the reason? *Curr HIV Res* 2009;**7**:254.
148. Kreidenweiss A, Kremsner PG, Mordmuller B. Comprehensive study of proteasome inhibitors against Plasmodium falciparum laboratory strains and field isolates from Gabon. *Malar J* 2008;**7**:187.
149. Gandotra S, Schnappinger D, Monteleone M, Hillen W, Ehrt S. In vivo gene silencing identifies the Mycobacterium tuberculosis proteasome as essential for the bacteria to persist in mice. *Nat Med* 2007;**13**:1515–20.
150. Lin G, Li D, de Carvalho LP, Deng H, Tao H, Vogt G, et al. Inhibitors selective for mycobacterial versus human proteasomes. *Nature* 2009;**461**:621–6.
151. Lonial S, Waller EK, Richardson PG, Jagannath S, Orlowski RZ, Giver CR, et al. Risk factors and kinetics of thrombocytopenia associated with bortezomib for relapsed, refractory multiple myeloma. *Blood* 2005;**106**:3777–84.
152. Richardson PG, Briemberg H, Jagannath S, Wen PY, Barlogie B, Berenson J, et al. Frequency, characteristics, and reversibility of peripheral neuropathy during treatment of advanced multiple myeloma with bortezomib. *J Clin Oncol* 2006;**24**:3113–20.
153. Ding Q, Dimayuga E, Markesbery WR, Keller JN. Proteasome inhibition induces reversible impairments in protein synthesis. *FASEB J* 2006;**20**:1055–63.
154. Cavaletti G, Gilardini A, Canta A, Rigamonti L, Rodriguez-Menendez V, Ceresa C, et al. Bortezomib-induced peripheral neurotoxicity: a neurophysiological and pathological study in the rat. *Exp Neurol* 2007;**204**:317–25.
155. Casafont I, Berciano MT, Lafarga M. Bortezomib induces the formation of nuclear poly(A) RNA granules enriched in Sam68 and PABPN1 in sensory ganglia neurons. *Neurotox Res* 2010;**17**:167–78.
156. Cata JP, Weng HR, Burton AW, Villareal H, Giralt S, Dougherty PM. Quantitative sensory findings in patients with bortezomib-induced pain. *J Pain* 2007;**8**:296–306.
157. Argyriou AA, Iconomou G, Kalofonos HP. Bortezomib-induced peripheral neuropathy in multiple myeloma: a comprehensive review of the literature. *Blood* 2008;**112**:1593–9.
158. Badros A, Goloubeva O, Dalal JS, Can I, Thompson J, Rapoport AP, et al. Neurotoxicity of bortezomib therapy in multiple myeloma: a single-center experience and review of the literature. *Cancer* 2007;**110**:1042–9.

159. Corso A, Mangiacavalli S, Varettoni M, Pascutto C, Zappasodi P, Lazzarino M. Bortezomib-induced peripheral neuropathy in multiple myeloma: a comparison between previously treated and untreated patients. *Leuk Res* 2010;**34**:471–4.
160. Richardson PG, Xie W, Mitsiades C, Chanan-Khan AA, Lonial S, Hassoun H, et al. Single-agent bortezomib in previously untreated multiple myeloma: efficacy, characterization of peripheral neuropathy, and molecular correlations with response and neuropathy. *J Clin Oncol* 2009;**27**:3518–25.
161. El-Cheikh J, Stoppa AM, Bouabdallah R, de Lavallade H, Coso D, de Collela JM, et al. Features and risk factors of peripheral neuropathy during treatment with bortezomib for advanced multiple myeloma. *Clin Lymphoma Myeloma* 2008;**8**:146–52.
162. Richardson PG, Sonneveld P, Schuster MW, Stadtmauer EA, Facon T, Harousseau JL, et al. Reversibility of symptomatic peripheral neuropathy with bortezomib in the phase III APEX trial in relapsed multiple myeloma: impact of a dose-modification guideline. *Br J Haematol* 2009;**144**:895–903.
163. Miyakoshi S, Kami M, Yuji K, Matsumura T, Takatoku M, Sasaki M, et al. Severe pulmonary complications in Japanese patients after bortezomib treatment for refractory multiple myeloma. *Blood* 2006;**107**:3492–4.
164. Ohri A, Arena FP. Severe pulmonary complications in African-American patient after bortezomib therapy. *Am J Ther* 2006;**13**:553–5.
165. Boyer JE, Batra RB, Ascensao JL, Schechter GP. Severe pulmonary complication after bortezomib treatment for multiple myeloma. *Blood* 2006;**108**:1113.
166. Gotoh A, Ohyashiki K, Oshimi K, Usui N, Hotta T, Dan K, et al. Lung injury associated with bortezomib therapy in relapsed/refractory multiple myeloma in Japan: a questionnaire-based report from the "lung injury by bortezomib" joint committee of the Japanese society of hematology and the Japanese society of clinical hematology. *Int J Hematol* 2006;**84**:406–12.
167. Narimatsu H, Hori A, Matsumura T, Kodama Y, Takita M, Kishi Y, et al. Cooperative relationship between pharmaceutical companies, academia, and media explains sharp decrease in frequency of pulmonary complications after bortezomib in Japan. *J Clin Oncol* 2008;**26**:5820–3.
168. Uy GL, Goyal SD, Fisher NM, Oza AY, Tomasson MH, Stockerl-Goldstein K, et al. Bortezomib administered pre-auto-SCT and as maintenance therapy post transplant for multiple myeloma: a single institution phase II study. *Bone Marrow Transplant* 2009;**43**:793–800.
169. Chanan-Khan A, Sonneveld P, Schuster MW, Stadtmauer EA, Facon T, Harousseau JL, et al. Analysis of herpes zoster events among bortezomib-treated patients in the phase III APEX study. *J Clin Oncol* 2008;**26**:4784–90.
170. Perfetti V, Palladini G, Brunetti L, Sgarella A, Brugnatelli S, Gobbi PG, et al. Bortezomib-induced paralytic ileus is a potential gastrointestinal side effect of this first-in-class anticancer proteasome inhibitor. *Eur J Gastroenterol Hepatol* 2007;**19**:599–601.
171. Moon SJ, Min CK, Lee DG, Lee S, Lee JW, Min WS, et al. Pseudomembranous colitis following bortezomib therapy in a myeloma patient. *Acta Haematol* 2007;**117**:211–4.
172. Stringer AM, Gibson RJ, Bowen JM, Logan RM, Yeoh AS, Keefe DM. Chemotherapy-induced mucositis: the role of gastrointestinal microflora and mucins in the luminal environment. *J Support Oncol* 2007;**5**:259–67.
173. Gibson RJ, Keefe DM. Cancer chemotherapy-induced diarrhoea and constipation: mechanisms of damage and prevention strategies. *Support Care Cancer* 2006;**14**:890–900.
174. Siniscalchi A, Tendas A, Ales M, Fratoni S, Cupelli L, Dentamaro T, et al. Bortezomib-related colon mucositis in a multiple myeloma patient. *Support Care Cancer* 2009;**17**:325–7.
175. Knoops L, Jacquemain A, Tennstedt D, Theate I, Ferrant A, Van den Neste E. Bortezomib-induced Sweet syndrome. *Br J Haematol* 2005;**13**:142.

176. Murase JE, Wu JJ, Theate I, Cole GW, Barr RJ, Dyson SW. Bortezomib-induced histiocytoid Sweet syndrome. *J Am Acad Dermatol* 2009;**60**:496–7.
177. Maseda D, Meister S, Neubert K, Herrmann M, Voll RE. Proteasome inhibition drastically but reversibly impairs murine lymphocyte development. *Cell Death Differ* 2008;**15**:600–12.
178. Hernandez-Espinosa D, Minano A, Martinez C, Ordonez A, Perez-Ceballos E, de Arriba F, et al. Inhibition of proteasome by bortezomib causes intracellular aggregation of hepatic serpins and increases the latent circulating form of antithrombin. *Lab Invest* 2008;**88**:306–17.
179. Herrmann J, Saguner AM, Versari D, Peterson TE, Chade A, Olson M, et al. Chronic proteasome inhibition contributes to coronary atherosclerosis. *Circ Res* 2007;**101**:865–74.
180. Chade AR, Herrmann J, Zhu X, Krier JD, Lerman A, Lerman LO. Effects of proteasome inhibition on the kidney in experimental hypercholesterolemia. *J Am Soc Nephrol* 2005;**16**:1005–12.
181. Arastu-Kapur S, Anderl JL, Kraus M, Parlati F, Shenk KD, Lee SJ, et al. Nonproteasomal targets of the proteasome inhibitors bortezomib and carfilzomib: a link to clinical adverse events. *Clin Cancer Res* 2011;**17**:2734–43.
182. Chen D, Frezza M, Schmitt S, Kanwar J, Dou QP. Bortezomib as the first proteasome inhibitor anticancer drug: current status and future perspectives. *Curr Cancer Drug Targets* 2011;**11**:239–53.
183. Kisselev AF, Goldberg AL. Proteasome inhibitors: from research tools to drug candidates. *Chem Biol* 2001;**8**:739–58.
184. Groll M, Huber R, Moroder L. The persisting challenge of selective and specific proteasome inhibition. *J Pept Sci* 2009;**15**:58–66.
185. Vivier M, Rapp M, Papon J, Labarre P, Galmier MJ, Sauziere J, et al. Synthesis, radiosynthesis, and biological evaluation of new proteasome inhibitors in a tumor targeting approach. *J Med Chem* 2008;**51**:1043–7.
186. Zhu Y, Zhao X, Zhu X, Wu G, Li Y, Ma Y, et al. Design, synthesis, biological evaluation, and structure-activity relationship (SAR) discussion of dipeptidyl boronate proteasome inhibitors, part I: comprehensive understanding of the SAR of alpha-amino acid boronates. *J Med Chem* 2009;**52**:4192–9.
187. Dorsey BD, Iqbal M, Chatterjee S, Menta E, Bernardini R, Bernareggi A, et al. Discovery of a potent, selective, and orally active proteasome inhibitor for the treatment of cancer. *J Med Chem* 2008;**51**:1068–72.
188. Agyin JK, Santhamma B, Nair HB, Roy SS, Tekmal RR. BU-32: a novel proteasome inhibitor for breast cancer. *Breast Cancer Res* 2009;**11**:R74.
189. Kupperman E, Lee EC, Cao Y, Bannerman B, Fitzgerald M, Berger A, et al. Evaluation of the proteasome inhibitor MLN9708 in preclinical models of human cancer. *Cancer Res* 2010;**70**:1970–80.
190. Zhou HJ, Aujay MA, Bennett MK, Dajee M, Demo SD, Fang Y, et al. Design and synthesis of an orally bioavailable and selective peptide epoxyketone proteasome inhibitor (PR-047). *J Med Chem* 2009;**52**:3028–38.
191. Coleman CS, Rocetes JP, Park DJ, Wallick CJ, Warn-Cramer BJ, Michel K, et al. Syringolin A, a new plant elicitor from the phytopathogenic bacterium Pseudomonas syringae pv. syringae, inhibits the proliferation of neuroblastoma and ovarian cancer cells and induces apoptosis. *Cell Prolif* 2006;**39**:599–609.
192. Groll M, Schellenberg B, Bachmann AS, Archer CR, Huber R, Powell TK, et al. A plant pathogen virulence factor inhibits the eukaryotic proteasome by a novel mechanism. *Nature* 2008;**452**:755–8.
193. Clerc J, Groll M, Illich DJ, Bachmann AS, Huber R, Schellenberg B, et al. Synthetic and structural studies on syringolin A and B reveal critical determinants of selectivity and potency of proteasome inhibition. *Proc Natl Acad Sci USA* 2009;**106**:6507–12.

194. Clerc J, Florea BI, Kraus M, Groll M, Huber R, Bachmann AS, et al. Syringolin A selectively labels the 20 S proteasome in murine EL4 and wild-type and bortezomib-adapted leukaemic cell lines. *Chembiochem* 2009;**10**:2638–43.
195. Groll M, Huber R, Potts BC. Crystal structures of Salinosporamide A (NPI-0052) and B (NPI-0047) in complex with the 20S proteasome reveal important consequences of beta-lactone ring opening and a mechanism for irreversible binding. *J Am Chem Soc* 2006;**128**:5136–41.
196. McGlinchey RP, Nett M, Eustaquio AS, Asolkar RN, Fenical W, Moore BS. Engineered biosynthesis of antiprotealide and other unnatural salinosporamide proteasome inhibitors. *J Am Chem Soc* 2008;**130**:7822–3.
197. Fenical W, Jensen PR, Palladino MA, Lam KS, Lloyd GK, Potts BC. Discovery and development of the anticancer agent salinosporamide A (NPI-0052). *Bioorg Med Chem* 2009;**17**:2175–80.
198. Kisselev AF, Callard A, Goldberg AL. Importance of the different proteolytic sites of the proteasome and the efficacy of inhibitors varies with the protein substrate. *J Biol Chem* 2006;**281**:8582–90.
199. Dick LR, Fleming PE. Building on bortezomib: second-generation proteasome inhibitors as anti-cancer therapy. *Drug Discov Today* 2010;**15**:243–9.
200. Milacic V, Chen D, Ronconi L, Landis-Piwowar KR, Fregona D, Dou QP. A novel anticancer gold(III) dithiocarbamate compound inhibits the activity of a purified 20S proteasome and 26S proteasome in human breast cancer cell cultures and xenografts. *Cancer Res* 2006;**66**:10478–86.
201. Milacic V, Ping QP. The tumor proteasome as a novel target for gold(III) complexes: implication for breast cancer therapy. *Coord Chem Rev* 2009;**253**:1649–60.
202. Zhang X, Frezza M, Milacic V, Ronconi L, Fan Y, Bi C, et al. Inhibition of tumor proteasome activity by gold-dithiocarbamato complexes via both redox-dependent and -independent processes. *J Cell Biochem* 2010;**109**:162–72.
203. Chen D, Frezza M, Shakya R, Cui QC, Milacic V, Verani CN, et al. Inhibition of the proteasome activity by gallium(III) complexes contributes to their anti prostate tumor effects. *Cancer Res* 2007;**67**:9258–65.
204. Yang H, Landis-Piwowar K, Chan TH, Dou QP. Green tea polyphenols as proteasome inhibitors: implication in chemoprevention. *Curr Cancer Drug Targets* 2011;**11**:296–306.
205. Cecarini V, Cuccioloni M, Mozzicafreddo M, Bonfili L, Angeletti M, Eleuteri AM. Targeting proteasomes with natural occurring compounds in cancer treatment. *Curr Cancer Drug Targets* 2011;**11**:307–24.
206. Landis-Piwowar KR, Huo C, Chen D, Milacic V, Shi G, Chan TH, et al. A novel prodrug of the green tea polyphenol (-)-epigallocatechin-3-gallate as a potential anticancer agent. *Cancer Res* 2007;**67**:4303–10.
207. Milacic V, Banerjee S, Landis-Piwowar KR, Sarkar FH, Majumdar AP, Dou QP. Curcumin inhibits the proteasome activity in human colon cancer cells in vitro and in vivo. *Cancer Res* 2008;**68**:7283–92.
208. Tiedemann RE, Schmidt J, Keats JJ, Shi CX, Zhu YX, Palmer SE, et al. Identification of a potent natural triterpenoid inhibitor of proteosome chymotrypsin-like activity and NF-κB with antimyeloma activity in vitro and in vivo. *Blood* 2009;**113**:4027–37.
209. Yang H, Chen D, Cui QC, Yuan X, Dou QP. Celastrol, a triterpene extracted from the Chinese "Thunder of God Vine", is a potent proteasome inhibitor and suppresses human prostate cancer growth in nude mice. *Cancer Res* 2006;**66**:4758–65.
210. Yang H, Shi G, Dou QP. The tumor proteasome is a primary target for the natural anticancer compound Withaferin A isolated from "Indian winter cherry" *Mol Pharmacol* 2007;**71**:426–37.
211. Baumann P, Muller K, Mandl-Weber S, Leban J, Doblhofer R, Ammendola A, et al. The peptide-semicarbazone S-2209, a representative of a new class of proteasome inhibitors,

induces apoptosis and cell growth arrest in multiple myeloma cells. *Br J Haematol* 2009;**144**:875–86.
212. Nickeleit I, Zender S, Sasse F, Geffers R, Brandes G, Sorensen I, et al. Argyrin A reveals a critical role for the tumor suppressor protein p27(kip1) in mediating antitumor activities in response to proteasome inhibition. *Cancer Cell* 2008;**14**:23–35.
213. Kazi A, Lawrence H, Guida WC, McLaughlin ML, Springett GM, Berndt N, et al. Discovery of a novel proteasome inhibitor selective for cancer cells over non-transformed cells. *Cell Cycle* 2009;**8**:1940–51.
214. Gustafsson L, Aits S, Onnerfjord P, Trulsson M, Storm P, Svanborg C. Changes in proteasome structure and function caused by HAMLET in tumor cells. *PLoS One* 2009;**4**:e5229.
215. Kristiansen M, Deriziotis P, Dimcheff DE, Jackson GS, Ovaa H, Naumann H, et al. Disease-associated prion protein oligomers inhibit the 26S proteasome. *Mol Cell* 2007;**26**:175–88.
216. Goldberg AL. On prions, proteasomes, and mad cows. *N Engl J Med* 2007;**357**:1150–2.
217. Chauhan D, Catley L, Li G, Podar K, Hideshima T, Velankar M, et al. A novel orally active proteasome inhibitor induces apoptosis in multiple myeloma cells with mechanisms distinct from bortezomib. *Cancer Cell* 2005;**8**:407–719.
218. Potts BC, Albitar MX, Anderson KC, Baritaki S, Berkers C, Bonavida B, et al. Marizomib, a proteasome inhibitor for all seasons: preclinical profile and a framework for clinical trials. *Curr Cancer Drug Targets* 2011;**11**:254–84.
219. Singh AV, Palladino MA, Lloyd GK, Potts BC, Chauhan D, Anderson KC. Pharmacodynamic and efficacy studies of the novel proteasome inhibitor NPI-0052 (marizomib) in a human plasmacytoma xenograft murine model. *Br J Haematol* 2010;**149**:550–9.
220. Roccaro AM, Leleu X, Sacco A, Jia X, Melhem M, Moreau AS, et al. Dual targeting of the proteasome regulates survival and homing in Waldenstrom macroglobulinemia. *Blood* 2008;**111**:4752–63.
221. Chauhan D, Singh A, Brahmandam M, Podar K, Hideshima T, Richardson P, et al. Combination of proteasome inhibitors bortezomib and NPI-0052 trigger in vivo synergistic cytotoxicity in multiple myeloma. *Blood* 2008;**111**:1654–64.
222. Chauhan D, Singh AV, Ciccarelli B, Richardson PG, Palladino MA, Anderson KC. Combination of novel proteasome inhibitor NPI-0052 and lenalidomide trigger in vitro and in vivo synergistic cytotoxicity in multiple myeloma. *Blood* 2010;**115**:834–45.
223. Ahn KS, Sethi G, Chao TH, Neuteboom ST, Chaturvedi MM, Palladino MA, et al. Salinosporamide A (NPI-0052) potentiates apoptosis, suppresses osteoclastogenesis, and inhibits invasion through down-modulation of NF-κB regulated gene products. *Blood* 2007;**110**: 2286–95.
224. Hideshima T, Ikeda H, Chauhan D, Okawa Y, Raje N, Podar K, et al. Bortezomib induces canonical nuclear factor-kappaB activation in multiple myeloma cells. *Blood* 2009;**114**: 1046–52.
225. Baritaki S, Yeung K, Palladino M, Berenson J, Bonavida B. Pivotal roles of snail inhibition and RKIP induction by the proteasome inhibitor NPI-0052 in tumor cell chemoimmunosensitization. *Cancer Res* 2009;**69**:8376–85.
226. Miller CP, Rudra S, Keating MJ, Wierda WG, Palladino M, Chandra J. Caspase-8 dependent histone acetylation by a novel proteasome inhibitor, NPI-0052: a mechanism for synergy in leukemia cells. *Blood* 2009;**113**:4289–99.
227. Cusack JC, Liu R, Xia L, Chao TH, Pien C, Niu W, et al. NPI-0052 enhances tumoricidal response to conventional cancer therapy in a colon cancer model. *Clin Cancer Res* 2006;**12**:6758–64.
228. Vlashi E, Mattes M, Lagadec C, Donna LD, Phillips TM, Nikolay P, et al. Differential Effects of the Proteasome Inhibitor NPI-0052 against Glioma Cells. *Transl Oncol* 2010;**3**:50–5.

229. Vlashi E, Kim K, Lagadec C, Donna LD, McDonald JT, Eghbali M, et al. *J Natl Cancer Inst* 2009;**101**:350–9.
230. Hamlin PA, Aghajanian C, Younes A, Hong DS, Palladino MA, Longenecker AM, et al. First-in-human phase I study of the novel structure proteasome inhibitor NPI-0052. *J Clin Oncol* 2009;**27**:3516.
231. Millward M, Price T, Mainwaring P, Spencer A, Longenecker A, Palladino MA, et al. Clinical trial of NPI-0052 in advanced malignancies including lymphoma and leukemia (advanced malignancies arm). *J Clin Oncol* 2009;**27**:3582.
232. Millward M, Price T, Townsend A, Sweeney C, Spencer A, Sukumaran S, et al. Phase 1 clinical trial of the novel proteasome inhibitor marizomib with the histone deacetylase inhibitor vorinostat in patients with melanoma, pancreatic and lung cancer based on in vitro assessments of the combination. *Invest New Drugs* 2011; in press. Epud ahead of print: http://www.ncbi.nlm.nih.gov/pubmed/22080430.
233. Demo SD, Buchholz TJ, Laidig GJ, Parlati F, Shenk KD, Smyth MS, et al. Biochemical and cellular characterization of the novel proteasome inhibitor PR-171. *Blood* 2005;**106**:1588.
234. Parlati F, Lee SJ, Aujay M, Suzuki E, Levitsky K, Lorens JB, et al. Carfilzomib can induce tumor cell death through selective inhibition of the chymotrypsin-like activity of the proteasome. *Blood* 2009;**114**:3439–47.
235. Demo SD, Kirk CJ, Aujay MA, Buchholz TJ, Dajee M, Ho MN, et al. Antitumor activity of PR-171, a novel irreversible inhibitor of the proteasome. *Cancer Res* 2007;**67**:6383–91.
236. Kuhn DJ, Chen Q, Voorhees PM, Strader JS, Shenk KD, Sun CM, et al. Potent activity of carfilzomib, a novel, irreversible inhibitor of the ubiquitin-proteasome pathway, against preclinical models of multiple myeloma. *Blood* 2007;**110**:3281–90.
237. Orlowski RZ, Stewart K, Vallone M, Molineaux C, Kunkel L, Gericitano J, et al. Safety and antitumor efficacy of the proteasome inhibitor carfilzomib (PR-171) dosed for five consecutive days in hematologic malignancies: phase I results. *Blood* 2007;**110**:409.
238. Alsina M, Trudel S, Vallone M, Molineaux C, Kunkel L, Goy A. Phase 1 single agent antitumor activity of twice weekly consecutive day dosing of the proteasome inhibitor carfitzomib (PR-171) in hematologic malignancies. *Blood* 2007;**110**:411.
239. O'Connor OA, Stewart AK, Vallone M, Molineaux CJ, Kunkel LA, Gerecitano JF, et al. A phase 1 dose escalation study of the safety and pharmacokinetics of the novel proteasome inhibitor carfilzomib (PR-171) in patients with hematologic malignancies. *Clin Cancer Res* 2009;**15**:7085–91.
240. Jagannath S, Vij R, Stewart K, Somlo G, Jakubowiak A, Trudel S, et al. Final results of PX-171-003-A0, part 1 of an open-label, single-arm, phase II study of carfilzomib (CFZ) in patients (pts) with relapsed and refractory multiple myeloma (MM). *J Clin Oncol* 2009;**27**:8504.
241. Wang L, Siegel D, Kaufman JL, Stewart AK, Jakubowiak AJ, Alsina M, et al. Updated results of bortezomib-naïve patients in PX-171-004, an ongoing open-label, phase II study of single-agent carfilzomib (CFZ) in patients with relapsed or refractory Myeloma (MM). *Blood* 2009;**114**:127–8.
242. Siegel D, Wang L, Orlowski RZ, Kaufman JL, Stewart AK, Kukreti V, et al. PX-171-004, an ongoing open-label, phase II study of single-agent carfilzomib (CFZ) in patients with relapsed or refractory myeloma (MM); updated results from the bortezomib-treated cohort. *Blood* 2009;**114**:128.
243. Jakubowiak A, Wang L, Orlowski RZ, Jagannath S, Siegel D, Stewart AK, et al. Influence of cytogenetics in patients with relapsed and refractory multiple myeloma (MM) treated with carfilzomib (CFZ). *Blood* 2009;**114**:725–6.
244. Chang H, Trieu Y, Qi X, Xu W, Stewart KA, Reece D. Bortezomib therapy response is independent of cytogenetic abnormalities in relapsed/refractory multiple myeloma. *Leuk Res* 2007;**31**:779–82.

245. Jagannath S, Richardson PG, Sonneveld P, Schuster MW, Irwin D, Stadtmauer EA, et al. Bortezomib appears to overcome the poor prognosis conferred by chromosome 13 deletion in phase 2 and 3 trials. *Leukemia* 2007;**21**:151–7.
246. Cavaletti G, Jakubowiak AJ. Peripheral neuropathy during bortezomib treatment of multiple myeloma: a review of recent studies. *Leuk Lymphoma* 2010;**51**:1178–87.
247. Piva R, Ruggeri B, Williams M, Costa G, Tamagno I, Ferrero D, et al. CEP-18770: a novel, orally active proteasome inhibitor with a tumor-selective pharmacologic profile competitive with bortezomib. *Blood* 2008;**111**:2765–75.
248. Sanchez E, Li M, Steinberg JA, Wang C, Shen J, Bonavida B, et al. The proteasome inhibitor CEP-18770 enhances the anti-myeloma activity of bortezomib and melphalan. *Br J Haematol* 2010;**148**:569–81.
249. Chauhan D, Tian Z, Zhou B, Kuhn D, Orlowski R, Raje N, et al. In vitro and in vivo selective antitumor activity of a novel orally bioavailable proteasome inhibitor MLN9708 against multiple myeloma cells. *Clin Cancer Res* 2011;**17**:5311–21.
250. Lee EC, Fitzgerald M, Bannerman B, Donelan J, Bano K, Terkelsen J, et al. Antitumor activity of the investigational proteasome inhibitor MLN9708 in mouse models of B-cell and plasma cell malignancies. *Clin Cancer Res* 2011;**17**:7313–23.
251. Dufour P, Lang JM, Giron C, Duclos B, Haehnel P, Jaeck D, et al. Sodium dithiocarb as adjuvant immunotherapy for high risk breast cancer: a randomized study. *Biotherapy* 1993;**6**:9–12.
252. Eneanya DI, Bianchine JR, Duran DO, Andresen BD. The actions and metabolic fate of disulfiram. *Annu Rev Pharmacol Toxicol* 1981;**21**:575–96.
253. Brar SS, Grigg C, Wilson KS, Holder WD, Dreau D, Austin C, et al. Disulfiram inhibits activating transcription factor/cyclic AMP-responsive element binding protein and human melanoma growth in a metal-dependent manner in vitro, in mice and in a patient with metastatic disease. *Mol Cancer Ther* 2004;**3**:1049–60.
254. Cvek B. Targeting malignancies with disulfiram (Antabuse): multidrug resistance, angiogenesis, and proteasome. *Curr Cancer Drug Targets* 2011;**11**:332–7.
255. Johansson B, Stankiewicz Z. Bis-(diethyldithiocarbamato) copper complex: a new metabolite of disulfiram? *Biochem Pharmacol* 1985;**34**:2989–91.
256. Suzuki Y, Fujii S, Tominaga T, Yoshimoto T, Yoshimura T, Kamada H. The origin of an EPR signal observed in dithiocarbamate-loaded tissues. Copper(II)-dithiocarbamate complexes account for the narrow hyperfine lines. *Biochim Biophys Acta* 1997;**1335**:242–5.
257. Cvek B, Dvorak Z. Targeting of nuclear factor-κB and proteasome by dithiocarbamate complexes with metals. *Curr Pharm Des* 2007;**13**:3155–67.
258. Chen D, Cui QC, Yang H, Dou QP. Disulfiram, a clinically used anti-alcoholism drug and copper-binding agent, induces apoptotic cell death in breast cancer cultures and xenografts via inhibition of the proteasome activity. *Cancer Res* 2006;**66**:10425–33.
259. Daniel KG, Chen D, Yan B, Dou QP. Copper-binding compounds as proteasome inhibitors and apoptosis inducers in human cancer. *Front Biosci* 2007;**12**:135–44.
260. Hindo SS, Frezza M, Tomco D, Heeg MJ, Hryhorczuk L, McGarvey BR, et al. Metals in anticancer therapy: copper(II) complexes as inhibitors of the 20S proteasome. *Eur J Med Chem* 2009;**44**:4353–61.
261. Cvek B, Milacic V, Taraba J, Dou QP. Ni(II), Cu(II), and Zn(II) diethyldithiocarbamate complexes show various activities against the proteasome in breast cancer cells. *J Med Chem* 2008;**51**:6256–8.
262. Verma R, Aravind L, Oania R, McDonald WH, Yates JR, Koonin EV, et al. Role of Rpn11 metalloprotease in deubiquitination and degradation by the 26S proteasome. *Science* 2002;**298**:611–5.

263. Gallery M, Blank JL, Lin Y, Gutierrez JA, Pulido JC, Rappoli D, et al. The JAMM motif of human deubiquitinase Poh1 is essential for cell viability. *Mol Cancer Ther* 2007;**6**:262–8.
264. Boguski MS, Mandl KD, Sukhatme VP. Drug discovery. Repurposing with a difference. *Science* 2009;**324**:1394–5.
265. Cvek B. Antabuse (disulfiram) as a pilot case of non-profit drug. *Int J Cancer* 2010;**127**:2486.
266. Schimmer AD. Clioquinol - a novel copper-dependent and independent proteasome inhibitor. *Curr Cancer Drug Targets* 2011;**11**:325–31.
267. Richards DA. The prophylactic value of clioquinol against travellers diarrhoea. *Lancet* 1971;**1**:44–5.
268. Woodward WE, Rahman AS. Trial of clioquinol in cholera. *Lancet* 1969;**2**:270.
269. Mao X, Li X, Sprangers R, Wang X, Venugopal A, Wood T, et al. Clioquinol inhibits the proteasome and displays preclinical activity in leukemia and myeloma. *Leukemia* 2009;**23**:585–90.
270. Yu H, Zhou Y, Lind SE, Ding WQ. Clioquinol targets zinc to lysosomes in human cancer cells. *Biochem J* 2009;**417**:133–9.
271. Yu H, Lou JR, Ding WQ. Clioquinol independently targets NF-κB and lysosome pathways in human cancer cells. *Anticancer Res* 2010;**30**:2087–92.
272. Zheng J, Benbrook DM, Yu H, Ding WQ. Clioquinol suppresses cyclin D1 gene expression through transcriptional and post-transcriptional mechanisms. *Anticancer Res* 2011;**31**:2739–41.
273. Daniel KG, Chen D, Orlu S, Cui QC, Miller FR, Dou QP. Clioquinol and pyrrolidine dithiocarbamate complex with copper to form proteasome inhibitors and apoptosis inducers in human breast cancer cells. *Breast Cancer Res* 2005;**7**:897–908.
274. Chen D, Cui QC, Yang H, Barrea RA, Sarkar FH, Sheng S, et al. Clioquinol, a therapeutic agent for Alzheimer's disease, has proteasome-inhibitory, androgen receptor-suppressing, apoptosis-inducing, and antitumor activities in human prostate cancer cells and xenografts. *Cancer Res* 2007;**67**:1636–44.
275. Diez M, Cerdan FJ, Arroyo M, Balibrea JL. Use of the copper/zinc ratio in the diagnosis of lung cancer. *Cancer* 1989;**63**:726–30.
276. Yoshida D, Ikeda Y, Nakazawa S. Quantitative analysis of copper, zinc and copper/zinc ratio in selected human brain tumors. *J Neurooncol* 1993;**16**:109–15.
277. Geraki K, Farquharson MJ, Bradley DA. Concentrations of Fe, Cu and Zn in breast tissue: a synchrotron XRF study. *Phys Med Biol* 2002;**47**:2327–39.
278. Zhai S, Yang L, Cui QC, Sun Y, Dou QP, Yan B. Tumor cellular proteasome inhibition and growth suppression by 8-hydroxyquinoline and clioquinol requires their capabilities to bind copper and transport copper into cells. *J Biol Inorg Chem* 2010;**15**:259–69.
279. Altun M, Galardy PJ, Shringarpure R, Hideshima T, LeBlanc R, Anderson KC, et al. Effects of PS-341 on the activity and composition of proteasomes in multiple myeloma cells. *Cancer Res* 2005;**65**:7896–901.
280. Dissemond J, Goette P, Moers J, Lindeke A, Goos M, Ferrone S, et al. Immunoproteasome subunits LMP2 and LMP7 downregulation in primary malignant melanoma lesions: association with lack of spontaneous regression. *Melanoma Res* 2003;**13**:371–7.
281. Heink S, Fricke B, Ludwig D, Kloetzel PM, Kruger E. Tumor cell lines expressing the proteasome subunit isoform LMP7E1 exhibit immunoproteasome deficiency. *Cancer Res* 2006;**66**:649–52.
282. Murakami Y, Kanda K, Yokota K, Kanayama H, Kagawa S. Prognostic significance of immuno-proteosome subunit expression in patients with renal-cell carcinoma: a preliminary study. *Mol Urol* 2001;**5**:113–9.
283. Atkins D, Breuckmann A, Schmahl GE, Binner P, Ferrone S, Krummenauer F, et al. MHC class I antigen processing pathway defects, ras mutations and disease stage in colorectal carcinoma. *Int J Cancer* 2004;**109**:265–73.

284. Meissner M, Reichert TE, Kunkel M, Gooding W, Whiteside TL, Ferrone S, et al. Defects in the human leukocyte antigen class I antigen processing machinery in head and neck squamous cell carcinoma: association with clinical outcome. *Clin Cancer Res* 2005;**11**:2552–60.
285. Mehta AM, Jordanova ES, Kenter GG, Ferrone S, Fleuren GJ. Association of antigen processing machinery and HLA class I defects with clinicopathological outcome in cervical carcinoma. *Cancer Immunol Immunother* 2008;**57**:197–206.
286. Cerruti F, Martano M, Petterino C, Bollo E, Morello E, Bruno R, et al. Enhanced expression of interferon-γ-induced antigen-processing machinery components in a spontaneously occurring cancer. *Neoplasia* 2007;**9**:960–9.
287. Kuhn DJ, Orlowski RZ, Bjorklund CC. Second generation proteasome inhibitors: carfilzomib and immunoproteasome-specific inhibitors (IPSIs). *Curr Cancer Drug Targets* 2011;**11**:285–95.
288. Ho YK, Bargagna-Mohan P, Wehenkel M, Mohan R, Kim KB. LMP2-specific inhibitors: chemical genetic tools for proteasome biology. *Chem Biol* 2007;**14**:419–30.
289. Kuhn DJ, Hunsucker SA, Chen Q, Voorhees PM, Orlowski M, Orlowski RZ. Targeted inhibition of the immunoproteasome is a potent strategy against models of multiple myeloma that overcomes resistance to conventional drugs and nonspecific proteasome inhibitors. *Blood* 2009;**113**:4667–76.
290. Muchamuel T, Basler M, Aujay MA, Suzuki E, Kalim KW, Lauer C, et al. A selective inhibitor of the immunoproteasome subunit LMP7 blocks cytokine production and attenuates progression of experimental arthritis. *Nat Med* 2009;**15**:781–7.
291. Nalepa G, Rolfe M, Harper JW. Drug discovery in the ubiquitin-proteasome system. *Nat Rev Drug Discov* 2006;**5**:596–613.
292. Hoeller D, Dikic I. Targeting the ubiquitin system in cancer therapy. *Nature* 2009;**458**:438–44.
293. Eldridge AG, O'Brien T. Therapeutic strategies within the ubiquitin proteasome system. *Cell Death Differ* 2010;**17**:4–13.
294. Yang Y, Kitagaki J, Dai RM, Tsai YC, Lorick KL, Ludwig RL, et al. Inhibitors of ubiquitin-activating enzyme (E1), a new class of potential cancer therapeutics. *Cancer Res* 2007;**67**:9472–81.
295. Lim HS, Archer CT, Kodadek T. Identification of a peptoid inhibitor of the proteasome 19S regulatory particle. *J Am Chem Soc* 2007;**129**:7750–1.
296. Lim HS, Cai D, Archer CT, Kodadek T. Periodate-triggered cross-linking reveals Sug2/Rpt4 as the molecular target of a peptoid inhibitor of the 19S proteasome regulatory particle. *J Am Chem Soc* 2007;**129**:12936–7.
297. Nakayama KI, Nakayama K. Ubiquitin ligases: cell-cycle control and cancer. *Nat Rev Cancer* 2006;**6**:369–81.
298. Soucy TA, Smith PG, Rolfe M. Targeting NEDD8-activated cullin-RING ligases for the treatment of cancer. *Clin Cancer Res* 2009;**15**:3912–6.
299. Soucy TA, Smith PG, Milhollen MA, Berger AJ, Gavin JM, Adhikari S, et al. An inhibitor of NEDD8-activating enzyme as a new approach to treat cancer. *Nature* 2009;**458**:732–6.
300. Daviet L, Colland F. Targeting ubiquitin specific proteases for drug discovery. *Biochimie* 2008;**90**:270–83.
301. Hussain S, Zhang Y, Galardy PJ. DUBs and cancer: the role of deubiquitinating enzymes as oncogenes, non-oncogenes and tumor suppressors. *Cell Cycle* 2009;**8**:1688–97.
302. Fontanini A, Foti C, Potu H, Crivellato E, Maestro R, Bernardi P, et al. The isopeptidase inhibitor G5 triggers a caspase-independent necrotic death in cells resistant to apoptosis: a comparative study with the proteasome inhibitor bortezomib. *J Biol Chem* 2009;**284**:8369–81.
303. D'Arcy P, Brnjic S, Olofsson MH, Fryknas M, Lindsten K, De Cesare M, et al. Inhibition of proteasome deubiquitinating activity as a new cancer therapy. *Nat Med* 2011;**17**:1636–40.

# Degradation of Damaged Proteins: The Main Function of the 20S Proteasome

Andrew M. Pickering and
Kelvin J. A. Davies

*Ethel Percy Andrus Gerontology Center of the Davis School of Gerontology and Division of Molecular & Computational Biology, Department of Biological Sciences, Dornsife College of Letters, Arts & Sciences: The University of Southern California, Los Angeles, California, USA*

I. Introduction ........ 228
II. Formation of Oxygen Radicals and Toxic Oxidants ........ 228
III. Removal of Oxygen Radicals and Toxic Oxidants by the Cell ........ 231
IV. The Role of the 20S Proteasome Versus the 26S Proteasome in Degrading Oxidized Proteins ........ 232
V. Mechanism of Selective Degradation of Damaged Proteins by the 20S Proteasome ........ 234
VI. The 20S Proteasome in Aging and the Removal of Age-related Protein Damage ........ 236
VII. The 20S Proteasome in Adaptation to Protein Damage ........ 238
VIII. Removal of Damaged Proteins by the 20S Proteasome in the Nucleus ........ 239
A. The Nuclear 20S Proteasome Regulator: PARP ........ 240
B. The Nuclear 20S/26S Proteasome Regulator: Pa200 ........ 240
C. The Nuclear 20S Proteasome Regulator: Pa28γ ........ 241
IX. Alternative Forms of the 20S Proteasome in the Cytoplasm and Other Non-nuclear Compartments and Their Role(s) in the Removal of Damaged Proteins ........ 241
A. Immunoproteasome and Oxidative Stress ........ 241
B. The Cytoplasmic Proteasome Activator Pa28αβ ........ 243
X. Summary ........ 243
References ........ 244

Cellular proteins are exposed to oxidative modification and other forms of damage through oxidative stress and disease, and as a consequence of aging. This oxidative damage results in loss and/or modification of protein function, which in turn compromises cell function and may even cause cell death. Therefore, the removal of damaged proteins is extremely important for the maintenance of normal cell function. The 20S proteasome functions primarily as a system for removal of such damaged proteins. Unlike the 26S proteasome, the 20S proteasome exhibits a high degree of selectivity in degrading the oxidized, or otherwise damaged, forms of cell proteins. The 20S proteasome is

Progress in Molecular Biology
and Translational Science, Vol. 109
DOI: 10.1016/B978-0-12-397863-9.00006-7

broadly distributed throughout the cell and has a range of specific functions in different organelles, which are controlled through a number of proteasome regulators. It is also activated, and its synthesis is induced, under conditions of enhanced oxidative stress, thus permitting greater removal of damaged proteins.

## I. Introduction

Protein damage is a natural and common aspect of all life forms, from multicellular endothermic mammals to single-celled prokaryotes. It can be caused in many different ways through a wide range of mechanisms. Such protein damage can occur as a result of external stimuli, including exposure to oxidants present in air pollution[1,2] and pesticides,[3] by ozone,[4,5] and by various other chemical agents. Protein damage can also be caused through exposure to radiation such as UV[6] or various forms of ionizing radiation.[7] In addition to external factors, protein damage can also be caused through internally generated oxidants and reactive species that are produced during metabolism[8] or immune responses.[9] Finally, damaged proteins can be formed through transcriptional or translational errors. Such synthesis errors can occur naturally but become significantly more prevalent with age due to a positive feedback loop from accumulating damage to transcriptional and translational machinery.[10]

Regardless of the mechanism through which it occurs, protein damage has many severe effects on cell function[11–15] and organismal viability.[16] Due to these severe outcomes, the rapid and efficient removal of damaged proteins is extremely important. In this chapter, we discuss the role of various forms of proteasome in the removal of damaged proteins. This chapter has a particular focus on the 20S proteasome, whose main function is thought to be the removal of damaged proteins. This chapter also briefly discusses the roles of the 26S proteasome and the immunoproteasome in the removal of damaged proteins.

## II. Formation of Oxygen Radicals and Toxic Oxidants

Oxygen is one of the greatest blessings, and perhaps curses, of complex life—this is what is often referred to as the "Oxygen Paradox".[13,17] Oxygen plays a vital role as the final acceptor in the electron transport chain. In the electron transport chain, a series of alternating hydrogen and electron carriers ensure that electrons are vectorially transported across the inner mitochondrial membrane such that protons are transported outward. The resulting electrochemical gradient is then used for the formation of ATP from ADP + P. In order to maintain electron flow in the inner mitochondrial membrane, electrons must also be removed from the electron transport chain by a terminal electron acceptor. This role is played by the cytochrome c oxidase complex (mitochondrial complex IV), which oxidizes electrons with oxygen to generate water, thereby "bleeding" electrons from the chain and allowing continual ATP production. The presence of oxygen is, thus, vital for mitochondrial function and overall energy balance in eucaryotes. While there are alternative processes to generate ATP that are independent of oxygen (such as fermentation), these processes are considerably less efficient than oxidative phosphorylation. As a result of the sizable difference in ATP production rates between oxygen-dependent and oxygen-independent pathways, it has been suggested by some groups that the increased efficiency of oxygen-dependent respiration over fermentation was one of the key factors that permitted the formation of complex multicellular life forms.[18]

While the role of oxygen as a final electron acceptor permits a greatly enhanced energy production rate, the reaction has the potential to generate a number of highly toxic by-products. In the electron transport chain, electrons pass through a series of electron acceptors, eventually terminating with cytochrome c oxidase, which catalyzes the formation of water. The termination of the electron transport chain seems to occur safely, with little or no detectable leakage of electrons; however, some electrons do leak out at earlier points in the chain.[19] Such leaked electrons can directly react with oxygen to form the superoxide anion radical ($O_2^{\bullet-}$), commonly referred to simply as superoxide. Free radicals such as $O_2^{\bullet-}$ possess an unpaired electron in their outer orbital (represented by the dot). These radicals have a stronger tendency to react with other molecules to either donate the extra electron or gain another, in either case restoring their electron pairing. In such oxidation/reduction (or redox) reactions, the electron donor is termed the reductant, while the electron acceptor is the oxidant. Naturally, both the oxidant and the reductant are modified, sometimes reversibly but often irreversibly, and both may go on to react further with surrounding molecules. In many cases, such free radical-driven, redox chain reactions produce species that cause serious damage to cellular constituents.[13]

Superoxide radicals are capable of reacting with iron or copper ions in the cell, thus converting them from $Fe^{3+}$ to $Fe^{2+}$ or $Cu^{2+}$ to $Cu^{+}$. This is harmful in itself as it results in the modification of metal ions, which form an important catalytic center of a wide range of cellular enzymes. In addition, the $Fe^{2+}$ or $Cu^{+}$ generated by this reaction is then capable of reacting (by the Fenton reaction[20]) with the mild oxidant hydrogen peroxide ($H_2O_2$) to form highly toxic hydroxyl radicals ($^{\bullet}OH$). This is termed the Haber–Weis reaction.[21]

Superoxide can instead react with two protons and another electron to form hydrogen peroxide. Hydrogen peroxide is a relatively mild oxidant, but it can cause modification of DNA, lipids, and proteins through reactions catalyzed by transition metals. Hydrogen peroxide also appears to be capable of modifications to some amino acid chains in proteins (e.g., amino acids containing thiol groups[22] or keto acids such as pyruvate[23]). More importantly, hydrogen peroxide is capable of reacting with additional electrons to form hydroxyl radicals, which are highly toxic and capable of reacting with a wide range of proteins as well as lipids and DNA.[24,25] Superoxide can also react with nitric oxide radicals ($NO^{\bullet}$), which are produced constantly as vasodilating agents and signaling molecules[26] to form peroxynitrite ($ONOO^{-}$). This is detrimental both through the reduction in nitric oxide signaling[27] and through the formation of peroxynitrite, which is a powerful oxidant (common free radical reactions are summarized in Fig. 1).

In addition to oxidative stress occurring through the formation of free radicals and oxidizing agents in metabolism, oxidative stress can occur through contact with a range of environmental factors. One of these environmental factors is air pollution. Air pollution contains a range of metals that can catalyze the formation of reactive oxidizing agents, such as iron, copper, chromium, and vanadium, through the Haber–Weis reaction.[28–30] Many particles in air pollution can also produce an inflammatory response, which in part involves the formation of oxygen radicals.[31] UV radiation is another source of oxidative

1. The four electron, univalent pathway for oxygen reduction

$$O_2 \xrightarrow{e^-} O_2{\bullet}^- \xrightarrow{e^- + 2H^+} H_2O_2 \xrightarrow{e^- + H^+} {\bullet}OH \xrightarrow{e^- + H^+} H_2O$$

(the third step also releases $H_2O$)

*Key:*
$O_2{\bullet}^-$ = Superoxide anion radical (or just Superoxide)
$H_2O_2$ = Hydrogen peroxide

${\bullet}OH$ = Hydroxyl radical

2. Superoxide dismutase can catalyze $H_2O_2$ formation from $O_2{\bullet}^-$

$$O_2{\bullet}^- + O_2{\bullet}^- + 2H^+ \longrightarrow H_2O_2 + O_2$$

3. Superoxide can reduce oxidized transition metal ions(such as iron and copper)

$$Fe^{3+} + O_2{\bullet}^- \longrightarrow Fe^{2+} + O_2$$

4. Reduced transition metal ions (such as iron and copper) can react with hydrogen peroxide in the Fenton reaction to catalyze hydroxyl radical production

$$Fe^{2+} + H_2O_2 \longrightarrow Fe^{3+} + {\bullet}OH + OH^-$$

5. Superoxide and transition metals can react together to drive Fenton chemistry in a doublet of steps termed the Haber-Weiss reaction.

$$O_2{\bullet}^- + Fe^{3+} \longrightarrow O_2 + Fe^{2+}$$
$$Fe^{2+} + H_2O_2 \longrightarrow Fe^{3+} + {\bullet}OH + OH^-$$

FIG. 1. Formation of oxygen radicals and reactive oxidants.

stress, the main source of which is solar irradiation, but there are also a range of lifestyle-related sources of exposure. The degree of solar exposure to UV radiation is dependent on a range of environmental factors such as cloud coverage and ozone level. UV radiation can cause the formation of both superoxide ($O_2^{\bullet -}$) and hydroxyl radicals ($^{\bullet}OH$). In addition, it can cause direct modification of proteins within the cell. Similar effects are also seen with infrared and ionizing radiation.[32]

## III. Removal of Oxygen Radicals and Toxic Oxidants by the Cell

Various oxygen and nitrogen radicals, as well as other reactive oxygen or nitrogen species, can (directly or indirectly) modify amino acids within proteins, damage vital prosthetic groups, or oxidize key transition metal centers. Although direct modification of amino acids or protein prosthetic groups typically increases local hydrophilicity (often changing or introducing charges), the net effect of protein oxidative modification is typically partial unfolding, which exposes the side chains of many hydrophobic amino acids that are normally buried in the interior of a properly folded protein. Thus, when free radicals, or related reactive species, modify a protein, they usually cause an increase in surface hydrophobicity and an increase in accessibility of protein hydrophobic groups.

While some mild modifications might be relatively harmless, other changes will reduce, inhibit, or even modify the function of the protein.[13,33,34] If severely oxidatively modified proteins are allowed to accumulate, cell function will be progressively compromised.[13] Thus, it is extremely important for any aerobic cell to have a system for removing oxidatively damaged proteins. It is worth noting that both DNA and lipid oxidation also present a serious challenge to maintenance of cell function, and aerobic cells have well-developed systems for combating these types of damage as well. For the purpose of this chapter, however, the focus is solely on protein oxidation.

The cells' first line of defense is the removal of oxygen radicals and the detoxification of reactive oxidants, to stop proteins becoming modified in the first place. Hydrogen peroxide is a relatively mild oxidant, although it can be reduced to generate hydroxyl radicals, which are considerably more reactive (Fig. 1). Because of this, it is important for hydrogen peroxide to be removed. This is done primarily through glutathione peroxidase,[35] which couples the reduction of hydrogen peroxide into water with the oxidation of glutathione (GSH). The oxidized glutathione will then react with another oxidized glutathione to form glutathione disulfide (GSSG). The glutathione disulfide is later reduced back to two glutathione molecules in a reaction catalyzed by glutathione reductase.[36] Catalase is another powerful enzyme that can remove hydrogen peroxide, but its actions seem limited to peroxisomes in eukaryotic cells.

In addition to hydrogen peroxide, superoxide is also capable of reacting with copper, iron, or nitric oxide radicals. As a result of this, even with efficient removal of hydrogen peroxide through glutathione peroxidase, there are other pathways by which protein damage can occur. In response to this, cells have evolved superoxide dismutase (SOD).[37] SOD catalyzes the formation of hydrogen peroxide from oxygen radicals, thus decreasing the chance of the superoxide reacting with copper, iron, or nitric oxide. Although it may seem counterintuitive for cells to produce an enzyme that generates hydrogen peroxide, it seems that the combination of SOD and glutathione peroxidase ensures that most of the superoxide produced is converted harmlessly to water without going through the hydroxyl radical step in the univalent pathway for oxygen reduction. Minimizing the production of hydroxyl radical, the most powerful oxygen radical of biological significance, seems to be the strategy that our cells employ.

In addition, there is a range of other antioxidant mechanisms, largely based on antioxidant compounds derived from fruits and vegetables in the diet, that are utilized by the cell. Ascorbic acid (vitamin C) is one of these; it is involved in a similar reaction to glutathione in which it functions as an electron acceptor for oxygen radicals, resulting in its conversion to dehydroascorbic acid (DHA). DHA may subsequently be reduced back to ascorbic acid by a reaction with glutathione in the endoplasmic reticulum.[38] Another example is vitamin E,[19] which functions as a scavenger of peroxyl radicals. Many other phenolic or polyphenolic dietary components have also been suggested as important biological antioxidants.

## IV. The Role of the 20S Proteasome Versus the 26S Proteasome in Degrading Oxidized Proteins

Although the first-line defenses of antioxidant enzymes and compounds described above are quite effective, they are not perfect, and some oxygen radicals and other reactive oxygen species do get through and cause protein damage. To combat this problem, the cell has mechanisms in place to remove damaged proteins and enable normal cell function to continue. Proteasome plays an important role in the removal of these damaged proteins. It has been shown that when a cellular protein is damaged (up to a point), it will be preferentially degraded, compared to a non-damaged protein.[39,40] When proteasome is removed from the cell, using immunoprecipitation, or depleted,

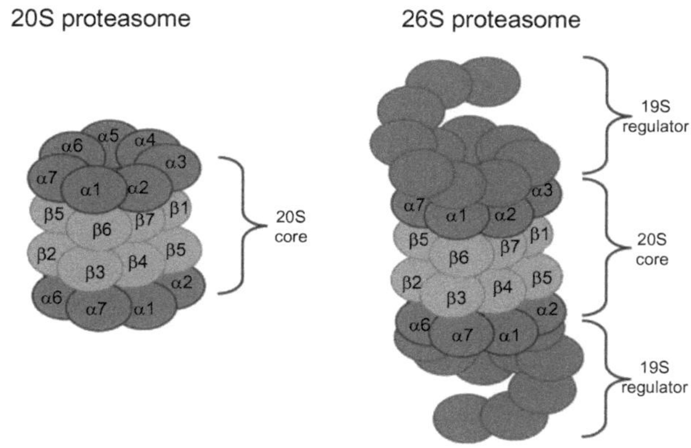


FIG. 2. Structures of the 20S and 26S proteasomes, based on diagrams published in McNaught et al. (2001).[42]

using antisense RNA or siRNA, most of the cells' ability to remove damaged proteins is lost, suggesting that proteasome is important in the removal of damaged cellular proteins.[39,41]

Proteasome comes in different forms, with many different potential activators or regulators. A strong candidate for the removal of damaged proteins is the 20S proteasome. The four-ringed 20S proteasome is the basic or core form of the proteasome, to which two 19S regulators can be added (one at either end of the four-ringed 20S "barrel or cylinder") to form the 26S proteasome (Fig. 2). The 20S proteasome is a 700-kDa barrel-shaped protein composed of four rings, each made up of seven subunits ranging from 20 to 35 kDa. The first ring contains alpha subunits 1, 2, 3, 4, 5, 6, and 7 while the second ring contains beta subunits 1, 2, 3, 4, 5, 6, and 7. The third ring again contains the beta 1–7 subunits, and the fourth ring again has the alpha 1–7 subunits. Thus, the 20S proteasome barrel or cylinder is symmetrical along its long axis, with the structure α ring, β ring, β ring, α ring. All the proteolytic activities of proteasome are found in the two β rings, where each of the β1, β2, and β5 subunits possesses specific proteolytic activities. In the 26S proteasome, 19S regulators are bound at either end of the 20S proteasome barrel (Fig. 2). The 19S regulator serves to feed proteins into the proteasome core and to restrict the proteins fed to the core to only those that possess a polyubiquitin tag. Degradation of polyubiquinated proteins by the 26S proteasome occurs in an ATP/$Mg^{2+}$-dependent manner. In the 20S proteasome, which is free of 19S regulatory caps, proteins are still capable of binding to the proteasome and of being fed into the core; however, degradation occurs in a manner that is ATP/ubiquitin-independent.[43,44]

Not only is the degradation of proteins by the 20S proteasome ATP/ ubiquitin-independent, but studies using purified proteasome have shown that the 20S proteasome will selectively degrade damaged or oxidized proteins over native proteins. The 20S proteasome shows a selective preference for degradation of oxidized over non-oxidized proteins for a wide range of proteins.[39,41,45] The 26S proteasome by comparison does not appear, in mammalian cells at least, to have any selective preference for degradation of oxidized proteins.[39] In addition to this, while depletion of the 20S proteasome results in an approximately 80% loss of capacity to degrade oxidized proteins, there is only a 20% reduction in capacity to degrade oxidized proteins, with depletion of the 26S proteasome from a cell.[41]

The 20S proteasome appears to be quite resistant to oxidative stress, in comparison with other proteins. In fact, in studies of its capacity to degrade oxidized proteins, under conditions of oxidative stress, its $I_{50}$ appears to be four times higher than the level of oxidation of a target protein that would result in optimum selective degradation.[45] This shows that the 20S proteasome can easily function under the non-toxic degrees of oxidative stress to which cells are normally exposed. In comparison, the $I_{50}$ of the purified 26S proteasome under oxidative stress appears to be 4–16 times lower than that of the 20S proteasome.[45] Similarly, in cell culture, the level of oxidation that completely blocks the ATP-stimulated 26S proteasome activity does not appear to significantly affect the 20S proteasome activity.[46] As a result, the 26S proteasome is actually inactivated under very mild degrees of oxidation compared to the 20S proteasome.[45,46]

In conclusion, both the 20S and 26S proteasomes are capable of degrading damaged proteins; however, the 26S proteasome does not appear to have any selectivity for damage over native proteins. Under conditions where protein damage occurs, the 26S proteasome is easily inactivated compared to the much hardier 20S proteasome, which is able to function at most relevant levels of cellular stress. Finally, if the 20S proteasome is depleted from the cell, the majority of the capacity to remove oxidized proteins is lost, compared to a very minor loss following the removal of the 26S proteasome. From these results, it appears that the 20S proteasome is the major system for removal of oxidized proteins, compared to the relatively minor role played by the 26S proteasome.

## V. Mechanism of Selective Degradation of Damaged Proteins by the 20S Proteasome

The 20S proteasome core is composed of an inner pair of rings of β subunits containing the proteolytic subunits, which are positioned within the lumen of the proteasome core (Fig. 3). Above and below the β rings are rings of α subunits which are not proteolytically active.[47] The α rings provide a

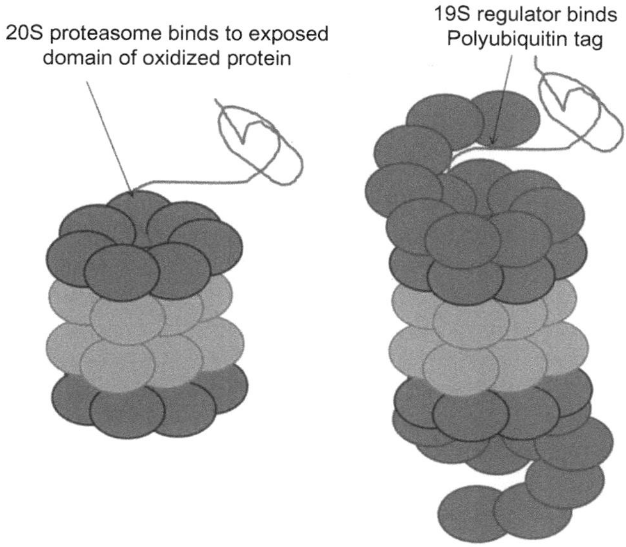


FIG. 3. Degradation of proteins by the 20S and 26S proteasomes.

scaffold for the proteolytic β rings, and are the binding sites for a variety of activators and regulators (including the 19S regulators). The α rings also function as gates to the proteasome, and these gates prevent correctly folded proteins from entering and becoming degraded.[48] In the case of the 26S proteasome, there are caps on each of the alpha rings known as 19S regulators. The 19S regulators act like magnets for ubiquitin-tagged proteins, whose polyubiquitin tails bind and initiate degradation of the tagged protein substrates. The 19S regulators remove the polyubiquitin tails, force the alpha rings open, and feed the substrate proteins into the proteasome core β rings where they are degraded.[49] The 26S proteasome is responsible for most of the turnover of normal proteins within cells. The key factor is the structure of each protein, which determines its susceptibility to polyubiquitinylation by so-called "E3" enzymes. Basically, proteins that are highly susceptible to ubiquitinylation are rapidly degraded by the 26S proteasome and, thus, have short half-lives. In contrast, proteins that are more resistant to ubiquitinylation will have longer cellular half-lives. As the 19S regulator is absent in the 20S proteasome, substrate proteins must be fed through in a different way.

As discussed earlier, most proteolysis of damaged or oxidized proteins is dependent on the 20S proteasome but independent of the 26S proteasome. While the 20S proteasome is not capable of degrading undamaged ubiquitin-tagged proteins as is the 26S proteasome, if the proteins are already unfolded, then the 20S proteasome appears to be just as capable as the 26S proteasome of degrading such substrates.[48] This indicates that proteins are fed into the 20S proteasome in the same way that they are into the 26S proteasome; however, the 20S proteasome does not have the structures possessed by the 26S proteasome to recognize and unfold polyubiquitinated proteins. The 20S proteasome does appear to have its own system for recognition and unfolding of target proteins; specifically, it will recognize and unfold damaged or oxidized proteins. As discussed earlier, protein damage results in the exposure of hydrophobic domains, which would not normally interact with the aqueous environment. These exposed residues are thought to act as sites of recognition for the 20S proteasome in the same way as polyubiquitin tags act as sites of recognition for the 26S proteasome (Fig. 3). These domains permit binding of the damaged protein to the 20S proteasome. The protein is then unfolded and fed into the core, where it may be degraded.[48] However, in the 26S proteasome, this selectivity for damaged proteins is lost because the outer proteasome α rings are covered by the 19S regulators, which bind polyubiquitin.

## VI. The 20S Proteasome in Aging and the Removal of Age-related Protein Damage

In the course of aging, there is a progressive decline in protein synthesis.[50] Despite this decline in synthesis, there is actually an increase in protein content with age. This discrepancy is the product of a buildup of damaged nonfunctioning proteins within cells.[50] This buildup of damaged proteins appears to derive from protein oxidation as well as a range of other modifications including glycation, methylation, ADP ribosylation, and deamidation.[50] The increase in damaged proteins in cells is extremely detrimental and decreases the functionality of all pathways in the cell. This forms part of the widely popular "Free Radical Theory of Aging." This theory suggests that many of the effects of aging are, at least in part, the product of accumulating oxidative damage within a cell, which results in deterioration of cell function and eventually death.[51]

Fortunately, much of the accumulated damage can be removed and the damaged proteins can be degraded and replaced by non-damaged ones. In fact, a mild degree of modification or damage to a protein makes it a better candidate for degradation by the 20S proteasome or other proteases.[33,39,41,52] However, if a protein becomes too heavily modified, it becomes a very poor candidate for degradation.[33,39,40] Thus, while many mildly oxidized proteins are readily degradable, at least some of the age-associated (or time-associated) accumulation of damaged proteins is due to proteins that are so highly modified that they are difficult or impossible to degrade. It has been argued that it is the buildup of these non-degradable damaged proteins that causes age-related effects; however, the accumulation of oxidized proteins in cells is exponential rather than linear over time, indicating that the rise in protein oxidation is not just a product of a buildup of indigestible proteins, but a potentially reversible change in cell function.[53]

The accumulation of oxidized or otherwise damaged proteins in cells during aging could be a product of either a rise in damaging conditions or a fall in the rate of removal of damaged proteins. It has been observed that with age there is a rise in the mitochondrial generation of oxidants.[54] In addition, it has been shown that there is a decline in protein turnover. This decline in protein turnover is, at least in part, the product of a sharp decrease in 20S proteasome function with age (which has been shown in a range of different tissues).[55–60] In addition to a decline in 20S proteasome function, there is also a drop in 20S proteasome levels in aging, which also reduces protein turnover.[61–63] The decline in 20S proteasome function is partly the product of an increase in modification or damage to the 20S proteasome in the course of aging.[55,60–62] For instance, it has been seen that the 20S proteasome isolated from old rats is 50% less proteolytically active than the 20S proteasome isolated from young rats.[64] As a result, not only is there a decrease in the amount of 20S proteasome present during the aging process but there is also a decrease in the ability of the remaining 20S proteasome to degrade the accumulating damaged proteins, thus resulting in an overall accumulation of damaged cellular proteins with age. As a result, older rats are less able to remove damaged proteins from their cells and tissues than are younger rats, which goes some way to explain the difference in the levels of protein damage found in the two animals.

In studies of long-lived animals, such as two long-lived bat species,[65] it has been argued that a major factor of their longer life span, compared to similar shorter-lived animals, is a product of higher 20S proteasome levels.[65,66] The higher levels of expression of the 20S proteasome result in lower levels of accumulation of oxidized proteins and greater resistance to oxidative stress, which has been argued as sufficient to account for the differences in life span.[65,66] Similarly, in studies with naked mole rats (which are the longest-living rodent species), there was also a very high expression of the 20S proteasome, resulting in greater oxidative stress resistance compared to similar rodents; this was suggested to potentially account for at least part of their extremely long life span.[67] Similarly, while mice show a progressive rise in accumulation of protein damage with age, there is almost no change in levels of protein damage in the course of aging in the naked mole rat, which is thought to be a product of their enhanced 20S proteasome levels and activity.[67] These results provide at least correlative and speculative evidence for an important role of the 20S proteasome as a determinant of life span through its ability to minimize the accumulation of damaged proteins in an organism as it ages.

## VII. The 20S Proteasome in Adaptation to Protein Damage

Cells or organisms are frequently exposed to a low level of oxidative stress or protein damage though, occasionally, they will experience much greater stress. Even for a short time, a large stress will result in greater accumulation of damaged proteins, which is highly damaging to the cell. To combat this accumulation, cells appear to have a transient response in which the capacity for degrading oxidized proteins is increased initially in a protein synthesis-independent manner (0.5–5 h after oxidative stress exposure). This is followed by protein synthesis-dependent increase in proteolytic capacity (5–48 h after oxidative stress exposure).[41,68] These responses make the cell much more resistant to normally toxic conditions of protein damage. The initial protein synthesis-independent (unaffected by cycloheximide) response is mostly blocked when proteasome inhibitors are used, indicating that proteasome plays an important role.[68] This increase in capacity to degrade oxidized proteins appears to coincide with a marked reduction in ATP-stimulated proteolytic capacity, which is restored 3–5 h later. It appears that this response is the product of a disassociation or disassembly of the 26S proteasome into free 19S regulators and 20S proteasomes. After the stress has abated and/or the protein synthesis-dependent response has begun, it is important for the 26S proteasome to be reformed so that normal cell function can return. Indeed, 3–5 h after the initial stress, the 26S proteasome appears to re-form from the free 19S regulator and 20S proteasome (Fig. 4). A number of chaperones appear to be involved in this process, and in the mammalian system, hsp70

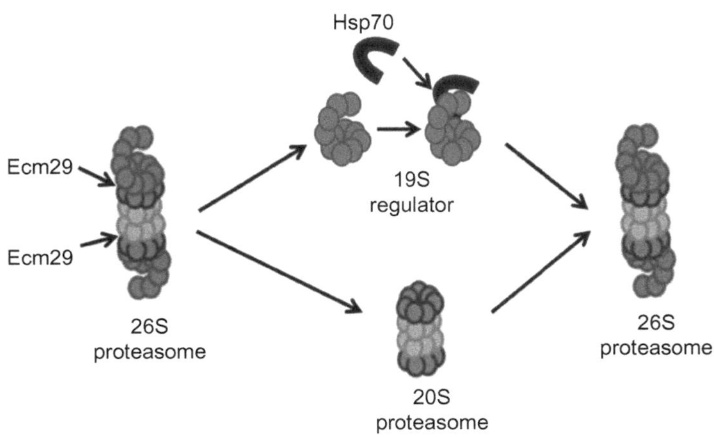


FIG. 4. Disassembly and reassembly of the 26S proteasome in response to oxidative stress.

has been shown to be important for the stabilization of the 19S regulator after dissociation from the 20S proteasome, and for the reassembly of functional 26S proteasomes.[68] In yeast, the Ecm29 protein has been shown to bind to the 19S regulator following dissociation (it may actually cause dissociation) and is required for the increase in survival and ability to degrade damaged proteins.[69] We suggest that Ecm29 may cause 26S disassembly, and hsp70 may protect 19S regulators and allow 26S reassembly in both yeast and mammals (Fig. 4), and we are currently testing this hypothesis.

In addition to the transient response of 26S proteasome dissociation/reassembly, there is also a slower adaptive response to a stress. Over the 5–48 h following oxidative stress, cellular capacity to remove damaged proteins and cellular tolerance to damaging conditions increase significantly.[41,68,70] This response, like the initial 0.5–5 h response described earlier, is largely blocked using proteasome inhibitors.[41,68] However, unlike the earlier response, this transient adaptive increase in stress resistance is accompanied by *de novo* synthesis of the 20S proteasome, and is blocked by siRNA against proteasome subunit genes, and by general protein synthesis inhibitors such as cycloheximide. Thus, it appears that the adaptive response is, at least in part, the product of an increase in *de novo* synthesis of the 20S proteasome. Such transient, and reversible, adaptive responses have also been called hormesis. Importantly, if the new 20S proteasome synthesis is blocked using siRNA, then the increase in the capacity to degrade oxidized proteins as well as the enhanced resistance to oxidative stress are blocked, indicating that the increase in the 20S proteasome is important for the adaptive response. In addition to the 20S proteasome, there is also an increase in synthesis of both the immunoproteasome and the Pa28$\alpha\beta$ proteasome regulator, both of which have also been shown to be important for adaptation.

## VIII. Removal of Damaged Proteins by the 20S Proteasome in the Nucleus

In addition to its presence in the cytoplasm, there is also a large amount of the 20S proteasome in the nucleus, which plays a very important role in stress resistance. As in the cytoplasm, the 20S proteasome in the nucleus is very important in the removal of oxidized or otherwise damaged proteins such that normal cell function can continue. Besides these roles, because of the specialist nature of many proteins in the nucleus, the nuclear 20S proteasome has additional specific functions. These nuclear-specific functions are regulated through proteasome regulators that are found only in the nucleus.

## A. The Nuclear 20S Proteasome Regulator: PARP

As in the cytoplasm, the nucleus appears to have a rapid response system to combat oxidative stress. In fact, in just 15 min following an oxidative stress, there is an 18-fold increase in proteasome-dependent proteolytic activity in the nucleus.[71] It has been shown that this is not due to *de novo* protein synthesis but due to modifications to the 20S proteasome to make it not only more proteolytically active but also more capable of specifically degrading oxidized histones.[71] The selective and very specific degradation of oxidized histones is very important for DNA repair during oxidative stress and, therefore, very important for cell survival and minimizing mutations. The histones around which DNA is wrapped play an important role in the control of DNA expression,[72] and so it is very important that any damage to the histone complexes in the chromatin be repaired as well as the DNA itself. In fact, damaged histones must actually be removed from the DNA before DNA repair can effectively begin, and proteasome is the enzyme that catalyzes this selective proteolytic removal of damaged histone proteins.

As noted earlier, following oxidative stress, there is a very large and rapid increase in proteasome-dependent proteolytic activity in the nucleus,[71] and this change in proteasome activity is regulated by another protein called poly-ADP ribose polymerase (PARP). Following stress, PARP functions by transferring ADP moieties off NAD+ onto other proteins.[73] PARP is believed to use this mechanism heavily as a means of signaling DNA damage and inducing its repair.[74] It appears that the 20S proteasome can also act as a substrate for modification by PARP. It was shown in studies using purified 20S proteasome that ADP ribosylation of 20S proteasome by PARP resulted in an increase in selectivity for the degradation of oxidized histones, which accounted for the changes in proteolytic selectivity and increased activity observed under oxidative stress of the nucleus. Furthermore, the increase in proteolytic activity in the nucleus under oxidative stress is lost with inhibition of PARP activity, indicating that PARP is a key regulator of proteasome in the nucleus.[71]

## B. The Nuclear 20S/26S Proteasome Regulator: Pa200

Another regulator of the nuclear 20S proteasome is the Pa200 (also known as PSME4) proteasome regulator. This is a 200-kDa activator of proteasome activity (hence Pa200) that is highly conserved throughout the eukaryotic domain.[75,76] Pa200 is a nuclear-localized regulator of proteasome. Like PARP, Pa200 appears to be involved in the repair of DNA damage. Unlike PARP, however, which appears to assist in DNA damage repair by activating the 20S proteasome to increase preferential degradation of damaged histones, the role that Pa200 plays is not entirely clear.[71,75] It has been seen in studies using the yeast homologue of Pa200 (Blm10) that mutation of Blm10 results in cells becoming much less resistant to DNA damage.[77] Similarly, in mammalian culture, the depletion of Pa200 results in greater DNA damage susceptibility to ionizing radiation.[78] Interestingly, although Pa200 is referred to as a proteasome activator because of its ability to enhance the 20S proteasome's ability to degrade short peptides, it does not appear to enhance the ability of the 20S proteasome to degrade whole proteins.[75] In at least a partial elucidation of its role, it appears that under cell damage through ionizing radiation or hydrogen peroxide treatment, there is a rapid formation of foci of Pa200 on the DNA chromatin. The formation of these DNA foci correlates with an increase in Pa200 binding to either the 20S proteasome or the so-called hybrid proteasome. This hybrid proteasome is proteasome on which there is a 19S regulator at one end and the Pa200 regulator at the other end.[75,78] It is unclear whether Pa200 functions mainly as a regulator of the 20S proteasome or as a regulator of the 26S proteasome through the formation of hybrid proteasomes. It is also unclear whether Pa200 functions primarily through degradation of damaged proteins in the chromatin or by triggering some form of DNA repair mechanism. It is clear, however, that whatever its actual role, it is a highly conserved regulator of proteasome in response to oxidation or other damage in the nucleus.

### C. The Nuclear 20S Proteasome Regulator: Pa28γ

The final regulator of the 20S proteasome in the nucleus is the proteasome regulator Pa28γ (also known as REGγ, 11sγ, or PSME3). There is no evidence at present of a role for Pa28γ in the removal of damaged proteins. The Pa28γ regulator forms a homoheptameric ring at either end of the 20S proteasome. It is nuclear-localized and is a genetic ortholog of the better-studied Pa28αβ regulator.[79] It weakly enhances the 20S proteasome's ability to degrade short peptides. Its main function, however, is largely unknown. Pa28γ has been shown to be important in cell cycle progression as well as apoptosis.[80] It has also been shown to be a critical regulator of p53 expression during a stress response[81]; however, its full role is largely unknown.

## IX. Alternative Forms of the 20S Proteasome in the Cytoplasm and Other Non-nuclear Compartments and Their Role(s) in the Removal of Damaged Proteins

### A. Immunoproteasome and Oxidative Stress

Another form of proteasome which has been suggested to be involved in the proteolytic removal of oxidized proteins is the immunoproteasome (Fig. 5), described in more detail in other chapters. This form of proteasome differs from the 20S proteasome in that it has three alternative subunits (β1i, β2i, and β5i), sometimes known as (lmp2, mecl-1, and lmp7). These alternative subunits

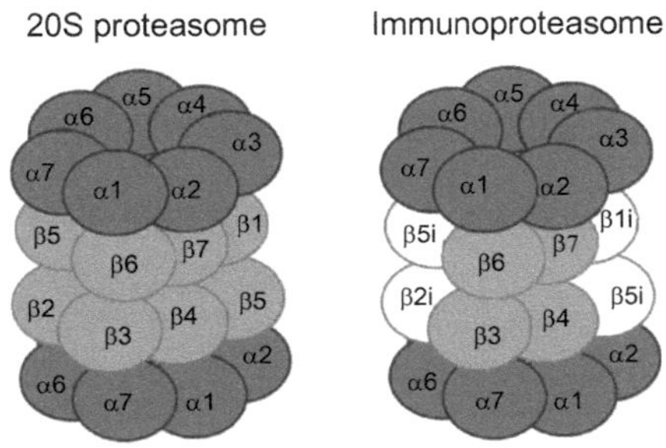


FIG. 5. Subunit composition of the 20S proteasome and the immunoproteasome.

result in differences in the sites of cleavage of proteins degraded by the immunoproteasome, compared to the 20S proteasome, which will give a different range of peptides.[82,83] The peptides generated by the immunoproteasome, because of their composition and structure, are favored as peptides for MHC class 1 antigen presentation.[84,85] Expression of the immunoproteasome is strongly induced by the regulator interferon-γ. Interferon-γ is a cell-signaling molecule that is important in both innate and adaptive immune response.[86] This along with the localization of the immunoproteasome to the endoplasmic reticulum has led to the belief that it functions primarily as a means of generating MHC class 1 peptides for cell surface presentation.[84,85] This theory of the immunoproteasome's role being primarily that of immune response has, however, come into question recently with the observation that interferon-γ itself produces an oxidative stress response. It is entirely possible then that interferon-γ-induced synthesis of the immunoproteasome might actually be an oxidative stress response rather than an immune response.[87]

In addition to a hypothesized role for the immunoproteasome in immune response, it appears that it plays a role in the degradation of oxidized proteins. The immunoproteasome has been shown to be as capable as the 20S proteasome of selectively degrading oxidized proteins[41] and its expression does increase under conditions of protein damage.[41,56,88,89] It has been hypothesized that protein oxidation might function as a universal marker for selection of proteins for MHC antigen presentation. This "PrOxI hypothesis"[84] suggests a potential explanation of, and reason for, the immunoproteasome's capacity to selectively degrade oxidized proteins. It also seems that the immunoproteasome can, like the 20S proteasome, play a role in the removal of damaged proteins as part of cellular damage repair. In support of this, it has been shown that the immunoproteasome, like the 20S proteasome, is inducible during oxidative stress and in aging. Furthermore, if the immunoproteasome is removed or depleted in certain cells, then those cells are more susceptible and less adaptable to oxidative stress.[41,90,91] As a result, while the role of the immunoproteasome remains somewhat controversial, several studies have provided highly suggestive evidence that the immunoproteasome plays a role in the selective removal of oxidized proteins from the cell.

## B. The Cytoplasmic Proteasome Activator Pa28αβ

The proteasome regulator Pa28αβ, also known as REGαβ, 11Sαβ or PSME1, and PSME2, is an activator of 20S proteasome activity. Pa28αβ is localized in the cytoplasm and forms a heteroheptameric ring composed of three alpha subunits and four beta subunits.[92] This ring can fit on either end of the core of the 20S proteasome. When associated with the 20S proteasome, it induces a large increase in the capacity to degrade short peptides.[79] Pa28αβ has been shown to be important in the process of oxidative stress adaptation. Like the 20S proteasome, there is a strong increase in the production of Pa28αβ when cells are exposed to an adaptive dose of an oxidant.[41] It has also been shown, using mutants of Pa28αβ and using siRNA to block Pa28αβ expression, that the adaptive response to oxidative stress is notably weaker and there is much less of an increase in the capacity of cells to degrade oxidized proteins without the expression of functional Pa28αβ. This indicates that Pa28αβ plays an important role both in oxidative stress adaptation and in stimulating the degradation of oxidized proteins.[41] There is also a strong suggestion for a potential role for Pa28αβ in MHC class I antigen presentation. Like the immunoproteasome, Pa28αβ is regulated by interferon-γ.[93] This has led to some suggesting that the two might act together, with Pa28αβ forming a cap on the immunoproteasome to enable or assist MHC class I antigen presentation of self and nonself antigens.[84] Additionally, it has been shown that increases in Pa28αβ expression result in enhanced recognition of viral antigens by cytotoxic T-cells.[94]

# X. Summary

Cells accumulate protein damage through exposure to environmental toxins, physiological stress, protein synthesis errors, and as a product of age. These damaged proteins reduce cell function and, in some cases, can even threaten cell viability. To allow normal cell function to return and prevent toxicity, the 20S proteasome functions to remove such damaged proteins and minimize their tendency to aggregate, cross-link, and accumulate. Unlike the 26S proteasome, the 20S proteasome is capable of selectively degrading the oxidatively damaged forms of cellular proteins. It is also more resilient to the conditions where damage would occur (oxidative stress) and where the removal or inhibition of 20S proteasome cells appears poorly equipped to remove damaged proteins. As a result, while the 26S proteasome is highly important in the regulation of normal cell function, it does not appear to play a major role in the removal of damaged proteins.

The removal of damaged proteins by the 20S proteasome, as well as being important for maintenance of normal cell function, appears to have an important role in aging. It seems that part of the decline in cell and organismal function as a product of age is through an increase in protein damage due to diminished 20S proteasome levels and activity. In addition to its role in aging and normal cell maintenance, the 20S proteasome appears to be important in transient and reversible adaptation to stress. Thus, 20S proteasome activity is not static but is directly activated by oxidants, and its synthesis is inducible by (oxidant) signaling molecules such as hydrogen peroxide, resulting in an enhanced resistance to protein damage and enhanced proteolytic removal of damaged proteins.

There are also several regulators of the 20S proteasome that are important for specific functions in different compartments. PARP binds to the 20S proteasome in response to oxidative stress and enhances the removal of damaged histones. Pa200, which also binds to proteasome in the nucleus, appears to have some role in response to DNA damage but its function is unclear. Pa28αβ appears to enhance the proteasome's proteolytic capacity in the cytoplasm and is important in oxidative stress response. Similar to the 20S proteasome, the immunoproteasome has also been found to selectively degrade oxidized proteins. The immunoproteasome is also transiently and reversibly induced by oxidative stress, and we suggest that it is also important in the overall degradation of damaged proteins.

## References

1. Halliwell B, Hu ML, Louie S, Duvall TR, Tarkington BK, Motchnik P, et al. Interaction of nitrogen dioxide with human plasma. Antioxidant depletion and oxidative damage. *FEBS Lett* 1992;**313**:62–6.
2. Menzel DB. The toxicity of air pollution in experimental animals and humans: the role of oxidative stress. *Toxicol Lett* 1994;**72**:269–77.
3. Abdollahi M, Ranjbar A, Shadnia S, Nikfar S, Rezaie A. Pesticides and oxidative stress: a review. *Med Sci Monit* 2004;**10**:RA141–7.
4. Cross CE, Motchnik PA, Bruener BA, Jones DA, Kaur H, Ames BN, et al. Oxidative damage to plasma constituents by ozone. *FEBS Lett* 1992;**298**:269–72.
5. Cross CE, Reznick AZ, Packer L, Davis PA, Suzuki YJ, Halliwell B. Oxidative damage to human plasma proteins by ozone. *Free Radic Res Commun* 1992;**15**:347–52.
6. Hu ML, Tappel AL. Potentiation of oxidative damage to proteins by ultraviolet-A and protection by antioxidants. *Photochem Photobiol* 1992;**56**:357–63.

7. Leach JK, Van Tuyle G, Lin PS, Schmidt-Ullrich R, Mikkelsen RB. Ionizing radiation-induced, mitochondria-dependent generation of reactive oxygen/nitrogen. *Cancer Res* 2001; **61**:3894–901.
8. Kappus H. Oxidative stress in chemical toxicity. *Arch Toxicol* 1987;**60**:144–9.
9. Baeuerle PA, Rupec RA, Pahl HL. Reactive oxygen intermediates as second messengers of a general pathogen response. *Pathol Biol (Paris)* 1996;**44**:29–35.
10. Orgel LE. The maintenance of the accuracy of protein synthesis and its relevance to ageing. *Proc Natl Acad Sci USA* 1963;**49**:517–21.
11. Davies KJA, Delsignore ME, Lin SW. Protein damage and degradation by oxygen radicals. II. Modification of amino acids. *J Biol Chem* 1987;**262**:9902–7.
12. Davies KJA, Delsignore ME. Protein damage and degradation by oxygen radicals. III. Modification of secondary and tertiary structure. *J Biol Chem* 1987;**262**:9908–13.
13. Davies KJA. Oxidative stress: the paradox of aerobic life. *Biochem Soc Symp* 1995;**61**:1–31.
14. Davies KJA. Protein damage and degradation by oxygen radicals. I. general aspects. *J Biol Chem* 1987;**262**:9895–901.
15. Davies KJA, Lin SW, Pacifici RE. Protein damage and degradation by oxygen radicals. IV. Degradation of denatured protein. *J Biol Chem* 1987;**262**:9914–20.
16. Copeland JM, Cho J, Lo Jr. T, Hur JH, Bahadorani S, Arabyan T, et al. Extension of Drosophila life span by RNAi of the mitochondrial respiratory chain. *Curr Biol* 2009;**19**:1591–8.
17. Davies KJA, Ursini F. *The oxygen paradox*. Padova: Cleup University Press; 1995.
18. Raymond J, Segre D. The effect of oxygen on biochemical networks and the evolution of complex life. *Science* 2006;**311**:1764–7.
19. Halliwell B, Gutteridge MC. *Free radicals in biology and medicine*. 3rd ed., Oxford University Press, Oxford, UK 2000.
20. Fenton HJH. Oxidation of tartaric acid in presence of iron. *J Chem Soc Trans* 1894;**65**:899–910.
21. Haber F, Weiss J. Über die Katalyse des Hydroperoxydes. *Die Naturwissenschaften* 1932;**20**:948–50.
22. Brodie AE, Reed DJ. Reversible oxidation of glyceraldehyde 3-phosphate dehydrogenase thiols in human lung carcinoma cells by hydrogen peroxide. *Biochem Biophys Res Commun* 1987;**148**:120–5.
23. Holleman MAF. Notice sur l'action de l'eau oxygénée sur les acides a-cétoniques et sur les dicétones. *Recl Trav Chim Pays-Bas Belg* 1904;**23**:169–71.
24. Cadet J, Delatour T, Douki T, Gasparutto D, Pouget JP, Ravanat JL, et al. Hydroxyl radicals and DNA base damage. *Mutat Res* 1999;**424**:9–21.
25. Wolff SP, Garner A, Dean RT. Free radicals, lipids and protein degradation. *Trends Biochem Sci* 1986;**11**:27–31.
26. Garthwaite J, Boulton CL. Nitric oxide signaling in the central nervous system. *Annu Rev Physiol* 1995;**57**:683–706.
27. Moncada S, Higgs EA. Molecular mechanisms and therapeutic strategies related to nitric oxide. *FASEB J* 1995;**9**:1319–30.
28. Lloyd DR, Carmichael PL, Phillips DH. Comparison of the formation of 8-hydroxy-2'-deoxyguanosine and single- and double-strand breaks in DNA mediated by fenton reactions. *Chem Res Toxicol* 1998;**11**:420–7.
29. Ghio AJ, Pritchard RJ, Lehmann JR, Winsett DW, Hatch GE. Lung inflammation after exposure to nonfibrous silicates increases with chelatable [Fe3+]. *J Toxicol Environ Health* 1996;**49**:11–28.
30. Ghio AJ, Stonehuerner J, Dailey LA, Carter JD. Metals associated with both the water-soluble and insoluble fractions of an ambient air pollution particle catalyze an oxidative stress. *Inhal Toxicol* 1999;**11**:37–49.

31. Becker S, Dailey LA, Soukup JM, Grambow SC, Devlin RB, Huang YC. Seasonal variations in air pollution particle-induced inflammatory mediator release and oxidative stress. *Environ Health Perspect* 2005;**113**:1032–8.
32. Schröder P, Krutmann J. Environmental oxidative stress—environmental sources of ROS. In: *The handbook of environmental chemistry*, vol. 2. Berlin, Heidelberg: Springer-Verlag; 2005. pp. 19–31.
33. Davies KJA. Protein modification by oxidants and the role of proteolytic enzymes. *Biochem Soc Trans* 1993;**21**:346–53.
34. Davies KJA. An overview of oxidative stress. *IUBMB Life* 2000;**50**:241–4.
35. Mills GC. The purification and properties of glutathione peroxidase of erythrocytes. *J Biol Chem* 1959;**234**:502–6.
36. Carlberg I, Mannervik B. Glutathione reductase. *Methods Enzymol* 1985;**113**:484–90.
37. McCord JM, Fridovich I. Superoxide dismutase. An enzymic function for erythrocuprein (hemocuprein). *J Biol Chem* 1969;**244**:6049–55.
38. Nishikimi M, Yagi K. Biochemistry and molecular biology of ascorbic acid biosynthesis. *Subcell Biochem* 1996;**25**:17–39.
39. Davies KJA. Degradation of oxidized proteins by the 20S Proteasome. *Biochimie* 2001;**83**:301–10.
40. Davies KJA, Goldberg AL. Proteins damaged by oxygen radicals are rapidly degraded in extracts of red blood cells. *J Biol Chem* 1987;**262**:8227–34.
41. Pickering AM, Koop AL, Teoh CY, Ermak G, Grune T, Davies KJA. The Immunoproteasome, the 20S Proteasome and the PA28alphabeta Proteasome regulator are oxidative-stress-adaptive proteolytic complexes. *Biochem J* 2010;**432**:585–94.
42. McNaught KS, Olanow CW, Halliwell B, Isacson O, Jenner P. Failure of the ubiquitin-proteasome system in Parkinson's disease. *Nat Rev Neurosci* 2001;**2**:589–94.
43. Coux O, Tanaka K, Goldberg AL. Structure and functions of the 20S and 26S Proteasomes. *Annu Rev Biochem* 1996;**65**:801–47.
44. Ciechanover A. The ubiquitin-Proteasome proteolytic pathway. *Cell* 1994;**79**:13–21.
45. Reinheckel T, Sitte N, Ullrich O, Kuckelkorn U, Davies KJA, Grune T. Comparative resistance of the 20S and 26S Proteasome to oxidative stress. *Biochem J* 1998;**335**(Pt 3):637–42.
46. Reinheckel T, Ullrich O, Sitte N, Grune T. Differential impairment of 20S and 26S Proteasome activities in human hematopoietic K562 cells during oxidative stress. *Arch Biochem Biophys* 2000;**377**:65–8.
47. Unno M, Mizushima T, Morimoto Y, Tomisugi Y, Tanaka K, Yasuoka N, et al. Structure determination of the constitutive 20S Proteasome from bovine liver at 2.75 A resolution. *J Biochem* 2002;**131**:171–3.
48. Liu CW, Corboy MJ, DeMartino GN, Thomas PJ. Endoproteolytic activity of the Proteasome. *Science* 2003;**299**:408–11.
49. Navon A, Goldberg AL. Proteins are unfolded on the surface of the ATPase ring before transport into the Proteasome. *Mol Cell* 2001;**8**:1339–49.
50. Rattan SI. Synthesis, modifications, and turnover of proteins during aging. *Exp Gerontol* 1996;**31**:33–47.
51. Harman D. Aging: a theory based on free radical and radiation chemistry. *J Gerontol* 1956;**11**:298–300.
52. Davies KJA. Oxidative stress, antioxidant defenses, and damage removal, repair, and replacement systems. *IUBMB Life* 2000;**50**:279–89.
53. Stadtman ER. Protein oxidation and aging. *Science* 1992;**257**:1220–4.
54. Sohal RS, Dubey A. Mitochondrial oxidative damage, hydrogen peroxide release, and aging. *Free Radic Biol Med* 1994;**16**:621–6.

55. Chondrogianni N, Gonos ES. Proteasome dysfunction in mammalian aging: steps and factors involved. *Exp Gerontol* 2005;**40**:931–8.
56. Ferrington DA, Husom AD, Thompson LV. Altered Proteasome structure, function, and oxidation in aged muscle. *FASEB J* 2005;**19**:644–6.
57. Husom AD, Peters EA, Kolling EA, Fugere NA, Thompson LV, Ferrington DA. Altered Proteasome function and subunit composition in aged muscle. *Arch Biochem Biophys* 2004;**421**:67–76.
58. Viteri G, Carrard G, Birlouez-Aragon I, Silva E, Friguet B. Age-dependent protein modifications and declining Proteasome activity in the human lens. *Arch Biochem Biophys* 2004;**427**:197–203.
59. Ponnappan U, Zhong M, Trebilcock GU. Decreased Proteasome-mediated degradation in T cells from the elderly: a role in immune senescence. *Cell Immunol* 1999;**192**:167–74.
60. Carrard G, Dieu M, Raes M, Toussaint O, Friguet B. Impact of ageing on Proteasome structure and function in human lymphocytes. *Int J Biochem Cell Biol* 2003;**35**:728–39.
61. Keller JN, Huang FF, Markesbery WR. Decreased levels of Proteasome activity and Proteasome expression in aging spinal cord. *Neuroscience* 2000;**98**:149–56.
62. Bulteau AL, Petropoulos I, Friguet B. Age-related alterations of Proteasome structure and function in aging epidermis. *Exp Gerontol* 2000;**35**:767–77.
63. Petropoulos I, Conconi M, Wang X, Hoenel B, Bregegere F, Milner Y, et al. Increase of oxidatively modified protein is associated with a decrease of Proteasome activity and content in aging epidermal cells. *J Gerontol A Biol Sci Med Sci* 2000;**55**:B220–7.
64. Conconi M, Friguet B. Proteasome inactivation upon aging and on oxidation-effect of HSP 90. *Mol Biol Rep* 1997;**24**:45–50.
65. Salmon AB, Leonard S, Masamsetti V, Pierce A, Podlutsky AJ, Podlutskaya N, et al. The long lifespan of two bat species is correlated with resistance to protein oxidation and enhanced protein homeostasis. *FASEB J* 2009;**23**:2317–26.
66. Chondrogianni N, Gonos ES. Proteasome function determines cellular homeostasis and the rate of aging. *Adv Exp Med Biol* 2011;**694**:38–46.
67. Perez VI, Buffenstein R, Masamsetti V, Leonard S, Salmon AB, Mele J, et al. Protein stability and resistance to oxidative stress are determinants of longevity in the longest-living rodent, the naked mole-rat. *Proc Natl Acad Sci USA* 2009;**106**:3059–64.
68. Grune T, Catalgol B, Licht A, Ermak G, Pickering AM, Ngo JK, et al. HSP70 mediates dissociation and reassociation of the 26S Proteasome during adaptation to oxidative stress. *Free Radic Biol Med* 2011;**51**:1355–64.
69. Wang X, Yen J, Kaiser P, Huang L. Regulation of the 26S Proteasome complex during oxidative stress. *Sci Signal* 2010;**3**:ra88.
70. Wiese AG, Pacifici RE, Davies KJA. Transient adaptation of oxidative stress in mammalian cells. *Arch Biochem Biophys* 1995;**318**:231–40.
71. Ullrich O, Reinheckel T, Sitte N, Hass R, Grune T, Davies KJA. Poly-ADP ribose polymerase activates nuclear Proteasome to degrade oxidatively damaged histones. *Proc Natl Acad Sci USA* 1999;**96**:6223–8.
72. Jenuwein T, Allis CD. Translating the histone code. *Science* 2001;**293**:1074–80.
73. Banasik M, Komura H, Shimoyama M, Ueda K. Specific inhibitors of poly(ADP-ribose) synthetase and mono(ADP-ribosyl)transferase. *J Biol Chem* 1992;**267**:1569–75.
74. Piskunova TS, Yurova MN, Ovsyannikov AI, Semenchenko AV, Zabezhinski MA, Popovich IG, et al. Deficiency in Poly(ADP-ribose) Polymerase-1 (PARP-1) Accelerates Aging and Spontaneous Carcinogenesis in Mice. *Curr Gerontol Geriatr Res* 2008;**2008**:754190.
75. Ustrell V, Hoffman L, Pratt G, Rechsteiner M. PA200, a nuclear Proteasome activator involved in DNA repair. *EMBO J* 2002;**21**:3516–25.

76. Baumeister W, Walz J, Zuhl F, Seemuller E. The Proteasome: paradigm of a self-compartmentalizing protease. *Cell* 1998;**92**:367–80.
77. Schmidt M, Haas W, Crosas B, Santamaria PG, Gygi SP, Walz T, et al. The HEAT repeat protein Blm10 regulates the yeast Proteasome by capping the core particle. *Nat Struct Mol Biol* 2005;**12**:294–303.
78. Blickwedehl J, Agarwal M, Seong C, Pandita RK, Melendy T, Sung P, et al. Role for Proteasome activator PA200 and postglutamyl Proteasome activity in genomic stability. *Proc Natl Acad Sci USA* 2008;**105**:16165–70.
79. Li J, Rechsteiner M. Molecular dissection of the 11S REG (PA28) Proteasome activators. *Biochimie* 2001;**83**:373–83.
80. Rechsteiner M, Hill CP. Mobilizing the proteolytic machine: cell biological roles of Proteasome activators and inhibitors. *Trends Cell Biol* 2005;**15**:27–33.
81. Zhang Z, Zhang R. Proteasome activator PA28 gamma regulates p53 by enhancing its MDM2-mediated degradation. *EMBO J* 2008;**27**:852–64.
82. Fruh K, Gossen M, Wang K, Bujard H, Peterson PA, Yang Y. Displacement of housekeeping Proteasome subunits by MHC-encoded LMPs: a newly discovered mechanism for modulating the multicatalytic proteinase complex. *EMBO J* 1994;**13**:3236–44.
83. Belich MP, Glynne RJ, Senger G, Sheer D, Trowsdale J. Proteasome components with reciprocal expression to that of the MHC-encoded LMP proteins. *Curr Biol* 1994;**4**:769–76.
84. Teoh CY, Davies KJA. Potential roles of protein oxidation and the Immunoproteasome in MHC class I antigen presentation: the 'PrOxI' hypothesis. *Arch Biochem Biophys* 2004;**423**:88–96.
85. Tanaka K, Kasahara M. The MHC class I ligand-generating system: roles of immunoProteasomes and the interferon-gamma-inducible Proteasome activator PA28. *Immunol Rev* 1998;**163**:161–76.
86. Fruh K, Yang Y. Antigen presentation by MHC class I and its regulation by interferon gamma. *Curr Opin Immunol* 1999;**11**:76–81.
87. Watanabe Y, Suzuki O, Haruyama T, Akaike T. Interferon-gamma induces reactive oxygen species and endoplasmic reticulum stress at the hepatic apoptosis. *J Cell Biochem* 2003;**89**:244–53.
88. Ferrington DA, Hussong SA, Roehrich H, Kapphahn RJ, Kavanaugh SM, Heuss ND, et al. ImmunoProteasome responds to injury in the retina and brain. *J Neurochem* 2008;**106**:158–69.
89. Kotamraju S, Matalon S, Matsunaga T, Shang T, Hickman-Davis JM, Kalyanaraman B. Upregulation of immunoProteasomes by nitric oxide: potential antioxidative mechanism in endothelial cells. *Free Radic Biol Med* 2006;**40**:1034–44.
90. Ethen CM, Hussong SA, Reilly C, Feng X, Olsen TW, Ferrington DA. Transformation of the Proteasome with age-related macular degeneration. *FEBS Lett* 2007;**581**:885–90.
91. Hussong SA, Kapphahn RJ, Phillips SL, Maldonado M, Ferrington DA. ImmunoProteasome deficiency alters retinal Proteasome's response to stress. *J Neurochem* 2010;**113**:1481–90.
92. Zhang Z, Krutchinsky A, Endicott S, Realini C, Rechsteiner M, Standing KG. Proteasome activator 11S REG or PA28: recombinant REG alpha/REG beta hetero-oligomers are heptamers. *Biochemistry* 1999;**38**:5651–8.
93. Fabunmi RP, Wigley WC, Thomas PJ, DeMartino GN. Interferon gamma regulates accumulation of the Proteasome activator PA28 and immunoProteasomes at nuclear PML bodies. *J Cell Sci* 2001;**114**:29–36.
94. Groettrup M, Soza A, Eggers M, Kuehn L, Dick TP, Schild H, et al. A role for the Proteasome regulator PA28alpha in antigen presentation. *Nature* 1996;**381**:166–8.

# Changes of the Proteasomal System During the Aging Process

Martin A. Baraibar and Bertrand Friguet

*Laboratoire de Biologie Cellulaire du Vieillissement, UR4-IFR83, Université Pierre et Marie Curie–Paris 6, 4 place Jussieu, Paris Cedex 05, France*

I. Introduction .......... 249
II. Age-Related Decline of Proteasome Function in Organs and Tissues .......... 252
III. Proteasome Dysfunction During Cellular Aging .......... 255
IV. Modulation of Proteasome Function in Model Organisms .......... 257
V. Mechanisms Involved in the Age-Related Decline of Proteasome Function .... 260
VI. Rejuvenating Proteasome Function as an Antiaging Strategy .......... 264
References .......... 268

Accumulation of oxidized and damaged proteins is a hallmark of the aging process in different organs and tissues. Intracellular protein degradation is normally the most efficient mechanism to prevent toxicity associated with the accumulation of altered proteins without affecting the cellular reserves of amino acids. Protein degradation by the proteasomal system is a key process for the maintenance of cellular protein homeostasis and has come into the focus of aging research during the last decade. During the last few years, several lines of evidence have indicated that proteasome function is impaired during aging, suggesting that this decreased activity might be causally related to the aging process and the occurrence of age-associated diseases. This chapter reviews the proteasome status in organs, tissues, cells, and model organisms during aging as well as the molecular mechanisms involved in the age-related decline of proteasome function. Finally, interventions aimed at rejuvenating proteasome function as a potential antiaging strategy are discussed.

## I. Introduction

The quality of the cellular proteome (i.e., the entire pool of cellular proteins located inside the cell and within the plasma membrane) is tightly controlled, insuring that proteins are properly synthesized to become functionally active and that damaged nonfunctional proteins are efficiently removed through degradation or repair. Accumulation of damaged proteins is a hallmark of the

Progress in Molecular Biology
and Translational Science, Vol. 109
DOI: 10.1016/B978-0-12-397863-9.00007-9


1877-1173/12 $35.00

aging process in different organs and tissues. Protein oxidation by reactive oxygen species (ROS) and related modifications represent a central component of the aging process, which is implicit in the free radical theory of aging reported by Harman.[1,2] The harmful effect of the accumulation of oxidized proteins and their contribution to increased frailty and morbidity have been recognized since the early days of the formulation of this and other related theories. The toxic effect of altered proteins is currently known as "proteotoxicity," and the complex cellular systems that contribute to the preservation of protein homeostasis are referred to as the "proteostasis network."[3,4] This proteostasis network detects the altered proteins inside the cell and coordinates their folding, repair, or elimination from the cells by different proteolytic pathways (Fig. 1). The consequences of poor quality control depend on the type of protein alteration, the location of the altered protein, and on the defective step in quality control.

Intracellular protein degradation is normally the most efficient mechanism to prevent toxicity associated with the accumulation of altered proteins without affecting the cellular reserves of amino acids.[5–7] Cells can elicit alternative defense mechanisms when the load of proteins destined for degradation overrides the capacity of the proteolytic systems or when there is a primary failure in the function of these systems. For example, formation of large protein inclusions has been proposed to be used by cells in certain instances to protect themselves from the toxic effect associated with irreversible oligomeric species of pathogenic proteins.[8,9]

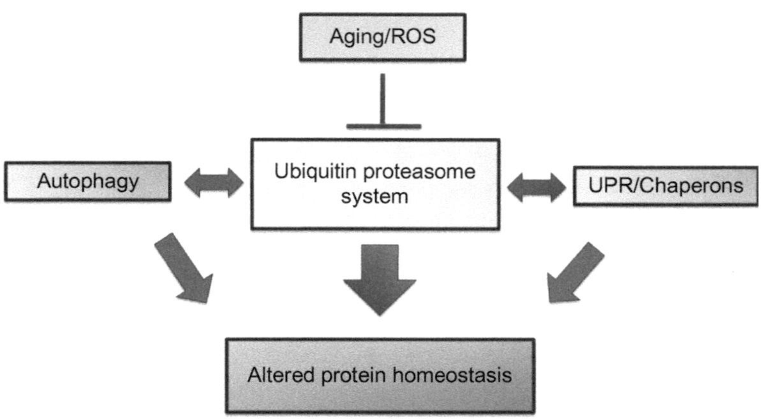


FIG. 1. Schematic representation of proteostasis network alteration during aging. Factors such as increased production of reactive oxygen species (ROS) and other toxic metabolites, accumulation of aberrant and/or modified proteins, failure of the ubiquitin–proteasome system, chaperone overload and impairment of the unfolded protein response (UPR), and decline in autophagy have been implicated to some extent in the deregulation of protein homeostasis during aging. (See Color Insert.)

The proteasomal and lysosomal pathways are the two main proteolytic machineries by which intracellular proteins are degraded. Protein degradation by the proteasome is a key process for the maintenance of cellular protein homeostasis and has come into the focus of aging research during the last decade.[10–13] Proteasomal-mediated protein degradation can be fully distinguished from lysosomal-mediated proteolysis by the fact that it occurs at neutral pH, preferentially degrades short-lived proteins, occurs in a protein complex, does not involve intracellular compartmentalization, and generates peptides instead of breaking proteins down to the amino acid level. Importantly, the proteasome is mainly involved in the degradation of abnormal, oxidized, and damaged proteins.[6,14–17] Moreover, the rapid, precise, and timely processing of a vast extent of cellular proteins by the ubiquitin–proteasome system (UPS) allows tight control of critical cellular functions such as DNA repair, cell cycle progression, development, apoptosis, gene transcription, signal transduction, senescence, immune response, metabolism, and protein quality control.

During the last few years, several lines of evidence have indicated that proteasome function is impaired during the aging process, suggesting that this decreased activity might be causally related to aging and age-associated diseases,[10,12,17,18] although some studies have shown that this decline may not be universal.[19,20] Pioneering studies from our group and those of Walter Ward indicated that proteasome proteolytic activity is compromised with aging.[21–23] Proteasome impairment has since been reported at different levels, including decreased transcription of certain proteasomal subunits in mice,[24] dissociation of the holocomplex in Drosophila,[25] and reduced proteolytic capacity in different aged mammalian tissues and organs.[11,26] In contrast, centenarians who represent an interesting example of successful aging and the long-lived naked mole rats were found to exhibit elevated proteasome levels and activity.[27,28]

A decrease in proteasome peptidase activity has been reported in aged tissues, such as liver,[21–23,29] spinal cord,[30] lens,[31] heart,[32] and retina,[33] of other mammals (mouse, rat, and bovine). Furthermore, an age-related decline of proteasome activity has also been shown ex vivo in human lymphocytes,[34,35] keratinocytes, and fibroblasts,[36,37] and in human primary cell cultures undergoing replicative senescence.[38–42] Impairment of proteasomal activity has also been reported during aging in model organisms such as *Drosophila melanogaster*[25,43] and *Caenorhabditis elegans*.[44]

The main emphasis has long been placed on preventing protein damage (i.e., interest in antioxidants as an effort to prevent oxidative damage) rather than potentiating the mechanisms that normally handle these damaged products. However, several lines of evidence support the idea that the main problem is not the damage per se, but rather how the cell handles this damage. These include better understanding of the cellular mechanisms that contribute to protein quality control, the fact that some of the genes coding for the components of these systems have been implicated in life span extension, and the growing evidence supporting the theory that failure of protein homeostasis represents an early event in aging. In addition, recent studies have shown that the proteasome can be activated by genetic manipulations as well as by factors that affect either its conformation and stability or the expression of its subunits and the rate of proteasome assembly. Indeed, overexpression of the 20S β5 subunit extended the replicative life span of cultured human fibroblasts,[45] overexpression of 19S Rpn11 prolonged *D. melanogaster* life span,[43] while overexpression of the proteasome-related transcription factor Rpn4 enhanced the replicative life span and resistance to proteotoxic stress of *Saccharomyces cerevisiae*.[46]

## II. Age-Related Decline of Proteasome Function in Organs and Tissues

The first experimental evidence of decreased proteasome activity during aging refers to *ex vivo* studies from our and other groups in liver and then in other organs and tissues. Indeed, we first showed that purified 20S proteasome from rat liver of old animals exhibited a 50% decrease in the peptidylglutamyl–peptide hydrolase activity when compared with their young counterparts.[21] Walter Ward and colleagues reported during the same year a 40% decrease of the same peptidase activity when assayed in crude homogenates and activated by the detergent sodium dodecyl sulfate.[22,23] In addition, we also reported that dietary self-selection of nutrients, which results in decreased protein uptake, could compensate the age-related decreased liver proteasome activity observed in rats fed with a standard diet.[47] These findings suggest that the beneficial effects associated with dietary restriction in calories and proteins include not only decreased macromolecular damage and increased expression of antioxidant enzymes but also preservation of protein maintenance, leading to life span extension. However, the fate of proteasome function on dietary restriction has been shown to be tissue-dependent and to vary according to experimental conditions.[48,49]

Studies from Jeffrey Keller and colleagues showed a decreased chymotrypsin-like proteasome peptidase activity throughout the central nervous system (CNS) of aged rats. Decreased proteasome activity was reported within the cortex, hippocampus, and spinal cord of 12-month-old rats.[30,50] In contrast, no activity impairment was observed in either the brain stem or cerebellum.[50] Interestingly, brain regions that have age-related increases in protein oxidation overlap with brain regions which have age-related declines in proteasome activity.[51,52] Loss of proteasome activity was found sufficient to elevate the levels of intracellular protein oxidation,[51,53] consistent with a possible role for proteasome inhibition directly contributing to age-related increases in protein oxidation in the aging brain. Additionally, proteasome inhibition exacerbates protein oxidation in neural cells following exposure to the neurotransmitter dopamine.[54] It is important to note that elevations in protein oxidation may be induced by chronic low-level proteasome inhibition, which is not acutely toxic.[51] In these conditions, neural cells possess both increased protein oxidation and protein aggregation, with >5% total cellular protein exhibiting aggregation and ~0.5% of protein being severely cross-linked.[51] These same neural cells exhibit elevated levels of lipofuscin,[55] with multiple age-related changes in gene expression. Taken together, these data indicate that low-level proteasome inhibition is sufficient to induce multiple aspects of neuronal aging.

The gradual loss of muscle mass (often termed as "sarcopenia"), which is a major cause of frailty and morbidity among the elderly, is a characteristic feature of aged mammalian organisms.[56] Studies by Deborah Ferrington and colleagues showed decreased 20S proteasome-specific activity in slow- and fast-twitch skeletal muscle of rats.[57,58] Structural changes including upregulation of the immunoproteasome, increased proteasome subunit oxidation, and decreased content of the proteasome-activating proteins relative to the 20S were proposed to explain the decreased specific activity observed. The differential specific activities measured in homogenates (decreased activity) versus purified 20S proteasome (similar activity) suggest the presence of endogenous inhibitors in aged muscle. Other functional analyses of proteasome in aged muscle showed slower degradation of oxidized calmodulin, and inadequate activation of the 20S by PA700. The potential outcomes from these age-associated functional defects could include an accumulation of oxidized proteins, declines in 20S regulation by PA700 and PA28, and decreased degradation of ubiquitin-conjugated proteins. More recently, Altun *et al.* studied the effects of aging on the content and activity of the 26S proteasome in hind limb muscle from 30-month-old Sprague-Dawley rats.[19] In this study, immunoblot analysis showed a two- to threefold increase in levels of the β1 and β5 subunits of the 20S as well as in subunits Rpt5 and Rpt6 of the 19S regulatory complex. Proteasome peptidase activities in this case were not measured using fluorogenic peptides as substrates, but cell extracts were treated with the specific proteasome active site probe, dansyl-Ahx3L3VS, which modifies covalently the active site threonine residues of the catalytic β1, β2, and β5 subunits. This approach revealed a two- to threefold increase in proteasomes in the aged muscle. This enhanced capacity of the UPS in muscle appears to be a regulated response, because the increase in proteasome content with age was attenuated by dietary restriction. This enhanced capacity of the UPS was proposed to be a response to the increased generation of damaged polypeptides.[19]

Proteasome peptidase activities have been found to be reduced in the heart of aged rats.[32] Analysis of the 20S proteasome purified from the heart of 8-, 21-, and 26-month-old rats indicates that loss of proteasome activity as a function of age is due to both a decreased proteasome content and alterations in the subunit composition, and/or the occurrence of posttranslational modifications. Specifically, age-dependent declines in peptidylglutamyl–peptide hydrolase and trypsin-like activities as well as proteolysis of the model protein [14C] methyl casein were evident following purification of the 20S proteasome from the hearts of senescent versus adult rats.[32]

Decreased proteasome peptidase activities have also been described in the human lens[59] and retina.[33] The aging process undergone by the eye lens proteins includes many posttranslational protein modifications due to UV irradiation and free radical generation. The data obtained on water-insoluble and water-soluble proteins of pooled postmortem lenses from various age groups showed increased protein oxidation and glycoxidation as a function of age. A decrease in the free ubiquitin and high-molecular-weight ubiquitin–protein conjugates, possibly corresponding to the crystallin subunits, was found to occur with age. The finding that ubiquitinated proteins were present in the lenses of the youngest age group(s), but only among the high-molecular-weight proteins and in relatively small proportions, suggests that these proteins cannot be degraded by the proteasome. The decrease in proteasome content observed may partly reflect a decrease in the rate of damaged protein elimination. In addition, the efficiency of the proteasome activity also seems to be affected with age, at least for the peptidylglutamyl–peptide hydrolase activity. Using an anti-carboxymethyllysine immuno-affinity column and an anti-20S antibody, the presence of the 20S proteasome was detected in the bound fraction, which indicates that the proteasome subunits had undergone glycoxidation. This finding might therefore explain the alteration of its proteolytic activities since no such modification of the proteasome was observed in the sample from the youngest age group. Moreover, both high- and low-molecular-weight immunoreactive protein bands were detected in age groups older than 0–10 years, most likely revealing the occurrence of either covalent aggregation or proteolytic degradation of proteasome subunits.[59]

UV radiation from the sun also impacts other exposed organs, such as the skin, and is considered the major cause of skin cancer and premature skin aging, also called photoaging. Photoaging induces morphological changes in the skin, such as wrinkle formation, due to general alteration of skin and, in particular, the dermal components. Several ex vivo studies performed with human skin epidermal cells and dermal fibroblasts showed a decreased proteasomal activity or expression of at least certain proteasome subunits in keratinocytes and fibroblasts during both aging and replicative senescence.[36,38–42] This decreased proteasome activity may explain the well-documented age-related accumulation of oxidized proteins. A decrease in ubiquitin carboxy-terminal hydrolase isoenzyme L1 has been reported, suggesting a possible failure of ubiquitin targeting with age.[60] The major role of the ubiquitin system in signal transduction associated with aging, in skin in particular, through the control of retinoid and NF-κB pathways and the production of a new generation of active molecules targeting E3 ubiquitin ligases have been underlined.[61]

Proteasome activity is also compromised during adipose tissue aging.[62] In these studies performed in mice, the authors reported that the three major proteolytic activities of the 26S and 20S proteasomes are decreased during adipose tissue aging with a concomitant decrease in α7 and β5 proteasome subunit expression. These data suggest that both non-catalytic (α) and catalytic (β) subunits of the core proteolytic complex exhibit decreased expression during aging in adipose tissue. However, increased levels of 19S subunit expression and no significant changes in the 11S subunits of the proteasome complex were observed. These data are consistent with 26S cap-associated proteins of the proteasome being differentially affected by aging in adipose tissue. In addition, increased protein oxidation as well as proteins modified by the lipid peroxidation 4-hydroxy-2-nonenal (HNE) had also been shown.[62] Interestingly, the proteasome activity is believed to influence at least in part the development of obesity and insulin signaling in type 2 diabetes[63] and to regulate the degradation of proteins essential to adipose function. Insulin receptor protein 1 is involved in insulin signaling and is inactivated by degradation through the UPS.[64] Taken together, these data suggest that age-related impairments in the proteasome function may be particularly important to the genesis of age-related insulin resistance in adipose tissue, as well as other features of adipocyte pathogenesis during aging.[65]

## III. Proteasome Dysfunction During Cellular Aging

Studies on human keratinocytes,[36] human fibroblasts,[42] and rat cardiomyocytes[32] have shown an age-related decline in proteasome function and a decreased expression of proteasome subunits. Interestingly, fibroblasts from healthy centenarians exhibited proteasome activity and proteasome subunit expression levels closer to those of younger individuals than older ones, suggesting that sustained proteasome activity could have contributed to their successful aging.[27] Ex vivo studies of proteasomal activities in human dermal fibroblasts from individuals ranging from 20 to 78 years also showed an age-associated decrease in proteasome activities. All three proteasome activities dramatically decreased in human dermal fibroblasts from 20- to 50-year-old donors, but remained stable in donors aged 50–78 years.[66] This age-associated loss of proteasome activity was accompanied by the accumulation of oxidized and ubiquitinated proteins. Moreover, Boraldi *et al.*[60] compared the proteasome composition of *in vitro* cultured dermal fibroblasts from healthy subjects of different ages (i.e., 15, 41, and 82 years old). They observed a reduction of the proteasome subunit α3 and the ubiquitin carboxyl-terminal hydrolase isozyme L1 with aging. This is consistent with the hypothesis of a less efficient proteasomal system during senescence.

Recently, by using a previously unreported system to determine proteasome activity in living cells based on a destabilized version of GFP, Pidder Jansen-Dürr and colleagues confirmed this age-associated decreased proteasome activity in skin fibroblasts from middle-aged donors compared with young donors.[67] However, fibroblasts derived from the oldest donors displayed a high heterogeneity in this assay.

In the immune system, aging is associated with significant deficits in immune function, also known as immunosenescence. Studies on human lymphocytes derived from old and young donors showed a decreased chymotrypsin-like activity in the former, associated with an increased yield of posttranslational modifications of proteasome subunits, while proteasome content and subunit composition were found unchanged.[34,35] In particular, some catalytic and regulatory subunits of the 20S proteasome were preferentially modified with age. Decreased proteasomal activity would be expected to contribute not only to accumulation of oxidized proteins but also to the lower activation of transcription factors such as NF-κβ, and most importantly to the lower production of antigenic peptides by the immunoproteasome for binding to MHC class I molecules.

Comparison of trabecular meshwork (TM) cells from young and old donors showed that aging in the TM is characterized by accumulation of oxidized proteins, a decline in proteasome activity, and the acquisition of features characteristic of cellular senescence, including increased expression of the cyclin inhibitors p16INK4A and p21Waf1/Cip1, as well as accumulation of p53, phosphorylation of p38MAPK, a sharp decline in proliferative capacity of the cells, and enlarged cell morphology.[68] *In vitro* replicative senescence of TM cells from younger donors was also found to be associated with increased accumulation of oxidized proteins and a decline in proteasome activity.[68]

During proliferative senescence of human BJ fibroblasts, a decline in all proteasome peptidase activities (trypsin-like, chymotrypsin-like, and peptidylglutamyl–peptide hydrolase) was reported by Tilman Grune and colleagues.[39,40] In addition, the activity of the cytosolic proteasome declines dramatically during senescence in nondividing BJ fibroblasts, with the peptidylglutamyl–peptide hydrolase activity being particularly affected.[41] Interestingly, this decline in proteasome activity was not accompanied with a decrease in the actual cellular proteasome content. Taken together, these comparative studies of proliferating and nonproliferating human BJ fibroblasts have revealed a good correlation between the accumulation of oxidized/cross-linked proteins and the decline in proteasome activity, and the overall cellular protein turnover during *in vitro* senescence.

Using WI38 fibroblasts as a model for cellular replicative senescence, Efstathios Gonos and colleagues[42] observed reduced levels of proteasomal peptidase activities coupled with increased levels of both oxidized and ubiquitinated proteins in senescent cells. An exhaustive analysis of proteasome subunit expression in WI 38 human fibroblasts undergoing replicative senescence has indicated that only the expression of the catalytic β-subunits is decreased, and that less of the 20S proteasome is assembled while certain α-subunits are found in a free state. Several subunits of the 19S regulatory complex were also found to be downregulated in senescent cells, which was accompanied by decreased levels of both 20S and 26S proteasome complexes. Interestingly, partial inhibition of proteasomes in young cells, caused by treatment with the proteasome inhibitor epoxomycin, induced a senescence-like phenotype, thus demonstrating the fundamental importance of the proteasome for retaining cellular proliferation and protein homeostasis.[42]

In summary, proteasome inhibition appears to be significantly correlated with the onset of *in vitro* cellular senescence in multiple cell types. These data have been supporting a role for proteasome inhibition as a mediator of senescence. Studies with pharmacological inhibitors have strengthened this concept, demonstrating that proteasome inhibition is sufficient to induce senescence and multiple features of the senescent phenotype.[42] In addition, following partial proteasome inhibition, senescence was found to be dominant in human dermal fibroblasts with functional p53 and Rb molecules; cell death was induced with high p53 levels and defective Rb pathways, whereas stress recovery and restoration of normal cycling occurred in cells that lacked functional p53.[69] Thus, it appears that proteasome inhibition may play a direct role in mediating cellular senescence, although the molecular basis for such involvement remains to be further elucidated.

## IV. Modulation of Proteasome Function in Model Organisms

As mentioned earlier, the proteasome is not only involved in the degradation of misfolded, oxidized, and damaged proteins, but it also has regulatory functions, acting in a spatially and temporally controlled fashion to degrade specific protein targets implicated in biological processes. The decline in the proteasome ability to function as an efficient protein quality control system is thought to contribute to, and be aggravated by, the accumulation of damaged proteins. Studies in the past recent years using model organisms have suggested an important role for proteasome function in modulating longevity.

In the budding yeast *S. cerevisiae*, proteasome abundance and activity have been shown to correlate with life span.[46,70] Yeast cells with elevated proteasome capacity retain rapid growth under protein stress-inducing conditions and are able to turnover structurally unstable and aggregation-prone proteins at an enhanced rate.[46] These results suggest a direct relation between proteasome function and longevity, with reduced life span in cells with low proteasome abundance or activity and strong life span extension upon upregulation of the UPS through a mechanism that is at least partially independent of previously known yeast longevity-modulating pathways.[46] Although the precise mechanism has yet to be further elucidated, the results suggest that improved protein quality control and, thus, improved maintenance of protein homeostasis, might be causal for life span extension in cells with increased proteasome capacity. However, the possibility that increased proteasome function also enhances turnover of negative regulators of life span in *S. cerevisiae* cannot be ruled out definitively.

In the nematode *C. elegans*, the inactivation of general proteasome function by RNAi knockdown of proteasome components elicited a life span shortening in wild-type animals as well as in the long-lived insulin/insulin-like growth factor (IGF)-1 receptor homolog daf-2/IIS-receptor mutants, and the glp-1(e2141ts) mutant, the life span of which is extended by germ cell loss.[71,72] On the other hand, proper proteasome function is required for the extension of life span as was reported in some of the long-lived mutant worms.[73] The consistency of these effects across different genotypes suggests that the "housekeeping" function of the proteasome is required in the adult worms to prevent the accumulation of misfolded or oxidized proteins and/or to perform other essential functions.

Another study in *C. elegans* has suggested that aip-1, the worm homolog of AIRAP (a highly conserved arsenic-inducible gene encoding a component of the 19S proteasome regulatory cap), plays an important role in preserving the life span of worms and in buffering proteotoxicity in circumstances that are not associated with exposure to arsenic. AIRAP was previously shown to adapt the cellular protein degradation machinery to the threat of protein misfolding by arsenic.[74] Thus, a proteasomal adaptation induced by environmental stress has linked resistance to proteotoxic insults with a healthy life span in *C. elegans*.[72]

Expression of the E3 ubiquitin ligase, RLE-1, has been shown to regulate aging by determining degradation of DAF-16 by the UPS in *C. elegans*.[73] The forkhead transcription factor, DAF-16, a downstream target of the insulin/IGF-I signaling pathway in worms, is indispensable for both life span regulation and stress resistance. Disruption of RLE-1 expression resulted in increased life span in *C. elegans*.[73] Importantly, expression of RLE-1 accelerated DAF-16 degradation via the proteasome pathway as the proteasome-specific inhibitor,

MG132, blocked DAF-16 degradation. RLE-1 mediates DAF-16 degradation via the ubiquitin pathway because RLE-1 directly interacts with and induces ubiquitin conjugation of DAF-16.[73]

A significant decline in proteasome activity in old (43–47 days of age) flies (*D. melanogaster*) compared to younger (1–32 days of age) flies has been observed.[25] The authors found that the major proteasome form in old flies is the 20S core particle while in younger flies the 26S complex is the preponderant proteasome form. These findings have established that auto-inhibited 20S proteasomes prevail in old flies, whereas the fully assembled 26S proteasome is highly active in young flies. Hence, these results suggest an age-dependent disassembly of the 26S proteasome.[25] Conversely, Tonoki *et al.* reported that the overexpression of Rpn11 prevented the age-related reduction of the 26S proteasome activity, which suppressed the age-dependent accumulation of ubiquitinated proteins and extended life span.[43] On the other hand, the loss of function of Rpn11 enhanced the age-related reduction of 26S proteasome activity, leading to a shorter life span, a premature age-dependent accumulation of ubiquitinated proteins, and an early onset of a neurodegenerative phenotype. These findings imply that improving the amount and/or activity of the 26S proteasome by the overexpression of a lid subunit of the 19S proteasome, such as Rpn11, could provide an extension of the mean life span and prevent the age-dependent onset or progression of neurodegeneration in *D. melanogaster.*

In mammals, the naked mole rat (*Heterocephalus glaber*) is the longest-living rodent known, with a maximum life span of 28.3 years.[75] When compared with mice, the naked mole rat showed a 1.5-fold age-related increase in proteasome activity, suggesting the presence of a higher protein turnover, facilitating the removal of damaged proteins and the preservation of a functional proteome.[28] Although oxidative damage in this long-lived rodent at a young age is higher than that in mice, high steady-state levels of both reversible and irreversible oxidative damage are maintained throughout life. Mice, however, show pronounced age-related increases in protein damage and concomitant functional impairments. Taken together, these results suggest that enhanced protein homeostasis coupled with a tightly modulated protein redox state may contribute significantly to the exceptional longevity of the naked mole rat and may be a critical determinant for the observed abrogated aging of this species.[28]

Although accumulation of macromolecular damage has been observed during the life span of organisms ranging from yeast to humans, these organisms produce youthful progeny with low damage levels, implying that these molecular insults are not transmitted to the offspring. A central role of the proteasome has also been described in this process. Pioneering studies on unicellular organisms such as *S. cerevisiae* or *Schizosaccharomyces pombe* showed that they segregate toxic damage away from the offspring.[76,77] In these models, carbonylated proteins form aggregates that are segregated at the time of cytokinesis. Consequently, daughter cells are born free of oxidative products, which instead accumulate in the mother cell compartment. Recent work revealed that this segregation process encompasses a retrograde flow of aggregates of oxidatively damaged proteins from the daughter to the mother cell. This process involves a series of cytoskeletal components named the "polarisome"[78] and depends on the life span modulator Sir2p.[76–78] More recently, Zhou *et al.* showed that protein aggregates undergo random walk without directional bias.[79] Although these data do not disagree with a role for actin or cell polarity in aggregate segregation, *in silico* studies suggest that their asymmetric inheritance can be a predictable outcome of the aggregates′ slow diffusion and the geometry of yeast cells.[79]

A pivotal study performed during mice embryogenesis revealed that carbonylated proteins are also eliminated, suggesting that similar processes occur in mammals.[80] In this case, the process by which damaged proteins are effectively cleared from the cytoplasm appears to be distinct from what has been observed in yeast. First, the authors did not observe any decrease in the concentration of damaged proteins until the late stages of development. At the blastocyst stage, high levels of damaged proteins were observed in the inner mass, while cells that underwent differentiation harbored markedly reduced levels of carbonylated proteins.[80] Second, the authors observed that the activity of the 20S proteasome also increased abruptly at this stage,[80] although the mechanism involved in this process has not yet been fully elucidated. The authors hypothesized that these two processes might be causally linked, which correlates with the fact that the 20S proteasome is responsible for the degradation of the majority of these proteins. More recently, Goudeau and Aguilaniu showed that, similar to mice, proteasome activity in the germline is required for the resetting of carbonylated proteins during reproduction in *C. elegans.*[81] Thus, oxidatively damaged proteins are eliminated during reproduction in worms through the proteasome.

## V. Mechanisms Involved in the Age-Related Decline of Proteasome Function

The trend for a decline in proteasome function observed during aging may be related to different reasons (Fig. 2), some of them believed to be tissue-specific. Downregulated expression of proteasomal subunits,[30] unbalanced levels of catalytic subunits,[42] defective expression of regulatory subunits,[58] damaging modifications in critical proteasome subunits,[35,37] extensive protein oxidation and cross-linking of proteasome substrates[41,82,83] have all been

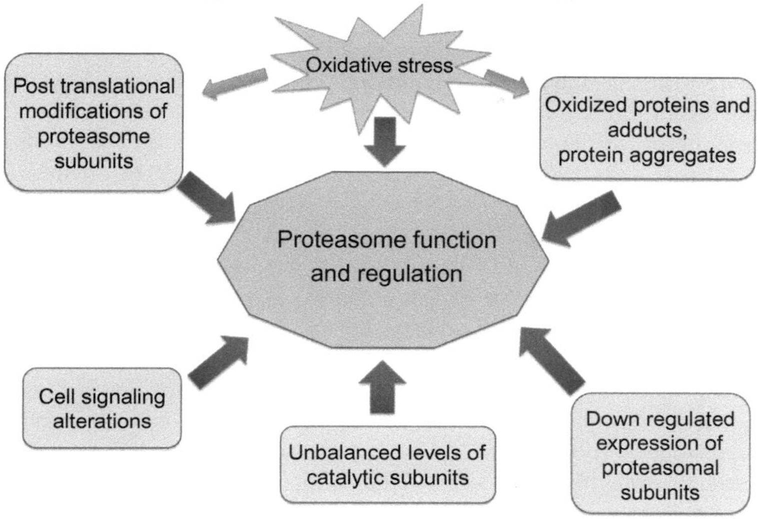


FIG. 2. Age-associated changes impacting on proteasome function and regulation. Proteasome dysfunction observed during aging has been related to different causes: increased posttranslational damaging modifications of proteasome subunits, formation of endogenous inhibitors such as extensively oxidized and/or cross-linked proteins, downregulation of proteasomal subunits, unbalanced levels of catalytic subunits, and altered cellular pathways of proteasome regulation. (See Color Insert.)

reported as possible causes of proteasome dysfunction. In addition, other factors such as the reduced ATP content in aging flies have been related to the reduced assembly of the 26S proteasome.[25]

ROS are routinely produced as a by-product of aerobic metabolism and oxidative phosphorylation.[84] Exposure to various environmental stressors (e.g., ionizing and nonionizing radiation, or certain chemical agents) can also result in the production of ROS.[85–88] In addition, ROS production and accumulation can be generated during disease pathogenesis or during the normal aging process.[84,89] Unneutralized ROS cause oxidative damage to lipids, proteins, and DNA, thus leading to aberrant molecular activities.[90–92] Protein oxidation is particularly detrimental as the resulting conformational changes to protein structures can render damaged proteins inactive or lead to functional abnormalities. Various types of protein oxidative modifications are induced directly by ROS or indirectly by reactions with secondary products of oxidative stress.[93] Cysteine and methionine residues in proteins are particularly prone to oxidative attack by almost all ROS. However, these modifications are not always linked to protein damage, because it has been shown that they can participate in cellular signaling events. Furthermore, the existence of repair mechanisms for certain cysteine and methionine oxidation products within proteins can reverse the damage. On the other hand, irreversible oxidation products of other residues result most frequently in their hydroxylation and carbonylation.

Compelling evidence indicates that the capacity of the proteasome to degrade proteins in cells or tissues is altered in response to environmental insults, including oxidative stress. It has been demonstrated in several cell types that exposure to mild oxidative stress results in a transient increase in the intracellular degradation of both short-lived and long-lived cellular proteins. However, extensive oxidative stress impairs the function of the proteolytic system, reducing intracellular protein degradation,[6,94–96] and inducing intracellular accumulation and aggregation of damaged or abnormal proteins. If modified proteins are no longer degradable by the proteasome, further oxidation can occur as well as covalent cross-linking to other proteins by products of lipid peroxidation such as HNE[82,97] and malondialdehyde,[98–100] two abundant bifunctional aldehydic compounds. This results in hydrophobic and insoluble protein aggregates, also termed "aggresomes",[101,102] preliminary stages of the so-called lipofuscin.[103,104] Hence, prevention of the accumulation of highly damaged proteins is the main function of the proteolytic degradation systems.

Several studies have shown that 19S and 20S proteasome subunits are themselves susceptible to oxidative modifications, including carbonylation, HNE modification, and glycation.[32,35,94,105] Treatment of purified proteasome with free radical-generating systems or the lipid peroxidation product 4-Hydroxynonenal (HNE) led to declines in peptidase activities.[21,106,107] Both trypsin-like and peptidylglutamyl–peptide hydrolase activities were susceptible to metal-catalyzed oxidation, while HNE inactivation was selective to trypsin-like activity.[21] Keller *et al.*[30] reported the formation of HNE adducts mainly in β-subunits in the spinal cord of Fisher rats after exposure to oxidative agents and suggested a relation between decreasing proteasomal activity and neuronal cell death. However, the constitutive proteasomal subunits were not the only subunits to be increasingly modified in aged cells; the β5i subunit was a target of HNE modification too.[30] Moreover, carbonylation of Rpt3 resulted in impaired Rpt3 ATPase activity and a subsequent decrease in ubiquitin/ATP-dependent proteolysis of the 26S proteasome.[108] In addition, administration of the pro-oxidant $Fe3^{+}$-nitriloacetate in mice induced a decrease in trypsin-like and peptidylglutamyl–peptide hydrolase activities of the proteasome. Interestingly, these activities recovered over time, coincident with the disappearance of HNE modification to the proteasome.[83]

Proteasome activity is also decreased *in vitro* and *in vivo* under conditions of chronic hyperglycemia. Intracellular hyperglycemia activates the formation of the reactive carbonyl compounds methylglyoxal (MGO) and glyoxal.[109] These reactive carbonyl compounds form covalent adducts in particular with lysine and arginine residues in proteins, a process also known as protein glycation, altering their function.[110] Glycation of the 20S proteasome subunits was observed and may explain its decreased catalytic activity.[111] This effect was also observed by exposure of endothelial cells to MGO, which induced glycation of the 20S proteasome subunits. Similar results were obtained using kidney from nondiabetic glyoxalase-1 knockdown mice, indicating that increased MGO alone is sufficient to cause proteasome modifications and dysfunction. In addition to decreasing chymotrypsin-like activity in the 20S proteasome, increased intracellular MGO caused a marked decrease in protein levels of the regulatory 19S component of the 26S proteasome.[111]

*S*-glutathiolation, the covalent attachment of glutathione (GSH) to protein thiol groups, is generated by two mechanisms: (i) the reaction of GSH with oxidized thiol groups (e.g., Cys-SOH or Cys-S-S-Cys), or (ii) the reaction of oxidized glutathione (GSSG) with reduced thiol residues (e.g., Cys-SH).[112] Although GSH is considered to have antioxidant function, by stabilizing oxidized protein thiol groups, preventing further, possibly irreversible thiol oxidation through *S*-glutathiolation, it can also modulate protein activity.[113] Upon $H_2O_2$-induced oxidative stress in yeast, *S*-glutathiolation of proteasome 20S subunits was observed both *in vitro* and *in vivo*.[114] Further functional studies determined that treatment of purified 20S proteasomes with GSH led to the inhibition of chymotrypsin-like and trypsin-like proteasome activities.[115] Mammalian proteasomes appear to have a biphasic response to *S*-glutathiolation, as low concentrations of GSH or GSSG increased the chymotrypsin-like activity of purified mammalian proteasomes whereas high levels of GSH or GSSG led to decreased activity.[115] Although *S*-glutathiolation of the 20S proteasome generally inhibits proteasome activity, the biphasic response observed for *S*-glutathiolation of mammalian proteasome suggests that *S*-glutathiolation acts as a redox-signaling trigger through which proteasome activity is regulated depending on the redox status of mammalian cells.[116]

In addition to oxidative and related modifications of particular proteasome subunits, other types of posttranslational modifications may be involved in modulating proteasome activities.[117] Poly [ADP-ribose] polymerase 1 (PARP1), a nuclear enzyme that transfers ADP-ribose moieties from $NAD^+$ to glutamic acid, aspartic acid, or lysine residues, is activated in response to oxidative stress.[118–120] Nuclear 20S proteasomes can be ADP-ribosylated by PARP1 in human hematopoietic K562 cells, resulting in increased chymotrypsin-like activity of the nuclear 20S proteasome.[117] Given the nuclear localization of PARP1 and its role in DNA repair,[121] ADP-ribosylation is likely unique to nuclear proteasomes and may function to enhance proteasomal degradation of oxidized nuclear proteins.[122] Phosphorylation of proteasome subunits has been linked also to the regulation of proteasome activity.[123–126] A recent study revealed that Rpt5 (19S subunit) is phosphorylated by human apoptosis signal-regulating kinase 1 (Ask1).[123] Phosphorylation resulted in the inhibition of Rpt5 ATPase activity and in the reduction of 26S proteasome proteolytic activities.[123] Additional kinases have been found to phosphorylate proteasome subunits including CK2 (formerly casein kinase II), cyclic AMP-dependent kinase (PKA), Ca2+/calmodulin-dependent kinase (CaM-K) II, AMP-activated protein kinase (AMPK), and *c-Abl* and *abl-related* gene (Arg) tyrosine kinases.[124,126] Proteasome subunit phosphorylation by these kinases appears to be involved in several proteasomal-related functions and regulations including proteasome assembly,[127,128] and proteolytic activities.[123] Liu *et al.* reported that c-Abl and Arg phosphorylation of α4 results in suppressed 20S and 26S proteasome proteolytic activities.[126]

Although proteasome phosphorylation by these kinases has not been directly linked to oxidative stress, activity modulation of CK2, PKA, CaM-KII, c-Abl, and Arg has been shown during oxidative stress.[129–131] Oxidative stress-mediated gene regulation is a known component of the defense mechanism for cellular responses to proteotoxic stress.[132,133] In yeast, much of the oxidative stress-driven transcriptional activation is controlled by the redox reactive transcription factor Yap1.[134] Rpn4, the transcriptional activator for proteasome genes, is a Yap1-targeted gene.[135–138] Upon oxidative stress, transient Yap1-mediated Rpn4 mRNA upregulation[137] and Yap1-dependent expression of several yeast proteasome components[139] have been observed; however, the biological consequences of these changes have not been evaluated.

## VI. Rejuvenating Proteasome Function as an Antiaging Strategy

Sustained protein maintenance during life appears to be a mandatory requirement for a healthy life span. Strategies for testing and developing intervention, prevention, and modulation of aging incorporate effective means to minimize the occurrence and accumulation of macromolecular damage. In that sense, restoring and/or stimulating proteasome activity may represent not only an efficient way for increased protection against oxidative stress, but also for developing antiaging strategies. Over the last few years, several attempts to activate the proteasome have been made by genetic approaches or other means, such as modulation by chemical or natural compounds.

Concerning genetic approaches for proteasome activation, the pioneers' *in vitro* studies pointed toward the overexpression of $\beta_{5i}$ subunits in lymphoblasts and HeLa cells, which enhances the chymotrypsin-like and trypsin-like proteasome catalytic activities.[140] A follow-up study showed increased trypsin-like and caspase-like activities following β1i and β1 subunit overexpression in HeLa cells.[141] In addition, stable overexpression of β5 subunit in WI38/T and

HL60 cells resulted in the upregulation of other β-type subunits, thus implying a common regulatory loop.[42] The observed co-regulation of β-type subunits has also been confirmed in other cell types, such as the lens epithelial cells[142] or the murine neuroblastoma cells,[143] following overexpression of the β5 subunit. The transfected β1 and β5 stable clones exhibited enhanced levels of chymotrypsin-like and peptidylglutamyl–peptide hydrolase proteolytic activities as compared with vector-transfected cell lines, indicating that increased levels of β1 and β5 subunits cause an increase in proteasome activity. The transfected cells showed increased survival ability when subjected to different stresses. Importantly, increased proteasome catalytic activities correlated with enhanced cell viability and reduced accumulation of oxidized proteins have been observed following oxidative stress.[42] Moreover, β5 subunit overexpression in primary human embryonic fibroblasts (IMR90 cells) resulted in the extension of life span by ~15–20%.[45] Similarly, the restoration of the normal level of proteasome subunits in aged human fibroblasts reduced the levels of aging biomarkers such as β-galactosidase and P21.[66]

Overexpression of proteasome assembly protein UMP1 improves cell viability following exposure to various oxidants in yeast[70] and mammalian cells.[144] In the former study, the authors used a yeast model system to explore the role of the proteasome in regulating the cytotoxicity observed in response to oxidative stressors as well as during cellular aging. Cells were engineered to express elevated levels of Ump1, a protein responsible for assembling proteasome subunits into a functional proteasome complex, in order to elucidate whether elevating the levels of proteasome complexes was sufficient to ameliorate oxidative stress and age-related cytotoxicity. Cells expressing Ump1 exhibited increased levels of overall proteasome activity. In order to elucidate whether increased expression of Ump1 alters cellular viability, the authors subjected vector- and Ump1-transformed cells to a variety of oxidative stressors (hydrogen peroxide, menadione, HNE). Following the administration of oxidative stressors, there was a tremendous loss of cellular viability in vector-transformed cells that was significantly attenuated in Ump1-transformed cells. Experiments aimed at elucidating if the preservation of proteasome-mediated protein degradation is also associated with an attenuation of oxidized protein levels were performed. As expected, following the administration of an oxidative stressor (menadione), there was a rapid increase in the amount of protein carbonyls in both vector- and Ump1-transformed cells. Interestingly, the elevation in oxidized protein levels was nearly identical in vector- and Ump1-transformed cells. However, following the removal of oxidative stressors, there was a profound decrease in oxidized protein levels in Ump1-transformed cells. Together, these studies suggest that an increased 20S proteasome expression and assembly would enhance the capacity of yeast cells to cope with oxidative stress.

Alternatively, disruption of Rpn4-mediated proteasome induction leads to reduced viability in response to oxidative stress in *S. cerevisiae*,[145] demonstrating the critical role of proteasome gene regulation for combating oxidative insults. In this study, a yeast strain in which *pre1* encoding of one of the essential proteasome subunits is no longer induced by Rpn4 was used. The authors found that the active proteasome level is lower in this strain than in the wild-type counterpart. Cell-cycle analysis showed that downregulation of *pre1* delays G2/M exit. Moreover, they demonstrated that loss of Rpn4-induced proteasome expression sensitizes cells to different stresses. This study explicitly reveals the physiological function of Rpn4-induced proteasome expression.

Other approaches involve proteasome activation by natural or chemical compounds. Pioneering studies in this direction were performed in our group in order to prevent the decreased activity of the proteasome upon UV irradiation of keratinocytes *in vitro*.[146] On the basis of previous studies that small compounds such as sodium dodecyl sulfate and some unsaturated fatty acids can stimulate proteasome activities *in vitro*,[147] the capacity of plant and algae extracts to increase the proteolytic activities of the proteasome has been tested. In this study, an extract rich in fatty acids belonging to the algae *Phaeodactylum tricornutum* has been identified. Importantly, this algae extract was able to stimulate and protect the proteolytic activities of the proteasome within human keratinocytes exposed to UVA and UVB irradiation, thus lowering the extent of the irradiation-induced protein oxidative damage and protecting the keratinocytes after UV irradiation.[146] Another study performed by Katsiki *et al.* has shown a stimulatory impact on proteasome activities *in vitro* using oleuropein, the most abundant of the phenolic compounds in *Olea europaea* leaf extract, olive oil, and olives.[148] The increased activities promote cellular resistance to oxidants and confer extension of human fibroblast life span.[148] Oleuropein most likely acts through structural changes of the 20S $\alpha$-gated channel conformation in a similar manner to SDS, although its effects are considerably stronger.

Proteasome activation has also been achieved following treatment with indirect antioxidants, such as sulforaphane and dithiolethiones. These phytochemicals increase intracellular levels of the antioxidant glutathione, boost synthesis of reducing equivalents such as NADPH, and increase the expression of enzymes, such as glutathione S-transferase and NADPH dehydrogenase, that detoxify chemicals poised to generate oxidants. These phytochemicals are well-characterized cancer-preventive agents that increase expression of multiple antioxidant proteins.[149] Induction of multiple families of cytoprotective genes by these compounds is mediated through the transcription factor Nrf2. Subunits of the 26S proteasome system represent one of several inducible gene families regulated by the Nrf2 pathway. Concordantly, administration of dithiolethiones to mice increases peptidase activities of the proteasome in several tissues.[150,151] Increased protein levels of PSMB5 and PSMB6 were observed in the liver, lung, small intestine, and colon, while induction of PSMB7 in these tissues was marginal. Peptidase activities of the proteasome were increased in the liver and lung from treated mice compared to those from vehicle-treated mice.[151] The results suggest that coordinated transcriptional regulation of multiple subunits of the proteasome seems to be one mechanism for enhancing proteasome function. Interestingly, it has been recently shown that Nrf-2-mediated proteasome activation delays senescence in human fibroblasts[152]

An increasingly studied area of aging research is focused on the topic of hormesis, whereby low levels of stress can stimulate cells to increase protective processes, such as proteasome activity, which can subsequently protect cells from the toxicity associated with exposure to multiple stressors.[153,154] One such approach is that of mild stress-induced hormesis, which stimulates a complex series of maintenance and repair systems and strengthens the homeodynamic space of cells and organisms.[155] In organism models, such as *C. elegans*, brief thermal stress sufficient to induce thermo-tolerance causes statistically significant increases in life span.[156] Dose–response relationships for thermo-tolerance and longevity have been shown to be very similar.[157] Hormesis through mild heat shock, natural and synthetic hormetins, and other stressors brings about several antiaging effects at the cellular level also, for instance, in human fibroblasts, keratinocytes, and telomerase-immortalized bone marrow stem cells.[153] These results further support the view that increased stress resistance causes increased life span. Some studies have suggested that the proteasome may play an important role in hormesis by contributing to maintenance of protein turnover during periods of stress.[153] Jeffrey Keller and colleagues have demonstrated that neural cells respond to low levels of oxidant exposure by increasing the expression of inducible proteasome subunits.[51] More recently, Kelvin Davies and colleagues showed that the increased capacity of adapted cells to withstand a high $H_2O_2$ challenge is, at least, partly dependent upon the 20S proteasome, the immunoproteasome, and PA28αβ regulator induction.[158] These findings demonstrate the importance of the 20S proteasome, the immunoproteasome, and PA28αβ in the overall cellular adaptation to oxidative stress. However, it is important to note that several questions remain unanswered, especially with respect to determining the optimal levels of stress that are hormetic at various ages, the cell-type specificity of the various stresses, and the combined action of multiple stresses. Hence, despite the complexity of its regulation and function, the proteosomal system represents an already validated target for improving protein homeostasis. Therefore, it is expected that such beneficial effects, if applicable at advanced age, would help to prevent or delay the onset of those age-related disease that are linked to the alteration of the proteostasis network.

## Acknowledgments

The authors are very thankful to the FP7 EU-funded projects MYOAGE (No. 223576) and MARK-AGE (No. 200880).

## References

1. Harman D. Aging: a theory based on free radical and radiation chemistry. *J Gerontol* 1956;**11**:298–300.
2. Harman D. Free radical theory of aging. *Triangle* 1973;**12**:153–8.
3. Taylor RC, Dillin A. Aging as an event of proteostasis collapse. *Cold Spring Harb Perspect Biol* 2011;**3**. doi:10.1101/cshperspect.a004440.
4. Roth DM, Balch WE. Modeling general proteostasis: proteome balance in health and disease. *Curr Opin Cell Biol* 2011;**23**:126–34.
5. Friguet B. Protein repair and degradation during aging. *Sci World J* 2002;**2**:248–54.
6. Grune T, Reinheckel T, Davies KJ. Degradation of oxidized proteins in K562 human hematopoietic cells by proteasome. *J Biol Chem* 1996;**271**:15504–9.
7. Goldberg AL. Protein degradation and protection against misfolded or damaged proteins. *Nature* 2003;**426**:895–9.
8. Kopito RR. Aggresomes, inclusion bodies and protein aggregation. *Trends Cell Biol* 2000;**10**: 524–30.
9. Caughey B, Lansbury PT. Protofibrils, pores, fibrils, and neurodegeneration: separating the responsible protein aggregates from the innocent bystanders. *Annu Rev Neurosci* 2003;**26**:267–98.
10. Friguet B. Aging of proteins and the proteasome. *Prog Mol Subcell Biol* 2002;**29**:17–33.
11. Farout L, Friguet B. Proteasome function in aging and oxidative stress: implications in protein maintenance failure. *Antioxid Redox Signal* 2006;**8**:205–16.
12. Breusing N, Grune T. Regulation of proteasome-mediated protein degradation during oxidative stress and aging. *Biol Chem* 2008;**389**:203–9.
13. Sitte N, Huber M, Grune T, Ladhoff A, Doecke WD, Von Zglinicki T. Proteasome inhibition by lipofuscin/ceroid during postmitotic aging of fibroblasts. *FASEB J* 2000;**14**: 1490–8.
14. Jung T, Grune T. The proteasome and its role in the degradation of oxidized proteins. *IUBMB Life* 2008;**60**:743–52.
15. Friguet B, Bulteau AL, Chondrogianni N, Conconi M, Petropoulos I. Protein degradation by the proteasome and its implications in aging. *Ann N Y Acad Sci* 2000;**908**:143–54.
16. Weih M, Schmitt M, Gieche J, Harms C, Ruscher K, Dirnagl U, et al. Proteolysis of oxidized proteins after oxygen-glucose deprivation in rat cortical neurons is mediated by the proteasome. *J Cereb Blood Flow Metab* 2001;**21**:1090–6.
17. Davies KJ. Degradation of oxidized proteins by the 20S proteasome. *Biochimie* 2001;**83**: 301–10.
18. Shringarpure R, Davies KJ. Protein turnover by the proteasome in aging and disease. *Free Radic Biol Med* 2002;**32**:1084–9.
19. Altun M, Besche HC, Overkleeft HS, Piccirillo R, Edelmann MJ, Kessler BM, et al. Muscle wasting in aged, sarcopenic rats is associated with enhanced activity of the ubiquitin proteasome pathway. *J Biol Chem* 2010;**285**:39597–608.
20. Cook C, Gass J, Dunmore J, Tong J, Taylor J, Eriksen J, et al. Aging is not associated with proteasome impairment in UPS reporter mice. *PLoS One* 2009;**4**:e5888.

21. Conconi M, Szweda LI, Levine RL, Stadtman ER, Friguet B. Age-related decline of rat liver multicatalytic proteinase activity and protection from oxidative inactivation by heat-shock protein 90. *Arch Biochem Biophys* 1996;**331**:232–4.
22. Shibatani T, Nazir M, Ward WF. Alteration of rat liver 20S proteasome activities by age and food restriction. *J Gerontol A Biol Sci Med Sci* 1996;**51**:B316–22.
23. Shibatani T, Ward WF. Effect of age and food restriction on alkaline protease activity in rat liver. *J Gerontol A Biol Sci Med Sci* 1996;**51**:B175–8.
24. Huber N, Sakai N, Eismann T, Shin T, Kuboki S, Blanchard J, et al. Age-related decrease in proteasome expression contributes to defective nuclear factor-kappaB activation during hepatic ischemia/reperfusion. *Hepatology* 2009;**49**:1718–28.
25. Vernace VA, Arnaud L, Schmidt-Glenewinkel T, Figueiredo-Pereira ME. Aging perturbs 26S proteasome assembly in Drosophila melanogaster. *FASEB J* 2007;**21**:2672–82.
26. Chondrogianni N, Gonos ES. Proteasome dysfunction in mammalian aging: steps and factors involved. *Exp Gerontol* 2005;**40**:931–8.
27. Chondrogianni N, Petropoulos I, Franceschi C, Friguet B, Gonos ES. Fibroblast cultures from healthy centenarians have an active proteasome. *Exp Gerontol* 2000;**35**:721–8.
28. Perez VI, Buffenstein R, Masamsetti V, Leonard S, Salmon AB, Mele J, et al. Protein stability and resistance to oxidative stress are determinants of longevity in the longest-living rodent, the naked mole-rat. *Proc Natl Acad Sci USA* 2009;**106**:3059–64.
29. Hayashi T, Goto S. Age-related changes in the 20S and 26S proteasome activities in the liver of male F344 rats. *Mech Ageing Dev* 1998;**102**:55–66.
30. Keller JN, Huang FF, Markesbery WR. Decreased levels of proteasome activity and proteasome expression in aging spinal cord. *Neuroscience* 2000;**98**:149–56.
31. Shang F, Nowell Jr. TR, Taylor A. Removal of oxidatively damaged proteins from lens cells by the ubiquitin-proteasome pathway. *Exp Eye Res* 2001;**73**:229–38.
32. Bulteau AL, Szweda LI, Friguet B. Age-dependent declines in proteasome activity in the heart. *Arch Biochem Biophys* 2002;**397**:298–304.
33. Louie JL, Kapphahn RJ, Ferrington DA. Proteasome function and protein oxidation in the aged retina. *Exp Eye Res* 2002;**75**:271–84.
34. Ponnappan U, Zhong M, Trebilcock GU. Decreased proteasome-mediated degradation in T cells from the elderly: a role in immune senescence. *Cell Immunol* 1999;**192**:167–74.
35. Carrard G, Dieu M, Raes M, Toussaint O, Friguet B. Impact of ageing on proteasome structure and function in human lymphocytes. *Int J Biochem Cell Biol* 2003;**35**:728–39.
36. Petropoulos I, Conconi M, Wang X, Hoenel B, Bregegere F, Milner Y, et al. Increase of oxidatively modified protein is associated with a decrease of proteasome activity and content in aging epidermal cells. *J Gerontol A Biol Sci Med Sci* 2000;**55**:B220–7.
37. Bulteau AL, Petropoulos I, Friguet B. Age-related alterations of proteasome structure and function in aging epidermis. *Exp Gerontol* 2000;**35**:767–77.
38. Merker K, Grune T. Proteolysis of oxidised proteins and cellular senescence. *Exp Gerontol* 2000;**35**:779–86.
39. Sitte N, Merker K, Von Zglinicki T, Grune T, Davies KJ. Protein oxidation and degradation during cellular senescence of human BJ fibroblasts: part I—effects of proliferative senescence. *FASEB J* 2000;**14**:2495–502.
40. Sitte N, Merker K, Von Zglinicki T, Davies KJ, Grune T. Protein oxidation and degradation during cellular senescence of human BJ fibroblasts: part II—aging of nondividing cells. *FASEB J* 2000;**14**:2503–10.
41. Sitte N, Merker K, von Zglinicki T, Grune T. Protein oxidation and degradation during proliferative senescence of human MRC-5 fibroblasts. *Free Radic Biol Med* 2000;**28**:701–8.
42. Chondrogianni N, Stratford FL, Trougakos IP, Friguet B, Rivett AJ, Gonos ES. Central role of the proteasome in senescence and survival of human fibroblasts: induction of a senescence-

like phenotype upon its inhibition and resistance to stress upon its activation. *J Biol Chem* 2003;**278**:28026–37.
43. Tonoki A, Kuranaga E, Tomioka T, Hamazaki J, Murata S, Tanaka K, et al. Genetic evidence linking age-dependent attenuation of the 26S proteasome with the aging process. *Mol Cell Biol* 2009;**29**:1095–106.
44. Hamer G, Matilainen O, Holmberg CI. A photoconvertible reporter of the ubiquitin-proteasome system in vivo. *Nat Methods* 2010;**7**:473–8.
45. Chondrogianni N, Tzavelas C, Pemberton AJ, Nezis IP, Rivett AJ, Gonos ES. Overexpression of proteasome beta5 assembled subunit increases the amount of proteasome and confers ameliorated response to oxidative stress and higher survival rates. *J Biol Chem* 2005;**280**:11840–50.
46. Kruegel U, Robison B, Dange T, Kahlert G, Delaney JR, Kotireddy S, et al. Elevated proteasome capacity extends replicative lifespan in Saccharomyces cerevisiae. *PLoS Genet* 2011;**7**:e1002253.
47. Anselmi B, Conconi M, Veyrat-Durebex C, Turlin E, Biville F, Alliot J, et al. Dietary self-selection can compensate an age-related decrease of rat liver 20 S proteasome activity observed with standard diet. *J Gerontol A Biol Sci Med Sci* 1998;**53**:B173–9.
48. Li F, Zhang L, Craddock J, Bruce-Keller AJ, Dasuri K, Nguyen A, et al. Aging and dietary restriction effects on ubiquitination, sumoylation, and the proteasome in the heart. *Mech Ageing Dev* 2008;**129**:515–21.
49. Zhang L, Li F, Dimayuga E, Craddock J, Keller JN. Effects of aging and dietary restriction on ubiquitination, sumoylation, and the proteasome in the spleen. *FEBS Lett* 2007;**581**:5543–7.
50. Keller JN, Hanni KB, Markesbery WR. Possible involvement of proteasome inhibition in aging: implications for oxidative stress. *Mech Ageing Dev* 2000;**113**:61–70.
51. Ding Q, Reinacker K, Dimayuga E, Nukala V, Drake J, Butterfield DA, et al. Role of the proteasome in protein oxidation and neural viability following low-level oxidative stress. *FEBS Lett* 2003;**546**:228–32.
52. Keller JN, Gee J, Ding Q. The proteasome in brain aging. *Ageing Res Rev* 2002;**1**:279–93.
53. Abd El Mohsen MM, Iravani MM, Spencer JP, Rose S, Fahim AT, Motawi TM, et al. Age-associated changes in protein oxidation and proteasome activities in rat brain: modulation by antioxidants. *Biochem Biophys Res Commun* 2005;**336**:386–91.
54. Keller JN, Huang FF, Dimayuga ER, Maragos WF. Dopamine induces proteasome inhibition in neural PC12 cell line. *Free Radic Biol Med* 2000;**29**:1037–42.
55. Sullivan PG, Dragicevic NB, Deng JH, Bai Y, Dimayuga E, Ding Q, et al. Proteasome inhibition alters neural mitochondrial homeostasis and mitochondria turnover. *J Biol Chem* 2004;**279**:20699–707.
56. Evans WJ. What is sarcopenia? *J Gerontol A Biol Sci Med Sci* 1995;**50**:5–8 Spec No.
57. Husom AD, Peters EA, Kolling EA, Fugere NA, Thompson LV, Ferrington DA. Altered proteasome function and subunit composition in aged muscle. *Arch Biochem Biophys* 2004;**421**:67–76.
58. Ferrington DA, Husom AD, Thompson LV. Altered proteasome structure, function, and oxidation in aged muscle. *FASEB J* 2005;**19**:644–6.
59. Viteri G, Carrard G, Birlouez-Aragon I, Silva E, Friguet B. Age dependent protein modifications and declining proteasome activity in the human lens. *Arch Biochem Biophys* 2004;**427**:197–203.
60. Boraldi F, Bini L, Liberatori S, Armini A, Pallini V, Tiozzo R, et al. Proteome analysis of dermal fibroblasts cultured in vitro from human healthy subjects of different ages. *Proteomics* 2003;**3**:917–29.
61. Bregegere F, Milner Y, Friguet B. The ubiquitin-proteasome system at the crossroads of stress-response and ageing pathways: a handle for skin care? *Ageing Res Rev* 2006;**5**:60–90.

62. Dasuri K, Zhang L, Ebenezer P, Fernandez-Kim SO, Bruce-Keller AJ, Szweda LI, et al. Proteasome alterations during adipose differentiation and aging: links to impaired adipocyte differentiation and development of oxidative stress. *Free Radic Biol Med* 2011;**51**:1727–35.
63. Chang TL, Chang CJ, Lee WY, Lin MN, Huang YW, Fan K. The roles of ubiquitin and 26S proteasome in human obesity. *Metabolism* 2009;**58**:1643–8.
64. Lee AV, Gooch JL, Oesterreich S, Guler RL, Yee D. Insulin-like growth factor I-induced degradation of insulin receptor substrate 1 is mediated by the 26S proteasome and blocked by phosphatidylinositol 3'-kinase inhibition. *Mol Cell Biol* 2000;**20**:1489–96.
65. Sakamoto K, Sato Y, Sei M, Ewis AA, Nakahori Y. Proteasome activity correlates with male BMI and contributes to the differentiation of adipocyte in hADSC. *Endocrine* 2010;**37**:274–9.
66. Hwang JS, Chang I, Kim S. Age-associated decrease in proteasome content and activities in human dermal fibroblasts: restoration of normal level of proteasome subunits reduces aging markers in fibroblasts from elderly persons. *J Gerontol A Biol Sci Med Sci* 2007;**62**:490–9.
67. Koziel R, Greussing R, Maier AB, Declercq L, Jansen-Durr P. Functional interplay between mitochondrial and proteasome activity in skin aging. *J Invest Dermatol* 2011;**131**:594–603.
68. Caballero M, Liton PB, Challa P, Epstein DL, Gonzalez P. Effects of donor age on proteasome activity and senescence in trabecular meshwork cells. *Biochem Biophys Res Commun* 2004;**323**:1048–54.
69. Chondrogianni N, Trougakos IP, Kletsas D, Chen QM, Gonos ES. Partial proteasome inhibition in human fibroblasts triggers accelerated M1 senescence or M2 crisis depending on p53 and Rb status. *Aging Cell* 2008;**7**:717–32.
70. Chen Q, Thorpe J, Dohmen JR, Li F, Keller JN. Ump1 extends yeast lifespan and enhances viability during oxidative stress: central role for the proteasome? *Free Radic Biol Med* 2006;**40**:120–6.
71. Ghazi A, Henis-Korenblit S, Kenyon C. Regulation of Caenorhabditis elegans lifespan by a proteasomal E3 ligase complex. *Proc Natl Acad Sci USA* 2007;**104**:5947–52.
72. Yun C, Stanhill A, Yang Y, Zhang Y, Haynes CM, Xu CF, et al. Proteasomal adaptation to environmental stress links resistance to proteotoxicity with longevity in Caenorhabditis elegans. *Proc Natl Acad Sci USA* 2008;**105**:7094–9.
73. Li W, Gao B, Lee SM, Bennett K, Fang D. RLE-1, an E3 ubiquitin ligase, regulates C. elegans aging by catalyzing DAF-16 polyubiquitination. *Dev Cell* 2007;**12**:235–46.
74. Stanhill A, Haynes CM, Zhang Y, Min G, Steele MC, Kalinina J, et al. An arsenite-inducible 19S regulatory particle-associated protein adapts proteasomes to proteotoxicity. *Mol Cell* 2006;**23**:875–85.
75. Buffenstein R, Jarvis JU. The naked mole rat—a new record for the oldest living rodent. *Sci Aging Knowledge Environ* 2002;**2002**:pe7.
76. Aguilaniu H, Gustafsson L, Rigoulet M, Nystrom T. Asymmetric inheritance of oxidatively damaged proteins during cytokinesis. *Science* 2003;**299**:1751–3.
77. Erjavec N, Cvijovic M, Klipp E, Nystrom T. Selective benefits of damage partitioning in unicellular systems and its effects on aging. *Proc Natl Acad Sci USA* 2008;**105**:18764–9.
78. Liu B, Larsson L, Caballero A, Hao X, Oling D, Grantham J, et al. The polarisome is required for segregation and retrograde transport of protein aggregates. *Cell* 2010;**140**:257–67.
79. Zhou C, Slaughter BD, Unruh JR, Eldakak A, Rubinstein B, Li R. Motility and segregation of hsp104-associated protein aggregates in budding yeast. *Cell* 2011;**147**:1186–96.
80. Hernebring M, Brolen G, Aguilaniu H, Semb H, Nystrom T. Elimination of damaged proteins during differentiation of embryonic stem cells. *Proc Natl Acad Sci USA* 2006;**103**:7700–5.
81. Goudeau J, Aguilaniu H. Carbonylated proteins are eliminated during reproduction in C. elegans. *Aging Cell* 2010;**9**:991–1003.
82. Friguet B, Szweda LI. Inhibition of the multicatalytic proteinase (proteasome) by 4-hydroxy-2-nonenal cross-linked protein. *FEBS Lett* 1997;**405**:21–5.

83. Okada K, Wangpoengtrakul C, Osawa T, Toyokuni S, Tanaka K, Uchida K. 4-Hydroxy-2-nonenal-mediated impairment of intracellular proteolysis during oxidative stress. Identification of proteasomes as target molecules. *J Biol Chem* 1999;**274**:23787–93.
84. Finkel T, Holbrook NJ. Oxidants, oxidative stress and the biology of ageing. *Nature* 2000;**408**:239–47.
85. Vile GF, Tanew-Ilitschew A, Tyrrell RM. Activation of Nf-kappa B in human skin fibroblasts by the oxidative stress generated by UVa Radiation. *Photochem Photobiol* 1995;**62**:463–8.
86. Leach JK, Van Tuyle G, Lin PS, Schmidt-Ullrich R, Mikkelsen RB. Ionizing radiation-induced, mitochondria-dependent generation of reactive oxygen/nitrogen. *Cancer Res* 2001;**61**:3894–901.
87. Chou AP, Li S, Fitzmaurice AG, Bronstein JM. Mechanisms of rotenone-induced proteasome inhibition. *Neurotoxicology* 2010;**31**:367–72.
88. Watanabe Y, Suzuki O, Haruyama T, Akaike T. Interferongamma induces reactive oxygen species and endoplasmic reticulum stress at the hepatic apoptosis. *J Cell Biochem* 2003;**89**:244–53.
89. Leutner S, Eckert A, Muller WE. Ros generation, lipid peroxidation and antioxidant enzyme activities in the aging brain. *J Neural Transm* 2001;**108**:955–67.
90. Sedelnikova OA, Redon CE, Dickey JS, Nakamura AJ, Georgakilas AG, Bonner WM. Role of oxidatively induced DNA lesions in human pathogenesis. *Mutat Res* 2010;**704**:152–9.
91. Adibhatla RM, Hatcher JF. Lipid oxidation and peroxidation in cns health and disease: from molecular mechanisms to therapeutic opportunities. *Antioxid Redox Signal* 2010;**12**:125–69.
92. Bochkov VN, Oskolkova OV, Birukov KG, Levonen AL, Binder CJ, Stockl J. Generation and biological activities of oxidized phospholipids. *Antioxid Redox Signal* 2010;**12**:1009–59.
93. Stadtman ER. Protein oxidation and aging. *Science* 1992;**257**:1220–4.
94. Baraibar MA, Hyzewicz J, Rogowska-Wrzesinska A, Ladouce R, Roepstorff P, Mouly V, et al. Oxidative stress-induced proteome alterations target different cellular pathways in human myoblasts. *Free Radic Biol Med* 2011;**51**:1522–32.
95. Grune T, Reinheckel T, Joshi M, Davies KJA. Proteolysis in cultured liver epithelial cells during oxidative stress: role of multicatalytic proteinase complex, proteasome. *J Biol Chem* 1995;**270**:2344–51.
96. Wu M, Bian Q, Liu Y, Fernandes AF, Taylor A, Pereira P, et al. Sustained oxidative stress inhibits NF-κB activation partially via inactivating the proteasome. *Free Radic Biol Med* 2009;**46**:62–9.
97. Cohn JA, Tsai L, Friguet B, Szweda LI. Chemical characterization of a protein-4-hydroxy-2-nonenal cross-link: immunochemical detection in mitochondria exposed to oxidative stress. *Arch Biochem Biophys* 1996;**328**:158–64.
98. Voitkun V, Zhitkovich A. Analysis of DNA-protein crosslinking activity of malondialdehyde in vitro. *Mutat Res* 1999;**424**:97–106.
99. Montine TJ, Huang DY, Valentine WM, Amarnath V, Saunders A, Weisgraber KH, et al. Crosslinking of apolipoprotein E by products of lipid peroxidation. *J Neuropathol Exp Neurol* 1996;**55**:202–10.
100. Burcham PC, Kuhan YT. Diminished susceptibility to proteolysis after protein modification by the lipid peroxidation product malondialdehyde: inhibitory role for crosslinked and non-crosslinked adducted proteins. *Arch Biochem Biophys* 1997;**340**:331–7.
101. Amidi F, French BA, Chung D, Halsted CH, Medici V, French SW. M-30 and 4HNE are sequestered in different aggresomes in the same hepatocytes. *Exp Mol Pathol* 2007;**83**:296–300.
102. Wong ES, Tan JM, Soong WE, Hussein K, Nukina N, Dawson VL, et al. Autophagy-mediated clearance of aggresomes is not a universal phenomenon. *Hum Mol Genet* 2008;**17**:2570–82.

103. Harman D. Lipofuscin and ceroid formation: the cellular recycling system. *Adv Exp Med Biol* 1989;**266**:3–15.
104. Brunk UT, Terman A. The mitochondrial-lysosomal axis theory of aging: accumulation of damaged mitochondria as a result of imperfect autophagocytosis. *Eur J Biochem* 2002;**269**:1996–2002.
105. Ahmed EK, Rogowska-Wrzesinska A, Roepstorff P, Bulteau AL, Friguet B. Protein modification and replicative senescence of WI-38 human embryonic fibroblasts. *Aging Cell* 2010;**9**:252–72.
106. Reinheckel T, Sitte N, Ullrich O, Kuckelkorn U, Davies KJ, Grune T. Comparative resistance of the 20S and 26S proteasome to oxidative stress. *Biochem J* 1998;**335**:637–42.
107. Reinheckel T, Grune T, Davies KJ. The measurement of protein degradation in response to oxidative stress. *Methods Mol Biol* 2000;**99**:49–60.
108. Ishii T, Sakurai T, Usami H, Uchida K. Oxidative modification of proteasome: identification of an oxidation-sensitive subunit in 26S proteasome. *Biochemistry* 2005;**44**:13893–901.
109. Brownlee M. Biochemistry and molecular cell biology of diabetic complications. *Nature* 2001;**414**:813–20.
110. Thornalley PJ. Dicarbonyl intermediates in the maillard reaction. *Ann N Y Acad Sci* 2005;**1043**:111–7.
111. Queisser MA, Yao D, Geisler S, Hammes HP, Lochnit G, Schleicher ED, et al. Hyperglycemia impairs proteasome function by methylglyoxal. *Diabetes* 2010;**59**:670–8.
112. Pompella A, Visvikis A, Paolicchi A, De Tata V, Casini AF. The changing faces of glutathione, a cellular protagonist. *Biochem Pharmacol* 2003;**66**:1499–503.
113. Shackelford RE, Heinloth AN, Heard SC, Paules RS. Cellular and molecular targets of protein s-glutathiolation. *Antioxid Redox Signal* 2005;**7**:940–50.
114. Demasi M, Shringarpure R, Davies KJ. Glutathiolation of the proteasome is enhanced by proteolytic inhibitors. *Arch Biochem Biophys* 2001;**389**:254–63.
115. Demasi M, Silva GM, Netto LE. 20 S proteasome from Saccharomyces cerevisiae is responsive to redox modifications and is S-glutathionylated. *J Biol Chem* 2003;**278**:679–85.
116. Aiken CT, Kaake RM, Wang X, Huang L. Oxidative stress-mediated regulation of proteasome complexes. *Mol Cell Proteomics* 2011;**10**(R110):006924.
117. Ullrich O, Reinheckel T, Sitte N, Hass R, Grune T, Davies KJ. Poly-Adp ribose polymerase activates nuclear proteasome to degrade oxidatively damaged histones. *Proc Natl Acad Sci USA* 1999;**96**:6223–8.
118. Duan Y, Gross RA, Sheu SS. $Ca2^{+}$-dependent generation of mitochondrial reactive oxygen species serves as a signal for poly(Adp-ribose) polymerase-1 activation during glutamate excitotoxicity. *J Physiol* 2007;**585**:741–58.
119. Banasik M, Komura H, Shimoyama M, Ueda K. Specific inhibitors of poly(Adp-Ribose) synthetase and mono(Adp-Ribosyl)transferase. *J Biol Chem* 1992;**267**:1569–75.
120. Burkle A. Poly(Adp-Ribose). The most elaborate metabolite of $Nad^{+}$. *FEBS J* 2005;**272**:4576–89.
121. Satoh MS, Lindahl T. Role of poly(Adp-Ribose) formation in DNA repair. *Nature* 1992;**356**:356–8.
122. Catalgol B, Wendt B, Grimm S, Breusing N, Ozer NK, Grune T. Chromatin repair after oxidative stress: role of parp-mediated proteasome activation. *Free Radic Biol Med* 2010;**48**:673–80.
123. Um JW, Im E, Park J, Oh Y, Min B, Lee HJ, et al. Ask1 negatively regulates the 26s proteasome. *J Biol Chem* 2010;**285**:36434–46.
124. Djakovic SN, Schwarz LA, Barylko B, DeMartino GN, Patrick GN. Regulation of the proteasome by neuronal activity and calcium/calmodulin-dependent protein kinase Ii. *J Biol Chem* 2009;**284**:26655–65.

125. Moreno D, Viana R, Sanz P. Two-hybrid analysis identifies Psmd11, a non-ATPase subunit of the proteasome, as a novel interaction partner of AMP-activated protein kinase. *Int J Biochem Cell Biol* 2009;**41**:2431–9.
126. Liu X, Huang W, Li C, Li P, Yuan J, Li X, et al. Interaction between C-Abl and Arg tyrosine kinases and proteasome subunit Psma7 regulates proteasome degradation. *Mol Cell* 2006;**22**:317–27.
127. Castaño JG, Mahillo E, Arizti P, Arribas J. Phosphorylation of C8 and C9 subunits of the multicatalytic proteinase by casein kinase Ii and identification of the C8 phosphorylation sites by direct mutagenesis. *Biochemistry* 1996;**35**:3782–9.
128. Bose S, Stratford FL, Broadfoot KI, Mason GG, Rivett AJ. Phosphorylation of 20s proteasome alpha subunit C8 (Alpha7) stabilizes the 26s proteasome and plays a role in the regulation of proteasome complexes by gamma-interferon. *Biochem J* 2004;**378**:177–84.
129. Humphries KM, Pennypacker JK, Taylor SS. Redox regulation of Camp-dependent protein kinase signaling: kinase versus phosphatase inactivation. *J Biol Chem* 2007; **282**:22072–9.
130. Takahashi M, Ko LW, Kulathingal J, Jiang P, Sevlever D, Yen SH. Oxidative stress-induced phosphorylation, degradation and aggregation of alpha-synuclein are linked to upregulated Ck2 and cathepsin D. *Eur J Neurosci* 2007;**26**:863–74.
131. Murtaza I, Wang HX, Feng X, Alenina N, Bader M, Prabhakar BS, et al. Down-regulation of catalase and oxidative modification of protein kinase Ck2 lead to the failure of apoptosis repressor with caspase recruitment domain to inhibit cardiomyocyte hypertrophy. *J Biol Chem* 2008;**283**:5996–6004.
132. Allen RG, Tresini M. Oxidative Stress and Gene Regulation. *Free Radic Biol Med* 2000;**28**:463–99.
133. Sone H, Akanuma H, Fukuda T. Oxygenomics in environmental stress. *Redox Rep* 2010;**15**:98–114.
134. Delaunay A, Isnard AD, Toledano MB. H2O2 Sensing through oxidation of the Yap1 transcription factor. *EMBO J* 2000;**19**:5157–66.
135. Mannhaupt G, Schnall R, Karpov V, Vetter I, Feldmann H. Rpn4p acts as a transcription factor by binding to Pace, a nonamer box found upstream of 26s proteasomal and other genes in yeast. *FEBS Lett* 1999;**450**:27–34.
136. Dohmen RJ, Willers I, Marques AJ. Biting the hand thatfeeds: Rpn4-dependent feedback regulation of proteasome function. *Biochim Biophys Acta* 2007;**1773**:1599–604.
137. Hahn JS, Neef DW, Thiele DJ. A stress regulatory network for co-ordinated activation of proteasome expression mediated by yeast heat shock transcription factor. *Mol Microbiol* 2006;**60**:240–51.
138. Owsianik G, Balzi IL, Ghislain M. Control of 26s proteasome expression by transcription factors regulating multidrug resistance in Saccharomyces cerevisiae. *Mol Microbiol* 2002;**43**:1295–308.
139. Lee J, Godon C, Lagniel G, Spector D, Garin J, Labarre J, et al. Yap1 and Skn7 control two specialized oxidative stress response regulons in yeast. *J Biol Chem* 1999;**274**:16040–6.
140. Gaczynska M, Rock KL, Spies T, Goldberg AL. Peptidase activities of proteasomes are differentially regulated by the major histocompatibility complex-encoded genes for LMP2 and LMP7. *Proc Natl Acad Sci USA* 1994;**91**:9213–7.
141. Gaczynska M, Goldberg AL, Tanaka K, Hendil KB, Rock KL. Proteasome subunits X and Y alter peptidase activities in opposite ways to the interferon-gamma-induced subunits LMP2 and LMP7. *J Biol Chem* 1996;**271**:17275–80.
142. Liu Y, Liu X, Zhang T, Luna C, Liton PB, Gonzalez P. Cytoprotective effects of proteasome beta5 subunit overexpression in lens epithelial cells. *Mol Vis* 2007;**13**:31–8.

143. Kwak MK, Cho JM, Huang B, Shin S, Kensler TW. Role of increased expression of the proteasome in the protective effects of sulforaphane against hydrogen peroxide-mediated cytotoxicity in murine neuroblastoma cells. *Free Radic Biol Med* 2007;**43**:809–17.
144. Chondrogianni N, Gonos ES. Overexpression of hUMP1/POMP proteasome accessory protein enhances proteasome-mediated antioxidant defence. *Exp Gerontol* 2007;**42**:899–903.
145. Wang X, Xu H, Ju D, Xie Y. Disruption of Rpn4-induced proteasome expression in Saccharomyces cerevisiae reduces cell viability under stressed conditions. *Genetics* 2008;**180**:1945–53.
146. Bulteau AL, Moreau M, Saunois A, Nizard C, Friguet B. Algae extract-mediated stimulation and protection of proteasome activity within human keratinocytes exposed to UVA and UVB irradiation. *Antioxid Redox Signal* 2006;**8**:136–43.
147. Dahlmann B, Rutschmann M, Kuehn L, Reinauer H. Activation of the multicatalytic proteinase from rat skeletal muscle by fatty acids or sodium dodecyl sulphate. *Biochem J* 1985;**228**:171–7.
148. Katsiki M, Chondrogianni N, Chinou I, Rivett AJ, Gonos ES. The olive constituent oleuropein exhibits proteasome stimulatory properties in vitro and confers life span extension of human embryonic fibroblasts. *Rejuvenation Res* 2007;**10**:157–72.
149. Fahey JW, Talalay P. Antioxidant functions of sulforaphane: a potent inducer of Phase II detoxication enzymes. *Food Chem Toxicol* 1999;**37**:973–9.
150. Kwak MK, Wakabayashi N, Greenlaw JL, Yamamoto M, Kensler TW. Antioxidants enhance mammalian proteasome expression through the Keap1-Nrf2 signaling pathway. *Mol Cell Biol* 2003;**23**:8786–94.
151. Kwak MK, Huang B, Chang H, Kim JA, Kensler TW. Tissue specific increase of the catalytic subunits of the 26S proteasome by indirect antioxidant dithiolethione in mice: enhanced activity for degradation of abnormal protein. *Life Sci* 2007;**80**:2411–20.
152. Kapeta S, Chondrogianni N, Gonos ES. Nuclear erythroid factor 2-mediated proteasome activation delays senescence in human fibroblasts. *J Biol Chem* 2010;**285**:8171–84.
153. Rattan SI, Ali RE. Hormetic prevention of molecular damage during cellular aging of human skin fibroblasts and keratinocytes. *Ann N Y Acad Sci* 2007;**1100**:424–30.
154. Rattan SI. Targeting the age-related occurrence, removal, and accumulation of molecular damage by hormesis. *Ann N Y Acad Sci* 2010;**1197**:28–32.
155. Le Bourg E, Rattan SIS, editors. *Mild stress and healthy aging: applying hormesis in aging research and interventions*. Dordrecht, The Netherlands: Springer; 2008.
156. Lithgow GJ, White TM, Melov S, Johnson TE. Thermotolerance and extended life-span conferred by single-gene mutations and induced by thermal stress. *Proc Natl Acad Sci USA* 1995;**92**:7540–4.
157. Cypser JR, Johnson TE. Multiple stressors in Caenorhabditis elegans induce stress hormesis and extended longevity. *J Gerontol A Biol Sci Med Sci* 2002;**57**:B109–14.
158. Pickering AM, Koop AL, Teoh CY, Ermak G, Grune T, Davies KJ. The immunoproteasome, the 20S proteasome and the PA28alphabeta proteasome regulator are oxidative-stress-adaptive proteolytic complexes. *Biochem J* 2010;**432**:585–94.

# Proteasome and Cancer

Betul Catalgol

*Department of Biochemistry, Faculty of Medicine, Genetic and Metabolic Diseases Research Center (GEMHAM), Marmara University, Haydarpasa, Istanbul, Turkey*

Cancer is one of the most frightful diseases mostly resulting in mortality; it has recently become more possible to overcome with the help of new therapies. In this direction, carcinogenesis is defined as a complicated process that can include several different factors that contribute to its progress. Proteasome is implicated in cancer studies as it is the main degradation system for oxidatively damaged proteins and also for several proteins playing a role in the cell cycle and transcription, which are important for cancer improvement. Because of this crucial role of proteasome in cancer development, myriad *in vitro* and *in vivo* studies have focused on the proteasome in different cancer cases. In this chapter, the involvement of proteasome in the degradation of cancer-related proteins is explained with the results of representative studies. Related to these proteins, the use of proteasome inhibitors in cancer treatment is reviewed.

Carcinogenesis is associated with a wide range of factors and it is known to be a process of transformation of normal cells into cancer cells. Changes in the cellular and genetic levels direct the progression of cancer that reprograms cells for uncontrolled cell division. Tumor growth has been associated with genetic alterations in cell-cycle regulators such as oncogenes and tumor suppressors. Abnormal responses to cytokines and growth factors play an important role in the progress. On the other hand, cancer growth and spread can be induced by the altered degradation of cell-cycle proteins such as p21, p27, p53, and cyclins.

Cancer has a wide incidence in the body including all the organs and therefore many cancer types have been identified such as leukemia, multiple myeloma (MM), lymphoma, glioblastoma, and lung, prostate, ovarian, gastric, pancreatic, and colorectal cancers. Leukemia is a cancer of blood cells and lymphoma is a cancer of lymphatic cells in the immune system, mainly classified according to cell types as B, T, and NK cell tumors (also classified as Hodgkin's and non-Hodgkin's lymphoma). The latest classification by the WHO (2008) lists 70 different forms of lymphoma divided into four broad groups.[1] MM is the second most common hematologic malignancy after non-Hodgkin's lymphoma. MM is a B-cell malignancy of the plasma cells that typically attacks the bone marrow, causing anemia, thrombocytopenia, and

Progress in Molecular Biology
and Translational Science, Vol. 109
DOI: 10.1016/B978-0-12-397863-9.00008-0

neutropenia. In MM, the bone and soft tissues may be attacked, potentially causing pain, hypercalcemia, and paralysis.[2] Another important type of cancer, glioblastoma, is the most common form of central nervous system cancers, which represents a challenge for the therapy.

Protein oxidation, mainly proteasome, is involved in cancer development as the ubiquitin proteasomal system (UPS) is a key regulator in many molecular pathways. Proteasome not only removes abnormal proteins that may be misfolded, aged, or damaged by oxidation, but also regulates the half-life of the short-lived regulatory proteins such as cyclins that are involved in the control of cell-cycle and transcription regulators. This role of proteasome in the degradation of the bulk of proteins such as oncogenes, tumor suppressor proteins, transcription factors, and signaling molecules makes it extremely important in cancer studies[3] (Table I). Many of the studies have shown that, in relation to high oxidative stress levels, cancer cells have higher levels of proteasome compared to their normal counterparts.[4,5] Several molecules have also been identified that affect the structure and function of proteasome in cancer cells such as interferon-$\gamma$, tumor necrosis factor alpha (TNF-$\alpha$), and autocrine interleukin-3 (IL-3).[6,7]

Apoptosis is an important process for the progression and spread of cancer. The cell cycle is shown to be regulated by proteasome at G1/S and G2/M transitions,[8–10] and cyclin-dependent kinase inhibitors p21 and p27 are degraded by proteasome and thus stabilized by proteasome inhibition.[11] In this direction, proteasome inhibition was shown to cause apoptosis in various tumor cell types such as MM,[12,13] leukemia,[14] Hodgkin's[15] and non-Hodgkin's lymphoma,[16] glioblastoma,[17] lung cancer,[18] prostate cancer,[19] ovarian cancer,[20] gastric cancer,[21] pancreatic cancer[22] and colorectal cancer.[23] Apoptosis-inducing factor (AIF) plays a role in the caspase-independent form of apoptosis, resulting in large-scale fragmented DNA (50 kBp). AIF is thought to leak

TABLE I

PROTEINS REGULATED BY PROTEASOMAL DEGRADATION, WHICH PLAY A ROLE IN CARCINOGENESIS

| Proteins | Functions in carcinogenesis |
|---|---|
| Cyclins, Cdk inhibitors(p21, p27) | Cell-cycle regulation |
| Apoptosis-inducing factor | Apoptosis (caspase-independent) |
| Bcl-2, Bax, Bik, tBid | Apoptosis |
| P53 | Negative regulator of cell growth, tumor supression |
| p14 (ARF), p16(INK4a) | Tumor supression |
| Activator protein-1 (c-Jun, c-Fos) | Cell proliferation, differentiation, apoptosis, metastasis |
| NF$\kappa$B | Tumor development, metastasis, cancer therapy |
| Wnt/$\beta$-catenin | Adherens junctions |
| RB/E2F1 | Cell-cycle regulation, apoptosis |

into the cytoplasm and be degraded by proteasome.[24] Furthermore, Bcl-2,[25] Bax,[26] Bik,[27] and tBid,[28] as pro- and antiapoptotic proteins, have been identified as proteasomal targets. Proteasome inhibition led to the mitochondrial accumulation of Bik and resulted in apoptosis in leukemia cells[27]; stabilization of tBid resulted in cytochrome c release in Hela cells[28] and cleavage of Bcl-2 by caspases in M-07e leukemia cells[29] and MIA-PaCa-2 pancreatic cells.[22] Bcl-2 overexpression was shown to inhibit cell death caused by lactacystin in Ewings sarcoma cells[30] and by peptide aldehyde proteasome inhibitor (PSI) in human myeloid leukemic cells.[31] Apoptosis caused by PSI treatment of cell lines was shown to be via caspase or caspase-like activation.[32] Cytochrome c-mediated activation of caspase 9 as an apoptosis initiator was shown to be induced by PSIs.[14] The death receptor-activated initiator caspase 8 was also activated by lactacystin in human glioma cells[17] and adult T-cell leukemia cells.[33] Proteasome blockage is able to lead to the inhibition of cell growth and, as mentioned earlier, the induction of apoptosis, thereby preventing cancer progression. But, on the other hand, proteasome inhibition was shown to inhibit docetaxel-induced apoptosis via p21/p27 accumulation and cdk2/cdc2 inhibition.[34] This observation seems to be important for the optimization of therapeutic combination strategies that include proteasome inhibition. Recently, TNF-related apoptosis-inducing ligand (TRAIL) and X-linked inhibitor of apoptosis protein (XIAP) have gained increased attention and have been the focus in the cancer phenomenon. Selective apoptotic death has been processed by TRAIL leaving normal human cells unaffected. The involvement of this protein holds great promise as a potential anticancer agent, although some tumors develop resistance to TRAIL.[35] Proteasome is shown to be involved in the regulation of TRAIL expression in primary human interleukin-2 (IL-2)-activated NK cells, and proteasome inhibition by bortezomib disrupts TRAIL expression and TRAIL-dependent and/or TRAIL-independent pathway-mediated killing of myeloma cells.[36] Hodgkin–Reed–Sternberg cells, which are the tumor cells of classical Hodgkin's lymphoma, show uniformly upregulated XIAP expression, causing robust resistance to apoptosis.[37] Related to XIAP, bortezomib was shown to induce marked XIAP downregulation at the posttranslational level in Hodgkin's lymphoma cell lines.[38]

Tumor suppressor protein p53 is implicated in many cancer types and is known to be a negative regulator of cell growth.[39] In tumors, p53 reactivation has served as a potential approach in the development of cancer drug therapy. The role of proteasome in the degradation of p53 is confirmed by several studies.[40,41] In the human papillomavirus-associated carcinogenesis, mainly in cervical cancer, proteasomal degradation of p53 was shown to be accelerated by the oncoprotein E6.[42] p53 degradation is mediated by the ubiquitin protein ligase E3 enzyme, and ubiquitination is mediated by UBC4 and E2-F1.[43] In this direction, RAMB-1 was found to inhibit ubiquitin-dependent protein degradation without compromising the catalytic activities of the 20S proteasome, a mechanism distinct from that of bortezomib. This compound resulted in an increase in p53 steady-state levels and triggered cell death in cervical cancer cells.[44] It has been shown that MDM2 binds to p53, and promotes degradation of p53 through ubiquitin–proteasome degradation, and two small MDM2 inhibitors JNJ-26854165 and RO5045337 are used in the clinical development for the stabilization of p53.[45] On the other hand, p53 plays a role in the expression of many genes such as p21/Waf-1/Cip-1, which causes cell-cycle arrest, and Gadd-45, which engages in DNA repair.[46] In addition, there are plenty of proapoptotic genes upregulated by p53 including APAF-1, SIAH-1, PTEN, Fas, DR5, PMP22, Bax, Noxa, p53AIP1, and PUMA. The results from Miyashita et al. demonstrate the existence of a negative response element in the Bcl-2 gene through which p53 may either directly or indirectly transcriptionally downregulate expression of this gene involved in the regulation of programmed cell death.[47] Aberrant accumulation of β-catenin in tumors is often associated with mutational inactivation of the p53 tumor suppressor.[48] p14(ARF) and p16(INK4a) are other important tumor suppressor proteins, and in Burkitt lymphoma, INK4a/ARF locus on chromosome 9p21 is considered as a safeguard locus as it encodes p14(ARF) and p16(INK4a). Complete inactivation of the locus is achieved by the improper turnover of INK4/ARF proteins by the UPS, as the PSI MG-132 blocks p14(ARF) degradation and induces a dramatic stabilization of the p16(INK4a) protein, which opens new prospects for the treatment of Burkitt lymphoma.[49]

Activator protein-1 (AP-1) is a transcription factor closely related to cell proliferation, differentiation, and apoptosis, which makes it important for cancer. AP-1 comprises a dimer of the subunits Jun (c-Jun, Jun B, and Jun D) and Fos (c-Fos, Fos B, Fra-1, and Fra-2), which are degraded mainly by UPS.[50,51] AP-1 binds to the promoter regions of several genes such as IL-2,[52] IL-5,[53] CD95L,[54] matrix metalloproteinase-1 (MMP-1),[55] and tumor growth factor-β (TGF-β).[56] The proteasomeal degradation of c-Jun and c-Fos is controlled by the δ domain which is required for ubiquitination.[50] The antiapoptotic molecule cellular-FLICE-inhibitory protein long isoform [c-FLIP(L)] is necessary and sufficient to maintain resistance to TRAIL-induced apoptosis. c-FLIP(L) is found to be negatively regulated by c-Fos by direct binding to the putative promoter region of the c-FLIP(L) gene, and PSI MG-132 sensitizes TRAIL-resistant prostate cancer cells by inducing c-Fos and repressing c-FLIP(L).[57] In HepG2 cells, bortezomib markedly increased AP-1 activity and the expression of its transcriptional targets such as c-Jun, FasL, and BimEL, which are involved in apoptosis.[58]

A number of genes encoding cytokines, chemokines, growth factors, cell adhesion molecules, surface receptors, and acute phase proteins are regulated by the transcription factor NFκB.[59] It is identified as a nuclear transcription factor binding to the B site of the intronic promoter of the Ig κ chain.[60] In normal situations, NFκB is sequestered in the cytoplasm as a complex consisting of DNA-binding subunits p50 and p65 and is bound to its inhibitor, IκB. They may be p50 or p65 homodimers or p50/p65 heterodimers. Following the stimulation of cells by cytokines, stress, or chemotherapy, signaling cascades are triggered that lead to activation of IκB kinase, a heterodimeric protein kinase that catalyzes IκB phosphorylation. IκB kinase phosphorylates two serine residues Ser32 and Ser36 in the amino terminal regulatory domain of IκB.[61] The phosphorylated sites on IκB are recognized by E3RS (IκB/b-TrCP), an SCF-type E3 ubiquitin ligase, leading to ubiquitination, and IκB is then degraded by the proteasomal pathway, releasing free, active p50–p65 dimers, which translocate to the nucleus and bind to promoter regions of several target genes, thereby triggering their transcription.[62] The p50 activation process is also controlled by the UPS. p50 is the product of a large cytoplasmic precursor called p105. While the amino terminal of p105 contains inactivated p50, the carboxy terminal contains a number of closely adjacent sequence repeats called ankyrin repeats required for the inhibition of p50 DNA binding. It is known that p50 requires a proteolytic event to be released from the carboxy terminus. Studies have revealed that processing of the p105 requires ubiquitin, E2–F1, and a novel species of E3. p105 is the first known molecule to be processed by the ubiquitin system rather than being completely destroyed. Besides the degradation of p105 by UPS, the p50 subunit is known to be generated by novel cotranslational biogenesis requiring the 26S proteasome.[63] As the transcription factor NFκB plays a role in the expression of many genes, it also plays a role in tumor growth and cancer cell survival, as well as in the resistance of cancer to existing therapies.[64–77] In some cancer cell lines, NFκB levels have been tested and found to be different compared to the untransformed cells. It was found to be highly expressed in human breast epithelial tumor cell lines.[74] The protein levels of the NFκB p65 subunit were shown to increase in the Hs294T melanoma cell line compared to normal retinal pigment epithelial cells.[78] Moreover, in human pancreatic adenocarcinoma cells,[79] and in papillary, anaplastic, and follicular tumor-derived thyroid cancer cell lines,[80] NFκB expressions were shown to be high. An increased DNA-binding activity of NFκB was determined in several bladder cancer cell lines.[81] Some types of cancer are characterized by constitutive NF-κB activity, whereas in others such activity is induced following chemotherapy. NF-κB-harboring tumors are generally resistant to chemotherapy and their eradication requires NFκB inhibition.[82] PSIs are generally used to block the degradation of IkBs, NFkB1/p105, or NF-kB2/p100. PS-341 treatment was shown to decrease NFκB DNA-binding activity by preventing IκBα degradation in human T-cell lymphotrophic virus I-infected T cells.[83] Another method of blocking the IκB–E3 interaction has been employed in numerous experiments both *in vivo* and *in vitro*: using the dominant IκB "super-repressor," which cannot be phosphorylated and ubiquitinated but maintains its NF-κB inhibitory capacity.[84] Recently, the introduction of this mutant to human squamous cell carcinoma halted tumor growth *in vivo*.[85] Moreover, expression of adenoviral dominant IκB sensitized fibrosarcomas initially resistant to TNF-α- or chemotherapy (CPT-11)-induced apoptosis.[86] However, the use of such super-repressors awaits further developments in the field of gene therapy. A direct approach for inhibiting IκB–E3 interaction could involve competitive inhibition. This may be achieved by using cell-penetrating IκB-phosphopeptides.[87] More effectively, some small-molecule inhibitors that structurally mimic the ligase recognition motif could specifically inactivate (through allosteric inhibition or disassembly) the SCF-type E3 complex.

Metastasis and angiogenesis are important processes for the damage of cancer in the body, and cell adhesion molecules including E-selectin, ICAM-1, and VCAM-1 play a role in these processes, which are regulated by NFκB and therefore by proteasome.[88] MMPs degrade extracellular matrix proteins, inducing the junctions to open and the tumor to spread. NFκB takes place in the expressions of MMPs, and blockage of the NFκB pathway by PSIs prevents TNF-dependent MMP-1, MMP-3, ICAM-1, and cyclo-oxygenase 2 expressions, which play an important role in preventing metastasis.[89] Roccaro et al. observed the antiangiogenic effect of the PSI bortezomib in MM via direct and indirect effects on endothelial cells. This effect was associated with the inhibition of MM patient-derived endothelial cell proliferation, inhibition of vascular endothelial growth factor (VEGF) and interleukin-6 (IL-6) secretions, and downregulation of VEGF, IL-6, insulin-like growth factor-1, angiopoietin 1 (Ang1), and Ang2 expressions. Moreover, in functional assays for angiogenesis, including chemotaxis, adhesion to fibronectin, capillary formation on Matrigel, and chick embryo chorioallantoic membrane assay, bortezomib induced a dose-dependent inhibition of angiogenesis.[90]

Wnt signaling is a critical pathway for cell proliferation and differentiation.[91] In the Wnt/β-catenin-dependent pathway, the accumulation and nuclear translocation of β-catenin is important for the adherens junctions.[92] The accumulation of β-catenin may be induced by dysregulation of Wnt pathways, mutations in β-catenin, and adenomatous polyposis coli. When it is not bound to cadherin, it can interact with several different proteins inside the cell such as axin. As a complex with axin, β-catenin is phosphorylated by glycogen synthase kinase-3 beta (GSK-3β), which creates a signal for the rapid degradation of β-catenin by proteasome.[93] GSK-3β phosphorylates β-catenin in three conserved Ser residues and one Thr residue in the N-terminal domain. Mutations in these residues resulted in increased stability and increased activity of the protein.[94] In the nucleus, β-catenin functions as a transcriptional co-activator of the T-cell factor/lymphoid enhancer factor family of DNA-binding transcription factors.[95,96] Several Wnt target genes including c-myc, cyclin D1, multidrug resistance protein 1 (MDR1), and VEGF, many of which are involved in tumorigenesis, are activated following the binding of this complex.[97–99] Thus, β-catenin is known to function as an oncogene and is produced in higher amounts in patients with basal cell carcinoma, leading to the increase in proliferation of related tumors. It is also thought that nuclear β-catenin has oncogenic potential in the initiation and progression of colon and esophagus carcinomas.[94,100,101] In addition, Hsp90 was shown to modulate the phosphorylation of β-catenin in the human MCF-7 epithelial breast cancer cell model.[102]

The retinoblastoma protein (RB)–E2F1 pathway has a central role in regulating the cell cycle. The E2F family plays an important role in regulating the G1/S transition in cell-cycle progression. E2F1 is one of the best-studied proteins of the E2F group. E2F1 is known as an oncogene and inducer of apoptosis in different conditions and is degraded by UPS.[103] Retinoblastoma tumor suppressor (Rb) protein binds to E2F1 and prevents ubiquitinylation and results in a marked stabilization. On the other hand, RB protein is also degraded by proteasome and it has been shown to be targeted for proteasome-dependent degradation by cellular and viral oncoproteins gankyrin, MDM2, human papillomavirus E7, Epstein–Barr virus nuclear antigen-3C, and human megalovirus pp71 protein.[104–108] PAX8 protein (one of the tissue-specific developmental regulators) interacts with the RB protein and regulates cell cycle. PAX8 expression was shown to increase in renal cell carcinoma, bladder, ovarian, and thyroid cancer cell lines, and silencing of PAX8 in cancer cell lines leads to a striking reduction in the expression of E2F1 and its target genes, as well as a proteasome-dependent destabilization of RB protein.[109]

The proteasomal pathway regulates also the processing of MHC class-I antigens[110] and metabolic enzymes such as tyrosine amino transferase and Cu/Zn superoxide dismutase[111,112] which may be important for cancer development.

The differences in proteasome levels and activities between cancer and normal cells suggest that proteasome may serve as a promising target for cancer therapy.[113] There are a myriad of PSIs used in studies and also in clinical trials. Bortezomib, designed by Adams et al. in 1999[113] as a highly specific PSI, is the first to have entered clinical trials.[114] It was shown to have direct antitumor effects and also to increase the sensitivity to radiotherapy.[113,115] Studies proved this PSI to be active across a 60-tumor cell line panel from the National Cancer Institute, and the potency of proteasome inhibition was correlated with growth-inhibitory effects.[113] The combination of bortezomib with standard chemotherapies demonstrated increased tumor cell apoptosis and decreased tumor cell proliferation.[116] Additionally, the cytotoxicity of many standard chemotherapeutics such as paclitaxel and cisplatin was shown to be enhanced by PSIs.[18,117] PSIs were shown to diminish the removal of cisplatin covalent adducts from

DNA of tumor cells by nucleotide excision repair. The mechanism for this was explained in two studies. In the first, proteasome inhibition caused a depletion in ubiquitinated histone H2A in nucleosomes. As the ubiquitinated form of histones promotes chromatin condensation and possibly interferes with the function of DNA damage recognition and enzyme repair, proteasome is thought to be effective in this regard.[118] In the second study, proteasome inhibition was shown to cause a decrease in the excision repair cross-complementation group 1 response to cisplatin.[26] In this direction, *p*-glycoprotein as a product of multidrug resistance gene-1 is responsible for the maturation and pumping function in the removal of many anticancer agents from the cytoplasm, and PSIs were shown to be effective on this protein in overcoming the failure of chemotherapy.[119]

Radiation therapy is a medical use of ionizing radiation for the control of malignant cells. The resistance of cancer cells to radiotherapy leads researchers to focus on the contribution of cellular and molecular factors. Cells in the late growth and mitosis (G2/M) phase of the cell cycle have been found to be more sensitive to radiation. Regulatory proteins for cell-cycle progress and cell death play significant roles in the cell's survival and resistance to radiation. In addition, constitutive and radiation-induced activation of NFκB was shown to contribute to tumor cell resistance in head and neck squamous carcinomas.[120] The PSI bortezomib increased the sensitivity of human Hodgkin's lymphoma cells to ionizing radiation.[15] Moreover, inhibition of radiation-induced NFκB activation by PS-341 resulted in increased apoptosis and decreased cell growth in LoVo human colon adenocarcinoma cells. Radiosensivity of these cells was shown to increase and tumor volume in LoVo xenografts was decreased by a combination of radiotherapy and PS-341 treatment.[121] MG-132 and lactacystin were also shown to induce apoptosis and sensitize the cells to radiation.[122]

There are thousands of studies showing the role of PSIs in MM. This cancer type becomes resistant to many existing therapeutic options and median survival duration is approximately 5 years after diagnosis.[123] Numerous signal transduction pathways play a role in the interaction of MM and the stroma. MM cell growth and survival is induced by the growth factors produced by adherent stroma and also by direct ligand interactions between myeloma cell surface and the stroma. Furthermore, new vessel formation is induced by direct and indirect interactions, and cytokines can supress the ability of a patient's immune system to recognize and kill MM. These complex mechanisms led to the idea of developing new drugs to target tumor cells and the environment.[124] Since then, new inhibitors, modeled according to the effects of bortezomib, are thought to have provided valuable conventional therapies for cancer. Besides the direct effects on the proteasome, E3 ligase and the ubiquitination system may be manipulated as targets for increasing efficiency. In MM, bortezomib influences several pathways besides the proteasome. NFκB is the main transcription factor targeted by bortezomib. It also disrupts adhesion between MM and the stroma by blocking cytokines such as TNF-$\alpha$, IL-6, and IL-1, and by blocking new vessel formation with the inhibition of VEGF, fibroblast growth factor, and perhaps through direct effects on activated endothelium.[124] Furthermore, the apoptotic pathway is facilitated via caspase-8, Bcl-2, and cytochrome c release. Caspase-8, caspase-3, and Fas upregulation is induced by the activation of c-Jun N-terminal kinase (JNK) by bortezomib and this leads to the activation of p53 following the phosphorylation from the serine 15 site and increases apoptosis. Hideshima et al. demonstrated that PS-341 mediates antimyeloma activity by inducing p53 and MDM2 protein expression, p53 phosphorylation at Ser15 site, c-Jun N-terminal kinase (JNK), caspase-8 and caspase-3 activation, DNA protein kinase catalytic subunit, and ATM and MDM2 cleavage.[125] Bortezomib induces FasL expression, probably because of increased c-myc expression that occurs as a result of proteasome inhibition. The expressions of insulin-like growth factor-1 and insulin-like growth factor-1 receptor were downregulated by bortezomib. It also suppresses growth in MM cell lines and primary MM cells via inhibition of IL-6-induced Ras/Raf/mitogen-activated protein kinase pathway activation.[126] On the other hand, the inhibition of cytoprotective proteins such as heat shock protein 90 (Hsp90), which is upregulated by bortezomib, may increase the sensitivity and overcome resistence to bortezomib.[127] In bortezomib-resistant mantle cell lymphoma, both BiP/Grp78 knockdown and cell pretreatment with the Hsp90 inhibitor IPI-504 were shown to lead to apoptotic cell death.[128] Serum IgM increase is the gold standard for monitoring MM, and bortezomib was shown to reduce the levels of clonal IgM protein is serum.[129]

After the phase I development of bortezomib in patients with advanced cancer, efficacy in MM was reported.[130] In addition, the completed phase II trial of bortezomib proved very beneficent in relapsed refractory myeloma. Bortezomib has been approved by the Food and Drug Administration in 2003 for the treatment of relapsed and refractory myeloma. Early results from phase I and II trials showed antitumor activity against several types of neoplasia which are refractory to standard antineoplastic agents such as thalidomide and dexamethasone. Responses have been seen after 1–2 cycles in most patients.[131] Richardson et al. tested bortezomib in 202 patients with relapsed, refractory MM in a phase II trial.[132] In this study, patients with measurable monoclonal protein levels who were refractory to chemotherapy were chosen, and in 193 patients who could be evaluated, the rate of response to bortezomib was found to be 35%, and those with a response included 7 patients in whom myeloma protein became undetectable and 12 in whom myeloma protein was detectable only by immunofixation. Grade 3 adverse events included thrombocytopenia (in 28% of patients), fatigue (in 12%), peripheral neuropathy (in 12%), and neutropenia (in 11%). An international phase III trial (Assessment of

Proteasome Inhibition for Extending Remissions: APEX) with 669 patients showed bortezomib to be superior to high-dose dexamethasone for the treatment of patients with MM who have had a relapse after one to three previous therapies.[133]

In the light of the information that proteasome is involved in almost all the important pathways for cancer development, progress, metastasis, and therapy, it is hoped that future studies will bring new ideas focusing on proteasome for cancer therapy.

## References


1. Turner JJ, Hughes AM, Kricker A, Milliken S, Grulich A, Kaldor J, et al. Use of the WHO lymphoma classification in a population-based epidemiological study. *Ann Oncol* 2004;**15**:631–7.
2. Hideshima T, Richardson P, Chauhan D, Palombella VJ, Elliott PJ, Adams J, et al. The proteasome inhibitor PS-341 inhibits growth, induces apoptosis, and overcomes drug resistance in human multiple myeloma cells. *Cancer Res* 2001;**61**:3071–6.
3. Coux O, Tanaka K, Goldberg AL. Structure and functions of the 20S and 26S proteasomes. *Annu Rev Biochem* 1996;**65**:801–47.
4. Shimbara N, Orino E, Sone S, Ogura T, Takashina M, Shono M, et al. Regulation of gene expression of proteasomes (multi-protease complexes) during growth and differentiation of human hematopoietic cells. *J Biol Chem* 1992;**267**:18100–9.
5. Kanayama H, Tanaka K, Aki M, Kagawa S, Miyaji H, Satoh M, et al. Changes in expressions of proteasome and ubiquitin genes in human renal cancer cells. *Cancer Res* 1991;**51**:6677–85.
6. Rock KL, York IA, Saric T, Goldberg AL. Protein degradation and the generation of MHC class I-presented peptides. *Adv Immunol* 2002;**80**:1–70.
7. Pallares-Trujillo J, Carbo N, Lopez-Soriano FJ, Argiles JM. Does the mechanism responsible for TNF-mediated insulin resistance involve the proteasome? *Med* Hypotheses 2000;**54**:565–9.
8. Machiels BM, Henfling ME, Gerards WL, Broers JL, Bloemendal H, Ramaekers FC, et al. Detailed analysis of cell cycle kinetics upon proteasome inhibition. *Cytometry* 1997;**28**:243–52.
9. Shah SA, Potter MW, McDade TP, Ricciardi R, Perugini RA, Elliott PJ, et al. 26S proteasome inhibition induces apoptosis and limits growth of human pancreatic cancer. *J Cell Biochem* 2001;**82**:110–22.
10. Seufert W, Futcher B, Jentsch S. Role of a ubiquitin-conjugating enzyme in degradation of S- and M-phase cyclins. *Nature* 1995;**373**:78–81.
11. Pagano M, Tam SW, Theodoras AM, Beerromero P, Delsal G, Chau V, et al. Role of the ubiquitin-proteasome pathway in regulating abundance of the cyclin-dependent kinase inhibitor P27. *Science* 1995;**269**:682–5.
12. Martinelli G, Tosi P, Ottaviani E, Soverini S, Tura S. Molecular therapy for multiple myeloma. *Haematologica* 2001;**86**:908–17.
13. Hideshima T, Chauhan D, Podar K, Schlossman RL, Richardson P, Anderson KC. Novel therapies targeting the myeloma cell and its bone marrow microenvironment. *Semin Oncol* 2001;**28**:607–12.

14. Almond JB, Snowden RT, Hunter A, Dinsdale D, Cain K, Cohen GM. Proteasome inhibitor-induced apoptosis of B-chronic lymphocytic leukaemia cells involves cytochrome c release and caspase activation, accompanied by formation of an similar to 700 kDa Apaf-1 containing apoptosome complex. *Leukemia* 2001;**15**:1388–97.
15. Pajonk F, Pajonk K, McBride WH. Apoptosis and radiosensitization of Hodgkin cells by proteasome inhibition. *Int J Radiat Oncol Biol Phys* 2000;**47**:1025–32.
16. Schenkein D. Proteasome inhibitors in the treatment of B-cell malignancies. *Clin Lymphoma* 2002;**3**:49–55.
17. Tani E, Kitagawa H, Ikemoto H, Matsumoto T. Proteasome inhibitors induce Fas-mediated apoptosis by c-Myc accumulation and subsequent induction of FasL message in human glioma cells. *FEBS Lett* 2001;**504**:53–8.
18. Oyaizu H, Adachi Y, Okumura T, Okigaki M, Oyaizu N, Taketani S, et al. Proteasome inhibitor 1 enhances paclitaxel-induced apoptosis in human lung adenocarcinoma cell line. *Oncol Rep* 2001;**8**:825–9.
19. Herrmann JL, Briones F, Brisbay S, Logothetis CJ, McDonnell TJ. Prostate carcinoma cell death resulting from inhibition of proteasome activity is independent of functional Bcl-2 and p53. *Oncogene* 1998;**17**:2889–99.
20. Mimnaugh EG, Yunmbam MK, Li QD, Bonvini P, Hwang SG, Trepel J, et al. Prevention of cisplatin-DNA adduct repair and potentiation of cisplatin-induced apoptosis in ovarian carcinoma cells by proteasome inhibitors. *Biochem Pharmacol* 2000;**60**:1343–54.
21. Fan XM, Wong BCY, Wang WP, Zhou XM, Cho CH, Yuen ST, et al. Inhibition of proteasome function induced apoptosis in gastric cancer. *Int J Cancer* 2001;**93**:481–8.
22. Bold RJ, Virudachalam S, McConkey DJ. Chemosensitization of pancreatic cancer by inhibition of the 26S proteasome. *J Surg Res* 2001;**100**:11–7.
23. Lind DS, Hochwald SN, Malaty J, Rekkas S, Hebig P, Mishra G, et al. Nuclear factor-kappa B is upregulated in colorectal cancer. *Surgery* 2001;**130**:363–9.
24. Kroemer G. Mitochondrial control of apoptosis. *Bull Acad Natl Med* 2001;**185**:1135–43.
25. Dimmeler S, Breitschopf K, Haendeler J, Zeiher AM. Dephosphorylation targets Bcl-2 for ubiquitin-dependent degradation: a link between the apoptosome and the proteasome pathway. *J Exp Med* 1999;**189**:1815–22.
26. Li B, Dou QP. Bax degradation by the ubiquitin/proteasome-dependent pathway: involvement in tumor survival and progression. *Proc Natl Acad Sci U S A* 2000;**97**:3850–5.
27. Marshansky V, Wang X, Bertrand R, Luo H, Duguid W, Chinnadurai G, et al. Proteasomes modulate balance among proapoptotic and antiapoptotic Bcl-2 family members and compromise functioning of the electron transport chain in leukemic cells. *J Immunol* 2001;**166**:3130–42.
28. Breitschopf K, Zeiher AM, Dimmeler S. Ubiquitin-mediated degradation of the proapoptotic active form of bid. A functional consequence on apoptosis induction. *J Biol Chem* 2000;**275**:21648–52.
29. Zhang XM, Lin H, Chen C, Chen BD. Inhibition of ubiquitin-proteasome pathway activates a caspase-3-like protease and induces Bcl-2 cleavage in human M-07e leukaemic cells. *Biochem J* 1999;**340**(Pt 1):127–33.
30. Soldatenkov VA, Dritschilo A. Apoptosis of Ewing's sarcoma cells is accompanied by accumulation of ubiquitinated proteins. *Cancer Res* 1997;**57**:3881–5.
31. Soligo D, Servida F, Delia D, Fontanella E, Lamorte G, Caneva L, et al. The apoptogenic response of human myeloid leukaemia cell lines and of normal and malignant haematopoietic progenitor cells to the proteasome inhibitor PSI. *Br J Haematol* 2001;**113**:126–35.
32. Drexler HC. Activation of the cell death program by inhibition of proteasome function. *Proc Natl Acad Sci U S A* 1997;**94**:855–60.

33. Yamada Y, Sugahara K, Tsuruda K, Nohda K, Mori N, Hata T, et al. Lactacystin activates FLICE (caspase 8) protease and induces apoptosis in Fas-resistant adult T-cell leukemia cell lines. *Eur J Haematol* 2000;**64**:315–22.
34. Nawrocki ST, Sweeney-Gotsch B, Takamori R, McConkey DJ. The proteasome inhibitor bortezomib enhances the activity of docetaxel in orthotopic human pancreatic tumor xenografts. *Mol Cancer Ther* 2004;**3**:59–70.
35. Ge R, Wang Z, Zeng Q, Xu X, Olumi AF. F-box protein 10, an NF-kappaB-dependent anti-apoptotic protein, regulates TRAIL-induced apoptosis through modulating c-Fos/c-FLIP pathway. *Cell Death Differ* 2011;**18**:1184–95.
36. Feng X, Yan J, Wang Y, Zierath JR, Nordenskjold M, Henter JI, et al. The proteasome inhibitor bortezomib disrupts tumor necrosis factor-related apoptosis-inducing ligand (TRAIL) expression and natural killer (NK) cell killing of TRAIL receptor-positive multiple myeloma cells. *Mol Immunol* 2010;**47**:2388–96.
37. Kashkar H, Haefs C, Shin H, Hamilton-Dutoit SJ, Salvesen GS, Kronke M, et al. XIAP-mediated caspase inhibition in Hodgkin's lymphoma-derived B cells. *J Exp Med* 2003;**198**:341–7.
38. Kashkar H, Deggerich A, Seeger JM, Yazdanpanah B, Wiegmann K, Haubert D, et al. NF-kappaB-independent down-regulation of XIAP by bortezomib sensitizes HL B cells against cytotoxic drugs. *Blood* 2007;**109**:3982–8.
39. Levine AJ, Momand J, Finlay CA. The p53 tumour suppressor gene. *Nature* 1991;**351**:453–6.
40. Ciechanover A, DiGiuseppe JA, Bercovich B, Orian A, Richter JD, Schwartz AL, et al. Degradation of nuclear oncoproteins by the ubiquitin system in vitro. *Proc Natl Acad Sci U S A* 1991;**88**:139–43.
41. Maki CG, Huibregtse JM, Howley PM. In vivo ubiquitination and proteasome-mediated degradation of p53(1). *Cancer Res* 1996;**56**:2649–54.
42. Scheffner M, Whitaker NJ. Human papillomavirus-induced carcinogenesis and the ubiquitin-proteasome system. *Semin Cancer Biol* 2003;**13**:59–67.
43. Rolfe M, Beer-Romero P, Glass S, Eckstein J, Berdo I, Theodoras A, et al. Reconstitution of p53-ubiquitinylation reactions from purified components: the role of human ubiquitin-conjugating enzyme UBC4 and E6-associated protein (E6AP). *Proc Natl Acad Sci U S A* 1995;**92**:3264–8.
44. Anchoori RK, Khan SR, Sueblinvong T, Felthauser A, Iizuka Y, Gavioli R, et al. Stressing the ubiquitin-proteasome system without 20S proteolytic inhibition selectively kills cervical cancer cells. *PLoS One* 2011;**6**:e23888.
45. Yuan Y, Liao YM, Hsueh CT, Mirshahidi HR. Novel targeted therapeutics: inhibitors of MDM2, ALK and PARP. *J Hematol Oncol* 2011;**4**:16.
46. Raycroft L, Wu HY, Lozano G. Transcriptional activation by wild-type but not transforming mutants of the p53 anti-oncogene. *Science* 1990;**249**:1049–51.
47. Miyashita T, Harigai M, Hanada M, Reed JC. Identification of a p53-dependent negative response element in the bcl-2 gene. *Cancer Res* 1994;**54**:3131–5.
48. Sadot E, Geiger B, Oren M, Ben-Ze'ev A. Down-regulation of beta-catenin by activated p53. *Mol Cell Biol* 2001;**21**:6768–81.
49. Roberti A, Rizzolio F, Lucchetti C, de LL, Giordano A. Ubiquitin-mediated protein degradation and methylation-induced gene silencing cooperate in the inactivation of the INK4/ARF locus in Burkitt lymphoma cell lines. *Cell Cycle* 2011;**10**:127–34.
50. Jariel-Encontre I, Salvat C, Steff AM, Pariat M, Acquaviva C, Furstoss O, et al. Complex mechanisms for c-fos and c-jun degradation. *Mol Biol Rep* 1997;**24**:51–6.
51. Stancovski I, Gonen H, Orian A, Schwartz AL, Ciechanover A. Degradation of the proto-oncogene product c-Fos by the ubiquitin proteolytic system in vivo and in vitro: identification and characterization of the conjugating enzymes. *Mol Cell Biol* 1995;**15**:7106–16.

52. McMurray RW, Ndebele K, Hardy KJ, Jenkins JK. 17-beta-estradiol suppresses IL-2 and IL-2 receptor. *Cytokine* 2001;**14**:324–33.
53. Mori A, Kaminuma O, Ogawa K, Okudaira H, Akiyama K. Transcriptional regulation of IL-5 gene by nontransformed human T cells through the proximal promoter element. *Intern Med* 2000;**39**:618–25.
54. Melino G, Bernassola F, Catani MV, Rossi A, Corazzari M, Sabatini S, et al. Nitric oxide inhibits apoptosis via AP-1-dependent CD95L transactivation. *Cancer Res* 2000;**60**:2377–83.
55. Behren A, Simon C, Schwab RM, Loetzsch E, Brodbeck S, Huber E, et al. Papillomavirus E2 protein induces expression of the matrix metalloproteinase-9 via the extracellular signal-regulated kinase/activator protein-1 signaling pathway. *Cancer Res* 2005;**65**:11613–21.
56. Verrecchia F, Vindevoghel L, Lechleider RJ, Uitto J, Roberts AB, Mauviel A. Smad3/AP-1 interactions control transcriptional responses to TGF-beta in a promoter-specific manner. *Oncogene* 2001;**20**:3332–40.
57. Li W, Zhang X, Olumi AF. MG-132 sensitizes TRAIL-resistant prostate cancer cells by activating c-Fos/c-Jun heterodimers and repressing c-FLIP(L). *Cancer Res* 2007;**67**:2247–55.
58. Lauricella M, Emanuele S, D'Anneo A, Calvaruso G, Vassallo B, Carlisi D, et al. JNK and AP-1 mediate apoptosis induced by bortezomib in HepG2 cells via FasL/caspase-8 and mitochondria-dependent pathways. *Apoptosis* 2006;**11**:607–25.
59. Baeuerle PA, Henkel T. Function and activation of NF-kappa B in the immune system. *Annu Rev Immunol* 1994;**12**:141–79.
60. Sen R, Baltimore D. Multiple nuclear factors interact with the immunoglobulin enhancer sequences. *Cell* 1986;**46**:705–16.
61. Wu J. On the role of proteasomes in cell biology and proteasome inhibition as a novel frontier in the development of immunosuppressants. *Am J Transplant* 2002;**2**:904–12.
62. Brown K, Gerstberger S, Carlson L, Franzoso G, Siebenlist U. Control of I kappa B-alpha proteolysis by site-specific, signal-induced phosphorylation. *Science* 1995;**267**:1485–8.
63. Fan CM, Maniatis T. Generation of p50 subunit of NF-kappa B by processing of p105 through an ATP-dependent pathway. *Nature* 1991;**354**:395–8.
64. Schwartz SA, Hernandez A, Mark EB. The role of NF-kappaB/IkappaB proteins in cancer: implications for novel treatment strategies. *Surg Oncol* 1999;**8**:143–53.
65. Mosialos G. The role of Rel/NF-kappa B proteins in viral oncogenesis and the regulation of viral transcription. *Semin Cancer Biol* 1997;**8**:121–9.
66. Xiao G, Cvijic ME, Fong A, Harhaj EW, Uhlik MT, Waterfield M, et al. Retroviral oncoprotein Tax induces processing of NF-kappaB2/p100 in T cells: evidence for the involvement of IKKalpha. *EMBO J* 2001;**20**:6805–15.
67. Houldsworth J, Mathew S, Rao PH, Dyomina K, Louie DC, Parsa N, et al. REL proto-oncogene is frequently amplified in extranodal diffuse large cell lymphoma. *Blood* 1996;**87**:25–9.
68. Rao PH, Houldsworth J, Dyomina K, Parsa NZ, Cigudosa JC, Louie DC, et al. Chromosomal and gene amplification in diffuse large B-cell lymphoma. *Blood* 1998;**92**:234–40.
69. Neri A, Fracchiolla NS, Roscetti E, Garatti S, Trecca D, Boletini A, et al. Molecular analysis of cutaneous B- and T-cell lymphomas. *Blood* 1995;**86**:3160–72.
70. Cabannes E, Khan G, Aillet F, Jarrett RF, Hay RT. Mutations in the IkBa gene in Hodgkin's disease suggest a tumour suppressor role for IkappaBalpha. *Oncogene* 1999;**18**:3063–70.
71. Jungnickel B, Staratschek-Jox A, Brauninger A, Spieker T, Wolf J, Diehl V, et al. Clonal deleterious mutations in the IkappaBalpha gene in the malignant cells in Hodgkin's lymphoma. *J Exp Med* 2000;**191**:395–402.
72. Reuther JY, Reuther GW, Cortez D, Pendergast AM, Baldwin Jr. AS. A requirement for NF-kappaB activation in Bcr-Abl-mediated transformation. *Genes Dev* 1998;**12**:968–81.

73. Mukhopadhyay T, Roth JA, Maxwell SA. Altered expression of the p50 subunit of the NF-kappa B transcription factor complex in non-small cell lung carcinoma. *Oncogene* 1995;**11**:999–1003.
74. Sovak MA, Bellas RE, Kim DW, Zanieski GJ, Rogers AE, Traish AM, et al. Aberrant nuclear factor-kappaB/Rel expression and the pathogenesis of breast cancer. *J Clin Invest* 1997;**100**:2952–60.
75. Nair A, Venkatraman M, Maliekal TT, Nair B, Karunagaran D. NF-kappaB is constitutively activated in high-grade squamous intraepithelial lesions and squamous cell carcinomas of the human uterine cervix. *Oncogene* 2003;**22**:50–8.
76. Cogswell PC, Guttridge DC, Funkhouser WK, Baldwin Jr. AS. Selective activation of NF-kappa B subunits in human breast cancer: potential roles for NF-kappa B2/p52 and for Bcl-3. *Oncogene* 2000;**19**:1123–31.
77. Thornburg NJ, Pathmanathan R, Raab-Traub N. Activation of nuclear factor-kappaB p50 homodimer/Bcl-3 complexes in nasopharyngeal carcinoma. *Cancer Res* 2003;**63**:8293–301.
78. Shattuck-Brandt RL, Richmond A. Enhanced degradation of I-kappaB alpha contributes to endogenous activation of NF-kappaB in Hs294T melanoma cells. *Cancer Res* 1997; **57**:3032–9.
79. Wang W, Abbruzzese JL, Evans DB, Larry L, Cleary KR, Chiao PJ. The nuclear factor-kappa B RelA transcription factor is constitutively activated in human pancreatic adenocarcinoma cells. *Clin Cancer Res* 1999;**5**:119–27.
80. Visconti R, Cerutti J, Battista S, Fedele M, Trapasso F, Zeki K, et al. Expression of the neoplastic phenotype by human thyroid carcinoma cell lines requires NFkappaB p65 protein expression. *Oncogene* 1997;**15**:1987–94.
81. Sumitomo M, Tachibana M, Ozu C, Asakura H, Murai M, Hayakawa M, et al. Induction of apoptosis of cytokine-producing bladder cancer cells by adenovirus-mediated IkappaBalpha overexpression. *Hum Gene Ther* 1999;**10**:37–47.
82. Amit S, Ben-Neriah Y. NF-kappaB activation in cancer: a challenge for ubiquitination- and proteasome-based therapeutic approach. *Semin Cancer Biol* 2003;**13**:15–28.
83. Tan C, Waldmann TA. Proteasome inhibitor PS-341, a potential therapeutic agent for adult T-cell leukemia. *Cancer Res* 2002;**62**:1083–6.
84. Van Antwerp DJ, Martin SJ, Kafri T, Green DR, Verma IM. Suppression of TNF-alpha-induced apoptosis by NF-kappaB. *Science* 1996;**274**:787–9.
85. Duffey DC, Chen Z, Dong G, Ondrey FG, Wolf JS, Brown K, et al. Expression of a dominant-negative mutant inhibitor-kappaBalpha of nuclear factor-kappaB in human head and neck squamous cell carcinoma inhibits survival, proinflammatory cytokine expression, and tumor growth in vivo. *Cancer Res* 1999;**59**:3468–74.
86. Wang CY, Cusack Jr. JC, Liu R, Baldwin Jr. AS. Control of inducible chemoresistance: enhanced anti-tumor therapy through increased apoptosis by inhibition of NF-kappaB. *Nat Med* 1999;**5**:412–7.
87. Yaron A, Gonen H, Alkalay I, Hatzubai A, Jung S, Beyth S, et al. Inhibition of NF-kappa-B cellular function via specific targeting of the I-kappa-B-ubiquitin ligase. *EMBO J* 1997;**16**:6486–94.
88. Chen CC, Rosenbloom CL, Anderson DC, Manning AM. Selective inhibition of E-selectin, vascular cell adhesion molecule-1, and intercellular adhesion molecule-1 expression by inhibitors of I kappa B-alpha phosphorylation. *J Immunol* 1995;**155**:3538–45.
89. Pajonk F, McBride WH. The proteasome in cancer biology and treatment. *Radiat Res* 2001;**156**:447–59.
90. Roccaro AM, Hideshima T, Raje N, Kumar S, Ishitsuka K, Yasui H, et al. Bortezomib mediates antiangiogenesis in multiple myeloma via direct and indirect effects on endothelial cells. *Cancer Res* 2006;**66**:184–91.

91. Kikuchi A, Kishida S, Yamamoto H. Regulation of Wnt signaling by protein-protein interaction and post-translational modifications. *Exp Mol Med* 2006;**38**:1–10.
92. Wodarz A, Nusse R. Mechanisms of Wnt signaling in development. *Annu Rev Cell Dev Biol* 1998;**14**:59–88.
93. Miller JR, Moon RT. Signal transduction through beta-catenin and specification of cell fate during embryogenesis. *Genes Dev* 1996;**10**:2527–39.
94. Morin PJ, Sparks AB, Korinek V, Barker N, Clevers H, Vogelstein B, et al. Activation of beta-catenin-Tcf signaling in colon cancer by mutations in beta-catenin or APC. *Science* 1997;**275**:1787–90.
95. Behrens J, von Kries JP, Kuhl M, Bruhn L, Wedlich D, Grosschedl R, et al. Functional interaction of beta-catenin with the transcription factor LEF-1. *Nature* 1996;**382**:638–42.
96. Huber O, Korn R, McLaughlin J, Ohsugi M, Herrmann BG, Kemler R. Nuclear localization of beta-catenin by interaction with transcription factor LEF-1. *Mech Dev* 1996;**59**:3–10.
97. He TC, Sparks AB, Rago C, Hermeking H, Zawel L, da Costa LT, et al. Identification of c-MYC as a target of the APC pathway. *Science* 1998;**281**:1509–12.
98. Shtutman M, Zhurinsky J, Simcha I, Albanese C, D'Amico M, Pestell R, et al. The cyclin D1 gene is a target of the beta-catenin/LEF-1 pathway. *Proc Natl Acad Sci U S A* 1999;**96**:5522–7.
99. Tetsu O, McCormick F. Beta-catenin regulates expression of cyclin D1 in colon carcinoma cells. *Nature* 1999;**398**:422–6.
100. Bian YS, Osterheld MC, Bosman FT, Fontolliet C, Benhattar J. Nuclear accumulation of beta-catenin is a common and early event during neoplastic progression of Barrett esophagus. *Am J Clin Pathol* 2000;**114**:583–90.
101. Polakis P. The oncogenic activation of beta-catenin. *Curr Opin Genet Dev* 1999;**9**:15–21.
102. Cooper LC, Prinsloo E, Edkins AL, Blatch GL. Hsp90alpha/beta associates with the GSK3beta/axin1/phospho-beta-catenin complex in the human MCF-7 epithelial breast cancer model. *Biochem Biophys Res Commun* 2011;**413**:550–4.
103. Hofmann F, Martelli F, Livingston DM, Wang Z. The retinoblastoma gene product protects-E2F-1 from degradation by the ubiquitin-proteasome pathway. *Genes Dev* 1996;**10**:2949–59.
104. Boyer SN, Wazer DE, Band V. E7 protein of human papilloma virus-16 induces degradation of retinoblastoma protein through the ubiquitin-proteasome pathway. *Cancer Res* 1996;**56**:4620–4.
105. Higashitsuji H, Itoh K, Nagao T, Dawson S, Nonoguchi K, Kido T, et al. Reduced stability of retinoblastoma protein by gankyrin, an oncogenic ankyrin-repeat protein overexpressed in hepatomas. *Nat Med* 2000;**6**:96–9.
106. Kalejta RF, Shenk T. Proteasome-dependent, ubiquitin-independent degradation of the Rb family of tumor suppressors by the human cytomegalovirus pp 71 protein. *Proc Natl Acad Sci U S A* 2003;**100**:3263–8.
107. Knight JS, Sharma N, Robertson ES. Epstein-Barr virus latent antigen 3C can mediate the degradation of the retinoblastoma protein through an SCF cellular ubiquitin ligase. *Proc Natl Acad Sci U S A* 2005;**102**:18562–6.
108. Sdek P, Ying H, Chang DL, Qiu W, Zheng H, Touitou R, et al. MDM2 promotes proteasome-dependent ubiquitin-independent degradation of retinoblastoma protein. *Mol Cell Biol* 2005;**20**:699–708.
109. Li CG, Nyman JE, Braithwaite AW, Eccles MR. PAX8 promotes tumor cell growth by transcriptionally regulating E2F1 and stabilizing RB protein. *Oncogene* 2011;**30**:4824–34.
110. Rock KL, Gramm C, Rothstein L, Clark K, Stein R, Dick L, et al. Inhibitors of the proteasome block the degradation of most cell proteins and the generation of peptides presented on MHC class I molecules. *Cell* 1994;**78**:761–71.

111. Hoffman EK, Wilcox HM, Scott RW, Siman R. Proteasome inhibition enhances the stability of mouse Cu/Zn superoxide dismutase with mutations linked to familial amyotrophic lateral sclerosis. *J Neurol Sci* 1996;**139**:15–20.
112. Gross-Mesilaty S, Hargrove JL, Ciechanover A. Degradation of tyrosine aminotransferase (TAT) via the ubiquitin-proteasome pathway. *FEBS Lett* 1997;**405**:175–80.
113. Adams J, Palombella VJ, Sausville EA, Johnson J, Destree A, Lazarus DD, et al. Proteasome inhibitors: a novel class of potent and effective antitumor agents. *Cancer Res* 1999;**59**:2615–22.
114. Adams J. Development of the proteasome inhibitor PS-341. *Oncologist* 2002;**7**:9–16.
115. Pervan M, Pajonk F, Sun JR, Withers HR, McBride WH. Molecular pathways that modify tumor radiation response. *Am J Clin Oncol* 2001;**24**:481–5.
116. Teicher BA, Ara G, Herbst R, Palombella VJ, Adams J. The proteasome inhibitor PS-341 in cancer therapy. *Clin Cancer Res* 1999;**5**:2638–45.
117. Li QQ, Yunmbam MK, Zhong X, Yu JJ, Mimnaugh EG, Neckers L, et al. Lactacystin enhances cisplatin sensitivity in resistant human ovarian cancer cell lines via inhibition of DNA repair and ERCC-1 expression. *Cell Mol Biol (Noisy-le-grand)* 2001;**47**:OL61–72 [Online Pub].
118. Mimnaugh EG, Chen HY, Davie JR, Celis JE, Neckers L. Rapid deubiquitination of nucleosomal histones in human tumor cells caused by proteasome inhibitors and stress response inducers: effects on replication, transcription, translation, and the cellular stress response. *Biochemistry* 1997;**36**:14418–29.
119. Ohkawa K, Asakura T, Takada K, Sawai T, Hashizume Y, Okawa Y, et al. Calpain inhibitor causes accumulation of ubiquitinated P-glycoprotein at the cell surface: possible role of calpain in P-glycoprotein turnover. *Int J Oncol* 1999;**15**:677–86.
120. Kato T, Duffey DC, Ondrey FG, Dong G, Chen Z, Cook JA, et al. Cisplatin and radiation sensitivity in human head and neck squamous carcinomas are independently modulated by glutathione and transcription factor NF-kappaB. *Head Neck* 2000;**22**:748–59.
121. Russo SM, Tepper JE, Baldwin Jr. AS, Liu R, Adams J, Elliott P, et al. Enhancement of radiosensitivity by proteasome inhibition: implications for a role of NF-kappaB. *Int J Radiat Oncol Biol Phys* 2001;**50**:183–93.
122. Kurland JF, Meyn RE. Protease inhibitors restore radiation-induced apoptosis to Bcl-2-expressing lymphoma cells. *Int J Cancer* 2001;**96**:327–33.
123. Wolf J, Richardson PG, Schuster M, LeBlanc A, Walters IB, Battleman DS. Utility of bortezomib retreatment in relapsed or refractory multiple myeloma patients: a multicenter case series. *Clin Adv Hematol Oncol* 2008;**6**:755–60.
124. Richardson P. New therapies for the treatment of multiple myeloma. *Clin Adv Hematol Oncol* 2005;**3**:345–6.
125. Hideshima T, Mitsiades C, Akiyama M, Hayashi T, Chauhan D, Richardson P, et al. Molecular mechanisms mediating antimyeloma activity of proteasome inhibitor PS-341. *Blood* 2003;**101**:1530–4.
126. Mitsiades N, Mitsiades CS, Poulaki V, Chauhan D, Richardson PG, Hideshima T, et al. Biologic sequelae of nuclear factor-kappaB blockade in multiple myeloma: therapeutic applications. *Blood* 2002;**99**:4079–86.
127. Mitsiades CS, Mitsiades NS, McMullan CJ, Poulaki V, Kung AL, Davies FE, et al. Antimyeloma activity of heat shock protein-90 inhibition. *Blood* 2006;**107**:1092–100.
128. Roue G, Perez-Galan P, Mozos A, Lopez-Guerra M, Xargay-Torrent S, Rosich L, et al. The Hsp90 inhibitor IPI-504 overcomes bortezomib resistance in mantle cell lymphoma in vitro and in vivo by down-regulation of the prosurvival ER chaperone BiP/Grp78. *Blood* 2011;**117**:1270–9.
129. Jagannath S, Durie BG, Wolf J, Camacho E, Irwin D, Lutzky J, et al. Bortezomib therapy alone and in combination with dexamethasone for previously untreated symptomatic multiple myeloma. *Br J Haematol* 2005;**129**:776–83.

130. Anderson KC. Targeted therapy of multiple myeloma based upon tumor-microenvironmental interactions. *Exp Hematol* 2007;**35**:155–62.
131. Badros A, Gahres N. Bortezomib, thalidomide, and dexamethasone for relapsed multiple myeloma: add it up and wait. *Clin Adv Hematol Oncol* 2005;**3**:916–7.
132. Richardson PG, Barlogie B, Berenson J, Singhal S, Jagannath S, Irwin D, et al. A phase 2 study of bortezomib in relapsed, refractory myeloma. *N Engl J Med* 2003;**348**:2609–17.
133. Richardson PG, Sonneveld P, Schuster MW, Irwin D, Stadtmauer EA, Facon T, et al. Bortezomib or high-dose dexamethasone for relapsed multiple myeloma. *N Engl J Med* 2005;**352**:2487–98.

# The Ubiquitin–Proteasome System and Cardiovascular Disease

Saul R. Powell,[*] Joerg Herrmann,[†] Amir Lerman,[†] Cam Patterson,[‡,§,¶] and Xuejun Wang[||]

*[*]Center for Heart and Lung Research, The Feinstein Institute for Medical Research, Manhasset, New York, USA*

*[†]Department of Internal Medicine, Division of Cardiovascular Diseases, Mayo Clinic, Rochester, Minnesota, USA*

*[‡]Department of Medicine, The McAllister Heart Institute, University of North Carolina Chapel Hill, North Carolina, USA*

*[§]Department of Pharmacology, The McAllister Heart Institute, University of North Carolina Chapel Hill, North Carolina, USA*

*[¶]Department of Cell and Developmental Biology, The McAllister Heart Institute, University of North Carolina Chapel Hill, North Carolina, USA*

*[||]Protein Quality Control and Degradation Research Center, Division of Basic Biomedical Sciences, Sanford School of Medicine, University of South Dakota, Vermillion, South Dakota, USA*

I. Introduction ........ 296
II. Regulation of the Cardiac Proteasome ........ 297
A. Heterogeneity as a Basis for Regulation ........ 297
B. Posttranslational Modification and Associating Partners as a Basis for Regulation ........ 298
III. Involvement of the UPS in atherosclerosis ........ 300
A. The UPS in the Initiation Stage of Atherosclerosis ........ 301
B. The UPS and Progression Stage of Atherosclerosis ........ 303
C. The UPS in the Complication Stage of Atherosclerosis ........ 306
D. Proteasome Inhibition and Atherosclerosis: For Better or for Worse? .. 308
E. Summary: The UPS and Atherosclerosis ........ 312
IV. The Role of Cardiac E3 Ligases ........ 312
A. Pathologic Cardiac Hypertrophy ........ 312
B. Protection Against Ischemic Heart Disease ........ 315

Progress in Molecular Biology
and Translational Science, Vol. 109
DOI: 10.1016/B978-0-12-397863-9.00009-2

C. The UPS in Familial Hypertrophic Cardiomyopathy: Proteasomal Involvement Without an Identified Cardiac E3 Ligase .................... 317
D. Summary: Cardiac E3 Ligases .................... 318
V. The UPS in PQC and Cardiac Proteinopathy .................... 319
A. Cardiac PQC .................... 319
B. Proteasome Functional Insufficiency in Cardiac Proteinopathy .................... 321
C. PFI Contributes to the Genesis of Proteinopathy .................... 322
D. Potential Mechanisms Underlying the Pathogenic Role of PFI .................... 323
E. Improving PQC as a Strategy to Treat Cardiac Proteinopathy .................... 324
VI. The UPS in Heart Failure .................... 325
A. Alterations in UPS in Early Failure .................... 325
B. Alterations in UPS Function in Late Failure .................... 326
VII. The UPS in Myocardial Ischemia/Reperfusion Injury .................... 327
A. Potential Mechanisms Involved in Postischemic Proteasome Dysfunction .................... 328
B. The 11S-Activated Proteasome and Myocardial Ischemia .................... 329
C. The UPS in IPC .................... 331
D. Proteasome Inhibition in Myocardial Ischemia: for Better or for Worse? 333
VIII. Concluding Statement .................... 333
References .................... 334

Over the past decade, the role of the ubiquitin–proteasome system (UPS) has been the subject of numerous studies to elucidate its role in cardiovascular physiology and pathophysiology. There have been many advances in this field including the use of proteomics to achieve a better understanding of how the cardiac proteasome is regulated. Moreover, improved methods for the assessment of UPS function and the development of genetic models to study the role of the UPS have led to the realization that often the function of this system deviates from the norm in many cardiovascular pathologies. Hence, dysfunction has been described in atherosclerosis, familial cardiac proteinopathies, idiopathic dilated cardiomyopathies, and myocardial ischemia. This has led to numerous studies of the ubiquitin protein (E3) ligases and their roles in cardiac physiology and pathophysiology. This has also led to the controversial proposition of treating atherosclerosis, cardiac hypertrophy, and myocardial ischemia with proteasome inhibitors. Furthering our knowledge of this system may help in the development of new UPS-based therapeutic modalities for mitigation of cardiovascular disease.

## I. Introduction

The ubiquitin–proteasome system (UPS) is the major nonlysosomal pathway for intracellular degradation of proteins and plays major roles in regulating many cellular processes. Many of these processes have already been or will be described in the accompanying chapters. For the most part, the structure of the 26S proteasome, the immunoproteasome, and the 11S-activated proteasome have been described in detail in earlier chapters of this book. To avoid overlap, the sections in this chapter deal exclusively with UPS function and regulation as they pertain to the cardiovascular system. The chapter first examines regulation of the cardiac proteasome followed by an in-depth discussion of the role of the UPS and the controversial use of proteasome inhibitors in atherosclerosis. A description of the ubiquitin protein (E3) ligases follows with a discussion of their potential roles in cardiac physiology and pathophysiology. This leads up to the entire issue of protein quality control (PQC) in the cardiovascular system and how the UPS functions and malfunctions in both familial proteinopathies and idiopathic dilated cardiomyopathies (heart failure). The chapter ends with a discussion of proteasome dysfunction in myocardial ischemia, the potential role of the UPS in ischemic preconditioning (IPC), and a revisit of the controversial proposed use of proteasome inhibitors to treat some aspects of ischemia.

## II. Regulation of the Cardiac Proteasome

Given the importance of the UPS in cardiovascular physiology and pathology, it was only natural that the cardiac proteasome would attract the attention of one of the premier proteomics groups who have conducted an extensive characterization to provide some clues to the regulation of proteasome function. The following section is a discussion of the regulation of the UPS as it applies to the cardiac proteasome although it may be applicable to proteasomes in other cell types as well.

### A. Heterogeneity as a Basis for Regulation

Prior to 2005, the prevailing hypothesis was that both β rings of the 20S proteasome were homologous to each other, that is, both rings had the same distribution of constitutive versus immunoforms of β1, β2, or β5 subunits. Since then, studies from the Ping group at UCLA have revealed a rather surprising heterogeneity in at least the β rings.[1–4] These investigators have shown that the total cardiac 20S proteasome population can consist of multiple subpopulations containing different proportions of the catalytic β subunits, that is, the β1, β2, and β5 subunits may be replaced by the immunoform on one β ring, but not the other, or perhaps even a mixture of immunoform versus non-immunoform subunits. As predicted, the different subpopulations of the 20S proteasome exhibit unique proteolytic activities, and specificity and selectivity for various substrates, suggesting a role in regulation of intracellular proteolysis.[3–5] A potential scheme illustrating how proteasomes may be regulated by heterogeneous configurations is presented in Fig. 1. In regard to the 19S

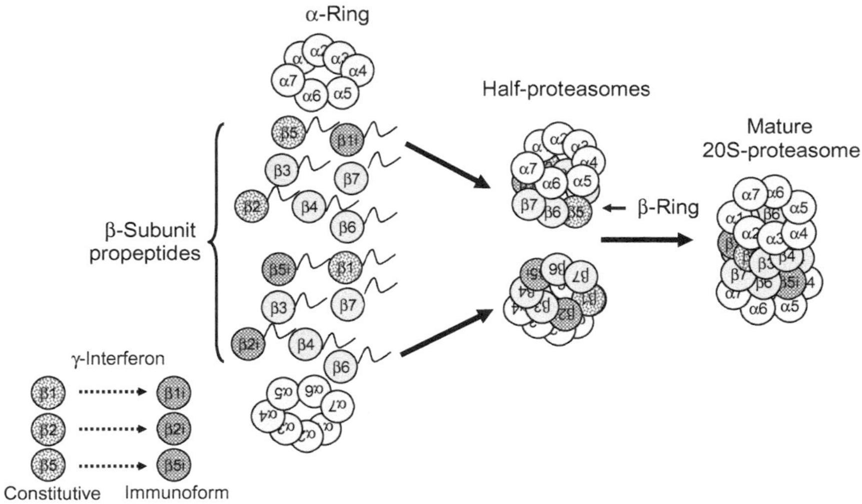


FIG. 1. Regulation of 20S proteasomes by assembly of heterogeneous configurations. Under the influence of γ-interferon, select β-subunits are replaced by their immunoforms, which are directed by the complete α-rings to form half-proteasomes with specific configurations. In the example shown earlier, the top half-proteasome has constitutive β2 and β5 subunits and the β1 subunit has been replaced by the immunoform. The bottom half-proteasome has a constitutive β1 subunit, but the β2 and β5 subunits have been replaced by their immunoforms. These two distinctly different half-proteasomes mate to form a mature 20S proteasome that has a heterogeneous configuration. For the sake of clarity, the involvement of the chaperones PAC1, PAC2, PAC3, and PAC4, and the maturation factor POMP, are not depicted here. Modified and reproduced from Ref. 6 with the permission of Walters Kluwer Health.

regulatory particle, alternate splicing of Rpn10 has been reported[2] and a recent study describes isolation of two distinct populations of the 19S regulatory particle from murine heart.[7] The presence of multiple proteasome subpopulations suggests the potential for directing assembly to effect selective degradation (or not) of a protein or class of proteins, possibly altering the outcome of a disease process.

## B. Posttranslational Modification and Associating Partners as a Basis for Regulation

These two topics are intimately related and are discussed together. Often, as is the case with phosphorylation, it is the associating partner of the proteasome that results in the posttranslational modification. Studies have suggested a dynamic link between the activity of the proteasome and metabolic demand as glycosylation of the 19S regulatory particle ATPase, Rpt2, can inhibit ATPase activity and shut down proteolytic activity.[8] Phosphorylation of the 19S regulatory particle ATPase, Rpt6, at $ser^{120}$ by cAMP-dependent protein kinase A (PKA) is associated with an increase in proteasome proteolytic activities. Dephosphorylation by the serine threonine protein phosphatase, PP1γ, reverses this.[9] Taken together, these findings suggest that hormones using the PKA system can act in concert with glucose metabolism to regulate proteasome function.

In regard to the heart, Ping and coworkers have published a series of studies that have characterized both the associating partners and posttranslational modifications of the murine cardiac proteasome and have provided much of the evidence linking PKA-mediated phosphorylation of multiple proteasome subunits with changes in proteolytic activity.[2,7,10–13] Multiple posttranslational modifications have been described, including N-terminal acetylation of Rpn1, Rpn5, Rpn6, Rpt3, and Rpt6 of the 19S regulatory particle, and α2, α5, α7, β3, and β4 subunits of the 20S proteasome; and myristolation of Rpt2. In addition, phosphorylation of the α1, α2, α3, α6, α7, β2, and β5 subunits was described and linked to the previously unknown proteasome-associating partner, PKA. Incubation with PKA increased all three proteolytic activities, which was reversed by incubation with the serine threonine protein phosphatase, PP2A.[2,11–13] Taken as a whole, these studies suggest that cAMP-dependent PKA plays a central role in regulating proteasome function and activity, which may be altered in the development of certain cardiac diseases, such as hypertrophy.[10] Based largely on these studies, it was proposed that the cardiac proteasome may actually be regulated by multiple kinases and may be a component of signal transduction pathways[6] (Fig. 2 presents one possible generic scenario).

For the most part, the cardiac 19S regulatory particle remains a mystery. However, as pointed out earlier, a recent study has reported isolation of two distinct populations of cardiac 19S regulatory particles, one of which associates with heat shock protein 90 (HSP90) and has a decreased functional potency.[7] Inhibition of the associating HSP90 led to an enhanced activating potential of this subpopulation of 19S regulatory particles on 20S proteasomes. While much work needs to be done on regulation of both the 20S proteasome and 19S regulatory particle, as well as other regulatory particles, it is becoming clearer that the cardiac proteasome is a dynamic structure under dynamic control, which probably allows it to respond to rapid changes in cardiac metabolism.

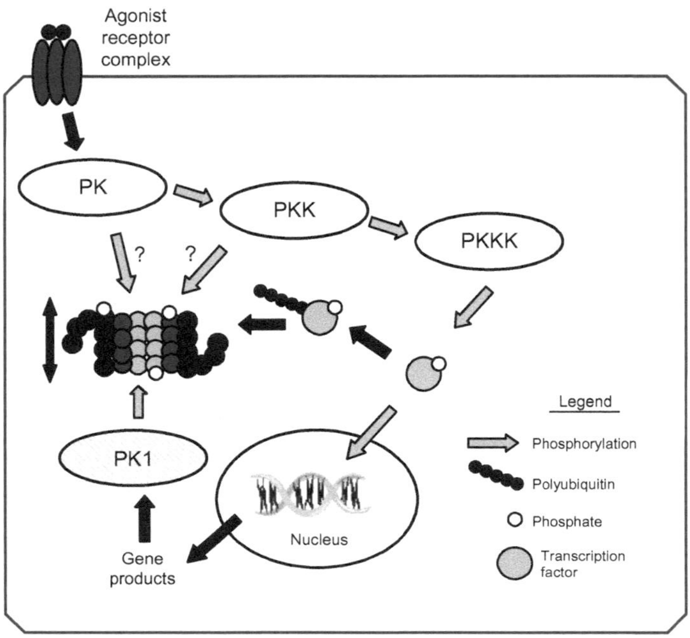


FIG. 2. Regulation of proteasome activity by kinase cascades. In this scenario, binding of an agonist with a cell surface receptor initiates a phosphorylation cascade with an end effector protein kinase kinase kinase (PKKK) phosphorylating some transcription factor which can then enter the nucleus and interact with DNA resulting in gene transcription. Phosphorylation of signaling intermediates, in this case the transcription factor, is often a signal for ubiquitination and targeting to the 26S proteasome for degradation. To amplify or dampen the effect of the signaling intermediate, any one of the protein kinases within the cascade might phosphorylate proteasome subunits resulting in increased or decreased peptidase activity, which would lessen or enhance availability of the signaling molecule accordingly. Another possibility would be for the gene product to activate a different protein kinase (PK1), which would phosphorylate proteasome, resulting in activation and thus decreased availability of the transcription factor, essentially a form of feedback control. Modified and reproduced from Ref. 6 with the permission of Walters Kluwer Health.

## III. Involvement of the UPS in atherosclerosis

Beyond rhetoric, one may wonder about an involvement of the leading cellular degenerative machinery in the leading human degenerative disease process. Over the past decade, a number of studies have explored this question and an integrative view is provided in this section. As the influence of the UPS varies by disease stage, its role in atherosclerosis is reviewed accordingly (Fig. 3).

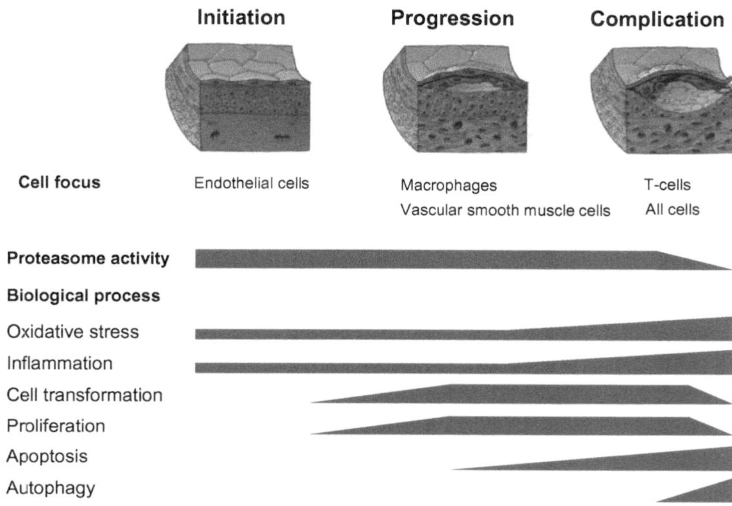


FIG. 3. Focus overview of the role of the ubiquitin–proteasome system in atherosclerosis. In general, the ubiquitin system is less vulnerable to changes and its activity remains preserved throughout the stages of atherosclerosis. On the contrary, proteasome activity seems to be increased in the initiation and progression stage to meet the increased substrate demand generated by oxidative stress, heightened metabolic activity, and cell transformation. Accumulation of oxidative stress and aging of the system likely lead to a decline in proteasome activity, which is notable in the complication stage along with apoptosis and autophagy.

## A. The UPS in the Initiation Stage of Atherosclerosis

It has been said that atherosclerosis is a disease of the intima of large to medium-sized arteries with hardening plaque formation. While it has now been realized that the entire arterial wall undergoes remodeling, traditional teaching still holds true in the sense that the disease process starts with the lining of the intima, that is, the endothelium. Endothelial dysfunction and activation presents the first and still reversible event in the atherosclerotic sequence.[14,15] On a molecular level, this disease stage is closely related to reduced bioavailability of nitric oxide (NO), which can be due to a reduction of NO production and/or an increase in NO consumption. Over the last few years, exciting discoveries have been made concerning the influence of the UPS, both under normal and pathological conditions, on the activity of the enzyme that generates NO in the vascular wall: endothelial NO synthase (eNOS).

Under normal conditions, eNOS is kept in an inactive state at the cell membrane by interaction with caveolin-1. Importantly, the posttranscriptional level of eNOS and hence the relative ratio to caveolin-1 is influenced by the proteasome and possibly favorably altered by proteasome inhibition.[16] In addition to release from caveolin-1 and dimerization, normal eNOS activity requires Akt-induced phosphorylation, particularly of its serine residue 1179, and dephosphorylation of this residue by protein phosphatase 2A leads to eNOS inactivation.[17] Interestingly, both Akt and protein phosphatase 2A are influenced in their activity by the UPS but seemingly in a different manner. While ubiquitination of Akt leads to its degradation by the proteasome complex, ubiquitination of protein phosphatase 2A leads to its translocation from the cytosol to the cell membrane, where it can then interact with eNOS.[17] These differences also highlight the various regulatory influences of the ubiquitin system on intracellular proteins.[18] NO generated by eNOS exerts a broad range of effects within endothelial and neighboring cells. On the level of the vascular smooth muscle cells (VSMCs) of the media, NO binds and activates soluble guanyl cyclase (sGC). This enzyme then generates cGMP from GTP, which mediates VSMC relaxation by various mechanisms. Importantly, cellular levels of sGC are negatively controlled by proteasomal degradation following ubiquitination by the co-chaperone/ubiquitin ligase carboxyl terminus of Hsc70 interacting protein (CHIP).[19] For this reason, while the positive effect of proteasome inhibition on ubiquitination of protein phosphatase 2A can prevail over other outlined effects, leading to a decrease in eNOS activity on endothelial cell level, whereas the overall scenario might be different when testing organs or organisms.[16,17]

Cardiovascular risk factors can downregulate eNOS activity and efficacy quite potently. One key mechanism is depletion of tetrahydro-L-biopterin (BH4), which serves as an essential cofactor for normal eNOS function. Very intriguingly, under hyperglycemic conditions, BH4 can become depleted by UPS-dependent degradation of guanosine 5′-triphosphate cyclohydrolase I.[20] Oxidative stress products such as 4-hydroxynonenal (4-HNE) generated in the setting of exposure to hypercholesterolemia or hypertension can accomplish the same.[21] In the absence of BH4, eNOS generates superoxide anions instead of NO (so-called eNOS uncoupling).[22] By this mechanism, the bioavailability of NO can be reduced quite quickly as NO is no longer produced and any remaining NO is consumed by superoxide anions to generate the potent cytotoxin, peroxynitrite. While superoxide anions are the epitome of oxidative stress in the arterial wall, a number of other reactive oxygen and nitrate species are generated and insufficiencies of antioxidant defense mechanisms have to be considered as well. One of these involves nuclear erythroid 2-related factor 2 (Nrf2). As long as the level of oxidative stress is low, this transcription factor is marked for proteasomal degradation by constitutive ubiquitination, which involves Keap-1 as a substrate adaptor for a Cul3-dependent E3 ubiquitin ligase complex.[23] Under circumstances of increased cellular oxidative stress, Keap-1 itself undergoes ubiquitination and proteasomal degradation, which stabilizes Nrf2 and allows it to bind to genome sequences with an antioxidant response element. This leads to the expression of genes encoding for proteins which have been linked to the amelioration of oxidative stress. Of note, Nrf2 also leads to an upregulation of proteasome subunits, and their overexpression via the Nrf2 pathway has been found to increase cellular resistance even against toxic misfolded proteins.[24] Likewise, proteasome inhibitors can "shield" against oxidative injury in an Nrf2-dependent manner.[25] However, this potential is not infinite and, at higher dosages, proteasome inhibitors can in fact worsen cellular oxidative stress.[26] Vice versa, a dual effect on proteasome function has been shown for oxidative stress products such as oxLDL and HNE.[27,28]

Interestingly, there are two other transcription factors whose activation pathways are stimulated by elevated intracellular levels of oxidative stress with involvement of the UPS.[29] The first one is nuclear factor kappa-B (NFκB), and gene sequences controlled by NFκB include those that encode for vascular cell adhesion molecule-1, E-selectin, and monocyte chemoattractant protein-1 (Fig. 4).[30] These molecules lead to the attraction and accumulation of inflammatory cells to and in the arterial wall, and hence NFκB is a pivotal factor for the development of atherosclerosis as an inflammatory disease from its early stages onward. NFκB also adds to the cell proliferation aspects of atherosclerosis, for instance, by regulating the expression of endothelin-1. The second transcription factor to be mentioned in this context is hypoxia-inducible factor (HIF)-1α (Fig. 5).[31,32] Importantly, HIF-1α adds to and stimulates NFκB activity but remains best known for facilitating vasa vasorum and then plaque neovascularization.

## B. The UPS and Progression Stage of Atherosclerosis

The dysfunctional endothelium with reduced barrier function allows cholesterol, especially in the form of so-called low-density lipoprotein (LDL) particles, to enter the subintimal space. In an environment marked by oxidative stress, these particles then undergo various modifications that lead to their entrapment and uptake by macrophages. An imbalance of cholesterol uptake and efflux along with suppression of apoptosis leads to the conversion of macrophages to foam cells. The expression of CD36, a member of the family of scavenger receptors that mediate the uptake of modified LDL particles, is regulated by protein kinase C alpha (PKCα). Importantly, ubiquitination of PKCα leads to the down-regulation of this enzyme, increase in CD36 levels, cholesterol accumulation, and development of atherosclerosis.[33] On the other side of the input–output equation, cholesterol efflux from macrophages depends on ATP-binding cassette transporter molecules (such as ABCA1 and ABCG1), and their expression level is likewise negatively controlled by the UPS.[34] As such, inhibition of the UPS can yield two opposing effects: increase of cholesterol uptake via CD36 and increase in reverse cholesterol transport via ABCA1 and ABCG1. A pathway by

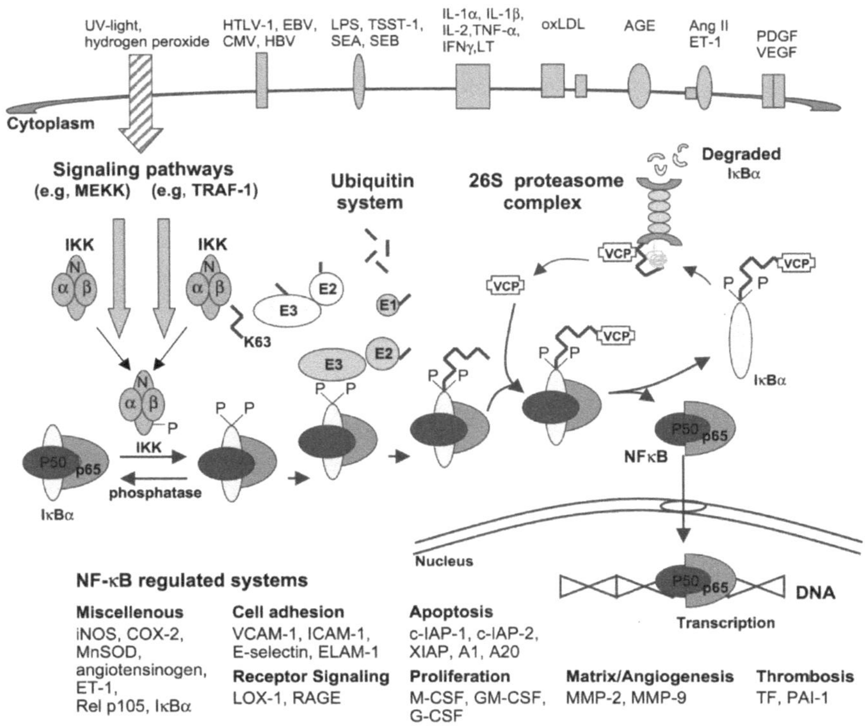


FIG. 4. The classical example of NFκB is the heterodimer of p50 and p65 and the classical activation pathway entails the release from its association with an inhibitory molecule of the IκB family. Upon exposure of the cell to various stimuli such as increase in oxidative stress, two specific serine residues are rapidly phosphorylated by the IKK1/2 kinases, whose activity is coupled to upstream signaling cascades by their association with a third protein called NFκB essential modulator (NEMO). Once phosphorylated, IκBs undergo degradation via the ubiquitin–proteasome pathway in the main route of NFκB activation. This is mediated by initial recognition of phosphorylated IκBs by the β-TrCP component of the ubiquitin ligase complex Skp1/Cul1/ROC1/F-box protein FWD 1 and subsequent covalent attachment of multiple moities of ubiquitin to the IκBs. The multi-ubiquitin chain allows subsequent association with the amino-terminal domain of valosin-containing protein (VCP), and thereby the targeting of IκBs to the 26S proteasome and the translocation of the transcriptionally active dimers to the nucleus. Worth noticing is the fact that the very same IKK/E3 combination is also involved in the generation of p50 (NFκB1) from its precursor molecule p105 by limited proteolysis, adding to its NFκB activating function. Modified and reproduced with permission from the European Society of Cardiology.[30]

which the UPS may contribute to foam cell maintenance is via adipose differentiation-related protein. This protein is associated with lipid droplets in various cell types including foam cells, and a functional UPS is required for the

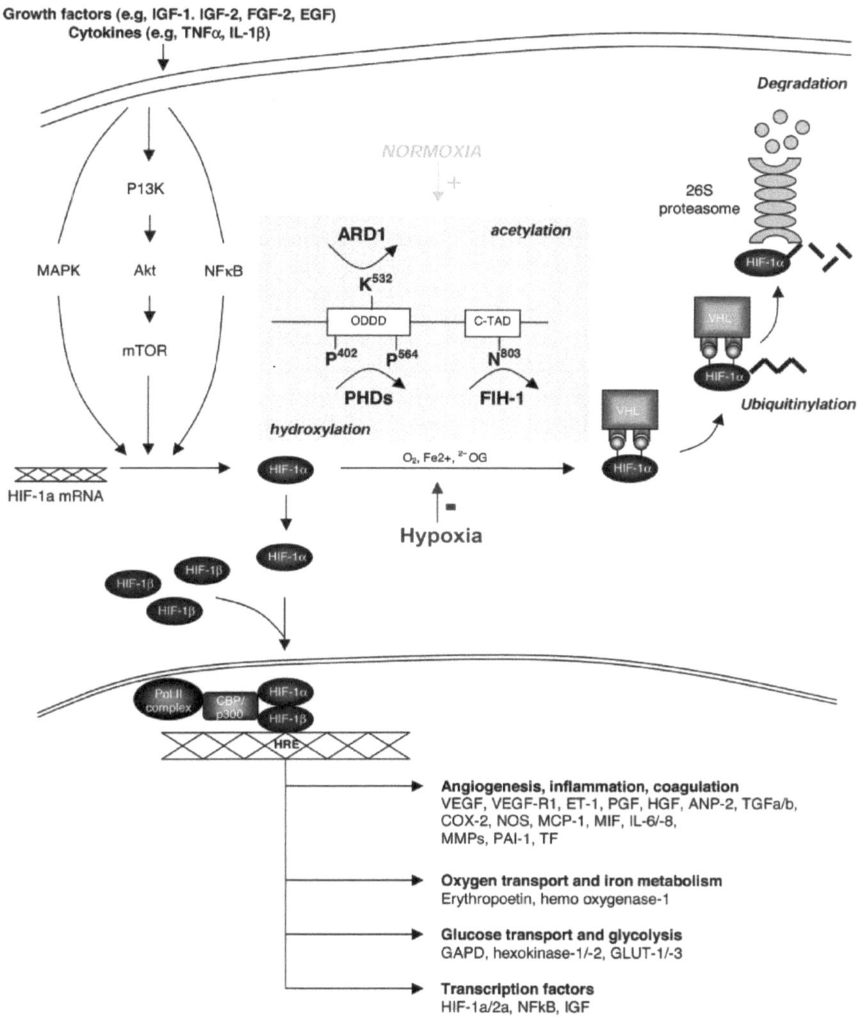


FIG. 5. Illustration of the hypoxia-inducible factor (HIF) activation pathway, highlighting hydroxylation and acetylation of HIF-1α depending on cellular oxygen tension, that is, normoxia and hypoxia. Hydroxylation of the proline residues 402 and 564 of HIF-1α allows recognition by the ubiquitin system, and thereby labeling for degradation by the proteasome complex. Modified and reproduced with permission from the American Heart Association.[31]

regression of these cells.[35] Finally, and in fact the very first link between the UPS and foam cells was provided by the observation that aggregated LDL particles stimulate the expression of a ubiquitin-conjugating enzyme that mediates the ubiquitination and degradation of p53, and thereby facilitates the suppression of apoptosis of lipid-bearing macrophages contributing to foam cell formation.[36]

As yet another reflection of the evolving disease process, VSMCs invade the subendothelial space from the media and transform into a proliferating phenotype. They also become metabolically active and generate the collagen-rich matrix that eventually overlies an atheromatous lipid pool. This transformation of smooth muscle cells involves extensive intracellular remolding and is prevented by proteasome inhibitors.[30] Myocardin has been identified as one specific molecular mediator of this process. This transcriptional coactivator controls the expression of genes for a contractile smooth muscle cell phenotype but is downregulated by CHIP-mediated ubiquitination and proteasomal degradation.[37] Likely mediated by upregulation of the p21 cyclin-dependent kinase inhibitor, proteasome inhibitors also inhibit migration and proliferation of VSMCs.[38,39] A shift in the balance in favor of proapoptotic molecules may also contribute to the antiproliferation effects of proteasome inhibitors.[39,40] Under normal conditions, NO may recapture these effects by inhibitory S-nitrosylation of proteolytic core proteins and modification of proteasome composition.[41] Furthermore, NO inhibits activation of NFκB in VSMCs by interfering with phosphorylation and thereby ubiquitination and proteasomal degradation of IκBα.[42] Importantly, the activation sequence of NFκB as a pivotal contributor to the proliferative response of the arterial wall can also be modified by deubiquitinating enzymes. Ubiquitin C-terminal hydroxylase L1 is a prominent example, expressed not only in endothelial cells but also and especially in neointimal VSMCs in an effort to downregulate NFκB activation by IκBα deubiquitination.[43] Cylindromatosis, another deubiquitinating enzyme expressed strongly in VSMCs, accomplishes the same by acting on TRAF2, a central molecule in the intracellular tumor necrosis factor (TNF)-α-receptor signaling cascade upstream from IκBs.[44]

## C. The UPS in the Complication Stage of Atherosclerosis

The developing atherosclerotic lesions can impair blood flow to the extent that it does not meet the demand of an end organ, leading to recurrent episodes of ischemia. While this in itself has clinical implications, the greater concern relates to acute (near-) closure of arteries as a consequence of acute plaque rupture causing acute ischemic events in various vascular territories. In a simplified equation, the vulnerability of a plaque toward this complication stage is determined by the relative content of collagen, produced by transformed VSMCs, and the volume of the lipid pool, generated by degenerating macrophages. Interestingly enough, the ratio of fibrous cap to necrotic core is influenced by inflammation and neovascularization at their very shoulder points.

T cells are of utmost importance for these tissue dynamics, and the UPS is, for one, uniquely involved in the regulation of NFκB activity in these cells.[15] Upon T cell stimulation, covalent attachment of regulatory ubiquitin chains to the regulatory protein Malt1 allows interaction with the IκB kinase (IKK) complex and its paracaspase activity cleaves A20, which catalyzes the removal of K63-linked ubiquitin chains from Malt1. As such, the balance between these two proteins regulates the strength and duration of the IKK/NFκB response upon T cell receptor (TCR)/CD28 costimulation.[45] The significance of this activation mode for atherosclerotic cardiovascular disease has been questioned by the fact that atherosclerotic plaque T cells often lack the costimulatory receptor CD28.[46] However, the significance of A20 for the attenuation of atherosclerosis has been confirmed in diabetic mice even though this was not specifically related to the TCR.[47] Another negative feedback loop in T cells involves Bcl10, which promotes activation of the IKK complex that then will phosphorylate Bcl10 after T cell antigen receptor stimulation and causes its proteolysis via the beta-TrCP ubiquitin ligase/proteasome pathway.[48]

T cell activity is also potently regulated by proteasome-mediated activation and nuclear translocation of the nuclear factor of activated T cells (NFAT). This transcription factor controls the expression of the activation-associated cell surface receptors CD25, CD28, CD120b, and CD134 as well as production of interferon (IFN)-γ, TNF-α, and interleukin (IL)-4 and IL-5.[49] Furthermore, T cells can be arrested in the G(1) phase by accumulation of cyclin-dependent kinase inhibitors p21(WAF1/CIP1) and p27(KIP1) and the disappearance of cyclin A, cyclin D2, and proliferating cell nuclear antigen.[49] These observations were made in CD4(+) T cells activated by dendritic cells (DCs), which is very relevant as DCs are particularly prevalent in the neovascularization-rich shoulder areas of complicated and symptomatic atherosclerotic plaques.[50] Importantly, the proteasome is crucial for the maturation and function of DCs. As a consequence, DCs with impaired proteasome function fail to stimulate allogeneic CD4(+) and CD8(+) T cells and autologous CD4(+) T cells sufficiently enough to regulate innate and adaptive immunity.[51,52] Moreover, inhibition of proteasome function also induces apoptosis of DCs via the mitochondrial pathway.[53] Similarly, with prolonged inhibition of proteasome function, T cells undergo apoptosis via the mitochondrial p53 pathway.[54,55] Of utmost significance, oxLDL, highly prevalent in human atherosclerotic plaques, can induce apoptosis of CD4+/CD25+ regulatory T cells in a time- and concentration-dependent manner.[56] Apoptosis of this regulatory T cell population removes a vital stabilizing force from the atherosclerotic plaque as these cells downregulate T cell responses to foreign and self-antigens.[57]

Recently, the immunoproteasome was identified as a potential link between inflammation and apoptosis of plaque cells.[58] In the presence of IFN-γ, fibrous cap VSMCs are sensitized to apoptosis via the Fas/Fas ligand pathway, which is prevented by interference with the inducible β5 subunit of the immunoproteasome. As it was subsequently realized, a switch to the immunoproteasome allowed proteolytic processing of myeloid cell leukemia (Mcl)-1, thereby removing a potent inhibitor (sequester) of proapoptotic molecules.[59] Intriguingly, this mode of regulation is in addition to the possible degradation of Mcl-1 via the conventional UPS.[60] These observations indicate that even subtle nuances in proteasome composition and activity may be important for the phenotype of biological and pathobiological processes.

In addition to apoptosis, autophagic cell death has been reported in atherosclerotic plaques with potentially detrimental consequences.[61] Autophagy is activated under circumstances of impaired proteasome function likely as a consequence of endoplasmic reticulum (ER) stress and as a compensatory effort to remove polyubiquitinated protein aggregates.[62] However, a number of other mechanisms can induce autophagy in the atherosclerotic plaque, not related to impairment of proteasome function.[61] Furthermore, autophagy can also have cytoprotective effects (particularly if the autophagic activity is not excessive). The ultimate biological consequences for the atherosclerotic plague therefore depend on the type of cell involved and the extent of stimulation of autophagic activity.[61]

## D. Proteasome Inhibition and Atherosclerosis: For Better or for Worse?

Utilizing a murine carotid collar of atherosclerosis, a 4-week treatment course with the proteasome inhibitor bortezomib, given once every 3 days at either a low- or high-dose regimen (corresponding to nearly 40% and 60% proteasome inhibition in the aorta, respectively), did not influence the composition of early plaques but led to a significant increase in necrotic core area and apoptotic cells as well as a decrease in VSMCs and collagen content in advanced atherosclerotic plaques (Fig. 6).[63] Most intriguingly, compared with carotid artery plaques from patients who did not have symptoms of focal cerebral ischemia within 6 months of carotid endarterectomy, plaques from symptomatic patients were characterized by lower collagen content and increased rate of apoptosis.[64] In correlation with the latter, the level of ubiquitination was higher and the level of proteasome activity was lower in symptomatic plaques (Fig. 7). Demonstrating that proteasome function can intrinsically decline in atherosclerosis with the potential to propagate plaque dynamics toward an unstable phenotype, these observations provided further

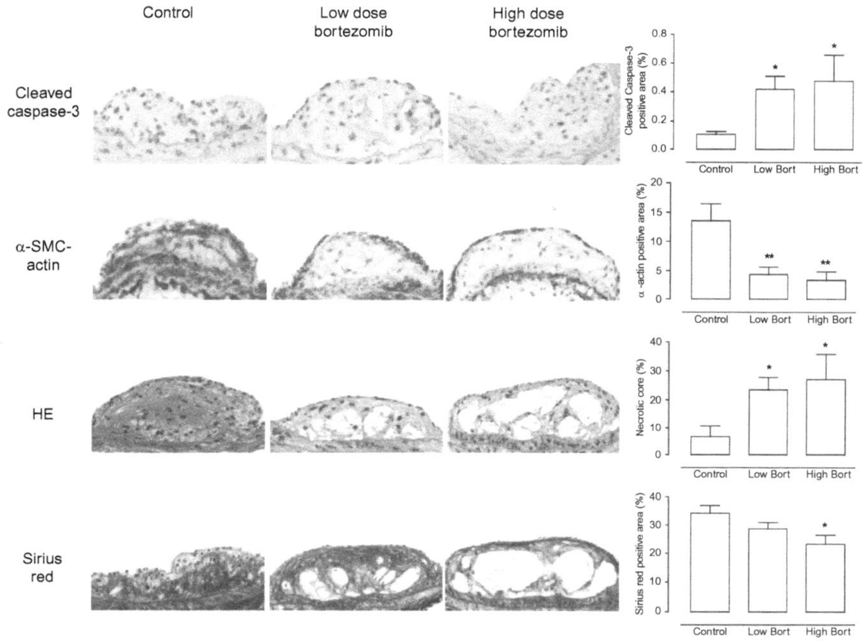


FIG. 6. Illustration of the effects of the boronate-type proteasome inhibitor, bortezomib, at low and high dose on the composition of advanced atherosclerotic plaques in ApoE-deficient mice including an increase in immunoreactivity for cleaved caspase-3 and a decrease for α-SMC-actin +cells, a decrease in collagen content by Sirius red staining, and an increase in necrotic core size by hematoxylin eosin (HE) staining; $^*p<0.05$, $^{**}p<0.01$ versus control; scale bar = 50 μm. Modified and reproduced with permission from Springer Verlag.[63]

insight into the functional significance of the original observation of elevated ubiquitin levels in advanced and complicated coronary artery plaques of patients with acute myocardial infarction (Fig. 8).[66] It has to be acknowledged though that some studies demonstrated an increase in proteasome activity in high-risk plaques, macrophages extracted from these plaques, and even peripheral monocytes.[67–69] These contradictory results may be explained by differences in the composition of the study population, as subsequent studies did indicate that proteasome activity was lower in atherosclerotic plaques of patients older than 60 years of age.[70] Intriguingly, an age-dependent decrease of proteasome function has been shown in other organs including the heart.[71,72] Hence, a decline in proteasome function may become a prevailing phenomenon over time with the consequences outlined above.

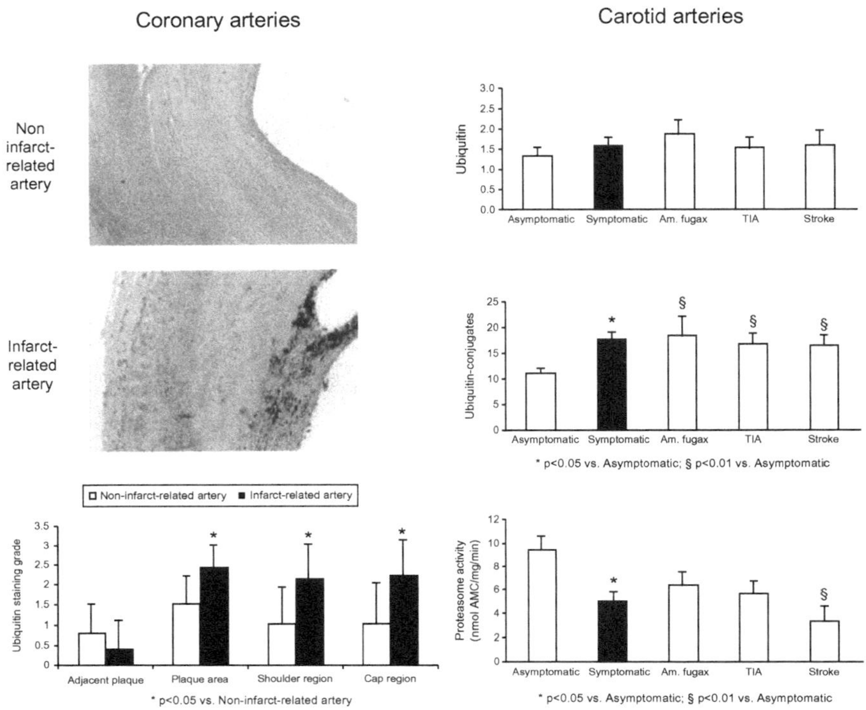


FIG. 7. In patients with fatal myocardial infarction, an increase in ubiquitin immunoreactivity is evident in the complicated plaques of the infarction-related coronary arteries compared with advanced plaques in the noninfarction-related coronary arteries. The difference in ubiquitin immunoreactivity is notable in the shoulder and fibrous cap areas (left panels). In patients undergoing carotid endarterectomy, immunoblotting-determined levels of ubiquitin–protein conjugates, but not free ubiquitin, are higher and chymotrypsin-like proteasome activity is lower in plaques from patients with symptoms of transient ischemic attack (TIA), stroke, or amaurosis (Am.) fugax within 6 months of surgery (right panel). Reproduced with permission from Elsevier.[65]

Similar to the debate in advanced atherosclerotic plaques, experimental studies in early atherosclerosis have yielded conflicting results. In agreement with the detrimental effect of the UPS on BH4 levels and NFκB activation, proteasome inhibitors or drugs that have proteasome-inhibitory potential such as aspirin were found to improve endothelial function and structural alterations of atherosclerosis in hypercholesterolemic and uremic rabbits as well as diabetic mice.[16,20,73,74] On the contrary, we noted a remarkable deterioration of endothelial function and aggravation of atherosclerosis in pigs on a normal or a high-cholesterol diet undergoing chronic proteasome inhibitor treatment (Fig. 8).[18] One readily notable difference between these studies is that the

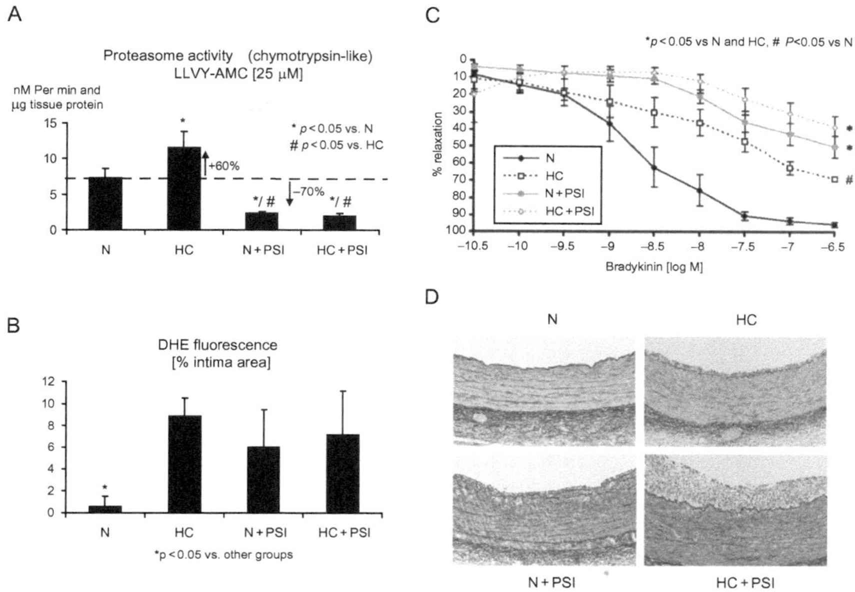


FIG. 8. Illustration of the effects of boronate-type proteasome inhibitor treatment (PSI) in addition to normal (N) or high-cholesterol (HC) for 12 weeks in domestic pigs. Compared with N, a 60% increase in coronary artery chymotrypsin-like proteasome activity is notable in HC and a 70% decrease in N + PSI and HC + PSI (panel A). This level of proteasome inhibition causes oxidative stress on the level of the intima, measurable, for instance, by DHE fluorescence, indicating in situ superoxide anion production similar to the degree observed in HC (panel B). Concomitantly, there is a decrease in the vasorelaxation response to the endothelium-dependent agonist bradykinin although more pronounced in PSI pigs (panel C). Along with endothelial dysfunction, intimal thickening develops in N + PSI similar to HC (panel B). Yet, even with a similar level of endothelial dysfunction, intimal thickening is greatly aggravated in HC + PSI. These observations suggest that, while chronic proteasome inhibition can be harmful under otherwise normal conditions, the consequences are much worse when it entails blunting a compensatory increase in proteasome activity. Modified and reproduced with permission from the American Heart Association.[18]

boronate-type proteasome inhibitor used in our study is more proteasome-specific and selectively inhibits chymotrypsin-like activity whereas the peptide aldehyde MG132 is not as proteasome and proteasome subunit specific. Final conclusions on the functional significance (beneficial or detrimental) of the upregulation of proteasome activity under pathological conditions which has been consistently outlined in all of these studies therefore remains confounded by the inherent limitations of using various proteasome inhibitors. Currently, there is agreement that proteasome inhibitors can be "poisons or remedies" depending on the context in which they are used.[75]

## E. Summary: The UPS and Atherosclerosis

This section reviewed the current and most up-to-date knowledge of the UPS in atherosclerosis with the following key points: First, the UPS is variably involved in the regulation of eNOS activity and modulation of oxidative stress, which is highly relevant for the initiation stage of atherosclerosis and beyond. Second, the UPS plays a central role in the activation pathway of NFκB in various cells, mediating adhesion molecule expression, cytokine release, and proliferation. Moreover, it can influence foam cell formation and maintenance, both of which influence atherosclerosis progression. Third, contrary to the initiation and progression stage, the complication stage of atherosclerosis seems to be linked to a reduction of proteasome activity mediated by accumulating levels of oxidative stress in aging cells over time. In this, atherosclerosis shares similarities with neurodegenerative disorders and protein quality diseases. Indeed, the UPS and particularly the proteasome complex is a vital component of PQC mechanisms as it prevents the accumulation of dysfunctional proteins, which can be toxic to the cell and its environment (Fig. 9).[36] As such, preservation rather than inhibition of proteasome function should be the goal while acknowledging some clear-cut exceptions and nuances. For instance, transient and reversible low-level proteasome inhibition has clearly been shown to exert beneficial effects such as increasing the resistance of cells against oxidative stress. The key is that this type of proteasome inhibition can be understood as a conditioning rather than an inhibition regimen similar to the protective effects of cardiac preconditioning.[26] Otherwise, a more local and focal mode of proteasome inhibition is to be favored such as the use in drug-eluting stents. Particularly advanced plaques may become vulnerable with systemic general proteasome inhibitor administration. In the coming years, further details of the function and significance of the UPS in atherosclerosis will likely be revealed and that will hopefully lead to the development of novel therapeutic strategies.

# IV. The Role of Cardiac E3 Ligases

## A. Pathologic Cardiac Hypertrophy

The role of the UPS, and specifically cardiac E3 ligases, in the development of pathologic cardiac hypertrophy is now well appreciated. Pathologic left ventricular hypertrophy is induced by a variety of factors (such as ischemia and volume or pressure overload) and results in a concomitant increase in both protein synthesis and proteolysis, the latter of which is regulated in large part by the UPS.[76,77] Following the induction of pathologic cardiac hypertrophy, the mRNA levels of ubiquitin, E1, E2, and E3s (specifically atrogin-1 and MuRF1

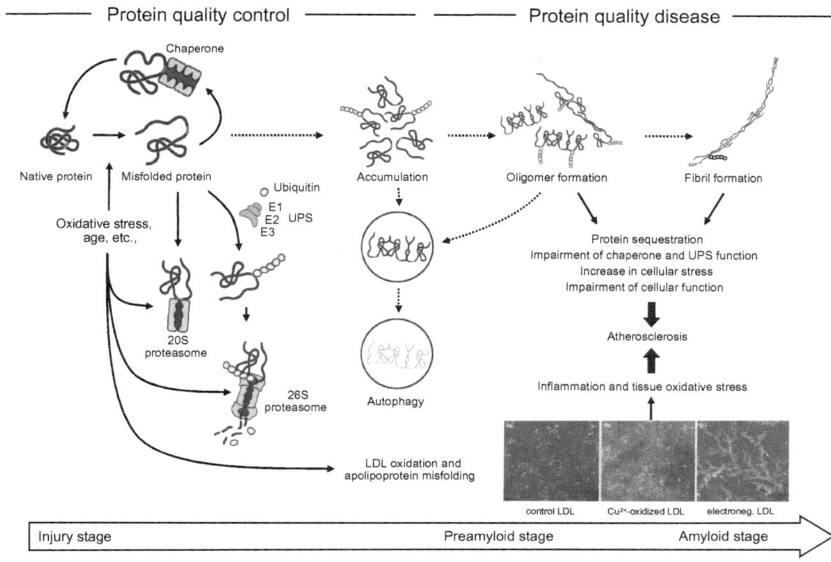


FIG. 9. Illustration of the role of the ubiquitin–proteasome system in protein quality aspects of atherosclerosis. In the initial phase of cardiovascular risk factor exposure, compensatory upregulation of chaperones and the ubiquitin–proteasome system prevents the overwhelming intracellular accumulation of damaged and dysfunctional proteins. The ubiquitin–proteasome system also contributes to the classical activation pathway of nuclear factor kappa-B and thereby to inflammation and cell proliferation. With the formation and growth of a metabolically active atherosclerotic plaque, there is further production of misfolded and damaged proteins in the progression phase. Once the classical protein quality mechanisms are overwhelmed and fail, these dysfunctional proteins accumulate (and aggregate) and autophagy remains the final clearance pathway. The accumulating proteins can undergo further oxidation, ubiquitination, and cross-linking. Furthermore, beta-pleated sheets can be formed and hence amyloid fibrils. In addition to intracellular proteins, proteins in the extracellular matrix can undergo conformational changes. For instance, oxidation and phospholipid hydrolysis of low-density lipoprotein (LDL) produces oxidatively modified and electronegative particles with unfolding of the apolipoprotein components. The generation of amyloid-like structures in this process serves as a potent "key" to the uptake of these modified LDLs by macrophages via scavenger receptors. Recognition of amyloid-like fibrils by CD36 leads to the production of reactive oxygen species, chemokines, and cytokines, which contributes further to the atherosclerotic disease process, including its complication phase. Reproduced with permission from Elsevier.[65]

(muscle ring finger 1)) increase.[78] A similar increase in transcriptional upregulation of atrogin-1 and MuRF1 occurs in an *in vivo* model of chronic heart failure (CHF), indicating that cardiac E3 ligases are involved in both the development and end-stage pathology associated with pathologic cardiac hypertrophy. Although there is still some doubt as to whether or not these increases in expression of atrogin-1 and MuRF1 are beneficial in the defense against pathologic cardiac hypertrophy, as will be detailed in the following sections, the evidence suggesting that these cardiac E3 ligases are pivotal in regulating the development of cardiac hypertrophy is accumulating.

### 1. Atrogin-1

Cardiac atrogin-1 (also known as MAFbx1) is a cardiac and skeletal muscle-specific F-box protein first identified as a crucial participant in skeletal muscle atrophy programs.[79,80] The expression level of atrogin-1 is tightly regulated at the transcriptional level in both skeletal myocytes and cardiomyocytes by members of the Forkhead family of transcription factors.[81,82] As with all cardiac E3 ligases elucidated thus far, atrogin-1 has generalized antihypertrophic activity, and is able to inhibit cardiac hypertrophy induced by either pathologic or physiologic stimuli. In the case of pathologic cardiac hypertrophy, atrogin-1 works as a cardiac E3 ligase and targets the "pro-hypertrophy" protein calcineurin for proteasomal degradation.[83] Calcineurin is a calcium/calmodulin-activated serine/threonine phosphatase that is a critical component of the signaling pathway that promotes pathologic hypertrophic growth of cardiomyocytes.[84] Calcineurin initiates nuclear signaling by dephosphorylating the NFAT transcription factor, which in turn activates transcription of hypertrophy-associated genes. Atrogin-1 inhibits this calcineurin-dependent transactivation and downstream NFATc4 translocation by serving as an adaptor for calcineurin to associate with Skp1, Cul1, and Roc1 (the common components of SCF ubiquitin ligase complexes) to assemble an SCF (atrogin-1) E3 complex that ubiquitinates calcineurin and targets it for proteasome-dependent degradation.[85] By targeting calcineurin for proteasomal degradation (and thereby inhibiting NFAT transcriptional activation of hypertrophy-associated genes), atrogin-1 disrupts a critical signaling pathway involved in pathologic hypertrophy.[83,86] In addition to its ability to inhibit pathologic cardiac hypertrophy, atrogin-1 can also decrease cardiac hypertrophy induced by physiologic stimuli. In this case, however, atrogin-1 uses a completely different mechanism of action. In response to the induction of physiologic cardiac hypertrophy, atrogin-1 targets the transcription factors Foxo1 and Foxo3 for ubiquitination.[87] Surprisingly though, instead of targeting the Foxo proteins for proteasomal degradation, atrogin-1-mediated ubiquitination results in enhanced Forkhead transcriptional activity and subsequent inhibition of Akt-dependent induction of hypertrophy.[88] Atrogin-1's ability to stimulate Foxo activity through ubiquitination is by virtue of the fact that atrogin-1 mediates the noncanonical addition of ubiquitin molecules on the Foxo proteins via lysine-63-linked chains, as opposed to the traditional canonical lysine-48.[87] Previous reports have documented the ability of lysine-63-linked chains to modulate protein function[89,90]; however, the atrogin-1-mediated non-canonical ubiquitination of the Forkhead transcription factors is the first report of this form of regulatory mechanism working within the heart.

#### 2. MuRF Proteins

The MuRF family of proteins comprises three members: MuRF1, MuRF2, and MuRF3. All these proteins have ubiquitin ligase activity and localize to the M-line and Z-disc of the sarcomere[91–93] where they function, in part, as mechanical stress sensors via their direct and indirect interactions with the giant protein titin.[85,94,95] Although all three MuRF proteins appear to play an important role in various aspects of myogenic responses and contractile regulation,[96] only MuRF1 appears to be involved in the regulation of skeletal muscle atrophy and inhibition of pathologic cardiac hypertrophy.[97] Although the mechanism by which MuRF1 mediates cardiac atrophy has not yet been fully elucidated, recent evidence suggests that MuRF1 may be instrumental in the initial proteolytic breakdown of the myofibril,[98] which then exposes the individual myofibrillar component proteins to subsequent proteolytic degradation. In cardiac muscle *in vitro*, MuRF1 exerts E3 ligase activity by ubiquitinating the sarcomeric protein cardiac troponin I, thereby contributing to its proteasomal degradation.[99] However, there are other ways in which MuRF1 may also contribute to the regulation of cardiac atrophy. For example, in cultured cardiomyocytes, MuRF1 inhibits PKCε activity through interactions with RACK1, the receptor for activated protein kinase C protein, which in turn suppresses focal adhesion kinase and ERK1/2 in cardiomyocytes, resulting in the suppression of cardiomyocyte hypertrophy.[100] It has also been suggested that MuRF1, through its association with the transcription serum response factor, is involved in a circuit that links mechanical force generation to transcriptional responses in cardiomyocytes, resulting in hypertrophic growth.[97] *In vivo*, MuRF1 has been linked to the regulation of cardiac hypertrophy in both mouse models and human disease.[101] MuRF1 protein levels are increased in human cardiac tissue isolated from patients undergoing therapeutic atrophy following placement of a left ventricular assist device (LVAD) as a treatment to decrease pressure overload associated with heart failure.[101]

## B. Protection Against Ischemic Heart Disease

The cardiac pathology profile associated with ischemic injury centers around the induction of cardiomyocyte apoptosis. A major player in this ischemia-related cardiomyocyte apoptosis is the transcription factor p53. p53 target genes include PUMA, Bax, and Apaf1, each of which plays a distinct yet complementary role in the apoptotic signaling pathway. Numerous reports have now demonstrated that p53 is a target substrate of multiple cardiac E3 ligases such as CHIP, murine double minute (MDM) 4, and MDM2. These cardiac E3 ligases ubiquitinate p53 and target it for proteasomal degradation, thereby decreasing cardiomyocyte apoptosis and resulting in an overall cardioprotective effect in response to ischemic injury.

### 1. CHIP AND MDM2

The E3 ligase/co-chaperone CHIP (carboxyl terminus of Hsp70-interacting protein) is highly expressed in cardiac muscle where it ubiquitinates specific proteins, targeting them for degradation by the 26S proteasome. CHIP is also characterized by its ability to bind the HSPs Hsc70–Hsp70 and Hsp90. Much of what is known about CHIP's cardioprotective function during ischemia/reperfusion (I/R) injury comes from studies carried out in CHIP-/-mice. CHIP-/-mice are far more susceptible to cardiac damage following an ischemic insult compared to wild-type mice, exhibiting a 50% larger infarct area compared to wild-type controls.[102] In addition, CHIP-/-mice exhibit a marked increase in frequency of arrhythmias following reperfusion, indicating an increased degree of cardiac damage compared to wild-type mice.[102] CHIP-deficient cardiomyocytes also undergo enhanced apoptosis following ischemia/reperfusion injury, indicating that CHIP inhibits apoptosis during cardiac stress. The mechanism through which CHIP exerts this antiapoptotic effect could be linked to the fact that CHIP interacts with both wild-type and mutant forms of p53. In response to stress, CHIP suppresses the aggregation and ubiquitinates both forms of p53, leading to their 26S proteasome-dependent degradation.[103] This in turn would decrease the ability of p53 to stimulate proapoptotic pathways.

MDM2 is another E3 ligase present in cardiac muscle that regulates the activity and degradation of proteins such as p53 and Apoptosis Repressor with Caspase recruitment domain (ARC), both of which are involved in cell growth and apoptosis. Similar to CHIP, MDM2 inhibits p53 transcriptional activity through ubiquitination and subsequent proteasomal degradation.[91,104,105] In cardiomyocytes, the decrease in p53 levels as a result of MDM2 ubiquitination leads to an inhibition of apoptosis during pathologic stress.[100] Conversely, if MDM2 is inactivated in cardiomyocytes, the level of p53 increases, resulting in enhanced I/R-induced apoptosis and a subsequent reduction in left ventricular function.[100] In addition to targeting p53, MDM2 also regulates ARC protein levels in cardiomyocytes.[106] MDM2-mediated ubiquitination of ARC leads to proteasomal degradation of ARC and subsequent inhibition of cardiomyocyte apoptosis. Recently, the MDM2 homologue MDM4 has been shown to also inhibit p53 in cardiomyocytes. Interestingly, a conditional MDM4 knockout mouse model demonstrated that decreased cardiac MDM4 levels result in the development of a dilated cardiomyopathy.[107] The cause of this dilated cardiomyopathy was linked back to a p53-dependent decrease in cardiomyocyte number, indicating the importance of MDM4 inhibition of p53 activity in fully differentiated cardiomyocytes. While MDM4 has been shown to have overlapping functions with MDM2, its role in ischemia/reperfusion injury has not yet been elucidated.

### 2. MuRF3

MuRF3 is another muscle-specific RING finger protein that was originally identified by its interaction with microtubules and its role in skeletal myoblast differentiation.[93] However, more recent work has expanded the role of MuRF3 to include activities within the heart. Mice lacking MuRF3 exhibit abnormal sarcomere structure and have a dramatically increased response to ischemia/reperfusion injury compared to wild-type mice. When these mice are challenged with ischemia/reperfusion injury, rapid left ventricular dilation, cardiac rupture, and myocyte degeneration occur within days of the initial insult.[108] Two substrates of MuRF3 have been identified as being γ-filamin and the four-and-a-half LIM domain protein (FHL2) and the expression of both these proteins is increased in the hearts of mice lacking MuRF3. Given the dramatic response of MuRF3-deficient mice to myocardial infarction, it has been suggested that MuRF3-dependent turnover of γ-filamin and FHL2 is essential to the integrity and function of the heart following acute ischemic events.[108] MuRF3 also plays a role in the regulation and turnover of sarcomeric contractile proteins including MHCIIa and beta/slow myosin heavy chain, a function that MuRF3 shares with MuRF1.[108] This cooperative relationship between MuRF1 and MuRF3 is illustrated by the pathological phenotype expressed by mice deficient in both proteins. MuRF1/MuRF3 knockout mice develop a skeletal muscle myopathy and hypertrophic cardiomyopathy in which myofiber fragmentation, MHC accumulation, and diminished muscle performance are also seen.[109] These findings reveal a critical role for MuRF1 and MuRF3 in proteasomal-dependent turnover of sarcomeric proteins and suggest a potential mechanism for the development of myosin storage myopathies.[108]

## C. The UPS in Familial Hypertrophic Cardiomyopathy: Proteasomal Involvement Without an Identified Cardiac E3 Ligase

While cardiac E3 ligases have been definitely linked to the regulation of cardiac hypertrophy and ischemia/reperfusion injury, there are other cardiovascular conditions that are known to involve the UPS, but for which a candidate ubiquitin ligase has yet to be discovered. One example is familial hypertrophic cardiomyopathy (FHC). FHC is an inherited cardiac disease that is transmitted as an autosomal-dominant trait. It is estimated that FHC affects 0.2% of the population and is the major cause of sudden death in young, healthy individuals.[110] Over 200 mutations associated with FHC have been identified in 14 different genes, with the majority of them involving genes encoding sarcomeric proteins.[111] Genetic analysis of families with FHC has revealed that most of the FHC-related mutations occur in the myosin-binding protein-C (cMyBP-C).[112] cMyBP-C is a critical constituent of sarcomeric thick filaments and has both regulatory and structural roles within the sarcomere. cMyBP-C interacts with myosin, titin, and actin, which, once again, are essential components of the sarcomeric apparatus.[110–113] Recently it has been demonstrated that the UPS rapidly targets naturally occurring mutations in cMyBP-C that result in a truncated protein for degradation. The increase in proteasomal workload associated with clearing of these mutated forms of cMyBP-C results in impairment of proteolytic capacity.[114] Frameshift mutant cMyBP-C proteins are expressed at markedly lower levels in muscle, despite the fact that the mRNA levels of these mutant forms of cMyBP-C match the wild-type cMyBP-C mRNA levels,[114] in both humans[115,116] and transgenic mice.[117,118] Experimentally, if the 26S proteasome is inhibited, expression of the truncated cMyBP-C mutant protein increases to that of wild-type cMyBP-C levels, supporting a role for proteasomal degradation in the control of cMyBP-C protein expression levels. Paradoxically, although the UPS is activated to remove mutant cMyBP-C proteins from the cell, over time, the increased proteasomal workload associated with this quality control mechanism results in a breakdown in activity of the proteasome, ultimately leading to the accumulation of mutated cMyBP-C proteins.[114] As such, truncated cMyBP-Cs act as both substrates of the UPS and obstructers of the proteolytic process, a dichotomy that is shared with certain protein mutants that accumulate in Alzheimer's disease and other storage-related neurological conditions.[119] The specific E3 ligases that target either wild-type or mutant cMyBP-C are currently unknown. If these E3 ligases could be identified, however, it would be possible to use them as therapeutic targets to alter the severity and progression of FHCs. The role of the UPS in PQC as it relates to cardiac proteinopathies is covered in detail in the next section.

## D. Summary: Cardiac E3 Ligases

As highlighted in this section, cardiac E3 ligases such as atrogin-1, CHIP, MDM2/4, and MuRF proteins play a pivotal role in regulating the molecular pathways underlying cardiovascular conditions such as cardiac hypertrophy and ischemic heart disease. Furthermore, proteasomal involvement has been reported in a number of other cardiovascular conditions, such as familial cardiac hypertrophy, although the identity of the participating cardiac E3 ligases remains unknown. However, despite the potential that cardiac E3 ligases have as therapeutic agents, more data need to be gathered in order to understand the extent to which these ligases control various cellular processes, both beneficial and detrimental. Such focus will allow identification of which cardiac E3 ligases might serve as therapeutic targets in the future.

## V. The UPS in PQC and Cardiac Proteinopathy

### A. Cardiac PQC

For a polypeptide to fulfill its normal duty in the cell, it must attain and maintain its native conformation via proper folding. A misfolded or unfolded protein is not only unable to perform its normal function, but it is also likely to be detrimental to the cell. PQC is a set of cellular mechanisms to keep the level of misfolded/unfolded proteins in the cell to the minimum. All proteins targeted to the secretory pathway go through the ER and are folded in the ER, while the folding of other proteins, such as cytoskeletal and cytosolic proteins, does not involve the ER. Hence, PQC is often divided into ER-associated PQC with which the unfolded protein response is associated, and ER-independent PQC. Given that myofibrils take the majority of space in the cardiomyocyte and that myofibrillar proteins are not folded via ER, ER-independent PQC takes care of more proteins in cardiomyocytes than the ER-associated PQC. Unfortunately, ER-independent PQC is much less understood than ER-associated PQC. Nevertheless, ER-associated and ER-independent PQC share the same general principle. Both are carried out by an intricate collaboration between molecular chaperones and targeted protein degradation. The latter is primarily performed by the UPS.[113] Both genetic mutations and environmental stress can give rise to misfolded protein molecules. In the cell, misfolded/unfolded proteins, which inevitably expose hydrophobic surfaces, will be sequestered by molecular chaperones to prevent aberrant aggregation and, in some cases, to help the misfolded/unfolded proteins to refold. If the refolding is unsuccessful, the misfolded proteins will then be known as terminally misfolded proteins and removed by proteasome-mediated degradation. Misfolded proteins that escape from the surveillance of chaperones and the UPS form aberrant aggregates. Recent evidence suggests that the autophagic–lysosomal pathway may also contribute to the removal of misfolded proteins, especially the aggregated form of misfolded proteins (Fig. 9).[120]

With a few exceptions, misfolded polypeptides must be polyubiquitinated before they can be recognized and degraded by the proteasome. A mature degradation signal, known as a degron, on the protein molecule is required to trigger the polyubiquitination and subsequent proteasome-mediated degradation of the protein. For a native protein, the maturation of its degradation signal is highly regulated and determines when it will be ubiquitinated; therefore, the degradation of a native protein is primarily determined by the ubiquitination step. Protein misfolding can either expose encrypted degrons or form a degron de novo (e.g., hydrophobic surfaces)[121]; therefore, it is conceivable that terminally misfolded proteins "automatically" activate the ubiquitination system and the rate-limiting step likely resides in the proteasome-mediated degradation step. Consistent with this notion, the accumulation of ubiquitinated proteins appears to be a common feature in the cells or tissues with increased production of misfolded proteins, and enhancing proteasome function facilitates the degradation of misfolded, but not native, proteins in the cell.[122,123] Therefore, measures to evaluate proteasome functional sufficiency, as well as means to enhance proteasome function, in a cell or organ are vital to the investigation into the role of the UPS in PQC and in the genesis of disease caused by misfolded proteins.

Proteasome peptidase activity assays provide a simple method to estimate changes in the proteolytic activity intrinsic to either the 20S or arguably the 26S proteasome in a crude protein extract, in a proteasome-enriched preparation from cells or tissue specimens, or in intact cells or tissues.[123] However, these assays do not assess whether the proteasome proteolytic activity is sufficient or not for adequate PQC in a given cell, tissue, or organ. To circumvent this deficit, full-length fluorescence proteins modified by fusion with a degron have been successfully adopted as surrogate substrates for the proteasome. Stable cell lines and stable transgenic mouse models that constitutively express such surrogate protein substrates have been established and employed to probe UPS functional status in the cell or organ in situ.[124–126] Regarding the usefulness of these surrogate protein substrates in the research into PQC and proteinopathy, $GFP^{u}$/GFPu and GFPdgn are arguably the most interesting and widely used ones. Both GFPu and GFPdgn were engineered via carboxyl fusion of degron CL1 to the enhanced green fluorescence protein (GFP). Conventional GFP is not an effective substrate of the UPS, with a long half-life when expressed in mammalian cells. The fusion of degron CL1 renders GFP a specific substrate for the UPS.[124] Amino acid replacement experiments showed that degron CL1, a consensus ubiquitination signal sequence from fission yeast,[127] triggers ubiquitination likely by forming a hydrophobic surface in its helical structure.[128] Hence, GFPu and GFPdgn are surrogates for misfolded proteins because surface exposure of hydrophobic residues is a conformational signature of misfolded proteins.[120] Changes in GFPu/GFPdgn protein levels in the absence of alterations in their synthesis inversely reflect the capacity of the UPS to remove misfolded proteins.

Emerging evidence suggests that proteasomes with a certain composition may be more efficiently equipped to degrade misfolded proteins. For instance, it is well known that viral infection or IFN stimulation triggers the replacement of conventional peptidase subunits (β1, β2, and β5) in 20S proteasomes with the inducible subunits (β1i, β2i, and β5i) and thereby upregulates immunoproteasomes.[129] A recent study has demonstrated that immunoproteasomes help the cell to more effectively remove oxidized proteins formed by increased oxidative stress induced by IFN.[130] It was also reported at almost the same time that upregulation of 11S proteasomes via overexpression of PA28α enhances proteasome-mediated removal of misfolded proteins and protects against oxidative stress in cultured cardiomyocytes.[131]

## B. Proteasome Functional Insufficiency in Cardiac Proteinopathy

Proteinopathies are a family of diseases caused by misfolded proteins and featured by the presence of aberrant protein aggregates in the cell. The best-studied cardiac proteinopathy is perhaps desmin-related cardiomyopathy (DRC).[132] DRC is the cardiac manifestation of desmin-related myopathy (DRM). DRM is a group of heterogeneous myopathies with a hallmark of the presence of desmin-positive aberrant protein aggregates in muscle cells. Mutations in the desmin gene, the aB-crystallin gene (*CryAB*), and a few other genes whose products interact with the desmin filaments have been linked to DRM.[132] DRM can affect all three types of muscle: skeletal, cardiac, and smooth muscle. DRC is the main cause of death of DRM patients. In the past decade, a few DRC mouse models have been created through cardiac-specific transgenic expression of a DRM-linked mutated gene,[133,134] such as a missense (R120G) mutant of CryAB ($CryAB^{R120G}$) and a 7-amino-acid (R173 through E179) deletion mutant of the desmin gene (D7-des).[135,136] Both $CryAB^{R120G}$ and D7-des transgenic mice recapitulate the main aspects of pathology observed in human DRC, including the presence of desmin-positive protein aggregates in cardiomyocytes and cardiac dysfunction[133,134]; hence, these mouse models provide useful platforms for studying DRC pathogenesis and exploring therapeutic strategies.[113,132]

Studies with cross-breeding GFPdgn mice with the mouse models of DRC showed marked increases of GFPdgn protein levels but not GFPdgn transcript levels in mouse hearts overexpressing either $CryAB^{R120G}$ or D7-des,[137,138] indicating UPS functional insufficiency in DRC hearts. The abundance of total ubiquitinated proteins was significantly increased in the DRC hearts, suggesting that UPS insufficiency is primarily caused by proteasome functional insufficiency (PFI). Further analyses revealed that the insufficiency resides mainly in the step delivering ubiquitinated proteins into the 20S proteasome as decreased protein abundance of critical 19S proteasome subunits, namely Rpt3 and Rpt5, is concurrently associated with the PFI while the peptidase activities of the 20S proteasome are not decreased. Additional studies have determined that aberrant protein aggregation is in part responsible for PFI in DRC hearts.[137,139] These are highly significant discoveries because although DRC per se is not a common disease, the DRC mice represent valuable models for pathogenic elucidation and therapeutic exploration of cardiac proteinopathies. It is increasingly suggested that a large subset of common heart diseases likely share pathogenic mechanisms with cardiac proteinopathy. Aberrant protein aggregates are also observed and considered the trigger of autophagy activation in the pressure-overloaded mouse heart.[140] Significant increases in myocardial ubiquitinated proteins have invariably been reported in studies using this experimental method, although some of the reports dispute the direction of changes in the proteasome peptidase activities as well as the effect of systemic proteasome inhibition on the hypertrophic responses and function of pressure-overloaded hearts.[141,142] Both the accumulation of aberrant protein aggregates and elevated levels of ubiquitinated proteins are consistent with PFI but the sufficiency of proteasome function in a pressure-overloaded or ischemic heart has not been directly assessed yet. A very high prevalence of aberrant protein aggregates in the form of preamyloid oligomers in human hearts with end-stage heart failure of common etiologies has been recently reported.[143,144] The coexistence of increased ubiquitinated proteins with aberrant protein aggregation signifies PFI and serves potentially as a cause of proteasome impairment. Hence, these observations place a large subset of CHF into the category of cardiac proteinopathy. A better understanding of the molecular pathogenesis of cardiac proteinopathy should shed light on the molecular mechanisms by which these heart diseases progress to CHF.

## C. PFI Contributes to the Genesis of Proteinopathy

PFI has been observed or implicated in a variety of diseases, such as neural degenerative diseases, cardiac proteinopathy, and more recently, a large subset of CHF. However, until very recently, a proteasome activator was unavailable, making it difficult to directly test whether PFI plays a role in the development of any of these diseases. The recent discovery of a benign measure to enhance proteasome function has instantly changed this situation. Using cultured neonatal rat ventricular myocytes and human embryonic kidney cells, it was found that overexpressing PA28α (PA28αOE) stabilizes PA28α and thereby upregulates 11S proteasomes, which destabilizes GFPu, a surrogate misfolded protein substrate for the UPS,[120] without affecting the protein levels of endogenous and ectopically overexpressed native proteins.[122] Therefore, this marks the establishment of the first benign measure to enhance proteasome proteolytic function in cultured cells. A biological relevance of this exciting new measure was further demonstrated by testing the effect of PA28αOE on oxidative stress. Enhancing proteasome function by PA28αOE decreased the abundance of protein carbonyls, a widely used marker of oxidized proteins, and the prevalence of apoptosis in cultured cardiomyocytes triggered by the treatment of hydrogen peroxide.[131] These findings suggest that PFI plays an important role in injury by oxidative stress. The latter represents a major cause of myocardial I/R injury. Additionally, PA28α overexpression and knockdown were shown to enhance and delay the degradation of $CryAB^{R120G}$ in cultured neonatal rat ventricular myocytes.

To verify whether PA28αOE enhances proteasome function in intact animals, transgenic mouse lines were created to achieve temporally controllable cardiomyocyte-restricted (CR-PA28αOE),[122] using the "Tet-off" system reengineered by Sanbe *et al.*[145] Mice with CR-PA28αOE initiated at the perinatal period do not display remarkable changes in the expression of the fetal gene program, heart function, or cardiac morphology till at least 1 year of age, the longest time observed. Introducing GFPdgn into the CR-PA28αOE mice via cross-breeding, it was validated that CR-PA28αOE facilitates the degradation of GFPdgn, a surrogate misfolded protein.[126] This is evidenced by a marked decrease in GFPdgn protein level and the unaltered abundance and unchanged translational efficiency of GFPdgn transcripts. The effect on GFPdgn protein degradation disappears upon turning off CR-PA28αOE via administration of doxycycline.[122] These experiments mark the establishment of the first genetic method and the first animal model to benignly enhance proteasome function in the heart, or any other organs. Taking advantage of this model, Li *et al.* have further demonstrated that overexpression of PA28α significantly attenuates aberrant protein aggregation and cardiac hypertrophy and delays the premature death of mice with transgenic overexpression of $CryAB^{R120G}$ in cardiomyocytes, in a bona fide mouse model of cardiac proteinopathy.[122] Moreover, we demonstrated that CR-PA28αOE protects against myocardial I/R injury induced by coronary ligation and release in intact mice.[122] These studies prove for the first time that PFI plays an important pathogenic role in myocardial I/R injury and cardiac proteinopathy in both cell culture and intact animals. It is anticipated that these exciting discoveries, including the invention of new ways to benignly enhance proteasome function in a cell and an organ, will facilitate research into the pathogenic roles of UPS dysfunction and promote the development of new therapeutic strategies to treat a large subset of diseases, including heart diseases.

## D. Potential Mechanisms Underlying the Pathogenic Role of PFI

Conceivably, PFI can impair the heart via a number of mechanisms, such as compromising PQC and disturbing signaling pathways. Sarcomeres are the most fundamental units of mechanical function in cardiac and skeletal muscles. An extremely vigorous stoichiometry among different sarcomeric proteins is followed to form sarcomeres and maintained for proper function of the sarcomere. The replacement of a damaged sarcomeric protein molecule with a new one must be accurately coupled with the removal of the damaged one. The

UPS is responsible for the degradation of virtually all myofibrillar proteins. Hence, PFI in cardiomyocytes will leave defective proteins in the sarcomere that are not replaced in a timely fashion, and which will undoubtedly reduce the functioning of the sarcomere and ultimately the mechanical performance of the heart. In this regard, PFI may yield the same consequence as the genetic mutations of the sarcomeric proteins that are known to cause inherited cardiomyopathies,[146] a major cause of CHF. PFI inevitably accumulates misfolded proteins and permits aberrant protein aggregation. In both humans and mice, expression of misfolded proteins and resultant aberrant protein aggregation are sufficient to cause cardiomyopathy and CHF.[147–149] In addition to physical impedance to contraction and relaxation and structural disruption, aberrant protein aggregates can also be detrimental to multiple cellular processes. Aggregation-prone proteins can directly interact and inhibit proteasome proteolytic subunits.[150,151] In both cultured cardiomyocytes and intact mouse hearts, aberrant protein aggregation has been shown to impair proteasome proteolytic function,[117,137–139,152,153] thereby forming a vicious cycle.

Pathological changes in mitochondrial morphology and function as well as programmed cell death are often associated with CHF. Mitochondrial malfunction and increased cell death were observed in the mouse hearts of CryAB$^{R120G}$-based DRC that displays severe PFI before other cardiac pathology becomes discernible.[137,154] Pharmacologically induced proteasome inhibition has been shown to induce cell death in cultured cardiomyocytes.[125,155] The calcineurin–NFAT pathway is an important signaling pathway that transcriptionally regulates pathological remodeling of the heart.[156] Interestingly, proteasome inhibition can activate the calcineurin–NFAT pathway and promote maladaptive growth of cultured cardiomyocytes. NFAT transactivation is increased in the D7-des mouse heart and proteasome inhibition facilitates the progression of a stressed heart to heart failure.[142] Similar to PFI, the loss of an important chaperone (e.g., HSPB2/HSPB5 double knockout) can also activate the calcineurin–NFAT pathway in the heart,[142,157] suggesting that PQC inadequacy may play a role in the activation of this pathway during pathological cardiac remodeling.

## E. Improving PQC as a Strategy to Treat Cardiac Proteinopathy

No effective therapies are currently available for treating cardiac proteinopathies. However, by targeting the molecular pathogenesis of proteinopathy, recent experimental studies have shown some promise. As described earlier, enhancing proteasome function can protect against cardiac proteinopathy in mice.[122] Additionally, pharmacological agents that behave as chaperones (e.g., high dose of doxycycline) or as inducers of HSPs (e.g., geranylgeranylacetone) have been shown to suppress aberrant protein aggregation in CryAB$^{R120G}$-based DRC mouse hearts, to improve heart function, and to delay premature death of the DRC mice,[158,159] providing strong evidence that pharmacological chaperones are promising therapeutic candidates for clinical trials to treat cardiac proteinopathies.

## VI. The UPS in Heart Failure

Heart failure is defined "as a complex clinical syndrome that can result from any structural or functional cardiac disorder that impairs the ability of the ventricle to fill with, or eject, blood."[160] The causes and risk factors of heart failure are many, and include hypertension, ischemic heart disease, diabetes, obesity, valvular heart disease, and exposure to cardiotoxic agents.[160] As discussed in the previous section, some cases of heart failure are familial and associated with mutations in genes encoding for various myocardial proteins (cardiac proteinopathies), leading to issues with PQC. However, the majority of cases are idiopathic, where no one cause can be identified. This section describes alterations in the UPS as it relates to this latter group. As will become apparent, changes in activity of the proteasome are biphasic and seem to correlate with the severity of the process and thus can be described as early, studied primarily in animal models, versus late, studied using explants from human hearts.

### A. Alterations in UPS in Early Failure

One of the early manifestations of heart failure is development of cardiac hypertrophy secondary to ventricular remodeling. This early phase of heart failure development has been modeled in experimental animals using a variety of initiating events, including transverse aortic constriction (TAC), permanent left anterior descending artery (LAD) occlusion, and chronic exposure to hypertrophic agents, such as isoproteronol (ISO). Perhaps the earliest study was by Depre *et al.*,[161] who used a canine TAC model to show that, as the heart hypertrophies, the UPS is activated. Increased proteasome activity was associated with increased expression of proteasome subunits. This result was confirmed in subsequent studies in a murine ISO-induced hypertrophy model[162] and in cardiomyocytes exposed to hypertrophic agents.[163] This is not to say that these results are not without some controversy, as at least one study observed that proteasome activities are depressed prior to the onset of cardiac dysfunction in pressure-overloaded hearts of mice.[155] An interesting observation of these and other studies was that inhibition of the proteasome with any one of an assortment of proteasome inhibitors decreased the amount of hypertrophy associated with the initiating event.[161–164] Activation of the proteasome was associated with alterations in the stress-response chaperone H11 Kinase/Hsp22.[165] In addition, in mice treated with ISO, cardiac hypertrophy was associated with increased sensitivity of the proteasome to PKA-mediated activation, as well as an altered proteasome phosphoproteome.[10]

While the exact mechanism of proteasome activation is unclear at this time, the involvement of the UPS in cardiac hypertrophy and remodeling cannot be denied. This can be inferred from the skeletal muscle literature where it is clear that the proteasome and associated ubiquitination machinery mediate the breakdown of muscle protein (remodeling) necessary for exercise-induced physiologic hypertrophy.[166–169] Although not studied to the same extent, it is becoming increasingly clear from experimental models that pathological cardiac remodeling is also associated with changes in the ubiquitin protein ligases that target myofibrillar and other cardiac proteins, including atrogin-1 (MAFbx) and MuRf-1.[94,170,171] A recent study has shown that knockout of atrogin-1 is protective in both TAC and β-adrenergic-induced cardiac hypertrophy.[171] There is a key question that remains to be answered: What is the difference between the remodeling associated with cardiac hypertrophy that becomes pathologic and leads to heart failure and the physiologic remodeling associated with exercise that leads to strengthening and optimized function?

## B. Alterations in UPS Function in Late Failure

The availability of human heart tissues to study alterations in UPS function in early heart failure is extremely limited. This is not the case when studying late, or so-called end-stage, heart failure, where unfortunately there are ample supplies of explanted human hearts harvested at transplant. Often, these studies are limited by the availability of control human heart tissues, which generally are donor hearts originally destined for transplant but not used for a variety of reasons. Two of the earliest studies investigated for the presence of ubiquitinated proteins or substrates, indirect measures of UPS function at best, and observed these to be increased, suggesting dysfunction.[172,173] Subsequently, we published the first comprehensive examination of proteasome function in human end-stage heart failure and confirmed that proteasome activity is diminished in end-stage failure associated with increases in ubiquitinated proteins.[174] Unloading the heart via the use of an LVAD partially reversed the proteasome dysfunction.[174] This is an observation also reported by another pre- and post-LVAD study which showed enhanced expression of proteasome subunits.[175] It is of interest that similar results were obtained with markers of autophagy which were also downregulated following mechanical unloading via LVAD.[176] Autophagic and proteasome-mediated degradation are often considered to be parallel proteolytic systems that respond inversely to each other, that is, when proteasome activity is decreased, autophagy is increased, or vice versa.[123,177] However, in end-stage heart failure, both appear to be diminished.[174,176]

As part of our study, increased protein oxidation products that prompted an examination for oxidative damage to proteasome subunits were observed.[174] Protein carbonylation of at least one subunit of the 19S regulatory particle, Rpt5, was observed, suggesting that the dysfunction may be secondary to oxidative damage to the proteasome itself, although one rather intriguing study suggests that humoral auto-antibodies to proteasomes may be increased in dilated cardiomyopathy.[178] In all fairness, not all studies of human heart explants report the same dysfunction. One study reports increased expression of proteasomes in samples from patients with dilated cardiomyopathy referred for left ventriculoplasty[179]; however, proteasome activity was not assessed. Nonetheless, if one extrapolates from the animal and human studies, a relationship between progression of heart failure and function of the proteasome can be constructed in which activity initially increases during remodeling but then diminishes below baseline in end-stage heart failure.

## VII. The UPS in Myocardial Ischemia/Reperfusion Injury

Ischemia is defined as reduced tissue perfusion resulting in oxygen deprivation accompanied by diminished removal of metabolites. From a historical perspective, many of the earliest studies suggesting that ischemia may result in dysfunction of the UPS actually derive from the brain literature. These earlier studies report increases in insoluble ubiquitin conjugates in the mitochondria and decreased 26S proteasome activity in the gerbil cortex and hippocampus following transient forebrain ischemia.[180,181] Other studies have confirmed these results and provide a link between UPS dysfunction and oxidative stress.[182,183] Since the publication of earlier brain studies, there have been several studies examining UPS dysfunction following myocardial ischemia.

Dysfunctional postischemic proteasome in the heart was first described by Bulteau *et al.*,[184] who reported decreased proteasome chymotryptic, caspase-like, and trypsin-like activity accompanied by increased ubiquitinated proteins following 30 min of *in vivo* LAD artery occlusion. Despite the decrease in tissue proteasome activity, following purification only the decrease in trypsin-like activity was observed. In a series of studies, Powell *et al.*[185,186] confirmed this observation in the isolated perfused heart model and also demonstrated that ATP-dependent proteasome activity is preferentially affected. This suggested defects in 26S proteasome function which were consistent with increases in myocardial ubiquitinated proteins. One rather intriguing study[187] suggested that short periods of ischemia/reperfusion may result in selective proteasome dysfunction affecting turnover of specific proteins. This study took the indirect approach of assessing proteasome function by examining the effects of proteasome inhibitors on turnover of signaling proteins known to be subject to UPS-mediated degradation.[187] This study must be interpreted with caution as lack of studies of the ubiquitinated homologues of these signaling proteins makes it difficult to determine the role of the proteasome. Nonetheless, given the earlier discussion on regulation of the cardiac UPS and presence of proteasome subpopulations, the differential regulation of protein degradation proposed in this study is conceivable.

## A. Potential Mechanisms Involved in Postischemic Proteasome Dysfunction

Various mechanisms have been proposed to explain the dysfunction; however, two potential explanations have recently received credible experimental support.

### 1. Ischemia-Induced ATP Depletion

This mechanism is based on the known requirement for ATP for 20S proteasome and 19S regulatory particle docking and activation of the proteasome, as well as for protein ubiquitination.[188,189] A recent study suggests that intracellular ATP levels regulate proteasome activity[190] in a biphasic manner; thus, it is certainly conceivable that depletion of ATP during ischemia[191] could be partially responsible for decreased proteasome activity in the ischemic heart. In fact, this mechanism has been invoked to explain proteasome dysfunction in both *in vivo* canine and *ex vivo* isolated rat models of ischemia.[192,193] Since both of these studies measured proteasome activity in the presence of ATP, essentially adding back the nucleotide, the dysfunction should have been reversed, yet this was not the observed result. At least one study has shown that, even when measured in the presence of optimal concentrations of ATP, postischemic proteasome activity is diminished.[186] Taken as a whole, these studies suggest that the dysfunction measured using the standard fluorogenic peptide assays is not related to ATP depletion; however, this does not rule out the possibility that ATP depletion can still be a contributing factor to proteasome dysfunction during the ischemic period. It does, however, suggest a defect in the utilization of ATP by the postischemic proteasome. The presence of oxidative stress and oxidized cytosolic, myofibrillar, and mitochondrial proteins in the postischemic heart[194–197] has prompted us and others to hypothesize that the many subunits of the 20S proteasome and its regulatory particles may be targets for oxidative attack.

### 2. Oxidative Damage to Proteasome and/or Regulatory Subunits

Oxidative modification of proteins affects their secondary and tertiary structures resulting in unfolding, which can lead to loss of function and exposure of hydrophobic patches, which in turn can lead to increased susceptibility to degradation.[198,199] Studies of vulnerability of different populations of proteasomes (20S vs. 26S) to oxidative damage have shown that exposure of purified proteasome to oxidants, including peroxynitrite,[200,201] hypochlorite, hydrogen peroxide,[202] and 4-hydroxynonenal,[203,204] leads to inactivation with the 26S configuration being approximately 10-fold more sensitive.[202] In their original study, Bulteau *et al.*[184] report that several α-type subunits of the 20S proteasome are subject to 4-hydroxynonelation following ischemia/reperfusion though it was not clear that the dysfunction was related to the modification of these subunits. Subsequently, they reported that proteasome purified from rat heart seems to be somewhat resistant to inactivation requiring concentrations of 4-hydroxynonenal in excess of 100 μM.[27] The preferential effect of ischemia/reperfusion on the ATP-dependent activity of the proteasome[186,193] coupled with the reported higher vulnerability of the 26S proteasome to oxidative inactivation[205] has led to a suggestion that subunits of the 19S regulatory particle, which utilizes ATP, may be more sensitive to oxidative damage than those in the 20S proteasome. In fact, the Rpt3 (S6 ATPase) subunit has been reported to be sensitive to carbonylation reactions in SH-SY5Y cells exposed to an oxidative environment.[206] Consistent with this report, we have identified Rpt3/Rpt5 as 26S proteasome subunits significantly carbonylated during myocardial ischemia/reperfusion.[207] As indicated in the section on heart failure, this was the same subunit(s) that appeared to be oxidized in explants from human hearts from patients with end-stage heart failure.[174]

## B. The 11S-Activated Proteasome and Myocardial Ischemia

So far, this section has primarily dealt with the role of the 26S proteasome in myocardial ischemia. However, as described in an earlier chapter in this book as well as earlier in this chapter, the 20S proteasome can dock with and be regulated by an additional regulatory particle, the 11S activator ring. This regulatory particle can dock at either end or in combination with a 19S regulatory particle at one end (hybrid proteasome). Various forms of this alternate "zome" exist. If it contains one or more immunoforms of the catalytic subunits (β1*i*, β2*i*, β5*i*), it is often referred to as the immunoproteasome, which as described earlier is primarily involved with antigen presentation. If it is docked with a 20S that does not contain immunoforms, it is called the 11S-activated proteasome, which can remove oxidized proteins as studies have suggested.[208] This suggestion is given credence by the recent report that forced overexpression of the PA28α subunit in cardiomyocytes results in increased activity of the proteasome associated with enhanced resistance to an oxidative stress.[131] Moreover, a recent study has shown that cardiac-specific overexpression of the same subunit increases proteasome activity, improves postischemic function, and decreases infarct size.[132] Previously, we had shown that the

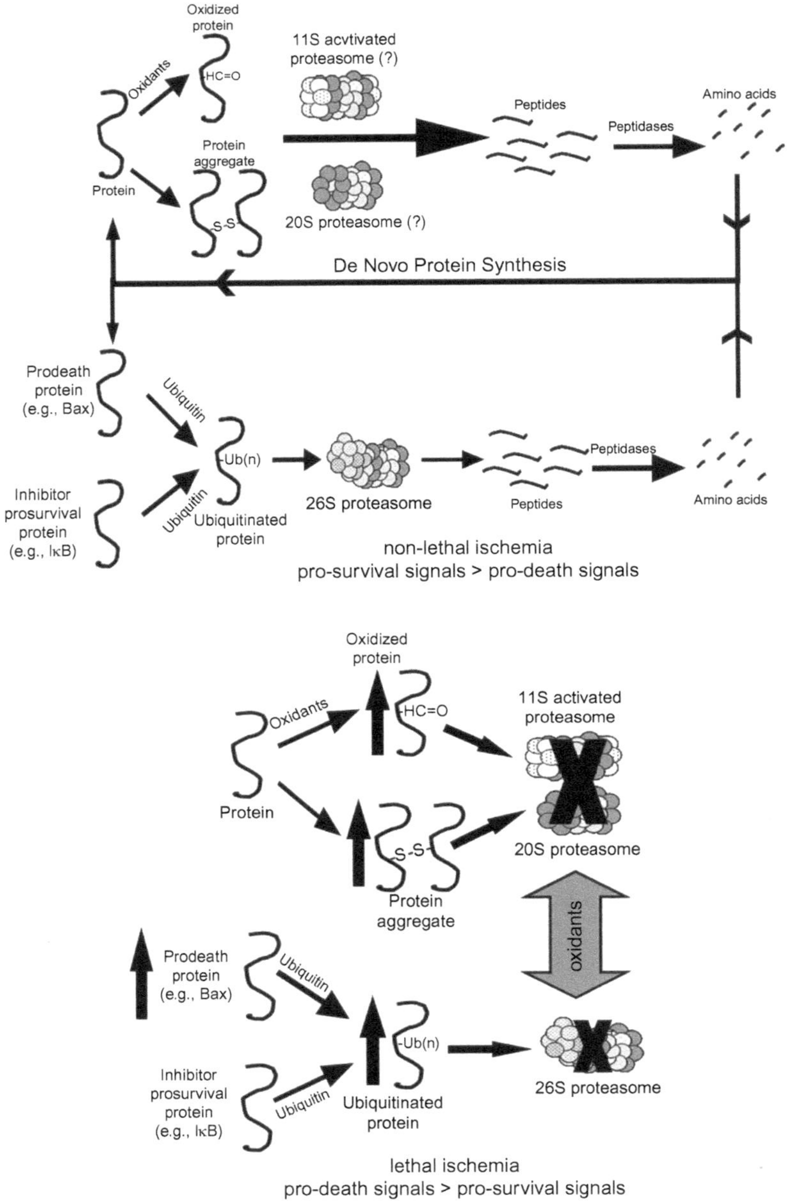


FIG. 10. Potential roles for the UPS and 11S-activated proteasome in short- and long-duration ischemia. In the nonischemic heart, oxidized, misfolded, and ubiquitinated proteins are degraded through both ubiquitin- and nonubiquitin-mediated pathways, recycling the constituent amino acids, and maintaining a dynamic balance between pro-survival and pro-death signals. During an ischemic insult resulting in cell death or dysfunction, UPS function is inhibited leading to accumulation of oxidized and ubiquitinated proteins. In addition, a condition known as dysregulation may occur in which normal UPS-mediated degradation of pro-death proteins is depressed. Reproduced with permission from Oxford University Press.[212]

proteasome was responsible for removal of proteins oxidized during myocardial ischemia and that this appeared not to require ubiquitin.[209] Given the oxidative stress associated with myocardial ischemia[197,210] as well as the reported upregulation of this form of the proteasome in a related cardiomyopathy also associated with oxidative stress,[211] it is conceivable that the 11S-activated proteasome plays a role in the removal of proteins oxidized during myocardial ischemia. A scheme describing the proposed functions of the 26S- and 11S-activated proteasomes is presented in Fig. 10.

## C. The UPS in IPC

As IPC is usually associated with myocardial ischemia, it is discussed in this section. IPC is a procedure to decrease vulnerability of the myocardium to ischemia/reperfusion injury via the prior exposure to short ischemic bursts. IPC tends to result in improved postischemic hemodynamic function and reduced markers of myocardial injury.[213] The current thinking is that IPC involves preischemic signaling changes that open the inward mitochondrial $K_{ATP}$ channels[214] and prevent opening of the mitochondrial permeability transition pore.[215] Three studies[192,193,207] have suggested that perhaps the UPS plays a role in IPC by facilitating some of the associated pre- or postischemic signaling changes.

Since the UPS may become dysfunctional during ischemia, by necessity IPC must in some way preserve postischemic proteasome function. As was the case with the ischemia studies, earlier studies of IPC protocols to protect the brain from ischemia/reperfusion had already examined this. These studies showed less accumulation of protein aggregates and proapoptotic proteins in the postischemic brain.[216–218] One of the earliest studies of the heart suggested that pharmacologic preconditioning of the myocardium with nicorandil affords some protection of postischemic function of the proteasome.[185] Nicorandil is an agent thought to open the inward mitochondrial $K_{ATP}$ channels[219] and thus mimic the effects of IPC. While all three of these recent studies provide convincing evidence that IPC preserves postischemic dysfunction of the UPS including improved peptidase activity and less accumulation of ubiquitinated or misfolded proteins,[192,193,207] they differ in their proposed mechanisms.

### 1. Preserved Function of the UPS Improves Degradation of δ Protein Kinase C During Reperfusion

Inhibition of the pro-death kinase, δ Protein Kinase C (δPKC), is reported to protect the heart from ischemic injury.[220] The pro-survival kinase, εPKC, is activated and translocated to cardiac mitochondria as a result of IPC.[221]

Churchill *et al.*[193] have shown that IPC improves postischemic UPS function and alters the ratios of these kinases to favor tissue survival. They have proposed that the UPS regulates the ratio between these two kinases because a proteasome inhibitor prevents the effects of IPC on postischemic function and the ratios of δPKC/εPKC. This is an attractive hypothesis; however, IPC alters the levels of many pro-survival and pro-death proteins thought to be regulated by the UPS including PTEN,[222,223] IκB,[187] and Bax.[207] As many diverse pathways are regulated by the UPS, however important it may be, it is likely that UPS-mediated changes in δPKC account for only a portion of the protective effects of IPC.

#### 2. Increased PKA-mediated activation of proteasome

Given the regulation of the cardiac proteasome by PKA (Section II.B above), this theory is highly credible. PKA phosphorylates several subunits of the 20S proteasome and the 19S regulatory particle[9,13] which activates proteasome peptidase activity. Asai *et al.*[192] present convincing evidence that PKA can enhance docking of the 20S proteasome and 19S regulatory particle(s), and suggest that IPC results in transient preischemic increases in PKA, which enhance assembly of the intact 26S proteasome. This then carries over to the postischemic period to explain the higher peptidase activity of the proteasome observed in the immediate postischemic period. Unfortunately, because assembly of intact proteasome at this time point was not assessed, the relationship between the improved function and assembly is not clear.

#### 3. Decreased postischemic oxidation of 19S regulatory particle subunits

As described earlier, the Rpt5 subunit, one of the ATPase subunits of the 19S regulatory particle, was recently described as subject to oxidative modification during myocardial ischemia.[207] One of the observations leading up to an examination of the vulnerability of 19S regulatory particle subunits to oxidative damage was the preferential effect of ischemia/reperfusion on the ATP-dependent activity of the proteasome.[193,207] Divald *et al.*[207] report decreased postischemic carbonylation of the Rpt5 subunit when the heart was subjected to an IPC protocol. As the Rpt5 subunit plays pivotal roles in the attachment of the "base" of the 19S regulatory particle to 20S proteasome α-rings[224] and the binding of the 19S particle "lid" to the base,[225] it is conceivable that by decreasing oxidation, IPC improves docking of these two complexes. However, as this was not examined, it is not clear whether the improved peptidase activity is related to the decreased oxidation of just this one subunit. Given the complexity of IPC, it is likely that no one single mechanism accounts for the protective effect of this procedure on postischemic proteasome function.

### D. Proteasome Inhibition in Myocardial Ischemia: for Better or for Worse?

The same controversy that surrounds the proposed use of proteasome inhibitors to treat atherosclerosis applies to myocardial ischemia. There have been a handful of studies in the literature suggesting that this strategy may be beneficial. As the use of proteasome inhibitors in myocardial ischemia has recently been discussed in depth,[226] little time will be spent on it here. Suffice it to say that given the proteasome dysfunction that may be present following myocardial ischemia, this strategy raises grave concerns[227] particularly in light of several studies[228–230] reporting unexpected cardiac toxicity associated with administration of the proteasome inhibitor bortizomeb (Velcade®) for the treatment of multiple myeloma. This is not to suggest that proteasome inhibitors could not be of value in treating some of the problems associated with myocardial ischemia. The problems with the first generation of proteasome inhibitors centers around their nonspecificity. However, it is conceivable that second-generation proteasome inhibitors with more selective effects, such as the immunoproteasome inhibitor ONYX-0914,[231] may be of value due to their potential anti-inflammatory properties. Likewise, we foresee altering degradation of specific proteins through targeting of specific ubiquitin protein ligases as a viable therapeutic intervention in the future.

## VIII. Concluding Statement

The last 15 years have witnessed a veritable explosion in research surrounding the various roles and functions of the UPS in cellular physiology and pathophysiology. While often described as simply a means for the removal of unwanted, damaged, or otherwise unneeded proteins, nothing could be further from the truth. In reality, in removing proteins the UPS plays a critical role in regulating many of the intracellular processes necessary for cell function and survival. Given the absolute requirement for the UPS, it was only a matter of time before this system was examined for its potential roles in cardiovascular physiology and pathophysiology. This chapter presented a synopsis of this research. Over the past decade, there have been many advances in our understanding of how the UPS functions in vascular and myocardial tissues and we have come to realize that many of the cardiovascular pathologies are associated with some degree of deviation of UPS function from the norm (increased or decreased). In many of these pathologies, it is not clear whether UPS dysfunction caused the primary disease process or is secondary to it, but it is clear that once the UPS becomes dysfunctional, deterioration is enhanced. However, in the familial proteinopathies, the dysfunctional proteasome plays a role not only in progression of the disease, but also, at least in experimental models, in the genesis of the proteinopathy. Although there is still a relative lack of knowledge, efforts to mitigate cardiovascular pathologies by modifying the function of the UPS are promising as recent studies with genetic models and development of activator and second-generation inhibitors have produced encouraging results. It is hoped that this chapter will spur research to further our knowledge of this system and lead to the development of new UPS-based therapeutic modalities for the mitigation of cardiovascular disease.

## Acknowledgments

This work was supported by National Institute of Health grants HL 92954, AG 31750 (to A. L.), HL 68936 (to S. R. P. and X. W.), HL72166 and HL85629 (to X. W.), GM061728 and HL065619 (to C. P.), and American Heart Asociation grant 0740025N (to X. W.).

## References

1. Gomes AV, Zong C, Edmondson RD, Berhane BT, Wang GW, Le S, et al. The murine cardiac 26S proteasome: an organelle awaiting exploration. *Ann N Y Acad Sci* 2005;**1047**:197–207.
2. Gomes AV, Zong C, Edmondson RD, Li X, Stefani E, Zhang J, et al. Mapping the murine cardiac 26S proteasome complexes. *Circ Res* 2006;**99**:362–71.
3. Drews O, Wildgruber R, Zong C, Sukop U, Nissum M, Weber G, et al. Mammalian proteasome subpopulations with distinct molecular compositions and proteolytic activities. *Mol Cell Proteomics* 2007;**6**:2021–31.
4. Gomes AV, Young GW, Wang Y, Zong C, Eghbali M, Drews O, et al. Contrasting proteome biology and functional heterogeneity of the 20 S proteasome complexes in mammalian tissues. *Mol Cell Proteomics* 2009;**8**:302–15.
5. Young GW, Wang Y, Ping P. Understanding proteasome assembly and regulation: importance to cardiovascular medicine. *Trends Cardiovasc Med* 2008;**18**:93–8.
6. Powell SR. The cardiac 26S proteasome. Regulating the regulator. *Circ Res* 2006;**99**:342–5.
7. Wang D, Zong C, Koag MC, Wang Y, Drews O, Fang C, et al. Proteome dynamics and proteome function of cardiac 19S proteasomes. *Mol Cell Proteomics* 2011;**10**:M110.
8. Zhang F, Su K, Yang X, Bowe DB, Paterson AJ, Kudlow JE. O-GlcNAc modification is an endogenous inhibitor of the proteasome. *Cell* 2003;**115**:715–25.

9. Zhang F, Hu Y, Huang P, Toleman CA, Paterson AJ, Kudlow JE. Proteasome function is regulated by cyclic AMP-dependent protein kinase through phosphorylation of Rpt6. *J Biol Chem* 2007;**282**:22460–71.
10. Drews O, Tsukamoto O, Liem D, Streicher J, Wang Y, Ping P. Differential regulation of proteasome function in isoproterenol-induced cardiac hypertrophy. *Circ Res* 2010;**107**: 1094–101.
11. Lu H, Zong C, Wang Y, Young GW, Deng N, Souda P, et al. Revealing the dynamics of the 20 S proteasome phosphoproteome: a combined CID and electron transfer dissociation approach. *Mol Cell Proteomics* 2008;**7**:2073–89.
12. Zong C, Young GW, Wang Y, Lu H, Deng N, Drews O, et al. Two-dimensional electrophoresis-based characterization of post-translational modifications of mammalian 20S proteasome complexes. *Proteomics* 2008;**8**:5025–37.
13. Zong C, Gomes AV, Drews O, Li X, Young GW, Berhane B, et al. Regulation of murine cardiac 20S proteasomes: role of associating partners. *Circ Res* 2006;**99**:372–80.
14. Hansson GK. Inflammation, atherosclerosis, and coronary artery disease. *N Engl J Med* 2005;**352**:1685–95.
15. Weber C, Zernecke A, Libby P. The multifaceted contributions of leukocyte subsets to atherosclerosis: lessons from mouse models. *Nat Rev Immunol* 2008;**8**:802–15.
16. Stangl V, Lorenz M, Meiners S, Ludwig A, Bartsch C, Moobed M, et al. Long-term up-regulation of eNOS and improvement of endothelial function by inhibition of the ubiquitin-proteasome pathway. *FASEB J* 2004;**18**:272–9.
17. Wei Q, Xia Y. Proteasome inhibition down-regulates endothelial nitric-oxide synthase phosphorylation and function. *J Biol Chem* 2006;**281**:21652–9.
18. Herrmann J, Saguner AM, Versari D, Peterson TE, Chade A, Olson M, et al. Chronic proteasome inhibition contributes to coronary atherosclerosis. *Circ Res* 2007;**101**:865–74.
19. Xia T, Dimitropoulou C, Zeng J, Antonova GN, Snead C, Venema RC, et al. Chaperone-dependent E3 ligase CHIP ubiquitinates and mediates proteasomal degradation of soluble guanylyl cyclase. *Am J Physiol Cell Physiol* 2007;**293**:H3080–7.
20. Xu J, Wu Y, Song P, Zhang M, Wang S, Zou MH. Proteasome-dependent degradation of guanosine 5′-triphosphate cyclohydrolase I causes tetrahydrobiopterin deficiency in diabetes mellitus. *Circulation* 2007;**116**:944–53.
21. Whitsett J, Picklo Sr. MJ, Vasquez-Vivar J. 4-Hydroxy-2-nonenal increases superoxide anion radical in endothelial cells via stimulated GTP cyclohydrolase proteasomal degradation. *Arterioscler Thromb Vasc Biol* 2007;**27**:2340–7.
22. Forstermann U, Munzel T. Endothelial nitric oxide synthase in vascular disease: from marvel to menace. *Circulation* 2006;**113**:1708–14.
23. Li W, Kong AN. Molecular mechanisms of Nrf2-mediated antioxidant response. *Mol Carcinog* 2009;**48**:91–104.
24. Park HM, Kim JA, Kwak MK. Protection against amyloid beta cytotoxicity by sulforaphane: role of the proteasome. *Arch Pharm Res* 2009;**32**:109–15.
25. Dreger H, Westphal K, Wilck N, Baumann G, Stangl V, Stangl K, et al. Protection of vascular cells from oxidative stress by proteasome inhibition depends on Nrf2. *Cardiovasc Res* 2010;**85**:395–403.
26. Bieler S, Meiners S, Stangl V, Pohl T, Stangl K. Comprehensive proteomic and transcriptomic analysis reveals early induction of a protective anti-oxidative stress response by low-dose proteasome inhibition. *Proteomics* 2009;**9**:3257–67.
27. Farout L, Mary J, Vinh J, Szweda LI, Friguet B. Inactivation of the proteasome by 4-hydroxy-2-nonenal is site specific and dependant on 20S proteasome subtypes. *Arch Biochem Biophys* 2006;**453**:135–42.

28. Vieira O, Escargueil-Blanc I, Jurgens G, Borner C, Almeida L, Salvayre R, et al. Oxidized LDLs alter the activity of the ubiquitin-proteasome pathway: potential role in oxidized LDL-induced apoptosis. *FASEB J* 2000;**14**:532–42.
29. Reuter S, Gupta SC, Chaturvedi MM, Aggarwal BB. Oxidative stress, inflammation, and cancer: how are they linked? *Free Radic Biol Med* 2010;**49**:1603–16.
30. Herrmann J, Ciechanover A, Lerman LO, Lerman A. The ubiquitin-proteasome system in cardiovascular diseases-a hypothesis extended. *Cardiovasc Res* 2004;**61**:11–21.
31. Herrmann J, Lerman LO, Mukhopadhyay D, Napoli C, Lerman A. Angiogenesis in atherogenesis. *Arterioscler Thromb Vasc Biol* 2006;**26**:1948–57.
32. Kietzmann T, Gorlach A. Reactive oxygen species in the control of hypoxia-inducible factor-mediated gene expression. *Semin Cell Dev Biol* 2005;**16**:474–86.
33. Bradshaw EL, Li XA, Guerin T, Everson WV, Wilson ME, Bruce-Keller AJ, et al. Nucleoside reverse transcriptase inhibitors prevent HIV protease inhibitor-induced atherosclerosis by ubiquitination and degradation of protein kinase C. *Am J Physiol Cell Physiol* 2006;**291**: C1271–8.
34. Ogura M, Ayaori M, Terao Y, Hisada T, Iizuka M, Takiguchi S, et al. Proteasomal inhibition promotes ATP-binding cassette transporter A1 (ABCA1) and ABCG1 expression and cholesterol efflux from macrophages in vitro and in vivo. *Arterioscler Thromb Vasc Biol* 2011;**31**:1980–7.
35. Masuda Y, Itabe H, Odaki M, Hama K, Fujimoto Y, Mori M, et al. ADRP/adipophilin is degraded through the proteasome-dependent pathway during regression of lipid-storing cells. *J Lipid Res* 2006;**47**:87–98.
36. Kikuchi J, Furukawa Y, Kubo N, Tokura A, Hayashi N, Nakamura M, et al. Induction of ubiquitin-conjugating enzyme by aggregated low density lipoprotein in human macrophages and its implications for atherosclerosis. *Arterioscler Thromb Vasc Biol* 2000;**20**:128–34.
37. Xie P, Fan Y, Zhang H, Zhang Y, She M, Gu D, et al. CHIP represses myocardin-induced smooth muscle cell differentiation via ubiquitin-mediated proteasomal degradation. *Mol Cell Biol* 2009;**29**:2398–408.
38. Meiners S, Laule M, Rother W, Guenther C, Prauka I, Muschick P, et al. Ubiquitin-proteasome pathway as a new target for the prevention of restenosis. *Circulation* 2002;**105**:483–9.
39. Barringhaus KG, Matsumura ME. The proteasome inhibitor lactacystin attenuates growth and migration of vascular smooth muscle cells and limits the response to arterial injury. *Exp Clin Cardiol* 2007;**12**:119–24.
40. Ying WZ, Zhang HG, Sanders PW. EGF receptor activity modulates apoptosis induced by inhibition of the proteasome of vascular smooth muscle cells. *J Am Soc Nephrol* 2007;**18**:131–42.
41. Kapadia MR, Eng JW, Jiang Q, Stoyanovsky DA, Kibbe MR. Nitric oxide regulates the 26S proteasome in vascular smooth muscle cells. *Nitric Oxide* 2009;**20**:279–88.
42. Katsuyama K, Shichiri M, Marumo F, Hirata Y. NO inhibits cytokine-induced iNOS expression and NF-kappaB activation by interfering with phosphorylation and degradation of IkappaB-alpha. *Arterioscler Thromb Vasc Biol* 1998;**18**:1796–802.
43. Takami Y, Nakagami H, Morishita R, Katsuya T, Cui TX, Ichikawa T, et al. Ubiquitin carboxyl-terminal hydrolase L1, a novel deubiquitinating enzyme in the vasculature, attenuates NF-kappaB activation. *Arterioscler Thromb Vasc Biol* 2007;**27**:2184–90.
44. Takami Y, Nakagami H, Morishita R, Katsuya T, Hayashi H, Mori M, et al. Potential role of CYLD (Cylindromatosis) as a deubiquitinating enzyme in vascular cells. *Am J Pathol* 2008;**172**:818–29.
45. Duwel M, Welteke V, Oeckinghaus A, Baens M, Kloo B, Ferch U, et al. A20 negatively regulates T cell receptor signaling to NF-kappaB by cleaving Malt1 ubiquitin chains. *J Immunol* 2009;**182**:7718–28.

46. Weyand CM, Younge BR, Goronzy JJ. T cells in arteritis and atherosclerosis. *Curr Opin Lipidol* 2008;**19**:469–77.
47. Shrikhande GV, Scali ST, da Silva CG, Damrauer SM, Csizmadia E, Putheti P, et al. O-glycosylation regulates ubiquitination and degradation of the anti-inflammatory protein A20 to accelerate atherosclerosis in diabetic ApoE-null mice. *PLoS One* 2010;**5**:e14240.
48. Lobry C, Lopez T, Israel A, Weil R. Negative feedback loop in T cell activation through IkappaB kinase-induced phosphorylation and degradation of Bcl10. *Proc Natl Acad Sci USA* 2007;**104**:908–13.
49. Berges C, Haberstock H, Fuchs D, Miltz M, Sadeghi M, Opelz G, et al. Proteasome inhibition suppresses essential immune functions of human CD4+ T cells. *Immunology* 2008;**124**:234–46.
50. Kawahara I, Kitagawa N, Tsutsumi K, Nagata I, Hayashi T, Koji T. The expression of vascular dendritic cells in human atherosclerotic carotid plaques. *Hum Pathol* 2007;**38**:1378–85.
51. Naujokat C, Berges C, Hoh A, Wieczorek H, Fuchs D, Ovens J, et al. Proteasomal chymotrypsin-like peptidase activity is required for essential functions of human monocyte-derived dendritic cells. *Immunology* 2007;**120**:120–32.
52. Straube C, Wehner R, Wendisch M, Bornhauser M, Bachmann M, Rieber EP, et al. Bortezomib significantly impairs the immunostimulatory capacity of human myeloid blood dendritic cells. *Leukemia* 2007;**21**:1464–71.
53. Nencioni A, Garuti A, Schwarzenberg K, Cirmena G, Dal Bello G, Rocco I, et al. Proteasome inhibitor-induced apoptosis in human monocyte-derived dendritic cells. *Eur J Immunol* 2006;**36**:681–9.
54. van der Heijden JW, Oerlemans R, Lems WF, Scheper RJ, Dijkmans BA, Jansen G. The proteasome inhibitor bortezomib inhibits the release of NFkappaB-inducible cytokines and induces apoptosis of activated T cells from rheumatoid arthritis patients. *Clin Exp Rheumatol* 2009;**27**:92–8.
55. Vaseva AV, Moll UM. The mitochondrial p53 pathway. *Biochim Biophys Acta* 2009;**1787**:414–20.
56. Meier P, Golshayan D, Blanc E, Pascual M, Burnier M. Oxidized LDL modulates apoptosis of regulatory T cells in patients with ESRD. *J Am Soc Nephrol* 2009;**20**:1368–84.
57. Taleb S, Tedgui A, Mallat Z. Regulatory T-cell immunity and its relevance to atherosclerosis. *J Intern Med* 2008;**263**:489–99.
58. Yang Z, Gagarin D, St Laurent III G, Hammell N, Toma I, Hu CA, et al. Cardiovascular inflammation and lesion cell apoptosis: a novel connection via the interferon-inducible immunoproteasome. *Arterioscler Thromb Vasc Biol* 2009;**29**:1213–9.
59. Yang-Yen HF. Mcl-1: a highly regulated cell death and survival controller. *J Biomed Sci* 2006;**13**:201–4.
60. Cuconati A, Mukherjee C, Perez D, White E. DNA damage response and MCL-1 destruction initiate apoptosis in adenovirus-infected cells. *Genes Dev* 2003;**17**:2922–32.
61. Martinet W, De Meyer GR. Autophagy in atherosclerosis. *Curr Atheroscler Rep* 2008;**10**:216–23.
62. Ding WX, Ni HM, Gao W, Yoshimori T, Stolz DB, Ron D, et al. Linking of autophagy to ubiquitin-proteasome system is important for the regulation of endoplasmic reticulum stress and cell viability. *Am J Pathol* 2007;**171**:513–24.
63. Van Herck JL, De Meyer GR, Martinet W, Bult H, Vrints CJ, Herman AG. Proteasome inhibitor bortezomib promotes a rupture-prone plaque phenotype in ApoE-deficient mice. *Basic Res Cardiol* 2010;**105**:39–50.
64. Versari D, Herrmann J, Gossl M, Mannheim D, Sattler K, Meyer FB, et al. Dysregulation of the ubiquitin-proteasome system in human carotid atherosclerosis. *Arterioscler Thromb Vasc Biol* 2006;**26**:2132–9.

65. Herrmann J, Soares SM, Lerman LO, Lerman A. Potential role of the ubiquitin-proteasome system in atherosclerosis: aspects of a protein quality disease. *J Am Coll Cardiol* 2008;**51**: 2003–10.
66. Herrmann J, Edwards WD, Holmes Jr. DR, Shogren KL, Lerman LO, Ciechanover A, et al. Increased ubiquitin immunoreactivity in unstable atherosclerotic plaques associated with acute coronary syndromes. *J Am Coll Cardiol* 2002;**40**:1919–27.
67. Marfella R, D'Amico M, Di Filippo C, Baldi A, Siniscalchi M, Sasso FC, et al. Increased activity of the ubiquitin-proteasome system in patients with symptomatic carotid disease is associated with enhanced inflammation and may destabilize the atherosclerotic plaque: effects of rosiglitazone treatment. *J Am Coll Cardiol* 2006;**47**:2444–55.
68. Marfella R, D'Amico M, Esposito K, Baldi A, Di Filippo C, Siniscalchi M, et al. The ubiquitin-proteasome system and inflammatory activity in diabetic atherosclerotic plaques: effects of rosiglitazone treatment. *Diabetes* 2006;**55**:622–32.
69. Marfella R, Di Filippo C, Baldi A, Siniscalchi M, Sasso FC, Crescenzi B, et al. The vascular smooth muscle cells apoptosis in asymptomatic diabetic carotid plaques: role of glycemic control. *J Am Coll Cardiol* 2006;**47**:2118–20.
70. Marfella R, Di Filippo C, Laieta MT, Vestini R, Barbieri M, Sangiulo P, et al. Effects of ubiquitin-proteasome system deregulation on the vascular senescence and atherosclerosis process in elderly patients. *J Gerontol A Biol Sci Med Sci* 2008;**63**:200–3.
71. Friguet B, Bulteau AL, Chondrogianni N, Conconi M, Petropoulos I. Protein degradation by the proteasome and its implications in aging. *Ann N Y Acad Sci* 2000;**908**:143–54.
72. Bulteau AL, Szweda LI, Friguet B. Age-dependent declines in proteasome activity in the heart. *Arch Biochem Biophys* 2002;**397**:298–304.
73. Tan C, Li Y, Tan X, Pan H, Huang W. Inhibition of the ubiquitin-proteasome system: a new avenue for atherosclerosis. *Clin Chem Lab Med* 2006;**44**:1218–25.
74. Feng B, Zhang Y, Mu J, Ye Z, Zeng W, Qi W, et al. Preventive effect of a proteasome inhibitor on the formation of accelerated atherosclerosis in rabbits with uremia. *J Cardiovasc Pharmacol* 2010;**55**:129–38.
75. Meiners S, Ludwig A, Stangl V, Stangl K. Proteasome inhibitors: poisons and remedies. *Med Res Rev* 2008;**28**:309–27.
76. Bao J, Sato K, Li M, Gao Y, Abid R, Aird W, et al. PR-39 and PR-11 peptides inhibit ischemia-reperfusion injury by blocking proteasome-mediated I kappa B alpha degradation. *Am J Physiol Heart Circ Physiol* 2001;**281**:H2612–8.
77. Gao Y, Lecker S, Post MJ, Hietaranta AJ, Li J, Volk R, et al. Inhibition of ubiquitin-proteasome pathway-mediated I kappa B alpha degradation by a naturally occurring antibacterial peptide. *J Clin Invest* 2000;**106**:439–48.
78. Luss H, Schmitz W, Neumann J. A proteasome inhibitor confers cardioprotection. *Cardiovasc Res* 2002;**54**:140–51.
79. Bodine SC, Latres E, Baumhueter S, Lai VK, Nunez L, Clarke BA, et al. Identification of ubiquitin ligases required for skeletal muscle atrophy. *Science* 2001;**294**:1704–8.
80. Gomes MD, Lecker SH, Jagoe RT, Navon A, Goldberg AL. Atrogin-1, a muscle-specific F-box protein highly expressed during muscle atrophy. *Proc Natl Acad Sci USA* 2001;**98**:14440–5.
81. Ni YG, Berenji K, Wang N, Oh M, Sachan N, Dey A, et al. Foxo transcription factors blunt cardiac hypertrophy by inhibiting calcineurin signaling. *Circulation* 2006;**114**:1159–68.
82. Sandri M, Sandri C, Gilbert A, Skurk C, Calabria E, Picard A, et al. Foxo transcription factors induce the atrophy-related ubiquitin ligase atrogin-1 and cause skeletal muscle atrophy. *Cell* 2004;**117**:399–412.
83. Pye J, Ardeshirpour F, McCain A, Bellinger DA, Merricks E, Adams J, et al. Proteasome inhibition ablates activation of NF-kappa B in myocardial reperfusion and reduces reperfusion injury. *Am J Physiol Heart Circ Physiol* 2003;**284**:H919–26.

84. Wilkins BJ, Dai YS, Bueno OF, Parsons SA, Xu J, Plank DM, et al. Calcineurin/NFAT coupling participates in pathological, but not physiological, cardiac hypertrophy. *Circ Res* 2004;**94**: 110–8.
85. Li HH, Kedar V, Zhang C, McDonough H, Arya R, Wang DZ, et al. Atrogin-1/muscle atrophy F-box inhibits calcineurin-dependent cardiac hypertrophy by participating in an SCF ubiquitin ligase complex. *J Clin Invest* 2004;**114**:1058–71.
86. Stansfield WE, Moss NC, Willis MS, Tang R, Selzman CH. Proteasome inhibition attenuates infarct size and preserves cardiac function in a murine model of myocardial ischemia-reperfusion injury. *Ann Thorac Surg* 2007;**84**:120–5.
87. Johnson SA, Hunter T. Kinomics: methods for deciphering the kinome. *Nat Methods* 2005;**2**:17–25.
88. Li HH, Willis MS, Lockyer P, Miller N, McDonough H, Glass DJ, et al. Atrogin-1 inhibits Akt-dependent cardiac hypertrophy in mice via ubiquitin-dependent coactivation of Forkhead proteins. *J Clin Invest* 2007;**117**:3211–23.
89. Coleman PS, Parmacek MS, Lesch M, Samarel AM. Protein synthesis and degradation during regression of thyroxine-induced cardiac hypertrophy. *J Mol Cell Cardiol* 1989;**21**:911–25.
90. Wilkinson KD, Ventii KH, Friedrich KL, Mullally JE. The ubiquitin signal: assembly, recognition and termination. Symposium on ubiquitin and signaling. *EMBO Rep* 2005;**6**:815–20.
91. Centner T, Yano J, Kimura E, McElhinny AS, Pelin K, Witt CC, et al. Identification of muscle specific ring finger proteins as potential regulators of the titin kinase domain. *J Mol Biol* 2001;**306**:717–26.
92. McElhinny AS, Kakinuma K, Sorimachi H, Labeit S, Gregorio CC. Muscle-specific RING finger-1 interacts with titin to regulate sarcomeric M-line and thick filament structure and may have nuclear functions via its interaction with glucocorticoid modulatory element binding protein-1. *J Cell Biol* 2002;**157**:125–36.
93. Spencer JA, Eliazer S, Ilaria Jr. RL, Richardson JA, Olson EN. Regulation of microtubule dynamics and myogenic differentiation by MURF, a striated muscle RING-finger protein. *J Cell Biol* 2000;**150**:771–84.
94. Adams V, Linke A, Wisloff U, Doring C, Erbs S, Krankel N, et al. Myocardial expression of Murf-1 and MAFbx after induction of chronic heart failure: effect on myocardial contractility. *Cardiovasc Res* 2007;**73**:120–9.
95. Razeghi P, Baskin KK, Sharma S, Young ME, Stepkowski S, Essop MF, et al. Atrophy, hypertrophy, and hypoxemia induce transcriptional regulators of the ubiquitin proteasome system in the rat heart. *Biochem Biophys Res Commun* 2006;**342**:361–4.
96. Parmacek MS, Magid NM, Lesch M, Decker RS, Samarel AM. Cardiac protein synthesis and degradation during thyroxine-induced left ventricular hypertrophy. *Am J Physiol* 1986;**251**: C727–36.
97. Willis MS, Ike C, Li L, Wang DZ, Glass DJ, Patterson C. Muscle ring finger 1, but not muscle ring finger 2, regulates cardiac hypertrophy in vivo. *Circ Res* 2007;**100**:456–9.
98. Cohen S, Brault JJ, Gygi SP, Glass DJ, Valenzuela DM, Gartner C, et al. During muscle atrophy, thick, but not thin, filament components are degraded by MuRF1-dependent ubiquitylation. *J Cell Biol* 2009;**185**:1083–95.
99. Frey N, Olson EN. Cardiac hypertrophy: the good, the bad, and the ugly. *Annu Rev Physiol* 2003;**65**:45–79.
100. Arya R, Kedar V, Hwang JR, McDonough H, Li HH, Taylor J, et al. Muscle ring finger protein-1 inhibits PKC{epsilon} activation and prevents cardiomyocyte hypertrophy. *J Cell Biol* 2004;**167**:1147–59.
101. Willis MS, Rojas M, Li L, Selzman CH, Tang RH, Stansfield WE, et al. Muscle ring finger 1 mediates cardiac atrophy in vivo. *Am J Physiol Heart Circ Physiol* 2009;**296**:H997–H1006.

102. Zhang C, Xu Z, He XR, Michael LH, Patterson C. CHIP, a cochaperone/ubiquitin ligase that regulates protein quality control, is required for maximal cardioprotection after myocardial infarction in mice. *Am J Physiol Heart Circ Physiol* 2005;**288**:H2836–42.
103. Hoshijima M. Mechanical stress-strain sensors embedded in cardiac cytoskeleton: Z disk, titin, and associated structures. *Am J Physiol Heart Circ Physiol* 2006;**290**:H1313–25.
104. Gregorio CC, Perry CN, McElhinny AS. Functional properties of the titin/connectin-associated proteins, the muscle-specific RING finger proteins (MURFs), in striated muscle. *J Muscle Res Cell Motil* 2005;**26**:389–400.
105. Mrosek M, Labeit D, Witt S, Heerklotz H, von Castelmur E, Labeit S, et al. Molecular determinants for the recruitment of the ubiquitin-ligase MuRF-1 onto M-line titin. *FASEB J* 2007;**21**:1383–92.
106. Foo RS, Chan LK, Kitsis RN, Bennett MR. Ubiquitination and degradation of the anti-apoptotic protein ARC by MDM2. *J Biol Chem* 2007;**282**:5529–35.
107. Xiong S, Van Pelt CS, Elizondo-Fraire AC, Fernandez-Garcia B, Lozano G. Loss of Mdm4 results in p53-dependent dilated cardiomyopathy. *Circulation* 2007;**115**:2925–30.
108. Fielitz J, van Rooij E, Spencer JA, Shelton JM, Latif S, van der Nagel R, et al. Loss of muscle-specific RING-finger 3 predisposes the heart to cardiac rupture after myocardial infarction. *Proc Natl Acad Sci USA* 2007;**104**:4377–82.
109. Fielitz J, Kim MS, Shelton JM, Latif S, Spencer JA, Glass DJ, et al. Myosin accumulation and striated muscle myopathy result from the loss of muscle RING finger 1 and 3. *J Clin Invest* 2007;**117**:2486–95.
110. Maron BJ. Triggers for sudden cardiac death in the athlete. *Cardiol Clin* 1996;**14**:195–210.
111. Bonne G, Carrier L, Richard P, Hainque B, Schwartz K. Familial hypertrophic cardiomyopathy: from mutations to functional defects. *Circ Res* 1998;**83**:580–93.
112. Richard P, Charron P, Carrier L, Ledeuil C, Cheav T, Pichereau C, et al. Hypertrophic cardiomyopathy: distribution of disease genes, spectrum of mutations, and implications for a molecular diagnosis strategy. *Circulation* 2003;**107**:2227–32.
113. Wang X, Robbins J. Heart failure and protein quality control. *Circ Res* 2006;**99**:1315–28.
114. Flavigny J, Robert P, Camelin JC, Schwartz K, Carrier L, Berrebi-Bertrand I. Biomolecular interactions between human recombinant beta-MyHC and cMyBP-Cs implicated in familial hypertrophic cardiomyopathy. *Cardiovasc Res* 2003;**60**:388–96.
115. Freiburg A, Gautel M. A molecular map of the interactions between titin and myosin-binding protein C. Implications for sarcomeric assembly in familial hypertrophic cardiomyopathy. *Eur J Biochem* 1996;**235**:317–23.
116. Gruen M, Gautel M. Mutations in beta-myosin S2 that cause familial hypertrophic cardiomyopathy (FHC) abolish the interaction with the regulatory domain of myosin-binding protein-C. *J Mol Biol* 1999;**286**:933–49.
117. Sarikas A, Carrier L, Schenke C, Doll D, Flavigny J, Lindenberg KS, et al. Impairment of the ubiquitin-proteasome system by truncated cardiac myosin binding protein C mutants. *Cardiovasc Res* 2005;**66**:33–44.
118. Squire JM, Luther PK, Knupp C. Structural evidence for the interaction of C-protein (MyBP-C) with actin and sequence identification of a possible actin-binding domain. *J Mol Biol* 2003;**331**:713–24.
119. Moolman JA, Reith S, Uhl K, Bailey S, Gautel M, Jeschke B, et al. A newly created splice donor site in exon 25 of the MyBP-C gene is responsible for inherited hypertrophic cardiomyopathy with incomplete disease penetrance. *Circulation* 2000;**101**:1396–402.
120. Wang X, Su H, Ranek MJ. Protein quality control and degradation in cardiomyocytes. *J Mol Cell Cardiol* 2008;**45**:11–27.
121. Ravid T, Hochstrasser M. Diversity of degradation signals in the ubiquitin-proteasome system. *Nat Rev Mol Cell Biol* 2008;**9**:679–90.

122. Li J, Horak KM, Su H, Sanbe A, Robbins J, Wang X. Enhancement of proteasomal function protects against cardiac proteinopathy and ischemia/reperfusion injury in mice. *J Clin Invest* 2011;**121**:3689–700.
123. Wang X, Li J, Zheng H, Su H, Powell SR. Proteasome functional insufficiency in cardiac pathogenesis. *Am J Physiol Heart Circ Physiol* 2011;**301**:H2207–19.
124. Bence NF, Sampat RM, Kopito RR. Impairment of the ubiquitin-proteasome system by protein aggregation. *Science* 2001;**292**:1552–5.
125. Dong X, Liu J, Zheng H, Glasford JW, Huang W, Chen QH, et al. In situ dynamically monitoring the proteolytic function of the ubiquitin-proteasome system in cultured cardiac myocytes. *Am J Physiol Heart Circ Physiol* 2004;**287**:H1417–25.
126. Kumarapeli RKA, Horak KM, Glasford JW, Li J, Chen Q, Liu J, et al. A novel transgenic mouse model reveals deregulation of the ubiquitin-proteasome system in the heart by doxorubicin. *FASEB J* 2005;**19**:2051–3.
127. Gilon T, Chomsky O, Kulka RG. Degradation signals for ubiquitin system proteolysis in Saccharomyces cerevisiae. *EMBO J* 1998;**17**:2759–66.
128. Gilon T, Chomsky O, Kulka RG. Degradation signals recognized by the Ubc6p-Ubc7p ubiquitin-conjugating enzyme pair. *Mol Cell Biol* 2000;**20**:7214–9.
129. Griffin TA, Nandi D, Cruz M, Fehling HJ, Kaer LV, Monaco JJ, et al. Immunoproteasome assembly: cooperative incorporation of interferon gamma (IFN-gamma)-inducible subunits. *J Exp Med* 1998;**187**:97–104.
130. Seifert U, Bialy LP, Ebstein F, Bech-Otschir D, Voigt A, Schroter F, et al. Immunoproteasomes preserve protein homeostasis upon interferon-induced oxidative stress. *Cell* 2010;**142**: 613–24.
131. Li J, Powell SR, Wang X. Enhancement of proteasome function by PA28alpha; overexpression protects against oxidative stress. *FASEB J* 2011;**25**:883–93.
132. McLendon PM, Robbins J. Desmin-related cardiomyopathy: an unfolding story. *Am J Physiol Heart Circ Physiol* 2011;**301**:H1220–8.
133. Wang X, Osinska H, Klevitsky R, Gerdes AM, Nieman M, Lorenz J, et al. Expression of R120G-alphaB-crystallin causes aberrant desmin and alphaB-crystallin aggregation and cardiomyopathy in mice. *Circ Res* 2001;**89**:84–91.
134. Wang X, Osinska H, Dorn GW, Nieman M, Lorenz JN, Gerdes AM, et al. Mouse model of desmin-related cardiomyopathy. *Circulation* 2001;**103**:2402–7.
135. Munoz-Marmol AM, Strasser G, Isamat M, Coulombe PA, Yang Y, Roca X, et al. A dysfunctional desmin mutation in a patient with severe generalized myopathy. *Proc Natl Acad Sci USA* 1998;**95**:11312–7.
136. Vicart P, Caron A, Guicheney P, Li Z, Prevost MC, Faure A, et al. A missense mutation in the alphaB-crystallin chaperone gene causes a desmin-related myopathy. *Nat Genet* 1998;**20**:92–5.
137. Chen Q, Liu JB, Horak KM, Zheng H, Kumarapeli AR, Li J, et al. Intrasarcoplasmic amyloidosis impairs proteolytic function of proteasomes in cardiomyocytes by compromising substrate uptake. *Circ Res* 2005;**97**:1018–26.
138. Liu J, Chen Q, Huang W, Horak KM, Zheng H, Mestril R, et al. Impairment of the ubiquitin-proteasome system in desminopathy mouse hearts. *FASEB J* 2006;**20**:362–4.
139. Liu J, Tang M, Mestril R, Wang X. Aberrant protein aggregation is essential for a mutant desmin to impair the proteolytic function of the ubiquitin-proteasome system in cardiomyocytes. *J Mol Cell Cardiol* 2006;**40**:451–4.
140. Tannous P, Zhu H, Nemchenko A, Berry JM, Johnstone JL, Shelton JM, et al. Intracellular protein aggregation is a proximal trigger of cardiomyocyte autophagy. *Circulation* 2008;**117**:3070–8.
141. Hedhli N, Depre C. Proteasome inhibitors and cardiac cell growth. *Cardiovasc Res* 2010;**85**: 321–9.

142. Tang M, Li J, Huang W, Su H, Liang Q, Tian Z, et al. Proteasome functional insufficiency activates the calcineurin-NFAT pathway in cardiomyocytes and promotes maladaptive remodelling of stressed mouse hearts. *Cardiovasc Res* 2010;**88**:424–33.
143. Gianni D, Li A, Tesco G, McKay KM, Moore J, Raygor K, et al. Protein aggregates and novel presenilin gene variants in idiopathic dilated cardiomyopathy. *Circulation* 2010;**121**:1216–26.
144. Sanbe A, Osinska H, Saffitz JE, Glabe CG, Kayed R, Maloyan A, et al. Desmin-related cardiomyopathy in transgenic mice: a cardiac amyloidosis. *Proc Natl Acad Sci USA* 2004;**101**:10132–6.
145. Sanbe A, Gulick J, Hanks MC, Liang Q, Osinska H, Robbins J. Reengineering inducible cardiac-specific transgenesis with an attenuated myosin heavy chain promoter. *Circ Res* 2003;**92**:609–16.
146. Marian AJ. Hypertrophic cardiomyopathy: from genetics to treatment. *Eur J Clin Invest* 2010;**40**:360–9.
147. Maloyan A, Gulick J, Glabe CG, Kayed R, Robbins J. Exercise reverses preamyloid oligomer and prolongs survival in alphaB-crystallin-based desmin-related cardiomyopathy. *Proc Natl Acad Sci USA* 2007;**104**:5995–6000.
148. Pattison JS, Sanbe A, Maloyan A, Osinska H, Klevitsky R, Robbins J. Cardiomyocyte expression of a polyglutamine preamyloid oligomer causes heart failure. *Circulation* 2008;**117**: 2743–51.
149. Sanbe A, Osinska H, Villa C, Gulick J, Klevitsky R, Glabe CG, et al. Reversal of amyloid-induced heart disease in desmin-related cardiomyopathy. *Proc Natl Acad Sci USA* 2005;**102**:13592–7.
150. Goldberg AL. On prions, proteasomes, and mad cows. *N Engl J Med* 2007;**357**:1150–2.
151. Kristiansen M, Deriziotis P, Dimcheff DE, Jackson GS, Ovaa H, Naumann H, et al. Disease-associated prion protein oligomers inhibit the 26S proteasome. *Mol Cell* 2007;**26**:175–88.
152. Bennett EJ, Bence NF, Jayakumar R, Kopito RR. Global impairment of the ubiquitin-proteasome system by nuclear or cytoplasmic protein aggregates precedes inclusion body formation. *Mol Cell* 2005;**17**:351–65.
153. Sanbe A, Yamauchi J, Miyamoto Y, Fujiwara Y, Murabe M, Tanoue A. Interruption of CryAB-amyloid oligomer formation by HSP22. *J Biol Chem* 2007;**282**:555–63.
154. Maloyan A, Sanbe A, Osinska H, Westfall M, Robinson D, Imahashi K, et al. Mitochondrial dysfunction and apoptosis underlie the pathogenic process in alpha-B-crystallin desmin-related cardiomyopathy. *Circulation* 2005;**112**:3451–61.
155. Tsukamoto O, Minamino T, Okada K, Shintani Y, Takashima S, Kato H, et al. Depression of proteasome activities during the progression of cardiac dysfunction in pressure-overloaded heart of mice. *Biochem Biophys Res Commun* 2006;**340**:1125–33.
156. Molkentin JD, Lu JR, Antos CL, Markham B, Richardson J, Robbins J, et al. A calcineurin-dependent transcriptional pathway for cardiac hypertrophy. *Cell* 1998;**93**:215–28.
157. Kumarapeli AR, Su H, Huang W, Tang M, Zheng H, Horak KM, et al. Alpha B-crystallin suppresses pressure overload cardiac hypertrophy. *Circ Res* 2008;**103**:1473–82.
158. Sanbe A, Daicho T, Mizutani R, Endo T, Miyauchi N, Yamauchi J, et al. Protective effect of geranylgeranylacetone via enhancement of HSPB8 induction in desmin-related cardiomyopathy. *PLoS One* 2009;**4**:e5351.
159. Zheng H, Tang M, Zheng Q, Kumarapeli AR, Horak KM, Tian Z, et al. Doxycycline attenuates protein aggregation in cardiomyocytes and improves survival of a mouse model of cardiac proteinopathy. *J Am Coll Cardiol* 2010;**56**:1418–26.
160. Hunt SA, Abraham WT, Chin MH, Feldman AM, Francis GS, Ganiats TG, et al. 2009 Focused update incorporated into the ACC/AHA 2005 Guidelines for the Diagnosis and Management of Heart Failure in Adults a Report of the American College of Cardiology Foundation/American Heart Association Task Force on Practice Guidelines Developed in

Collaboration With the International Society for Heart and Lung Transplantation. *J Am Coll Cardiol* 2009;**53**:e1–e90.
161. Depre C, Wang Q, Yan L, Hedhli N, Peter P, Chen L, et al. Activation of the cardiac proteasome during pressure overload promotes ventricular hypertrophy. *Circulation* 2006;**114**:1821–8.
162. Stansfield WE, Tang RH, Moss NC, Baldwin AS, Willis MS, Selzman CH. Proteasome inhibition promotes regression of left ventricular hypertrophy. *Am J Physiol Heart Circ Physiol* 2008;**294**:H645–50.
163. Meiners S, Dreger H, Fechner M, Bieler S, Rother W, Gunther C, et al. Suppression of cardiomyocyte hypertrophy by inhibition of the ubiquitin-proteasome system. *Hypertension* 2008;**51**:302–8.
164. Hedhli N, Lizano P, Hong C, Fritzky LF, Dhar SK, Liu H, et al. Proteasome inhibition decreases cardiac remodeling after initiation of pressure overload. *Am J Physiol Heart Circ Physiol* 2008;**295**:H1385–93.
165. Hedhli N, Wang L, Wang Q, Rashed E, Tian Y, Sui X, et al. Proteasome activation during cardiac hypertrophy by the chaperone H11 Kinase/Hsp22. *Cardiovasc Res* 2008;**77**: 497–505.
166. Urso ML, Clarkson PM, Hittel D, Hoffman EP, Thompson PD. Changes in ubiquitin proteasome pathway gene expression in skeletal muscle with exercise and statins. *Arterioscler Thromb Vasc Biol* 2005;**25**:2560–6.
167. Reid MB. Response of the ubiquitin-proteasome pathway to changes in muscle activity. *Am J Physiol Regul Integr Comp Physiol* 2005;**288**:R1423–31.
168. Nedergaard A, Vissing K, Overgaard K, Kjaer M, Schjerling P. Expression patterns of atrogenic and ubiquitin proteasome component genes with exercise: effect of different loading patterns and repeated exercise bouts. *J Appl Physiol* 2007;**103**:1513–22.
169. Murton AJ, Constantin D, Greenhaff PL. The involvement of the ubiquitin proteasome system in human skeletal muscle remodelling and atrophy. *Biochim Biophys Acta* 2008;**1782**:730–43.
170. Adams V, Linke A, Gielen S, Erbs S, Hambrecht R, Schuler G. Modulation of Murf-1 and MAFbx expression in the myocardium by physical exercise training. *Eur J Cardiovasc Prev Rehabil* 2008;**15**:293–9.
171. Usui S, Maejima Y, Pain J, Hong C, Cho J, Park JY, et al. Endogenous muscle atrophy F-box mediates pressure overload-induced cardiac hypertrophy through regulation of nuclear factor-kappaB. *Circ Res* 2011;**109**:161–71.
172. Hein S, Arnon E, Kostin S, Schonburg M, Elsasser A, Polyakova V, et al. Progression from compensated hypertrophy to failure in the pressure-overloaded human heart: structural deterioration and compensatory mechanisms. *Circulation* 2003;**107**:984–91.
173. Weekes J, Morrison K, Mullen A, Wait R, Barton P, Dunn MJ. Hyperubiquitination of proteins in dilated cardiomyopathy. *Proteomics* 2003;**3**:208–16.
174. Predmore JM, Wang P, Davis F, Bartolone S, Westfall MV, Dyke DB, et al. Ubiquitin proteasome dysfunction in human hypertrophic and dilated cardiomyopathies. *Circulation* 2010;**21**:997–1004.
175. Wohlschlaeger J, Sixt SU, Stoeppler T, Schmitz KJ, Levkau B, Tsagakis K, et al. Ventricular unloading is associated with increased 20s proteasome protein expression in the myocardium. *J Heart Lung Transplant* 2010;**29**:125–32.
176. Kassiotis C, Ballal K, Wellnitz K, Vela D, Gong M, Salazar R, et al. Markers of autophagy are downregulated in failing human heart after mechanical unloading. *Circulation* 2009;**120**: S191–7.
177. Zheng Q, Su H, Tian Z, Wang X. Proteasome malfunction activates macroautophagy in the heart. *Am J Cardiovasc Dis* 2011;**1**:214–26.

178. Voigt A, Bartel K, Egerer K, Trimpert C, Feist E, Gericke C, et al. Humoral anti-proteasomal autoimmunity in dilated cardiomyopathy. *Basic Res Cardiol* 2010;**105**:9–18.
179. Otsuka K, Terasaki F, Shimomura H, Tsukada B, Horii T, Isomura T, et al. Enhanced expression of the ubiquitin-proteasome system in the myocardium from patients with dilated cardiomyopathy referred for left ventriculoplasty: an immunohistochemical study with special reference to oxidative stress. *Heart Vessels* 2010;**25**:474–84.
180. Hayashi T, Takada K, Matsuda M. Post-transient ischemia increase in ubiquitin conjugates in the early reperfusion. *Neuroreport* 1992;**3**:519–20.
181. Kamikubo T, Hayashi T. Changes in proteasome activity following transient ischemia. *Neurochem Int* 1996;**28**:209–12.
182. Keller JN, Huang FF, Zhu H, Yu J, Ho YS, Kindy TS. Oxidative stress-associated impairment of proteasome activity during ischemia-reperfusion injury. *J Cereb Blood Flow Metab* 2000;**20**:1467–73.
183. Asai A, Tanahashi N, Qiu JH, Saito N, Chi S, Kawahara N, et al. Selective proteasomal dysfunction in the hippocampal CA1 region after transient forebrain ischemia. *J Cereb Blood Flow Metab* 2002;**22**:705–10.
184. Bulteau AL, Lundberg KC, Humphries KM, Sadek HA, Szweda PA, Friguet B, et al. Oxidative modification and inactivation of the proteasome during coronary occlusion/reperfusion. *J Biol Chem* 2001;**276**:30057–63.
185. Powell SR, Wang P, Katzeff HL, Shringarpure R, Teoh C, Khaliulin I, et al. Oxidized and ubiquitinated proteins may predict recovery of postischemic cardiac function. Essential role of the proteasome. *Antioxid Redox Signal* 2005;**7**:538–46.
186. Powell SR, Davies KJA, Divald A. Optimal determination of heart tissue 26S Proteasome activity requires maximal stimulating concentrations of ATP. *J Mol Cell Cardiol* 2007;**42**:265–9.
187. Gurusamy N, Goswami S, Malik G, Das DK. Oxidative injury induces selective rather than global inhibition of proteasomal activity. *J Mol Cell Cardiol* 2008;**44**:419–28.
188. Glickman MH, Ciechanover A. The ubiquitin-proteasome proteolytic pathway: destruction for the sake of construction. *Physiol Rev* 2002;**82**:373–428.
189. Horwitz AA, Navon A, Groll M, Smith DM, Reis C, Goldberg AL. ATP-induced structural transitions in PAN, the proteasome-regulatory ATPase complex in Archaea. *J Biol Chem* 2007;**282**:22921–9.
190. Huang H, Zhang X, Li S, Liu N, Lian W, McDowell E, et al. Physiological levels of ATP negatively regulate proteasome function. *Cell Res* 2010;**20**:1372–85.
191. Jennings RB, Steenbergen Jr. C. Nucleotide metabolism and cellular damage in myocardial ischemia. *Annu Rev Physiol* 1985;**47**:727–49.
192. Asai M, Tsukamoto O, Minamino T, Asanuma H, Fujita M, Asano Y, et al. PKA rapidly enhances proteasome assembly and activity in in vivo canine hearts. *J Mol Cell Cardiol* 2009;**46**:452–62.
193. Churchill EN, Ferreira JC, Brum PC, Szweda LI, Mochly-Rosen D. Ischaemic preconditioning improves proteasomal activity and increases the degradation of deltaPKC during reperfusion. *Cardiovasc Res* 2010;**85**:385–94.
194. Canton M, Neverova I, Menabo R, Van Eyk J, Di Lisa F. Evidence of myofibrillar protein oxidation induced by postischemic reperfusion in isolated rat hearts. *Am J Physiol Heart Circ Physiol* 2004;**286**:H870–7.
195. Khaliulin I, Schwalb H, Wang P, Houminer E, Grinberg L, Katzeff H, et al. Preconditioning improves postischemic mitochondrial function and diminishes oxidation of mitochondrial proteins. *Free Radic Biol Med* 2004;**37**:1–9.
196. Park Y, Kanekal S, Kehrer JP. Oxidative changes in hypoxic rat heart tissue. *Am J Physiol Heart Circ Physiol* 1991;**260**:H1395–405.

197. Powell SR, Gurzenda EM, Wahezi SE. Actin is oxidized during myocardial ischemia. *Free Radic Biol Med* 2001;**30**:1171–6.
198. Davies KJ, Lin SW, Pacifici RE. Protein damage and degradation by oxygen radicals. IV. Degradation of denatured protein. *J Biol Chem* 1987;**262**:9914–20.
199. Davies KJ, Delsignore ME. Protein damage and degradation by oxygen radicals. III. Modification of secondary and tertiary structure. *J Biol Chem* 1987;**262**:9908–13.
200. Amici M, Lupidi G, Angeletti M, Fioretti E, Eleuteri AM. Peroxynitrite-induced oxidation and its effects on isolated proteasomal systems. *Free Radic Biol Med* 2003;**34**:987–96.
201. Osna NA, Haorah J, Krutik VM, Donohue Jr. TM. Peroxynitrite alters the catalytic activity of rodent liver proteasome in vitro and in vivo. *Hepatology* 2004;**40**:574–82.
202. Reinheckel T, Sitte N, Ullrich O, Kuckelkorn U, Davies KJ, Grune T. Comparative resistance of the 20S and 26S proteasome to oxidative stress. *Biochem J* 1998;**335**:637–42.
203. Friguet B, Szweda LI. Inhibition of the multicatalytic proteinase (proteasome) by 4-hydroxy-2-nonenal cross-linked protein. *FEBS Lett* 1997;**405**:21–5.
204. Okada K, Wangpoengtrakul C, Osawa T, Toyokuni S, Tanaka K, Uchida K. 4-Hydroxy-2-nonenal-mediated impairment of intracellular proteolysis during oxidative stress. Identification of proteasomes as target molecules. *J Biol Chem* 1999;**274**:23787–93.
205. Reinheckel T, Ullrich O, Sitte N, Grune T. Differential impairment of 20S and 26S proteasome activities in human hematopoietic K562 cells during oxidative stress. *Arch Biochem Biophys* 2000;**377**:65–8.
206. Ishii T, Sakurai T, Usami H, Uchida K. Oxidative modification of proteasome: identification of an oxidation-sensitive subunit in 26 s proteasome. *Biochemistry* 2005;**44**:13893–901.
207. Divald A, Kivity S, Wang P, Hochhauser E, Roberts B, Teichberg S, et al. Myocardial ischemic preconditioning preserves postischemic function of the 26S proteasome through diminished oxidative damage to 19S regulatory particle subunits. *Circ Res* 2010;**106**:1829–38.
208. Pickering AM, Koop AL, Teoh CY, Ermak G, Grune T, Davies KJ. The immunoproteasome, the 20S proteasome and the PA28alphabeta proteasome regulator are oxidative-stress-adaptive proteolytic complexes. *Biochem J* 2010;**432**:585–94.
209. Divald A, Powell SR. Proteasome mediates removal of proteins oxidized during myocardial ischemia. *Free Radic Biol Med* 2006;**40**:156–64.
210. Powell SR, Hall D. Use of salicylate as a probe for •OH formation in isolated ischemic rat hearts. *Free Radic Biol Med* 1990;**9**:133–41.
211. Powell SR, Samuel SM, Wang P, Divald A, Thirunavukkarasu M, Koneru S, et al. Upregulation of myocardial 11S-activated proteasome in experimental hyperglycemia. *J Mol Cell Cardiol* 2008;**44**:618–21.
212. Powell SR, Divald A. The ubiquitin-proteasome system in myocardial ischaemia and preconditioning. *Cardiovasc Res* 2010;**85**:303–11.
213. Murry CE, Jennings RB, Reimer KA. New insights into potential mechanisms of ischemic preconditioning. *Circulation* 1991;**84**:442–5.
214. Gross GJ, Fryer RM. Sarcolemmal versus mitochondrial ATP-sensitive $K^+$ channels and myocardial preconditioning. *Circ Res* 1999;**84**:973–9.
215. Korge P, Honda HM, Weiss JN. Protection of cardiac mitochondria by diazoxide and protein kinase C: implications for ischemic preconditioning. *Proc Natl Acad Sci USA* 2002;**99**:3312–7.
216. Meller R, Cameron JA, Torrey DJ, Clayton CE, Ordonez AN, Henshall DC, et al. Rapid degradation of Bim by the ubiquitin-proteasome pathway mediates short-term ischemic tolerance in cultured neurons. *J Biol Chem* 2006;**281**:7429–36.
217. Pradillo JM, Romera C, Hurtado O, Cardenas A, Moro MA, Leza JC, et al. TNFR1 upregulation mediates tolerance after brain ischemic preconditioning. *J Cereb Blood Flow Metab* 2005;**25**:193–203.

218. Liu C, Chen S, Kamme F, Hu BR. Ischemic preconditioning prevents protein aggregation after transient cerebral ischemia. *Neuroscience* 2005;**134**:69–80.
219. Sato T, Sasaki N, O'Rourke B, Marban E. Nicorandil, a potent cardioprotective agent, acts by opening mitochondrial ATP-dependent potassium channels. *J Am Coll Cardiol* 2000;**35**: 514–8.
220. Inagaki K, Chen L, Ikeno F, Lee FH, Imahashi K, Bouley DM, et al. Inhibition of delta-protein kinase C protects against reperfusion injury of the ischemic heart in vivo. *Circulation* 2003;**108**:2304–7.
221. Baines CP, Song CX, Zheng YT, Wang GW, Zhang J, Wang OL, et al. Protein kinase Cepsilon interacts with and inhibits the permeability transition pore in cardiac mitochondria. *Circ Res* 2003;**92**:873–80.
222. Cai Z, Semenza GL. PTEN activity is modulated during ischemia and reperfusion: involvement in the induction and decay of preconditioning. *Circ Res* 2005;**97**:1351–9.
223. Cai Z, Shen Z, Kaer LV, Becker LC. Ischemic preconditioning-induced cardioprotection is lost in mice with immunoproteasome subunit low molecular mass polypeptide-2 deficiency. *FASEB J* 2008;**22**:4248–57.
224. Gillette TG, Kumar B, Thompson D, Slaughter CA, DeMartino GN. Differential role of the C-termini of AAA subunits of PA700 (19S regulator) in asymmetric assembly and activation of the 26S proteasome. *J Biol Chem* 2008;**283**:31813–22.
225. DeMartino GN. Purification of PA700, the 19S regulatory complex of the 26S proteasome. *Methods Enzymol* 2005;**398**:295–306.
226. Yu X, Kem DC. Proteasome inhibition during myocardial infarction. *Cardiovasc Res* 2010;**85**:312–20.
227. Powell SR. Proteasome inhibitors in myocardial ischemia, some concerns. *Ann Thorac Surg* 2008;**85**:1503–4.
228. Voortman J, Giaccone G. Severe reversible cardiac failure after bortezomib treatment combined with chemotherapy in a non-small cell lung cancer patient: a case report. *BMC Cancer* 2006;**6**:129.
229. Enrico O, Gabriele B, Nadia C, Sara G, Daniele V, Giulia C, et al. Unexpected cardiotoxicity in haematological bortezomib treated patients. *Br J Haematol* 2007;**138**:396–7.
230. Hacihanefioglu A, Tarkun P, Gonullu E. Acute severe cardiac failure in a myeloma patient due to proteasome inhibitor bortezomib. *Int J Hematol* 2008;**88**:219–22.
231. Muchamuel T, Basler M, Aujay MA, Suzuki E, Kalim KW, Lauer C, et al. A selective inhibitor of the immunoproteasome subunit LMP7 blocks cytokine production and attenuates progression of experimental arthritis. *Nat Med* 2009;**15**:781–7.

# Role of the Ubiquitin–Proteasome in Protein Quality Control and Signaling: Implication in the Pathogenesis of Eye Diseases

Fu Shang and Allen Taylor

*Laboratory for Nutrition and Vision Research, USDA Human Nutrition Research Center on Aging, Boston, Massachusetts, USA*

I. Introduction .......... 348
II. Role of UPP in Lens Protein Quality Control .......... 350
III. The Role of UPP in Regulating Lens Cell Proliferation .......... 355
IV. The Role of UPP in Regulating Lens Fiber Differentiation and Lens Development .......... 357
V. UPP Functions in Retinal Development .......... 360
VI. The UPP Regulates Signal Transduction: Implication in the Pathogenesis of Age-Related Macular Degeneration .......... 363
VII. UPP Function in Regulating the NF-κB Signaling Pathway .......... 363
VIII. UPP Function in Regulating Expression of Angiogenic Factors .......... 365
IX. The Role of UPP in Visual Signaling .......... 366
X. Redox Regulation of UPP Activity in the Eye .......... 367
XI. Interaction Between the UPP and Cellular Antioxidant Systems .......... 369
XII. Impairment or Dysregulation of the UPP and Age-Related Eye Diseases .......... 370
XIII. Interaction Between the UPP and ALP in RPE: Implications Regarding the Pathogenesis of AMD .......... 373
XIV. Protecting the UPP from Oxidative Stress May Be a Valid Strategy for Prolonging the Vision Function .......... 376
XV. Conclusions and Future Perspectives .......... 376
References .......... 377

The ubiquitin–proteasome pathway (UPP) plays important roles in many cellular functions, such as protein quality control, cell cycle control, and signal transduction. The selective degradation of aberrant proteins by the UPP is essential for the timely removal of potential cytotoxic damaged or otherwise abnormal proteins. Conversely, accumulation of the cytotoxic abnormal proteins in eye tissues is etiologically associated with many age-related eye diseases such as retina degeneration, cataract, and certain types of glaucoma. Age- or stress-induced impairment or overburdening of the UPP appears to contribute to the accumulation of abnormal proteins in eye tissues. Cell cycle and signal transduction are regulated by the conditional UPP-dependent degradation of the regulators of these processes. Impairment or overburdening of the


DOI: 10.1016/B978-0-12-397863-9.00010-9




1877-1173/12 $35.00

UPP could also result in dysregulation of cell cycle control and signal transduction. The consequences of the improper cell cycle and signal transduction include defects in ocular development, wound healing, angiogenesis, or inflammatory responses. Methods that enhance or preserve UPP function or reduce its burden may be useful strategies for preventing age-related eye diseases.

## I. Introduction

The ubiquitin–proteasome pathway (UPP) is a highly selective intracellular protein degradation system[1,2] that is involved in protein quality control, cell cycle regulation, and signal transductions. Proteins to be degraded are first tagged by covalent attachment of multiple ubiquitin molecules. Then the ubiquitin-tagged proteins are recognized and degraded to amino acids or small peptides by the 26S proteasome with the release of free and reusable ubiquitin.[3] In some cases, ubiquitin is degraded together with the tagged substrates by the proteasome.[4]

Ubiquitin is a highly conserved 76-amino acid polypeptide. Covalent attachment of ubiquitin to the protein substrate proceeds via a three-step cascade mechanism.[3] As illustrated in Fig. 1, ubiquitin is first activated by the ubiquitin-activating enzyme, E1, through the formation of a high-energy thiol ester bond between C-terminal glycine residue of ubiquitin and an internal cysteine residue of E1. Next, one of dozens of ubiquitin-conjugating enzymes, E2, transfers the activated ubiquitin, also via an E2~ubiquitin thiol ester intermediate, to the substrate that is specifically bound to a member of the ubiquitin–protein ligase family, E3. The E3 catalyzes the formation of a peptide/isopeptide bond between a carboxyl group at the C terminus of ubiquitin and an amine group of the substrate. There are two genes in the human genome that encode different isoforms of E1 and each form has a distinct preference for E2s.[5–8] There are at least 37 genes in the human genome that encode distinctive E2s.[9] The number of the genes encoding E3s is over 1000.[10,11] The diversity of E2s and E3s and the combinatorial possibilities of various E2s and E3s in various cellular contexts provide the molecular basis for the exquisite substrate specificity of the UPP.

For efficient targeting of the substrate to proteasome for degradation, multiple ubiquitins are often conjugated to the initial ubiquitin moiety to form a polyubiquitin chain.[12–14] The first ubiquitin is often conjugated to the ε-amino group of an internal lysine residue of the target protein. In the process of formation of ubiquitin chains, additional ubiquitins are conjugated to lysine

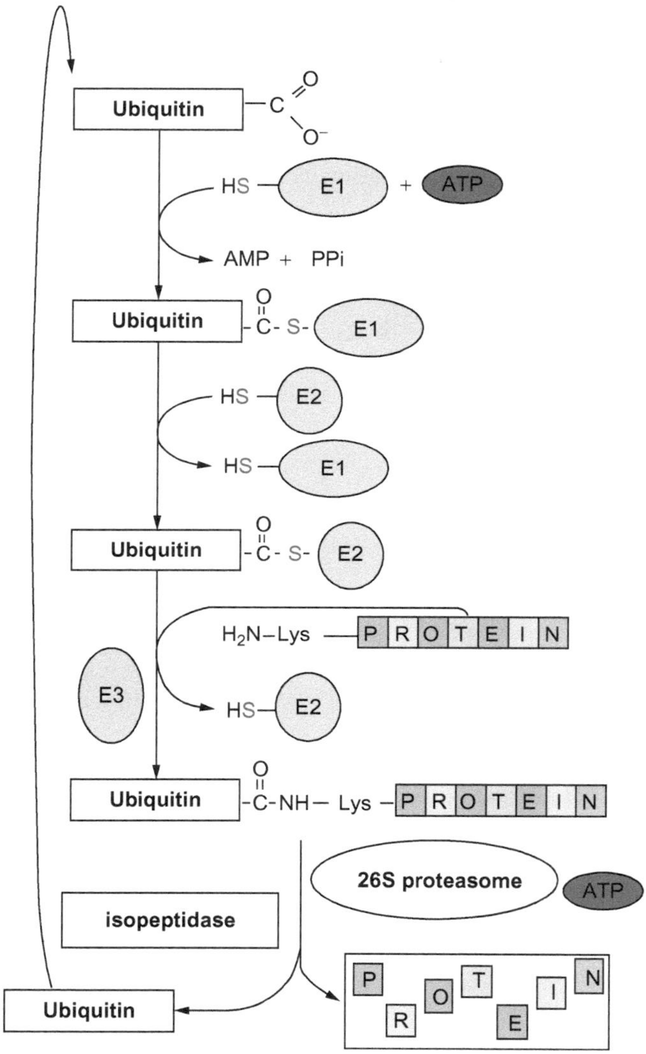


FIG. 1. Diagram of the ubiquitin–proteasome pathway. Ubiquitin is activated in an ATP-requiring reaction that results in a ubiquitin–"thiol ester" of E1. The ubiquitin is then transferred to E2 also as a thiol ester. In the presence of E3, substrate proteins have multiple ubiquitins attached. These conjugates are recognized and (usually) degraded in an ATP-dependent manner by the 26S proteasome. Isopeptidases and ubiquitin C-terminal hydrolases release ubiquitins. The ubiquitin is usually reused. (See Color Insert.)

residues or the N terminus of the previously conjugated ubiquitin molecule. Ubiquitin has seven internal lysine residues. Together with the amine group at the N terminus, there are eight possible positions for the second ubiquitin to attach. For the third ubiquitin, there are 15 possible positions of attachment. Thus, the structures of polyubiquitin chains in cells can be highly diverse and different linkages of the polyubiquitin chains may have specific functions.[15–19] For example, K48-linked ubiquitin chains often target substrates for proteasomal degradation[20,21] whereas substrates with K63-linked polyubiquitin chains are often involved in signal transduction and DNA repair.[22,23]

While some ubiquitinated proteins are not degraded by the proteasome, there are also proteins that are degraded by the proteasome independent of ubiquitination. Both the 26S proteasome and the 20S proteasome can degrade some proteins in the absence of ubiquitination.[24–28] We consider the proteasome-dependent but ubiquitin-independent protein degradation as a noncanonical UPP.[29–35]

The UPP is involved in many facets of cellular functions, including regulation of cell cycle, proliferation, development, signal transduction, transcriptional regulation, receptor downregulation, synaptic plasticity, and removal of damaged or obsolete proteins.[3,36–40] One of the best understood functions of the UPP is protein quality control by selective degradation of aberrant proteins.[41–44] Both canonical and noncanonical UPPs participate in the protein quality control process.[29–35,37,39,45–51] Another important function of the UPP is regulating the cell cycle or signal transduction pathway by conditional degradation of key regulators.[52–56] Efficient protein quality control and proper cell signaling are essential for the health of the eye. In this chapter, we discuss the role of UPP in protein quality control, cell cycle control, and signal transductions in the eyes and their implications in the pathogenesis of eye diseases.

## II. Role of UPP in Lens Protein Quality Control

The lens is a highly differentiated tissue that contains two types of cells: a single layer of epithelial cells underneath the anterior capsule and fiber cells that form the bulk of the lens. Lens fibers are formed by continuous differentiation from lens epithelial cells at the equatorial region of the lens.[57] Fiber differentiation involves extensive cellular remodeling, including expression of a large number of fiber-specific proteins and elimination of epithelial-specific proteins and organelles. In fully differentiated fiber cells, all organelles, including nuclei, mitochondria, and lysosomes, are eliminated.[58] The lens fibers stretch lengthwise from the posterior to the anterior poles and are arranged in concentric layers with the oldest fiber packed in the center and the newest differentiated fibers toward the outer surface.

Eye lenses contain the highest concentrations of proteins in the body. More than 90% of the dry materials in the lens are proteins. The majority of lens proteins are structural proteins called crystallins. The high concentrations of crystallins in cytoplasm provide the lens with a high refractive index and minimize light scattering at the membrane–cytoplasm interface. The proper packing of high concentrations of crystallins in the lens is essential for maintaining lens transparency and its ability to focus light onto the retina for clear vision.[59,60] Disruption of the proper arrangement of these crystallins by adverse modifications is causally related to cataractogenesis.[61–64] Upon aging and with environmental insults, lens proteins undergo various modifications, such as oxidation, glycation, deamidation, and truncations.[65–74] These modified or damaged proteins are usually unstable in aqueous solution and tend to aggregate and precipitate. The aggregation and precipitation of damaged/modified proteins accounts for a large portion of the age- and cataract-related increases in water-insoluble proteins in the lens.[75–82] Thus, efficient removal or repair of the damaged proteins before their aggregation and precipitation appears to be essential for lens transparency.

Some types of modifications, such as methionine oxidation, thiolation, and disulfide cross-link are reparable and lens cells contain enzymes that are capable of repairing these types of damaged proteins. Methionine sulfoxide reductases repair methionine oxidation.[83,84] Glutaredoxin and thioredoxin reduce proteins that are modified by thiolation and disulfide cross-links.[85,86] Lenses also contain molecular chaperones that refold and restore the functions of unfolded or misfolded proteins. However, proteins that are damaged beyond repair must be degraded by selective protein degradation systems. It was proposed that the UPP plays an important role in selective degradation of damaged proteins in the lens and loss of UPP activity may contribute to cataract formation.[48,87,88]

Consistent with a role for the UPP in lens protein quality control, all components of the UPP are retained in lens cells during the extensive cellular remodeling process. We have detected E1, E2s, E3s, isopeptidases, and the proteasome and a myriad of endogenous ubiquitin conjugates in both lens epithelium and lens fibers, including the oldest fibers in the core of the lenses.[7,37,39,49,89–94] The catalytic core of the 26S proteasome or the 20S proteasome was actually detected in the lens before the discovery of ubiquitin and the UPP.[95,96] The 20S proteasome was originally named neutral proteinase. Subsequently, it was renamed multicatalytic proteinase complex[97,98] or 20S multicatalytic protease and finally, the proteasome.[99,100] A functional 26S proteasome and its activity in UPP-mediated proteolysis were detected in lens epithelial cells and lens fibers.[37,94,101] All these components and functional ubiquitin-conjugating activity of endogenous substrates have been identified in all layers of the lens. However, the levels and activity of the UPP in the lens are not evenly distributed. The epithelial cells contain the highest activity and the fibers at the core contain the lowest activity.[94,102] In lens epithelial cells, immunodepletion experiments using antibody to E1 indicate that about 40–50% of intracellular proteolysis is ubiquitin-dependent.[103] In lens fibers, ~50% of the proteasome degradation was inhibited upon treatment with MG132, a potent proteasome inhibitor,[94,104] indicating that the UPP is a major proteolytic system in the lens.

Data from non-lens systems indicate crucial roles for the UPP in protein quality control and responses to various stresses. Disruption of the UPP in cells leads to hypersensitivity to various environmental insults, such as UV light, oxidation, heavy metals, and amino acid analogs.[47,105–111] All these insults are associated with accumulation of damaged proteins, and it is believed that the increased sensitivity to these insults is related to the failure to remove damaged proteins in a timely manner. It has long been hypothesized that damaged proteins in the lens are also selectively degraded by the UPP and that impairment of the UPP contributes to cataract formation.[48,90] Direct evidence in support of this hypothesis was collected during the last decade.

Oxidative stress is one of the major insults that are associated with cataract formation and selective degradation of oxidized proteins has been a topic of active investigation during the last few decades. $H_2O_2$ is a natural oxidant in the lens[112,113] and in aqueous humor in which the lens is bathed.[114–116] Exposure of lens proteins or cultured lens cells to $H_2O_2$ results in generation of protein carbonyls, disulfide-, and non-disulfide-linked aggregations.[50,103,111,117–119] We found that oxidized crystallins are preferentially degraded using the degradative systems present in extracts of lens cells as well as other degradation systems.[100,103,117,118] An increasing body of literature indicates that the proteasome is the major cytoplasmic proteolytic machinery for preferential degradation of damaged versus normal proteins. It has been demonstrated that various forms of oxidized proteins are degraded at faster rates than their native counterparts and inhibition of the proteasome stabilizes oxidized proteins in intact cells or in cell-free systems.[39,49,50,103,111,120–127] While there is consensus that the proteasome plays a key role in the selective degradation of oxidized proteins,[34,35,111,120–130] the extent to which ubiquitination is involved in the degradation of oxidized proteins remains controversial. Many publications suggest that oxidized proteins are degraded by the 20S proteasome independent of ubiquitin.[29,32,33,122,123,129,131–137] However, there is also solid evidence that supports the involvement of ubiquitination in degradation of some oxidized proteins.[45,50,103,111,124,138,139]

Evidence that supports the involvement of ubiquitin in the degradation of oxidized proteins includes the preferential ubiquitination of certain oxidized proteins[45,50,111,124,138] and the accumulation of ubiquitinated species of oxidized proteins upon proteasome inhibition.[50,111] The stabilization of ubiquitinated species of oxidized proteins by proteasome inhibitors indicates that these ubiquitin conjugates are en route to proteasomal degradation. A transient increase in levels of ubiquitinated proteins correlated with increased intracellular proteolysis during recovery from mild oxidative stress[39,49] also implies that the UPP is involved in restoring protein homeostasis during recovery from oxidative stress. A recent study provided direct evidence for the involvement of ubiquitination in proteasomal degradation of oxidized proteins.[139] In that study, the authors demonstrated that USP14, a proteasome-associated deubiquitinating enzyme, is a negative regulator of UPP by its trimming of the ubiquitin chain of ubiquitinated substrates. They further demonstrated that inhibition of USP14 not only promoted the degradation of classic substrates of the UPP, but also promoted the clearance of oxidized proteins and enhanced the resistance to oxidative stress,[139] indicating that ubiquitination is involved in targeting oxidized proteins for degradation.

Glutathiolation is another type of oxidative damage, which occurs when the sulfhydryl on glutathione and protein sulfhydryl form mixed disulfides in response to oxidative stress.[140] Glutathiolated γC-crystallin was degraded by the UPP whereas the unmodified crystallin was stable.[128] Consistent with the involvement of ubiquitination in the degradation of glutathiolated proteins, we found that addition of Ubc4, a ubiquitin-conjugating enzyme, to cell lysates promotes the degradation of glutathiolated γC-crystallin. Consistently, addition of a dominant negative ubiquitin to cell lysates inhibits the degradation of glutathiolated proteins.[128]

In addition to oxidation, calpain or other protease-mediated cleavage or truncation is a common protein modification in lens fibers.[141–148] Many of the partially cleaved or truncated proteins are unstable and they are prone to precipitation. Thus, they are often detected in the water-insoluble fractions of lens proteins. To test whether truncated proteins are also substrates for the UPP, we compared the susceptibilities of wt and three C-terminally truncated αA-crystallins.[104] We found that susceptibilities to degradation depend on the sites of truncations. Full-length αA-crystallin contains 173 amino acids. While wt αA-crystallin was degraded moderately in both lens fiber and reticulocyte lysates, $\alpha A_{1-168}$-crystallin was resistant to degradation. The susceptibility of $\alpha A_{1-163}$-crystallin to degradation was comparable to that of wt αA-crystallin. However, $\alpha A_{1-162}$-crystallin was much more susceptible than wt αA-crystallin to degradation in both lens fiber and reticulocyte lysates. The degradation of both wt and C-terminally truncated $\alpha A_{1-162}$-crystallins requires ATP and, like oxidized proteins, was stimulated by addition of Ubc4.[104] The degradation was substantially inhibited by the proteasome inhibitor (MG132) and a dominant negative mutant ubiquitin (K6W-Ub), indicating that at least part of the proteolysis was mediated by the UPP. Spectroscopic analyses of wt and C-terminally truncated αA-crystallins revealed that C-terminal truncation of αA-crystallin resulted in only subtle changes in secondary structures. However, C-terminal truncations resulted in significant changes in surface hydrophobicity and thermal stability.[104] It appears that the susceptibilities of the C-terminally truncated crystallins to UPP-mediated degradation were positively related to their surface hydrophobicity and negatively related to their thermal stabilities, indicating that exposure of hydrophobic patches may be a degradation signal for C-terminally truncated $\alpha A_{1-162}$-crystallin and other types of damaged proteins.

Deamidation is one the most common posttranslational modifications of lens proteins. Levels of deamidated proteins in the lens increase with age and accumulation of deamidated proteins contributes to lens opacity.[61,70,149–152] We recently determined the effects of deamidation on susceptibility to UPP-mediated degradation of β- and γ-crystallins.[153] The effects of deamidation on susceptibility to degradation by the UPP are substrate-dependent. γD-crystallin is resistant to UPP-mediated degradation and deamidation of γD-crystallin had no effects of its susceptibility to degradation. Wt βB1-crystallin was more susceptible to UPP-mediated degradation than wt γD- and βB2-crystallins.[153] Deamidation of βB1-crystallin had no effects or it decreased the degradation, depending on the sites of deamidation. In contrast, deamidation of βB2-crystallin increased its susceptibility to ubiquitination and degradation.[153] These data suggest that deamidation per se may not serve as a signal for degradation. Instead, deamidation-associated conformation changes may be the triggers of UPP-mediated degradation. Thus, the effects of deamidation on degradation are substrate and site specific.

Glycation is another posttranslational modification of lens proteins. Nonenzymatic attachment of sugars or their reactive metabolites, such as methylglyoxal, alters the activity and stability of lens proteins.[154–156] Glycated proteins progressively accumulate in the aging lens and the process is accelerated under diabetic conditions. Glycation induces irreversible changes in structural proteins, which lead to lens protein aggregation and formation of high-molecular-weight aggregates that scatter light and impede vision.[157] We determined the effects of glycation on degradation of lens proteins and found that glycated crystallins were resistant to degradation by the UPP.[156] Glycation even renders other forms of damaged proteins (glutathiolated or C-terminally truncated crystallins) resistant to degradation by the UPP.[156] It remains to be elucidated how glycation prevents the degradation of the UPP substrates. One of the possibilities is that glycation modifies lysine residues of the substrates and blocks their ubiquitination.[158] Even though glycation reduces the susceptibility of proteins to degradation by the UPP, inhibition of the UPP caused accumulation of glycated proteins in cells, particularly MGO-modified proteins,[156] suggesting that UPP is also involved in the degradation of glycated proteins, but is not as effective as other types of modified proteins. Inhibition of the proteasome in cells activates the autophagy pathway and drives glycated proteins into lysosomal fractions,[156] suggesting that the autophagic–lysosome pathway (ALP) was also involved in the removal of glycated proteins, particularly under conditions of UPP inhibition. The cross talk between UPP and ALP was implicated in the degradation of other forms of damaged proteins.[159,160]

Ubiquitination may also be involved in targeting mutant proteins for degradation in the lens. Both D140N mutant αB-crystallin and T5P-mutant γC-crystallin were degraded faster than the native proteins (our unpublished data). However, if the proteasome is overburdened or there is insufficient proteasome capacity, ubiquitinated proteins will accumulate and aggregate, which in turn triggers unfolded protein response and lens degeneration.[161] Accumulation of ubiquitinated proteins is also a common feature of many age-related degenerative diseases, probably due to functional insufficiency of the proteasome.

Consistent with the role of UPP in degradation of abnormal protein in lens, the activity of the UPP was induced upon expression of mutant proteins, such as mutant vimentin.[162] Expression of substoichiometric amounts of vimentin carrying the R113C point mutation disrupted the endogenous vimentin network and resulted in extensive protein aggregates in cortex fiber cells. Interestingly, postmitotic fibers became depleted of aggregates of the mutant protein, indicating that they were actively removed. Levels of ubiquitinated vimentin and proteasome activity were increased in cells that express mutant vimentin.[162]

## III. The Role of UPP in Regulating Lens Cell Proliferation

As in all other cell types, a functional UPP is required for the cell cycle of lens epithelial cells. Inhibition of the UPP by lactacystin also attenuates lens cell proliferation, as indicated by reduced incorporation of BrdU.[163]

Perturbation of the UPP in human lens epithelial cells in culture with proteasome inhibitor causes cell cycle arrest.[164] Consistent with a requirement for the UPP in multiple phases of the cell cycle, human lens epithelial cells are arrested in G0/G1 if the inhibitor is added before the cells enter the cell cycle. However, if the proteasome inhibitor is added in the S phase, cells are arrested in the G2/M phase.[165] Data also suggest that the G2/M phase is more susceptible to perturbation of the UPP. Expression of K6W-ubiquitin, a dominant negative inhibitor of the UPP[110] in human lens epithelial cells, retarded cell cycle in the G2/M phase, but not in the G0/G1 phase.[166] Expression of K6W-ubiquitin only partially inhibits the UPP functions in the cells.[110] It appears that the residual activity of the UPP in K6W-expressing cells is sufficient to allow the transition from the G1 to S phases. But the residual activity is not enough to support the M–G1 phase transition.[166] A similar phenomenon was observed in temperature-sensitive E1 mutant CHO cells, which were also arrested in the G2/M phase, rather than the G0/G1 phase at nonpermissive temperatures.[167]

Cell cycle regulation by UPP is accomplished by the degradation of specific cell cycle regulators, such as cyclin-dependent kinase activators (i.e., cyclin D and cyclin E), which promote the transition from the G1 to S phase, or Cdk inhibitors that control progress beyond the G1 phase. Cdk activators, such as cyclin A and cyclin B, promote transition through and beyond the G2/M checkpoint by binding to Cdc2 and Cdk2, and are required for progression through the S phase when associated with Cdk2.[168] Cdk inhibitors include $p21^{WAF}$, $p27^{Kip}$, and $p57^{Kip}$.[169–171] Inhibition of the proteasome in lens cells results in accumulation of $p21^{WAF}$ and $p27^{Kip}$.[164] The balance between positive and negative regulators of Cdks controls progress from one phase to another of the cell cycle.[19,53,169,172–178]

The ubiquitination of these cell cycle regulators is catalyzed by different E3s. Based on bioinformatic studies, human cells have between 600 and 1000 different E3s.[11] Many of these E3s contain a characteristic RING domain that interacts with ubiquitin-charged E2s and a wide variety of domains that interact with specific substrates. In the cullin family of RING E3 ligases (CRLs), a large cullin subunit provides a scaffold to bind both a RING-domain subunit and a substrate-recruitment factor. Several of these CRLs, including the Skp1–Cullin–F-box complex (SCF) and the anaphase-promoting complex (APC/C), have long been recognized as essential E3s required for cell cycle control.[179,180]

Although both SCF and APC/C target cell cycle regulators for degradation, they differ in important aspects. For example, by recruiting the Ubc3, the SCF E3s catalyze the ubiquitination of cyclin E, $p21^{Waf}$, $p27^{Kip}$, and other regulators that control the G1–S phase transition.[181] In contrast, the APC/C works with UbcH10 and Ube2S to catalyze the ubiquitination of cyclin A, cyclin B, securin, and other regulators that control the mitosis.[53] SCF E3s and APC/C E3s also form different types of polyubiquitin chains. Like many other E3s, the SCF E3s assemble K48-linked ubiquitin chains,[182] the canonical proteasomal targeting signal. In contrast, the APC/C E3s catalyze the formation of K11-linked ubiquitin chains.[19] The K11-linked ubiquitin chains also trigger the degradation of their substrates by the 26S proteasome.[183,184] The reasons why the APC/C assembles K11-linked ubiquitin chains to promote protein degradation are not known.

In addition to these differences in catalysis, the SCF and APC/C are controlled by distinct regulatory mechanisms. The SCF is active throughout all stages of the cell cycle, but usually recognizes its substrates only after these have been tagged for degradation through phosphorylation or related post-translational modifications.[185] The activity and stability of the APC/C is restricted to mitosis and G1. Phosphorylation of APC/C subunits[186] regulates these processes. Additionally, binding of several distinct APC/C activators or degradation of APC/C inhibitors and APC/C-specific E2s regulates the function of the APC.[176,177,187–190]

As in other cell types, levels of components and regulators of APC/C oscillate during the cell cycle of lens epithelial cells. For example, levels of UbcH10 were highest during mitosis and lowest during the G1–S phase transition.[191] Levels of UbcH7 also oscillate in a manner similar to UbcH10: highest in the G2/M phase and lowest in the S phase.[191,192] The levels of UbcH7 and UbcH10 are also regulated by ubiquitin-dependent degradation.[176,177,192] Although the substrates of UbcH7 are largely unknown, it appears that UbcH7 plays a role in controlling the cell cycle.[192,193] Knockdown of UbcH7 increases the proportion of cells in the S phase and doubles the time to traverse the S phase, whereas overexpression of UbcH7 reduces the proportion of cells in the S phase. UbcH7 knockdown was coincident with elevated levels of the checkpoint kinase Chk1, indicating that UbcH7 promotes S phase progression to G2 by modulating the intra-S phase checkpoint.[192,193]

Despite the constant levels of Ubc3 during the cell cycle, overexpression of Ubc3 in lens epithelial cells resulted in cell cycle delay in the metaphase.[165] The effect of Ubc3 on cell cycle delay is independent of its catalytic activity. Overexpression of wt and catalytic site mutant Ubc3 delayed the cell cycle in metaphase to a similar extent. In contrast, overexpression of catalytic site mutant Ubc2 had no effect on cell cycle of lens epithelial cells.[165] Expression of both wt and mutant Ubc3 stabilized M phase regulators and typical substrates of APC/C, such as cyclin A, cyclin B, and securin.[165] It appears that overexpression of Ubc3 may impair the function of APC/C and slow the degradation of APC/C substrates and thus the delay in metaphase. The mechanism of the inhibitory effects of Ubc3 on APC/C remains to be elucidated. Our unpublished data suggest that Ubc3 may compete with UbcH10 for binding to APC/C and interferes with the normal interaction between APC/C and UbcH10.

## IV. The Role of UPP in Regulating Lens Fiber Differentiation and Lens Development

Lens fiber cells are generated throughout life via continuous proliferation and differentiation of epithelial cells. Differentiation of lens cells involves dramatic metabolic, morphologic, and functional changes in cellular components. Using an *in vitro* differentiation model, we demonstrated that the ubiquitinating activity increases during the early stage (proliferating phase) of bFGF-induced differentiation and then declines in differentiated fibers.[93,194]

The upregulation of the UPP during fiber differentiation includes the increased expression of all three ubiquitin genes *(UBA, UBB, and UBC)*,[195] endogenous levels of ubiquitin conjugates and several ubiquitin-conjugating enzymes, such as Ubc1, Ubc2, Ubc4/5, and Ubc9, and altered expression of components of the proteasome.[93,194,196] The role of UPP in lens development was further demonstrated using transgenic mice that express K6W–ubiquitin specifically in the lens.[197] Expression of K6W–ubiquitin in the lens slows proliferation of epithelial cells and impedes the differentiation program, including expression of fiber-specific proteins and removal of organelles of matured fiber cells (Fig. 2). Using this system, we demonstrated that phosphorylation of nuclear lamin by the mitotic kinase is required for the removal of nuclei during the differentiation process. Disturbance of the UPP by expression of K6W–ubiquitin impedes the degradation of $p27^{Kip}$ and probably other inhibitors of Cdk1, and thus prevents the phosphorylation of nuclear lamin, a prerequisite for disassembly of the nuclear membrane. This precludes the entry of DNAase IIβ into the nucleus and degradation of DNA to nuclear

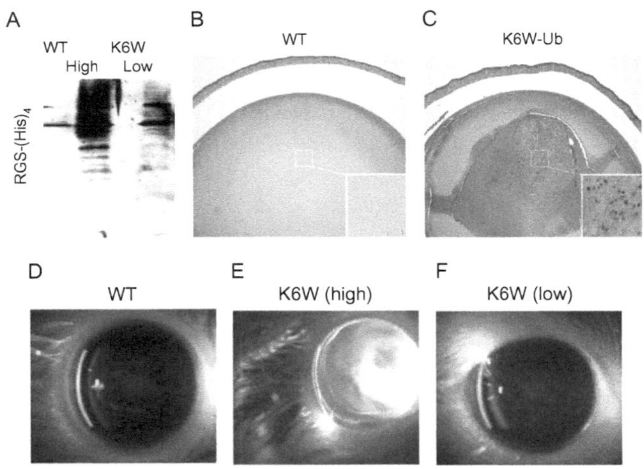


FIG. 2. Expression of K6W-ubiquitin in the lens causes defects in development and cataract in a dose-dependent manner. (Panel A) Western blotting shows the expression of K6W-ubiquitin in lenses of transgenic mice. (Panels B and C) Hematoxylin and eosin staining of sections of wild-type (wt) and K6W-ubiquitin-expressing lenses. Panel B shows the normal lens structure with elimination of nuclei and other organelles in the organelle-free zone (OFZ) in wt lenses. Panel C shows the retention of nuclei in the OFZ in lenses that express high levels of K6W-ubiquitin. (Panels D–F) Slit-lamp photographs of lenses of 3-month-old wt and transgenic mice. Lenses expressing high levels of K6W-ubiquitin show severe cataracts. Lenses from animals expressing low levels of K6W-ubiquitin are clear, comparable to wt lenses. (See Color Insert.)

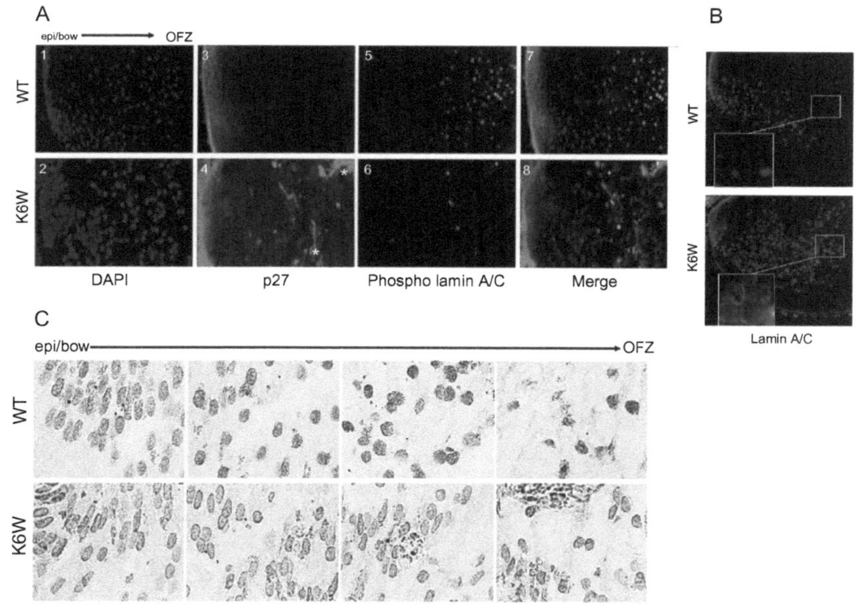


FIG. 3. Expression of K6W-ubiquitin in the lens interferes with lens differentiation and denucleation. (Panel A) Fluorescent micrographs of P2 K6W-ubiquitin-expressing lenses (panel A2) show retained nuclei (blue) in the organelle-free zone (OFZ) as compared to wild-type (wt) lenses (panel A1). Transgenic lenses (panel A4) show a higher distribution of p27 (red) along the fiber mass as compared to wt (panel A3) where it is only apparent at the bow region. Asterisks show nonspecific staining (panel A4). In addition, lenses that express K6W-ubiquitin (panel A6) show decreased levels of phosphorylated lamin A/C (green) at the edge of the core as compared to wt (panel A5). Panel B shows the retention of nuclei (blue) and nuclear envelopes (green) in the OFZ of K6W-ubiquitin-expressing lens as compared to wt lenses. Immunohistochemistry was done to detect levels of p27, phosphorylated lamin A/C, and lamin A/C using anti-p27, anti-lamin A/C (phospho Ser 392), and anti-lamin A/C, respectively. DAPI was used to stain nuclei. (Panel C) Light micrographs of P2 mice wt and K6W-ubiquitin transgenic lenses show the distribution of DNAse IIβ (dark brown). Comparable sections were photographed from the bow region of the lens to the edge of the OFZ. K6W-ubiquitin-expressing lenses show accumulation of DNAse IIβ around the nuclear envelope for all regions, whereas in wt lenses, the distribution of DNAse IIβ changes. Going from the bow toward the OFZ, DNAse IIβ enters the nucleus and does not accumulate around the nuclear envelope. Immunohistochemistry was done to detect the distribution of DNAse IIβ using anti-DNAse IIβ antibody. (See Color Insert.)

DNA (Fig. 3). It was also shown that removal of mitochondria proteins during lens fiber maturation also requires proteasome activity. Inhibition of the proteasome in the developing lens of the chicken eye slowed the degradation of succinate–ubiquinone oxidoreductase, an integral protein of the inner mitochondrial membrane.[198]

Coordinated differentiation and achievement of proper fiber cell morphology, including removal of nuclei and other organelles, are critical for lens transparency. These data indicating that maintaining a functional UPP is essential for cataract prevention.

## V. UPP Functions in Retinal Development

A growing body of data shows that the UPP plays critical roles in mammalian and fly eye differentiation and development. A role for ubiquitination in neurite outgrowth was presaged by experiments in PC-12 cells[199] and developing Xenopus retinal ganglion cells.[200] The roles for the UPP in axon guidance or steering are also established in mammalian systems.[36,201]

Controlled degradation of transcription factors or signaling molecules by the UPP is essential for proper retinal development. For example, UPP-mediated degradation of microphthalmia (Mitf) is required for retina/RPE differentiation. Mitf is a transcription factor that plays essential roles in proliferation, survival, and differentiation of neural crest-derived melanocytes during retinal development.[202,203] The degradation of Mitf depends on a specific degron called a PEST sequence.[203]

Formation of the retina has been most extensively studied in Drosophila. The fly eye is composed of ~750 unit eyes called ommatidia, arranged in a hexagonal pattern. Each ommatidium has eight photoreceptor neurons and six support cells. During larval life, the eye disc changes from an undifferentiated epithelial sheet of cells to a highly patterned neuroepithelium through a series of ordered differentiation steps. The first cells to be specified are the clusters of photoreceptors, which will each give rise to an ommatidial unit. During late larval and early pupal stages, the photoreceptors retreat below the apical surface of the eye as they recruit the next cell type, the four cone cells, each of which in turn recruit two primary pigment cells. The remaining cells will form the interommatidial lattice of secondary pigment cells, tertiary pigment cells and mechanosensory bristles that separate neighboring ommatidia and set the precise hexagonal pattern of the adult eye. In order to allow for proper alignment following primary pigment cell differentiation, approximately 30% of the secondary and tertiary pigment cells must be removed by apoptosis. If these excess cells are not eliminated, the adult eye appears roughened because ommatidial units are separated by a varying number of cells.

The apoptotic editing of these retinal cells is a highly regulated process. Many mutants which cause abnormal apoptosis and abnormal eyes are found to be related to aberrations in UPP function, indicating that the UPP plays an important role in regulating apoptotic pathways required for eye development. Some members of the inhibitors of apoptosis (IAP), an E3 protein family, provide a critical barrier to apoptosis by binding to and inactivating caspases. The drosophila IAP (DIAP) is itself an E3, which blocks apoptosis by promoting the ubiquitination and degradation of Dronc, an upstream caspase.[204] Caspase-mediated cleavage of DIAP1, at position 20, converts the stable, longer DIAP1 into the highly unstable, N-terminal Asn-bearing DIAP1.[205] An SCF is involved in degradation of DIAP and the F-box for this degradative process appears to be Drosophila Morgue.[206,207] Reaper (Rpr) and Grim, but not Hid, also promote the degradation of DIAP1 *in vivo*.[207] The degradation of DIAP1, which is mediated through sequential caspase activity and subsequent exposure of the N-end rule signal of the UPP, is essential for suppression of apoptosis through the coordinated mutual destruction of itself and associated active caspases.

Another family of E3s involved in retinal development includes the widely conserved seven in absentia (SINA) protein and its mammalian orthologs (seven in absentia homolog, Siah). These E3s facilitate degradation of a wide range of cellular proteins, including beta-catenin, a protein with dual roles in development. It is a cell junction protein thought to link the actin cytoskeleton to cell–cell adhesion, and also a nuclear protein that modulates transcription in response to the Wnt signaling pathway.[208,209] Overexpression of the Xenopus homolog Xsiah-2 during development results in a small eye characterized by a reduced size of the lens, the retina, and the pigmented epithelium.

As indicated earlier, ubiquitination can also be a signal for endocytosis. The Notch receptor is part of an intercellular signaling pathway that is required for proper eye development.[210] The only known ligand for Notch expressed in the pupal eye is Delta. Neuralized acts as an E3, triggering endocytosis of Delta, and inhibition of Drosophila neurogenesis.[211,212] Delta is expressed at high levels in cone cells and at lower levels in primary pigment cells and in some lattice cells.[213] While polyubiquitination usually results in the degradation of substrates, Delta appears to be monoubiquitinated at the cell surface.[211,212] Such monoubiquitination is consistent with Delta being endocytosed and delivered to lysosomes.[214] Planar cell polarity (PCP) is a common feature in many epithelia, reflected in cellular organization within the plane of an epithelium. In the Drosophila eye, Frizzled (Fz)/PCP signaling induces cell-fate specification of the R3/R4 photoreceptors through regulation of Notch activation in R4. Neuralized is asymmetrically expressed within the R3/R4 pair and the asymmetric expression is controlled in an Fz/PCP-dependent manner. The Fz/PCP-dependent neuralized expression in R3 ensures the proper R3/R4 specification.[215]

Several additional E3s appear to be involved in signaling pathways which control eye development. Hedgehog and Decapentaplegic pathways direct the progressive differentiation of the eye imaginal disc. Levels of Hedgehog and Decapentaplegic are controlled by the transcription factor Cubitus interruptus

(Ci). A HECT domain E3 regulates the levels of Hedgehog and Decapentaplegic by regulating levels of Ci.[216] Levels of Ci, along with Arm, and CycE are also regulated *in vivo* by the SCF E3, with a Cul1 backbone.[217] In cells anterior to the morphogenetic furrow, Ci proteolytic processing requires the activity of the Nedd8-modified, Cul1-based, $SCF^{Slimb}$ E3 complex. Nedd8 is a ubiquitin-like protein and is involved in regulating SCF E3 function by modifying SCF components or Ubc3. In cells posterior to the morphogenetic furrow, Ci degradation is controlled by a mechanism that requires the activity of Cul3, another member of the Cullin family. This posterior Ci degradation mechanism is activated by Hedgehog signaling. Cul3 suppresses Hedgehog signaling by downregulating the level of effector Ci. High-level Hedgehog signaling promotes Cul3-dependent Ci degradation, leading to the downregulation of Hedgehog signaling. This process is manifested in controlling cell proliferation during Drosophila retinal development. Consistent with the role of Cul3 in controlling levels of Ci and eye development, Cul3 mutation resulted in an accumulation of Ci and an increase in the population of interommatidial cells; moreover, this phenotype can be suppressed by depleting endogenous Ci.[218]

Deubiquitinating enzymes or isopeptidases also play critical roles in retinal development. The first deubiquitinating enzyme found in retina was a product of gene Uch-L1, also known as PGP 9.5.[219–221] Uch-L1 also exhibits a ubiquitin ligase activity.[222] The two opposing enzymatic activities of ubiquitination and deubiquitination control the degradation of α-synuclein and affect Parkinson's disease susceptibility.[223] However, the physiological role for Uch-L1 in retina remains to be defined.

Another deubiquitinating enzyme, Uch-L3, displays 52% amino acid identity to Uch-L1.[224] A murine Uch-L3 deletion mutant displays retinal degeneration, muscular degeneration, and mild growth retardation. In the normal retina, Uch-L3 was enriched in the photoreceptor inner segment. Curiously, the retina of Uch-L3-deficient mice showed no significant morphological abnormalities during retinal development. However, prominent retinal degeneration became manifested after 3 weeks of age with massive apoptosis of photoreceptors. Retina degeneration in Uch-L3-deficient mice was associated with an increase in immunoreactivities for manganese superoxide dismutase, cytochrome c oxidase I, and apoptosis-inducing factor in the inner segment, indicating that mitochondrial oxidative stress is involved.[225]

Deubiquinating enzymes are also involved in establishing the number of photoreceptors as well as in guiding the photoreceptor neuron projection to different synaptic layers in the brain. During eye development in Drosophila, axons from photoreceptor neurons 1–6 terminate in the lamina, while axons from photoreceptor neurons 7 and 8 pass through the lamina and stop in the medulla. As the photoreceptor neuron axons enter the lamina, they encounter both glial cells and neurons. Nonstop mutation of the ubiquitin hydrolase results in altered developmental cues from glial cells and disrupts the targeting of photoreceptor neurons 1–6. This particularly suggests that ubiquitin hydrolase activities in glial cells provide the initial stop signal promoting growth cone termination in the lamina.[226]

Fat facets is another essential deubiquitinating enzyme that limits the number of photoreceptor cells in each facet of the Drosophila compound eye to 8, by preventing the proteasomal degradation of specific substrates such as liquid facets (Lqf)[227–230] and beta-catenin. Lqf or epsin, the eukaryotic homolog, appears to bridge three components: the cell membrane, a transmembrane protein to be internalized, and the core endocytic complex during endocytosis.[231,232] Retention of Lqf facilitates endocytosis in order to generate an inhibitory signal that prevents ectopic photoreceptor determination. It remains to be determined whether neuralized and Lqf activities are related with respect to functions in endocytosis. The enhanced endocytosis due to deubiquitination by the fat facets protein is in contrast to the enhanced endocytosis noted upon ubiquitination of many plasma membrane proteins.[233] Interestingly, Fam[234] the mouse homolog of Fat facets, can substitute for Fat facets in all of its essential functions in Drosophila.[227] Biochemical experiments implicate Af-6 and beta-catenin as substrates for Fam.[235] Af-6 is a tight junction protein thought to mediate the effects of Ras signaling on cell–cell junctions.

# VI. The UPP Regulates Signal Transduction: Implication in the Pathogenesis of Age-Related Macular Degeneration

Efficient and proper cell signaling are essential for the health of the eye. One of the most important functions of the UPP is to regulate the intracellular signaling process by conditional degradation of the regulators of signaling pathways.[52–56] The signal pathways that are controlled by the UPP include, but are not limited to, the Notch pathway,[236,237] Wnt/PCP pathway,[238,239] NF-κB pathway,[54,240] STAT pathway,[241,242] MAP kinase signal pathways,[243–245] and the hypoxia-induced signaling pathway.[246,247] Proper regulation of these signal pathways is essential for development and visual functions. Dysregulation of these signal pathways may lead to retinal degeneration.

# VII. UPP Function in Regulating the NF-κB Signaling Pathway

Excessive ocular inflammation appears involved in the pathogenesis of age-related macular degeneration (AMD), and NF-κB is a master regulator of genes that control immune and inflammatory responses.[248] Relationships between the UPP and the NF-κB pathway have been extensively investigated. Actually the UPP is required for both activation and termination of this signal pathway.[56]

NF-κB proteins all share an approximately 300-amino acid Rel homology domain that regulates DNA binding, nuclear localization, and dimerization.[249] NF-κB family members form homo- and heterodimers, including p65/RelA, c-Rel, RelB, p105/p50, and p100/p52. NF-κB proteins are sequestered in the cytoplasm as latent complexes by inhibitory proteins, or I-κBs, that prevent NF-κB nuclear translocation and DNA binding.[250] Members of the I-κB family, consisting of I-κBα, I-κBβ, I-κBε, NF-κB1, NF-κB2, I-κBζ (also known as MAIL), and Bcl-3, all contain a series of ankyrin repeat domains that mediate binding with NF-κB subunits.[251] The NF-κB1 and NF-κB2 proteins, also known as p105 and p100, respectively, are precursor forms of the p50 and p52 NF-κB subunits.[243]

There are two unique NF-κB signaling pathways, termed canonical and noncanonical. These have distinct biological roles.[252] The canonical NF-κB pathway plays an important role in innate and adaptive immunity and cell survival. In this pathway, proinflammatory cytokines such as tumor necrosis factor alpha (TNF-α) or IL-1β, or pathogen-derived components such as bacterial lipopolysaccharide, trigger the rapid nuclear translocation of NF-κB subunits.[248,252] A wide variety of stimuli activate canonical NF-κB signaling, all of which converge at the IκB kinase (IKK), consisting of catalytic subunits IKKα and IKKβ and the regulatory subunit IKKγ (also known as NEMO).[253,254] IKK phosphorylates I-κB proteins, triggering their ubiquitination and degradation by the proteasome, thus allowing NF-κB to enter the nucleus and activate stimulus-specific gene programs.[255]

Ubiquitination is also involved in activation of IKK. For example, during TNF-α-induced NF-κB activation, TNF-α binds to TNF receptor 1 (TNFR1) and results in ubiquitination of receptor-interacting protein 1 (RIP1).[256,257] TNF-receptor-associated factors (TRAF) are a class of ring finger E3s that play an important role in IKK activation. Activated TRAFs function as ubiquitin ligases (E3) for RIP1 ubiquitination through the K63-linked polyubiquitin chain.[258,259] A20 acts as the deubiquitinating enzyme to disassemble the K63-linked polyubiquitin chain. The polyubiquitin chains of ubiquitinated RIP1 recruit IKKγ/Nemo in the IKK complex.[254,259,260] This ubiquitination requires the dimeric E2 enzyme complex Ubc13/Uev1a as well as the transforming growth factor receptor-β-activated kinase 1 (TAK1) and its regulatory subunits TAB1 and TAB2.[22]

TRAF6 also undergoes oligomerization and K63-linked polyubiquitination. This leads to the recruitment of the TAB1–TAB2–TAK1 complex and subsequent activation of TAK1, which in turn activates IKKβ by phosphorylation.[261,262] The activated IKK complex phosphorylates I-κBα at two sites (32 and 36),[263] triggering its ubiquitination and degradation.[264–266]

The UPP activity is also required to regulate the noncanonical pathway of NF-κB. The MAP 3 kinase MAP3K14, also known as NF-κB-inducing kinase (NIK), is a central player in the noncanonical pathway.[267] A multi-subunit E3 ubiquitin ligase complex consisting of TRAF2, TRAF3, cIAP1, and cIAP2 promotes the degradation of NIK, thus maintaining it at extremely low levels in most cell types.[268] If NIK is not degraded in time, it phosphorylates IKKα, and triggers the proteasome-dependent processing of NF-κB2.[269] NF-κB2 is processed by the proteasome to generate p52, which together with RelB regulates a distinct subset of target genes.[252]

In addition to regulating immune and inflammation responses, the NF-κB pathway is essential for survival. Many pro-survival genes, such as survivin, Bcl-2, and inhibitors of apoptosis, are under the control of NF-κB.[270–272] The activation of NF-κB in cultured ocular cells also requires UPP activity.[56,273–275] We found that sustained oxidative stress impairs UPP function and inhibits NF-κB activation.[275] Inhibition of NF-κB activation during recovery from transient oxidative stress reduces the cell viability.[275] Inhibition of the UPP in RPE also suppresses the expression of NF-κB-regulated genes, such as MCP-1 and IL-6.[274,276]

## VIII. UPP Function in Regulating Expression of Angiogenic Factors

While establishing and maintaining blood flow in the inner and outer retina are essential for retinal health, unscheduled angiogenesis is a major problem, particularly in the aging or diseased retina.

The hypoxia-inducible factor (HIF) signaling cascade controls vasculature adaptations to hypoxia, including the formation of new blood vessels. HIF is an αβ-heterodimer that was first recognized as a DNA-binding factor that mediates hypoxia-inducible activity of the erythropoietin 3′ enhancer.[277,278] In addition, the HIF signaling pathway is a key regulator of many other processes, including transcription of an array of pro-angiogenic factors, such as VEGF, TGF-β3, and various components of glucose transport and glycolysis. These are generally thought to be exploited to overcome vascular insufficiency.[279–281] Both the HIF-α and HIF-β subunits exist as a series of isoforms encoded by distinct genetic loci. HIF-β subunits are constitutive nuclear proteins, whereas HIF-α subunits are inducible by hypoxia. While regulation of HIF-α can occur at the mRNA level,[277] it is primarily regulated through posttranslational modification and UPP-mediated degradation of HIFα. Oxygen-dependent hydroxylation at two prolyl residues (Pro402 and Pro564 in human HIF-1α) mediates its interactions with E3, von Hippel–Lindau (VHL), which targets HIF-1α for rapid proteasomal degradation.[282–284] A recent study showed that, in addition to VHL, C terminus of Hsc70-interacting protein (CHIP), another E3, is also involved in degradation of HIF-1α in RPE. But CHIP-mediated ubiquitination and degradation of HIF-1α are independent of prolyl hydroxylation.[285]

Deubiquitinating enzymes are also involved in regulating the stability of HIF-1α by removing ubiquitin from the substrate. For example, VHL-interacting deubiquitinating enzyme 2 (VDU2) interacts with HIF-1α and specifically deubiquitinates and stabilizes HIF-1α.[286] Therefore, overexpression of VDU2 increases expression of HIF-1α targeted genes, such as VEGF. The balance between the VHL-mediated ubiquitination and VDU2-mediated deubiquitination of HIF-1α provides another level of control for HIF-1α stabilization.

The oxygen-regulated degradation of HIF-1α by the proteasome is important for controlling the balance of angiogenesis in the eye, as failure of degradation of HIF-1α increases the expression and secretion of VEGF and other pro-angiogenesis factors and abnormal angiogenesis, such as those observed in AMD and diabetic retinopathy.[274,287,288] The discoveries regarding HIF-1α signaling, together with new drugs that can control ubiquitination, deubiquitination, and degradation, provide new avenues for AMD therapy.[289]

## IX. The Role of UPP in Visual Signaling

The UPP also plays important roles in visual signaling. Light signals are transduced via GTP-binding proteins such as transducin. Transducin and other GTP-binding proteins are essential heterotrimeric (Tα, β, γ) signaling molecules, which dissociate into Tα-GTP and Tβγ upon light activation by heptahelical membrane receptors such as rhodopsin. Ubiquitination of Tγ occurs when it is dissociated from Tα. Although only Tγ is ubiquitinated, the entire Tβγ subunit complex is degraded.[290,291]

Ubiquitination of Tγ is selectively catalyzed by UbcH5 and UbcH7. Interestingly, Tβγ association with the photoreceptor-specific protein phosducin following light-induced transducin dissociation blocks Tβγ ubiquitination and its subsequent degradation. Conversely, inhibition of phosducin–Tβγ complex formation by phosphorylation of phosducin (by Ca2+/calmodulin-dependent protein kinase II) restores Tβγ ubiquitination and degradation. Thus, it appears that Tβγ is a substrate of the UPP and that phosducin serves to protect Tβγ following the light-dependent dissociation of Tα,β,γ.[291]

These observations suggest a UPP-dependent mechanism in which the cycle of phosducin phosphorylation/dephosphorylation regulates the light sensitivity of the photoreceptor by controlling the amount of transducin available for activation. As formation of the phosducin–T complex reduces the availability of free Tβγ for re-association with Tα GDP (a requirement for subsequent transducin activation), such a mechanism might be important for photoreceptor light adaptation, during which potentially saturating levels of light input to photoreceptor cells are countered by desensitization.[292] A potential problem with this hypothesis is that most phosducin is localized not in the outer segments of photoreceptors, where rhodopsin is present and where phototransduction occurs, but in the inner segments, which are primarily responsible for the metabolic functions of the cell. This hypothesis and these data may be reconciled by the observation that, upon continuous illumination, up to 90% of both Tα and Tβγ translocate from the photoreceptor outer segment to the inner segment.[293]

## X. Redox Regulation of UPP Activity in the Eye

The capacity of the UPP to degrade proteins in cells or tissues is altered in response to environmental insults, including oxidative stress. It has been demonstrated in several cell types that exposure to various forms of oxidative stress results in a transient increase in the intracellular degradation of both short-lived and long-lived cellular proteins. An increase in substrate availability is one of the reasons for the increased intracellular protein degradation in response to oxidative stress. Enhancement of ubiquitination and degradation capacities in response to oxidative stress is another reason for the increased intracellular protein degradation.[39,49,110,131,132,294,295] We demonstrated that both E1 and E2 activities of the ubiquitin-conjugation system increase in response to mild oxidative stress[39,50,296] and Pickering *et al.* demonstrated that the levels and activity of the proteasome, including the immunoproteasome, increased upon adaption to mild oxidative stress.[295] The AMD-related increase in the immunoproteasome in the retina might also be related to oxidative stress.[297]

While mild oxidative stress upregulates the activity of the UPP, all components of the UPP (E1, E2s, some E3s, proteasome, and deubiquitinating enzymes) can be targets of extensive oxidative insults because they all have sulfhydryl groups as critical features of their active sites.[38,296,298–304] Exposure to oxidative stress rapidly depletes reduced glutathione (GSH) and elevates the levels of oxidized glutathione (GSSG). GSSG reacts with the cysteine residues in the active site of E1 and E2, forming mixed disulfide bonds, and blocking their binding to ubiquitin.[38] In addition, other types of modifications, such as *S*-nitrosylation, can inactivate these enzymes.[299] When oxidants are removed from the medium and cells are allowed to recover, the GSH level is restored, and the levels of ubiquitin conjugates, together with the ability to form the conjugates, are also restored.[38,49,298] Moreover, after several hours of recovery from oxidation, there is hyperactivation of E1 and increased ubiquitination and proteolysis relative to levels found in untreated cells.[39,49,92] Thus, intracellular proteolysis can show a biphasic response to oxidative stress. Mild to moderate oxidative stress increases susceptibility of proteins to degradation and enhances the proteolytic capacity, therefore promoting intracellular protein degradation. In contrast, extensive but nonlethal oxidative stress impairs the functions of the proteolytic system, reducing intracellular protein degradation[38,39,49,131,132,275,298] and inducing intracellular accumulation and aggregation of damaged or abnormal, potentially cytotoxic, proteins.

The regulation of ubiquitin-conjugating activity by redox status was corroborated by treatment of RPE cells with diamide and these experiments also led to a new definition for an oxidative stress message. Diamide specifically oxidizes sulfhydryl groups and results in an elevation of GSSG/GSH ratios. It also causes dose-dependent inhibition of ubiquitination and ubiquitin-dependent protein degradation.[298] Thus, the GSSG/GSH ratio is the molecular redox message that regulates the cellular ubiquitin-conjugating activity. The downregulation of the ubiquitin-conjugating activity upon oxidative stress is due to reversible S-thiolation of E1 and E2 enzymes by increased levels of GSSG. The transient inhibition of the ubiquitin-conjugation system by reversible thiolation may also have a protective role: preventing ubiquitination and degradation of essential, reversibly modified proteins upon oxidative stress.[128] Conversely, it may also be perceived that glutathiolation of ubiquitin-conjugating enzymes is a protective mechanism, which blocks the cysteine residue from irreversible oxidation, allowing for recovery of activity when redox status is restored.

The proteasome is also a target of oxidative stress. Reactive oxygen species and reactive lipid peroxidation products, such as 4-hydroxynonenal (HNE), impair the proteasome.[134,300–303,305] It appears that the 26S proteasome is more susceptible than the 20S proteasome to oxidative inactivation.[306] The relative susceptibility of the ubiquitination system versus the proteasome to oxidative stress was also compared in RPE cells.[296] The proteasome is more susceptible to oxidative inactivation than the ubiquitination enzymes in RPE.[296] Sustained exposure to an $H_2O_2$ dose as low as 50 μM $H_2O_2$ resulted in 30–50% inactivation of the proteasome, whereas these levels of oxidative stress had no detectable effect on ubiquitin-conjugating enzymes and ubiquitination activity. Given the fact that the 26S proteasome is more susceptible to oxidative stress than the 20S proteasome,[306] the loss of the peptidase activity of the proteasome upon oxidative stress is most likely due to inactivation of the 26S proteasome. Oxidative inactivation of the 26S proteasome prior to inactivation of the ubiquitination system may explain why levels of endogenous ubiquitin conjugates increase in response to oxidative stress in various types of cells.[39,49,50,92,111,307–312] Age-related inactivation or inhibition of the proteasome[137,313] may also contribute to the increased levels of ubiquitin conjugates in old tissues.[307,308,314] Constant light exposure also resulted in an increase in levels of ubiquitin conjugates in retina,[315] probably due to impairment of the proteasome. Furthermore, extensive oxidative insults may result in nondegradable cross-linked protein aggregates and these may interfere with the functions of the proteasome indirectly in intact cells[316,317] or in cell-free systems.[305,318–321] This would result in insufficient proteasome capacity even if the proteasome per se was not inactivated by the oxidative stress.

## XI. Interaction Between the UPP and Cellular Antioxidant Systems

Cellular antioxidant systems provide the primary defense mechanisms for cells to cope with oxidative stress via directly quenching reactive oxygen species. Just as the UPP is affected by the cellular redox status, the UPP also regulates the cellular redox status. Regulation of the intracellular redox status by the UPP is mediated by degradation of nuclear factor-E2-related factor 2 (Nrf2), an important transcription factor involved in the transcriptional activation of antioxidant enzymes.[322–326] Nrf2 binds to the antioxidant-response element (ARE) and regulates ARE-mediated expression of antioxidant enzymes, such as catalase, superoxide dismutase, NAD(P)H:quinone oxidoreductase, certain forms of glutathione-S-transferases, and $\gamma$-glutamate cysteine ligase regulatory subunit.[327–329]

As with many other transcription factors, the abundance of Nrf2 is regulated by UPP-mediated degradation.[323] The Kelch-like ECH-associating protein (Keap1) is a cytosolic inhibitor of Nrf2.[330,331] The main function of Keap1 is to serve as an adapter for the cullin 3-dependent E3.[332,333] Keap1 binds to Cul3 via its N-terminal BTB/POZ domain and binds to Nrf2 via its C-terminal Kelch domain,[334] leading to the ubiquitination and degradation of Nrf2 through the 26S proteasome.[322,331,335,336]

Under normal cellular conditions, Nrf2 is constantly ubiquitinated by the cytosolic Keap1/Cul3/Rbx1 complex and degraded by the proteasome. When a cell is exposed to oxidative stress, Nrf2 dissociates from the Keap1/Cul3 complex and translocates into the nucleus, leading to activation of ARE-mediated gene expression.[337]

A recent study indicates that a specific ubiquitin-conjugating enzyme, UbcM2, acts as a redox sensor and a regulator of Nrf2 activation.[338] Oxidation of the cys-136 residue of UbcM2 increases its interaction with Nrf2 and results in Nrf2 stabilization. When the function of the UPP is impaired or inhibited, Nrf2 is accumulated and the ARE is activated.[335,339–341] The stabilization of Nrf2 with upregulation of antioxidant enzymes explains why partial inhibition of the UPP results in cytoprotection under certain conditions.[335,339–344] Nrf2 activation also enhances the expression of proteasome subunits and increases proteasome activity.[345] The activation of Nrf2 may explain why pretreatment of cultured neocortical neurons with low doses of proteasome inhibitors led to increased, rather than decreased, proteasome activity.[346]

In corroboration of essential roles for this antioxidant pathway in the eye, RPE are more susceptible to environmental stress in Nrf2 knockout mice[347,348] and to development of AMD-like lesions at a later stage of life.[349] Additional elucidation of the interrelationship between the UPP and antioxidant system will bring new drug targets for AMD and other oxidation-induced diseases.

## XII. Impairment or Dysregulation of the UPP and Age-Related Eye Diseases

Given the essential roles of the UPP in protein quality control and signaling transductions, it is anticipated that impairment or dysregulation of the UPP in ocular tissues would have severe consequences. Impairment of the UPP may contribute to accumulation and aggregation of various forms of damaged proteins, such as those observed in cataract lenses and subretina deposits. However, dysregulated induction of UPP activity may promote the degradation of functional proteins and impair vision functions. For example, inflammation-related degradation of rhodopsin causes deficiency of the functional proteins and has physiological ramifications.[350]

The architecture of the lens, with the oldest tissue in the interior and the most recently elaborated tissue in the epithelial surface, allowed us to demonstrate that all lens tissues have the capacity to catalyze the conjugation of ubiquitin to endogenous proteins; albeit, that capability is decreased in the more mature parts of a lens.[39,90,94,101] As with lens epithelial cells in culture, a brief exposure of whole young lens to peroxide elicits a ubiquitination response.[92] All developmental zones of the young lens increase their ability to form high-mass ubiquitin conjugates in response to mild oxidation. The extent of the ubiquitination response in young lenses is epithelium > cortical fiber > nuclear fibers 2012.[92] In these experiments, the oxidative stress was not associated with changes in masses of the bulk proteins, indicating that the increase in high-mass conjugates was due to polyubiquitination of existing proteins, rather than aggregation of the proteins. As in lens cells in culture, whole lenses from young animals showed a hyperactivation of E1 and increased ubiquitin thiol esters of various E2s upon oxidative stress. This indicates that E1 was rate-limiting of ubiquitination in response to stress. In contrast, in lenses taken from old animals, only epithelial and cortical tissue show increased ability to form ubiquitin conjugates in response to peroxide treatment. The nuclear fibers of old lenses lose the ability to mount a significant ubiquitin-dependent response to oxidation.[92] Furthermore, the ability to form ubiquitin conjugates de novo using exogenous ubiquitin is compromised in all layers in the old lens relative to younger lens.[90,92]

The dramatically reduced ubiquitin-conjugating activity in the inner regions of the bovine lens[90,92–94] appears to be due to a decline in the levels of a specific family of E2s, Ubc4/Ubc5, that are involved in degradation of damaged or obsolete proteins[104,128,351] as supplementation with Ubc4 or Ubc5 can partially restore the ubiquitin-conjugating activity in lysates prepared from the inner regions of these lenses.[94] The decline in levels and activities of these E2s in older lens tissue is also consistent with the hypothesis that the UPP is involved in the removal of damaged proteins, and that dysfunction of the UPP is usually related to the accumulation and aggregation of abnormal proteins in inner regions of older lenses. Accumulation of such damaged protein aggregates would exacerbate the situation as these aggregates inhibit the proteasome.[316]

Although the ability to form ubiquitin conjugates decreases with increasing age, the levels of endogenous ubiquitin conjugates in the human lens increase dramatically with age. A large proportion of these conjugates, particularly conjugates with masses $>250$ kDa, were found to accrue in the water-insoluble fraction.[352] Cataractous lenses, as compared with normal lenses of the same age, show more of these high-molecular-mass conjugates in the urea-soluble and urea-insoluble fractions of cortex and core. The accumulation of ubiquitin conjugates in cataractous lenses is consistent with the observation that the proteasome activity is decreased in cataractous as compared with age-matched clear human lenses[353] perhaps due to oxidative stress.[275,354,355] The vicious cycle of damage and inhibition of the 26S proteasome causing aggregation and precipitation continues as most of the conjugates in cataractous lenses contain many copies of ubiquitin and are of high molecular weight. If these conjugates are not degraded by the 26S proteasome or deubiquitinated by isopeptidases, they would be even more prone to precipitate and cause light scattering or cataract than the initial protein substrate. Inhibition of proteasome may also have additional sequels as proteasome inhibition in bovine lens epithelial cells causes cell death via apoptosis.[275,356]

Direct evidence of the essential role of the UPP for lens transparency was obtained by targeted expression of dominant negative ubiquitin in the lens. Expression of K6W-ubiquitin in the lens not only interrupted lens fiber differentiation, but also resulted in accumulation of protein aggregates in the fibers.[197] These aggregates also contain ubiquitin, consistent with the hypothesis that ubiquitinated proteins may be more prone to aggregation if not promptly processed by the proteasome or deubiqutinases.

Accumulation of ubiquitin or ubiquitin conjugates is also prominent in age-related sub-RPE deposits, that is, drusen and basal laminar deposits, and inclusion bodies in retinas of patients with ataxia type II,[315,357,358] as well as in ganglion cells of rats with chloroquine retinopathy.[359,360] The presence of ubiquitin in sub-RPE deposits—without active forms of ubiquitin-processing enzymes—raises the possibility that certain proteins become ubiquitinated within RPE without further degradation of the ubiquitin–protein complexes.[360–363] Several "amyloid diseases" are associated with accumulation rather than the timely degradation of ubiquitin conjugates derived from a myriad of, but as yet mostly uncharacterized, protein substrates. Lewy bodies are ubiquitin-containing inclusions in many premature neurological aging syndromes. Studies show that α-synuclein, a neuron-specific molecule of unknown functions, is a substrate for the UPP and is found within neurofibrillary deposits in the brains of patients with Parkinson–dementia and abnormal eye function.[364] This corroborates the concept that limited UPP activity is causally related to the "amyloid diseases." How these deposits are related to physiologic dysfunction is not known at this time, but it would appear that cataract and drusen-related AMD can be included among these age-related "amyloid diseases."

As in many other tissues, proteasome activity in retina decreases with aging.[365,366] The age-related decline in proteasome activity may be a consequence of oxidative stress, as the decline in proteasome activity was accompanied by increased levels of oxidatively modified proteins, including nitrotyrosine and 4-hydroxy-2-nonenal (HNE)-modified proteins.[365] In cell culture systems, we demonstrated that the proteasome in RPE is more susceptible to oxidative stress than the ubiquitin-conjugating system and that oxidative stress, including photooxidation, is often associated with accumulation of ubiquitinated proteins, in both neural retina and RPE.[38,55,56,296,315]

Impairment of the UPP in RPE not only resulted in accumulation of oxidatively modified proteins,[296] but also increased the expression and secretion of pro-angiogenesis and pro-inflammation factors.[55,56,274] Impairment of the UPP in RPE resulted in accumulation of HIF-1α and subsequently increased the expression of VEGF and other HIF-regulated genes.[274] The increased VEGF expression and secretion may trigger retinal and choroidal angiogenesis, such as those obtained in wet AMD and proliferative retinopathy. Prolonged inhibition of the UPP in RPE also triggers the expression of IL-8, a potent pro-inflammation and pro-angiogenesis factor.[55,56] Disruption or upregulation of NF-κB, p38 MAP kinase, or PI3K-signaling pathways appears to be involved in the increased expression of IL-8 upon inhibition of the UPP.[55,56] The inhibition of NF-κB signaling by proteasome inhibition is also related to the downregulation of MCP-1, an important monocyte chemotactic protein.[274,367] As MCP-1 deficiency in mice causes AMD-like lesions, such as accumulation of lipofuscin RPE, photoreceptor atrophy, and CNV.[368] The downregulation of the MCP-1 expression upon proteasome inhibition may be a factor for AMD pathogenesis. The relationship between impairment of the UPP and enhanced inflammation is not limited in the eyes. A recent study demonstrated that a mutation in the human proteasome subunit beta type 8 gene (PSMB8) that encodes the immunoproteasome subunit β5i was detected in patients with Nakajo-Nishimura syndrome, a disease with characteristics of chronic inflammation.[369] The β5i mutant is not efficiently incorporated during immunoproteasome biogenesis, resulting in reduced proteasome activity and accumulation of ubiquitinated and oxidized proteins within cells expressing immunoproteasomes.[369] As a result, the level of IL-6 and IFN-γ-inducible protein-10 in patient sera is markedly increased. Nuclear phosphorylated p38 and the secretion of IL-6 are increased in cells that harbor this mutant.

Direct evidence for the involvement of UPP impairment in AMD pathogenesis is still lacking. Ethen *et al.* compared proteasome activity in retina from donor eyes with different grades of AMD and found that the chymotrypsin-like activity of the proteasome increased in neurosensory retina with disease progression.[297] Increased activity correlated with a dramatic increase in the inducible subunits of the immunoproteasome suggest that the proteasome was transformed in the retina of AMD patients, probably due to local inflammation or oxidative stress. Compared to the standard proteasome, the immunoproteasome has higher chymotrypsin activity,[370] but the trypsin-like and caspase-like activities of the proteasome were comparable.[370] The transformation of the proteasome may be a mechanism for the increase in chymotrypsin-like activity in the retina of AMD patients. Induction of the immonoproteasome was also observed in the retina and brain upon cytotoxic T lymphocyte-induced damage.[371] The upregulation of the immunoproteasome in response to retinal and brain injury implies that the immonoproteasome may play a role in neuronal protection and/or repair of damage. Consistent with this notion, immunoproteasome deficiency rendered RPE more susceptible to oxidative stress.[370] All together, these data imply that age- or stress-related impairment of the UPP may contribute to pathogenesis of AMD and other age-related eye diseases.

## XIII. Interaction Between the UPP and ALP in RPE: Implications Regarding the Pathogenesis of AMD

Autophagy is a highly conserved housekeeping pathway that plays a critical role in the removal of damaged intracellular organelles or proteins that are not degraded by the UPP.[156,372,373] Proteins that are cleared by autophagic processes are encapsulated into autophagosomes and make their way into lysosomes for degradation.[374–379] Efficient autophagy is dependent on a balance between the formation and elimination of autophagosomes. Disruption of any part of this pathway will cause autophagic dysfunction. There are few reports regarding autophagy in retina diseases, but it appears that a decline in autophagy and lysosomal activity is associated with retinal aging and AMD.[380–382]

Both the UPP and ALP are involved in selective degradation of damaged proteins. While the UPP prefers to degrade soluble cytoplasmic proteins, the ALP mainly degrades damaged organelles or protein aggregates. However, there are substrate overlaps between the UPP and ALP. It appears that the UPP and ALP complement and provide backup for each other in the removal of damaged proteins. Such a backup mechanism was elucidated in a recent study that studied the degradation of MGO-modified proteins in cultured RPE.[156]

Consuming high glycemic diets is one of the major risk factors for development of AMD in humans.[383–387] Our recent studies demonstrate that feeding mice with a high glycemic index diet results in accumulation of glycated proteins, including MGO-modified proteins, in neural retina and RPE. Consuming high glycemic index diets also promotes the age-dependent development of early AMD-like lesions.[156,388] In cultured RPE, both the UPP and ALP are involved in the removal of MGO-modified proteins. Inhibition of the proteasome or lysosome alone only partially slowed the degradation of MGO-modified proteins.[156] When both the proteasome and lysosome were inhibited, there was greater accumulation of MGO-modified proteins. When the proteasome was inhibited, ubiquitin conjugates and MGO-modified proteins were enriched in the lysosome fraction, indicating that glycated and ubiquitinated proteins were delivered to the lysosome when the proteasome was not functional.[156] The interaction between the UPP and lysosome was also observed in lens epithelial cells.[389] We found that HNE-modified proteins were ubiquitinated, but degraded by the lysosome. In RPE, inhibition of the proteasome results in accumulation of perinuclear aggregates positive for Hsp70, ubiquitin–protein conjugates, and the lysosomal membrane protein LAMP-2.[390] Our work also indicates that impairment of the UPP in RPE activates the autophagic pathway as indicated by the increase in levels of LC3II upon inhibition of the proteasome, or expression of K6W-ubiquitin.[156] It appears that p62/sequestosome 1 as a mediator of proteasome inhibitor induced autophagy.[391] The p62/sequestosome-1 is a multifunctional protein containing several protein–protein interaction domains. P62/sequestosome 1 binds to ubiquitinated proteins through its UBA motif and the LC3 via the LC3-interacting region and mediates the autophagy of ubiquitinated proteins.[379] The interaction between UPP and ALP also provides an opportunity to manipulate the activities of these degradation pathways for healthy vision. For example, chemical induction of autophagy may help to prevent the accumulation of ubiquitin-containing aggregates. Enhancement of the UPP function may also reduce the burden of the ALP and thus promote the degradation of lipofuscin and dysfunctional organelles, such as damaged mitochondria by the ALP.[382]

The interaction between UPP and ALP and its physiologic significance is summarized in Fig. 4. This scheme proposes that under normal homeostatic conditions, the UPP and ALP function independently or in concert to maintain the proteome homeostasis (proteostasis). The proper functions of the UPP and ALP prevent the accumulation of abnormal proteins or adversely modified

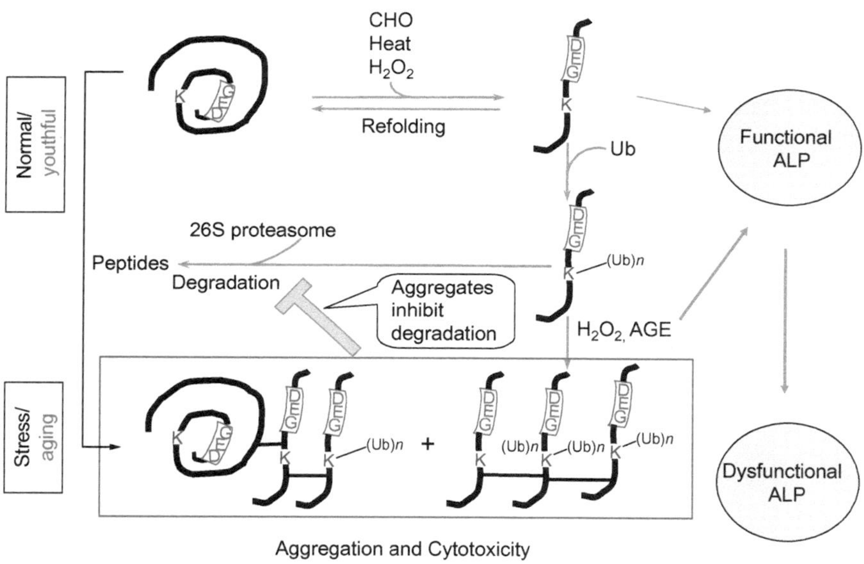


FIG. 4. Scheme of interaction between the ubiquitin–proteasome pathway (UPP) and autophagic–lysosome pathway (ALP) in protein quality control and the proposed ramifications of dysfunctions of these proteolytic pathways. Environmental stress, including consuming a high glycemic index diet, causes protein denaturation or unfolding, resulting in exposure of degradation signals (deg) and triggering of their ubiquitination and degradation. Majority of ubiquitinated proteins are degraded by the proteasome, but some ubiquitinated proteins are degraded by lysosome via autophagy, particularly when the function of the proteasome is impaired or overburdened. The rapid degradation of damaged proteins by the concerned action of the UPP and ALP prevents the accumulation of cytotoxic damaged proteins and maintains the homeostasis of the proteome, such as those observed in young and healthy tissues. When the levels of damaged proteins overwhelm the capacity of the UPP and ALP, damaged protein will accumulate. Some of the damaged protein may cross-link, forming the higher mass aggregates noted in the figures above. Accumulation of oligomerized damaged proteins may impair the proteolytic machinery, further compromising the proteome and resulting in diseases related to accumulation of damaged proteins, such as cataract and age-related macular degeneration. (See Color Insert.)

proteins, such as advanced glycation end proteins (AGEs). However, upon chronic stress and aging, these capacities of the UPP and ALP are insufficient to maintain the proteome or to protect themselves from becoming compromised, resulting thereby in accumulation of damaged proteins. Degrons, or amino acid residues that encode susceptibility for degradation on substrates, may be blocked by adverse modifications. This situation is exacerbated by the accumulation of damaged proteins, which may diminish UPP efficacy, setting up a vicious cycle which results in compromised cellular function. Environmental stress, including hyperglycemia, may also alter the interaction between the UPP and ALP and compromise the protein quality control mechanisms, triggering the development of AMD-like lesions.

## XIV. Protecting the UPP from Oxidative Stress May Be a Valid Strategy for Prolonging the Vision Function

Age- or stress-related impairment of the UPP appears to be a contributing factor for age-related eye diseases, such as cataract and AMD. Any means that can preserve the functions of the UPP could be beneficial. As oxidative stress is a major cause of age-related eye disease and dysfunction of the UPP, adequate dietary intake of nutritional antioxidants may protect the UPP from oxidative inactivation. Indeed, we recently demonstrated that supplementation of lutein, zeaxanthin, or vitamin E to cultured RPE can protect the proteasome from photooxidative inactivation (Bian, Qingning *et al.*, manuscript submitted). Lutein and zeaxanthin supplementation also attenuated photooxidation-induced accumulation of ubiquitin conjugates and alteration in inflammatory response, such as expression and secretion of MCP-1, complement factor H, and IL-8 (Bian *et al.*, manuscript submitted). The protection of the UPP from oxidative inactivation may be one of the mechanisms for the protective effects of lutein, zeaxanthin, and other nutritional antioxidants in reducing the risk of cataract and AMD.

## XV. Conclusions and Future Perspectives

The UPP is involved in regulating many cellular functions, from protein quality control to signal transduction and stress responses. A fully functional UPP is essential for eye development and maintaining the normal functions of ocular tissues. Aging or stress-related impairment of the UPP in the eye would result in accumulation of abnormal or damaged proteins, which would interfere with cellular function or cause light-scattering and cataract formation.

Impairment of the UPP would also disrupt signal transduction pathways, which in turn would alter inflammation response and angiogenesis, common features of AMD-related lesions. Any means that can protect the UPP from impairment or reduce the burden of the UPP would be beneficial for eye health. Sufficient intake of nutritional antioxidants, such as lutein, zeaxanthin, vitamins C, E, and zinc, may protect the proteasome from oxidative inactivation, preserve the UPP function in the eye, and reduce the risk for age-related eye diseases. Consuming low glycemic index diets would also help to preserve the function of UPP by decreasing the glycative/oxidative ratio and reducing the burdens of the UPP. The interaction between nutrition and UPP, as well as between UPP and other protective mechanisms, warrants comprehensive investigation.

## Acknowledgments

This work is partially supported by NIH grants EY 011717, EY 13250, USDA AFRI Award 2009-35200-05014, USDA contract 1950-510000-060-01A, AHAF grant M2010038, and Dennis L. Gierhart Charitable Gift Fund.

## References

1. Ciechanover A, Orian A, Schwartz AL. The ubiquitin-mediated proteolytic pathway: mode of action and clinical implications. *J Cell Biochem Suppl* 2000;**34**:40–51.
2. Glickman MH, Ciechanover A. The ubiquitin-proteasome proteolytic pathway: destruction for the sake of construction. *Physiol Rev* 2002;**82**:373–428.
3. Hershko A, Ciechanover A. The ubiquitin system. *Annu Rev Biochem* 1998;**67**:425–79.
4. Shabek N, Herman-Bachinsky Y, Ciechanover A. Ubiquitin degradation with its substrate provides clues to proteasome regulation. *Proc Natl Acad Sci USA* 2009;**106**(29):11907–12.
5. Handley-Gearhart PM, Stephen AG, Trausch-Azar JS, Ciechanover A, Schwartz AL. Human ubiquitin-activating enzyme, E1. Indication of potential nuclear and cytoplasmic subpopulations using epitope-tagged cDNA constructs. *J Biol Chem* 1994;**269**:33171–8.
6. Handley-Gearhart PM, Trausch-Azar JS, Ciechanover A, Schwartz AL. Rescue of complex temperature-sensitive phenotype of chinese hamster ovary E36ts20 cells by expression of human ubiquitin activating enzyme cDNA. *Biochem J* 1994;**304**:1015–20.
7. Shang F, Deng G, Obin M, Wu CC, Gong X, Smith D, et al. Ubiquitin-activating enzyme (E1) isoforms in lens epithelial cells: origin of translation, E2 specificity and cellular localization determined with novel site-specific antibodies. *Exp Eye Res* 2001;**73**:827–36.
8. Jin J, Li X, Gygi SP, Harper JW. Dual E1 activation systems for ubiquitin differentially regulate E2 enzyme charging. *Nature* 2007;**447**:1135–8.
9. Michelle C, Vourc'h P, Mignon L, Andres CR. What was the set of ubiquitin and ubiquitin-like conjugating enzymes in the eukaryote common ancestor? *J Mol Evol* 2009;**68**:616–28.
10. Pickart CM, Eddins MJ. Ubiquitin: structures, functions, mechanisms. *Biochim Biophys Acta* 2004;**1695**:55–72.
11. Deshaies RJ, Joazeiro CA. RING domain E3 ubiquitin ligases. *Annu Rev Biochem* 2009;**78**: 399–434.

12. Beal R, Deveraux Q, Xia G, Rechsteiner M, Pickart C. Surface hydrophobic residues of multiubiquitin chains essential for proteolytic targeting. *Proc Natl Acad Sci USA* 1996;**93**: 861–6.
13. Pickart CM. Targeting of substrates to the 26S proteasome. *FASEB J* 1997;**11**:1055–66.
14. Lam YA, Lawson TG, Velayutham M, Zweier JL, Pickart CM. A proteasomal ATPase subunit recognizes the polyubiquitin degradation signal. *Nature* 2002;**416**:763–7.
15. Pickart CM, Fushman D. Polyubiquitin chains: polymeric protein signals. *Curr Opin Chem Biol* 2004;**8**:610–6.
16. Peng J, Schwartz D, Elias JE, Thoreen CC, Cheng D, Marsischky G, et al. A proteomics approach to understanding protein ubiquitination. *Nat Biotechnol* 2003;**21**:921–6.
17. Xu P, Duong DM, Seyfried NT, Cheng D, Xie Y, Robert J, et al. Quantitative proteomics reveals the function of unconventional ubiquitin chains in proteasomal degradation. *Cell* 2009;**137**:133–45.
18. Komander D, Reyes-Turcu F, Licchesi JD, Odenwaelder P, Wilkinson KD, Barford D. Molecular discrimination of structurally equivalent Lys 63-linked and linear polyubiquitin chains. *EMBO Rep* 2009;**10**:466–73.
19. Wickliffe KE, Williamson A, Meyer HJ, Kelly A, Rape M. K11-linked ubiquitin chains as novel regulators of cell division. *Trends Cell Biol* 2011;**21**:656–63.
20. Chau V, Tobias JW, Bachmair A, Marriott D, Ecker DJ, Gonda DK, et al. A multiubiquitin chain is confined to specific lysine in a targeted short-lived protein. *Science* 1989;**243**:1576–83.
21. Thrower JS, Hoffman L, Rechsteiner M, Pickart CM. Recognition of the polyubiquitin proteolytic signal. *EMBO J* 2000;**19**:94–102.
22. Deng L, Wang C, Spencer E, Yang L, Braun A, You J, et al. Activation of the IkappaB kinase complex by TRAF6 requires a dimeric ubiquitin-conjugating enzyme complex and a unique polyubiquitin chain. *Cell* 2000;**103**:351–61.
23. Chiu RK, Brun J, Ramaekers C, Theys J, Weng L, Lambin P, et al. Lysine 63-polyubiquitination guards against translesion synthesis-induced mutations. *PLoS Genet* 2006;**2**:e116.
24. Murakami Y, Matsufuji S, Kameji T, Hayashi S-I, Igarashi K, Tamura T, et al. Ornithine decarboxylase is degraded by the 26S proteosome without ubiquitin. *Nature* 1992;**360**:597–9.
25. Tarcsa E, Szymanska G, Lecker S, O'Connor CM, Goldberg AL. Ca2+-free calmodulin and calmodulin damaged by in vitro aging are selectively degraded by 26 S proteasomes without ubiquitination. *J Biol Chem* 2000;**275**:20295–301.
26. Benaroudj N, Tarcsa E, Cascio P, Goldberg AL. The unfolding of substrates and ubiquitin-independent protein degradation by proteasomes. *Biochimie* 2001;**83**:311–8.
27. Zhang M, Pickart CM, Coffino P. Determinants of proteasome recognition of ornithine decarboxylase, a ubiquitin-independent substrate. *EMBO J* 2003;**22**:1488–96.
28. Matsuzawa S, Cuddy M, Fukushima T, Reed JC. Method for targeting protein destruction by using a ubiquitin-independent, proteasome-mediated degradation pathway. *Proc Natl Acad Sci USA* 2005;**102**:14982–7.
29. Pacifici RE, Salo DC, Davies KJA. Macroxyproteinase (M.O.P.): a 670 kDa proteinase complex that degrades oxidatively denatured proteins in red blood cells. *Free Rad Biol Med* 1989;**7**:521–36.
30. Salo DC, Pacifici RE, Lin SW, Giulivi C, Davies KJ. Superoxide dismutase undergoes proteolysis and fragmentation following oxidative modification and inactivation. *J Biol Chem* 1990;**265**:11919–27.
31. Dean RT, Geiseg S, Davies MJ. Reactive species and their accumulation on radical-damaged, proteins. *Trends Biochem Sci* 1993;**18**(11):437–41.
32. Shringarpure R, Grune T, Davies KJ. Protein oxidation and 20S proteasome-dependent proteolysis in mammalian cells. *Cell Mol Life Sci* 2001;**58**:1442–50.

33. Shringarpure R, Grune T, Mehlhase J, Davies KJ. Ubiquitin conjugation is not required for the degradation of oxidized proteins by proteasome. *J Biol Chem* 2003;**278**:311–8.
34. Ferrington DA, Sun H, Murray KK, Costa J, Williams TD, Bigelow DJ, et al. Selective degradation of oxidized calmodulin by the 20 S proteasome. *J Biol Chem* 2001;**276**:937–43.
35. Balog EM, Lockamy EL, Thomas DD, Ferrington DA. Site-specific methionine oxidation initiates calmodulin degradation by the 20S proteasome. *Biochemistry* 2009;**48**:3005–16.
36. Campbell DS, Holt CE. Chemotropic responses of retinal growth cones mediated by rapid local protein synthesis and degradation. *Neuron* 2001;**32**:1013–26.
37. Jahngen JH, Haas AL, Ciechanover A, Blondin J, Eisenhauer D, Taylor A. The eye lens has an active ubiquitin-protein conjugation system. *J Biol Chem* 1986;**261**:13760–7.
38. Jahngen-Hodge J, Obin M, Gong X, Shang F, Nowell T, Gong J, et al. Regulation of ubiquitin conjugating enzymes by glutathione following oxidative stress. *J Biol Chem* 1997;**272**: 28218–26.
39. Shang F, Gong X, Taylor A. Activity of ubiquitin dependent pathway in response to oxidative stress: ubiquitin activating enzyme (E1) is transiently upregulated. *J Biol Chem* 1997;**272**: 23086–93.
40. Taylor A. Nutritional and environmental influences on risk for cataract. In: Taylor A, editor. *Nutritional and environmental influences on the eye*. New York: CRC; 1999. pp. 53–93.
41. Goldberg AL. Protein degradation and protection against misfolded or damaged proteins. *Nature* 2003;**426**:895–9.
42. Shang F, Taylor A. Ubiquitin-proteasome pathway and cellular responses to oxidative stress. *Free Radic Biol Med* 2011;**51**:5–16.
43. Cyr DM, Hohfeld J, Patterson C. Protein quality control: U-box-containing E3 ubiquitin ligases join the fold. *Trends Biochem Sci* 2002;**27**:368–75.
44. McDonough H, Patterson C. CHIP: a link between the chaperone and proteasome systems. *Cell Stress Chaperones* 2003;**8**:303–8.
45. Hershko A, Heller H, Eytan E, Reiss Y. The protein substrate binding site of the ubiquitin-protein ligase system. *J Biol Chem* 1986;**261**:11992–9.
46. Hershko A, Eytan E, Ciechanover A. Immunochemical analysis of the turnover of ubiquitin-protein conjugates in intact cells. *J Biol Chem* 1982;**257**:13964–70.
47. Finley D, Ozkaynak E, Varshavsky A. The yeast polyubiquitin gene is essential for resistance to high temperature, starvation, and other stresses. *Cell* 1987;**48**:1035–46.
48. Taylor A, Davies KJA. Protein oxidation and loss of protease activity may lead to cataract formation in the aged lens. *Free Rad Biol Med* 1987;**3**:371–7.
49. Shang F, Taylor A. Oxidative stress and recovery from oxidative stress are associated with altered ubiquitin conjugating and proteolytic activities in bovine lens epithelial cells. *Biochem J* 1995;**307**:297–303.
50. Shang F, Nowell Jr. TR, Taylor A. Removal of oxidatively damaged proteins from lens cells by the ubiquitin-proteasome pathway. *Exp Eye Res* 2001;**73**:229–38.
51. Shang F, Taylor A. Function of the ubiquitin proteolytic pathway in the eye. *Exp Eye Res* 2004;**78**:1–14.
52. Frescas D, Pagano M. Deregulated proteolysis by the F-box proteins SKP2 and beta-TrCP: tipping the scales of cancer. *Nat Rev Cancer* 2008;**8**:438–49.
53. Wickliffe K, Williamson A, Jin L, Rape M. The multiple layers of ubiquitin-dependent cell cycle control. *Chem Rev* 2009;**109**:1537–48.
54. Chen ZJ. Ubiquitin signalling in the NF-kappaB pathway. *Nat Cell Biol* 2005;**7**:758–65.
55. Fernandes AF, Bian Q, Jiang JK, Thomas CJ, Taylor A, Pereira P, et al. Proteasome inactivation promotes p38 mitogen-activated protein kinase-dependent phosphatidylinositol 3-kinase activation and increases interleukin-8 production in retinal pigment epithelial cells. *Mol Biol Cell* 2009;**20**:3690–9.

56. Fernandes AF, Zhou J, Zhang X, Bian Q, Sparrow J, Taylor A, et al. Oxidative inactivation of the proteasome in retinal pigment epithelial cells. A potential link between oxidative stress and up-regulation of interleukin-8. *J Biol Chem* 2008;**283**:20745–53.
57. McAvoy JW, Chamberlain CG, de Iongh RU, Hales AM, Lovicu FJ. Lens development. *Eye (Lond)* 1999;**13**(Pt 3b):425–37.
58. Bassnett S. On the mechanism of organelle degradation in the vertebrate lens. *Exp Eye Res* 2009;**88**:133–9.
59. Delaye M, Tardieu A. Short-range order of crystallin proteins accounts for eye lens transparency. *Nature* 1983;**302**:415–7.
60. Ponce A, Sorensen C, Takemoto L. Role of short-range protein interactions in lens opacifications. *Mol Vis* 2006;**12**:879–84.
61. Gupta R, Asomugha CO, Srivastava OP. The common modification in alphaA-crystallin in the lens, N101D, is associated with increased opacity in a mouse model. *J Biol Chem* 2011;**286**: 11579–92.
62. Wilmarth PA, Tanner S, Dasari S, Nagalla SR, Riviere MA, Bafna V, et al. Age-related changes in human crystallins determined from comparative analysis of post-translational modifications in young and aged lens: does deamidation contribute to crystallin insolubility? *J Proteome Res* 2006;**5**:2554–66.
63. Andley UP. AlphaA-crystallin R49Cneo mutation influences the architecture of lens fiber cell membranes and causes posterior and nuclear cataracts in mice. *BMC Ophthalmol* 2009;**9**:4.
64. Banerjee PR, Pande A, Patrosz J, Thurston GM, Pande J. Cataract-associated mutant E107A of human gammaD-crystallin shows increased attraction to alpha-crystallin and enhanced light scattering. *Proc Natl Acad Sci USA* 2011;**108**:574–9.
65. Augusteyn RC. *Protein modification in cataract, possible oxidative mechanisms.* London: Academic Press; 1981.
66. Zhang Z, David LL, Smith DL, Smith JB. Resistance of human betaB2-crystallin to in vivo modification. *Exp Eye Res* 2001;**73**:203–11.
67. Hanson SR, Hasan A, Smith DL, Smith JB. The major in vivo modifications of the human water-insoluble lens crystallins are disulfide bonds, deamidation, methionine oxidation and backbone cleavage. *Exp Eye Res* 2000;**71**:195–207.
68. Feng J, Smith DL, Smith JB. Human lens beta-crystallin solubility. *J Biol Chem* 2000;**275**:11585–90.
69. Hanson SR, Smith DL, Smith JB. Deamidation and disulfide bonding in human lens gamma-crystallins. *Exp Eye Res* 1998;**67**:301–12.
70. Takemoto L, Boyle D. Increased deamidation of asparagine during human senile cataractogenesis. *Mol Vis* 2000;**6**:164–8.
71. Takemoto L. Deamidation of Asn-143 of gamma S crystallin from protein aggregates of the human lens. *Curr Eye Res* 2001;**22**:148–53.
72. Liu C, Pande J, Lomakin A, Ogun O, Benedek GB. Aggregation in aqueous solutions of bovine lens gamma-crystallins: special role of gamma(s). *Invest Ophthalmol Vis Sci* 1998;**39**: 1609–19.
73. Hood BD, Garner B, Truscott RJ. Human lens coloration and aging. Evidence for crystallin modification by the major ultraviolet filter, 3-hydroxy-kynurenine O-beta-D-glucoside. *J Biol Chem* 1999;**274**:32547–50.
74. Dickerson Jr. JE, Lou MF. A new mixed disulfide species in human cataractous and aged lenses. *Biochim Biophys Acta* 1993;**1157**:141–6.
75. Li LK, Roy D, Spector A. Changes in lens protein in concentric fractions from individual normal human lenses. *Curr Eye Res* 1986;**5**:127–53.
76. Cui XL, Lou MF. The effect and recovery of long-term H2O2 exposure on lens morphology and biochemistry. *Exp Eye Res* 1993;**57**:157–67.

77. Giblin FJ, Padgaonkar VA, Leverenz VR, Lin LR, Lou MF, Unakar NJ, et al. Nuclear light scattering, disulfide formation and membrane damage in lenses of older guinea pigs treated with hyperbaric oxygen. *Exp Eye Res* 1995;**60**:219–35.
78. Marcantonio JM. Susceptibility of the bovine lens 115 kDa beaded filament protein to degradation by calcium and calpain. *Curr Eye Res* 1992;**11**:103–8.
79. Fukiage C, Azuma M, Nakamura Y, Tamada Y, Shearer TR. Calpain-induced light scattering by crystallins from three rodent species. *Exp Eye Res* 1997;**65**:757–70.
80. Shearer TR, Shih M, Azuma M, David LL. Precipitation of crystallins from young rat lens by endogenous calpain. *Exp Eye Res* 1995;**61**:141–50.
81. Shearer TR, Shih M, Mizuno T, David LL. Crystallins from rat lens are especially susceptible to calpain-induced light scattering compared to other species. *Curr Eye Res* 1996;**15**:860–8.
82. Shih M, David LL, Lampi KJ, Ma H, Fukiage C, Azuma M, et al. Proteolysis by m-calpain enhances in vitro light scattering by crystallins from human and bovine lenses. *Curr Eye Res* 2001;**22**:458–69.
83. Brennan LA, Lee W, Giblin FJ, David LL, Kantorow M. Methionine sulfoxide reductase A (MsrA) restores alpha-crystallin chaperone activity lost upon methionine oxidation. *Biochim Biophys Acta* 2009;**1790**:1665–72.
84. Brennan LA, Lee W, Kantorow M. TXNL6 is a novel oxidative stress-induced reducing system for methionine sulfoxide reductase a repair of alpha-crystallin and cytochrome C in the eye lens. *PLoS One* 2010;**5**:e15421.
85. Lou MF. Redox regulation in the lens. *Prog Retin Eye Res* 2003;**22**:657–82.
86. Lofgren S, Fernando MR, Xing KY, Wang Y, Kuszynski CA, Ho YS, et al. Effect of thioltransferase (glutaredoxin) deletion on cellular sensitivity to oxidative stress and cell proliferation in lens epithelial cells of thioltransferase knockout mouse. *Invest Ophthalmol Vis Sci* 2008;**49**:4497–505.
87. Taylor A, Shang F, Obin M. Relationships between stress, protein damage, nutrition, and age-related eye diseases. *Mol Aspects Med* 1997;**18**:305–414.
88. Petersen A, Honarvar A, Zetterberg M. Changes in activity and kinetic properties of the proteasome in different rat organs during development and maturation. *Curr Gerontol Geriatr Res* 2010;**2010**:9 230697.
89. Jahngen JJ, Eisenhauer D, Taylor A. Lens proteins are substrates for the reticulocyte ubiquitin conjugation system. *Curr Eye Res* 1986;**5**:725–33.
90. Jahngen JH, Lipman RD, Eisenhauer DA, Jahngen EGE, Taylor A. Aging and cellular maturation causes changes in ubiquitin-eye lens protein conjugates. *Arch Biochem Biophys* 1990;**276**:32–7.
91. Jahngen-Hodge J, Laxman E, Zuliani A, Taylor A. Evidence for ATP and ubiquitin degradation of proteins in cultured bovine lens epithelial cells. *Exp Eye Res* 1991;**52**:41–7.
92. Shang F, Gong X, Palmer HJ, Nowell TR, Taylor A. Age-related decline in ubiquitin conjugation in response to oxidative stress in the lens. *Exp Eye Res* 1997;**64**:21–30.
93. Shang F, Gong X, McAvoy JW, Chamberlain C, Nowell TR, Taylor A. Ubiquitin-dependent pathway is up-regulated in differentiating lens cells. *Exp Eye Res* 1999;**68**:179–92.
94. Pereira P, Shang F, Girão H, Taylor A. Lens fibers have a fully functional ubiquitin-proteasome pathway. *Exp Eye Res* 2003;**76**:623–31.
95. van Heyningen R, Waley SG. Neutral proteinases in the lens. 2. Partial purification and properties. *Biochem J* 1963;**86**:92–101.
96. Trayhurn P, van Heyningen R. Neutral proteinase activity in the human lens. *Exp Eye Res* 1976;**22**:251–7.
97. Wagner BJ, Fu SCJ, Margolis JW, Fleshman KR. A synthetic endopeptidase substrate hydrolyzed by the bovine lens neutral proteinase preparation. *Exp Eye Res* 1984;**38**:477–83.

98. Wagner BJ, Margolis JW, Garland D, Roseman JE. Bovine lens neutral proteinase preferentially hydrolyses oxidatively modified glutamine synthetase. *Exp Eye Res* 1986;**43**:1141–3.
99. Ray K, Harris H. Purification of neutral lens endopeptidase; close similarity to a neutral proteinase in pituitary. *Proc Natl Acad Sci USA* 1985;**82**:7545–9.
100. Murakami K, Jahngen JH, Lin SW, Davies KJA, Taylor A. Lens proteosome shows enhanced rates of degradation of hydroxyl radical modified alpha-crystallin. *Free Rad Biol Med* 1990;**8**:217–22.
101. Huang LL, Jahngen-Hodge J, Taylor A. Bovine lens epithelial cells have a ubiquitin-dependent proteolysis system. *Biochim Biophys Acta* 1993;**1175**:181–7.
102. Girao H, Pereira P, Taylor A, Shang F. Subcellular redistribution of components of the ubiquitin-proteasome pathway during lens differentiation and maturation. *Invest Ophthalmol Vis Sci* 2005;**46**:1386–92.
103. Huang LL, Shang F, Nowell Jr. TR, Taylor A. Degradation of differentially oxidized α-crystallins in bovine lens epithelial cells. *Exp Eye Res* 1995;**61**:45–54.
104. Zhang X, Dudek EJ, Liu B, Ding L, Fernandes AF, Liang JJ, et al. Degradation of C-terminal truncated alpha A-crystallins by the ubiquitin-proteasome pathway. *Invest Ophthalmol Vis Sci* 2007;**48**:4200–8.
105. Johnson ES, Bartel B, Seufert W, Varshavsky A. Ubiquitin as a degradation signal. *EMBO J* 1992;**11**:497–505.
106. Spence J, Sadis S, Haas AL, Finley D. A ubiquitin mutant with specific defects in DNA repair and multiubiquitination. *Mol Cell Biol* 1995;**15**:1265–73.
107. Cheng L, Watt R, Piper PW. Polyubiquitin gene expression contributes to oxidative stress resistance in respiratory yeast (*Saccharomyces cerevisiae*). *Mol Gen Genet* 1994;**246**:358–62.
108. Jungmann J, Reins H-A, Schobert C, Jentsch S. Resistance to cadmium mediated by ubiquitin-dependent proteolysis. *Nature* 1993;**361**:369–71.
109. Tsirigotis M, Zhang M, Chiu RK, Wouters BG, Gray DA. Sensitivity of mammalian cells expressing mutant ubiquitin to protein damaging agents. *J Biol Chem* 2001;**11**:11.
110. Shang F, Deng G, Liu Q, Guo W, Haas AL, Crosas B, et al. Lys6-modified ubiquitin inhibits ubiquitin-dependent protein degradation. *J Biol Chem* 2005;**280**:20365–74.
111. Dudek EJ, Shang F, Valverde P, Liu Q, Hobbs M, Taylor A. Selectivity of the ubiquitin pathway for oxidatively modified proteins: relevance to protein precipitation diseases. *FASEB J* 2005;**19**:1707–9.
112. Devamanoharan PS, Ramachandran S, Varma SD. Hydrogen peroxide in the eye lens: radioisotopic determination. *Curr Eye Res* 1991;**10**:831–8.
113. Devamanoharan PS, Varma SD. H2O2 determination in rat lens: chemiluminescent versus radioisotopic methods. *Ophthalmic Res* 1995;**27**(Suppl 1):39–43.
114. Spector A, Garner WH. Hydrogen peroxide and human cataract. *Exp Eye Res* 1981;**33**:673–81.
115. Giblin FJ, McCready JP, Kodama T, Reddy VN. A direct correlation between the levels of ascorbic acid and H2O2 in aqueous humor. *Exp Eye Res* 1984;**38**:87–93.
116. Ramachandran S, Morris SM, Devamanoharan P, Henein M, Varma SD. Radio-isotopic determination of hydrogen peroxide in aqueous humor and urine. *Exp Eye Res* 1991;**53**:503–6.
117. Jahngen-Hodge J, Taylor A, Shang F, Huang LL, Mura C. Oxidative stress to lens crystallins. *Methods Enzymol* 1994;**233**:512–22.
118. Shang F, Huang L, Taylor A. Degradation of native and oxidized beta- and gamma-crystallins using bovine lens epithelial cell and rabbit reticulocyte extracts. *Curr Eye Res* 1994;**13**:423–31.
119. Taylor A, Hobbs M. The 2001 Assessment of Nutritional Influences on risk for cataract in Nutrition and Aging. In: Rosenberg I, Sastre A, editors. *Nutrition and aging*. Basel: Karger; 2002. pp. 163–92.

120. Blazquez H, Fominaya JM, Hofsteenge J. Oxidation of sulfhydryl group of ribonuclease inhibitor in epithelial cells is sufficient for its intracellular degradation. *J Biol Chem* 1996;**271**:18638–42.
121. Shang F, Taylor A. Oxidation and protein degradation in the lens. *Mol Aspects Med* 1997;**18**:339–51.
122. Grune T, Reinheckel T, Davies KJ. Degradation of oxidized proteins in mammalian cells. *FASEB J* 1997;**11**:526–34.
123. Grune T, Davies KJ. Breakdown of oxidized proteins as a part of secondary antioxidant defenses in mammalian cells. *Biofactors* 1997;**6**:165–72.
124. Iwai K, Drake SK, Wehr NB, Weissman AM, LaVaute T, Minato N, et al. Iron-dependent oxidation, ubiquitination, and degradation of iron regulatory protein 2: implications for degradation of oxidized proteins. *Proc Natl Acad Sci USA* 1998;**95**:4924–8.
125. Sitte N, Merker K, Grune T. Proteasome-dependent degradation of oxidized proteins in MRC-5 fibroblasts. *FEBS Lett* 1998;**440**:399–402.
126. Ullrich O, Grune T. Proteasomal degradation of oxidatively damaged endogenous histones in K562 human leukemic cells. *Free Radic Biol Med* 2001;**31**:887–93.
127. Friguet B. Oxidized protein degradation and repair in ageing and oxidative stress. *FEBS Lett* 2006;**580**:2910–6.
128. Zetterberg M, Zhang X, Taylor A, Liu B, Liang JJ, Shang F. Glutathiolation enhances the degradation of {gamma}C-crystallin in lens and reticulocyte lysates, partially via the ubiquitin-proteasome pathway. *Invest Ophthalmol Vis Sci* 2006;**47**:3467–73.
129. Sitte N, Merker K, Von Zglinicki T, Grune T, Davies KJ. Protein oxidation and degradation during cellular senescence of human BJ fibroblasts: part I—effects of proliferative senescence. *FASEB J* 2000;**14**:2495–502.
130. Sitte N, Merker K, von Zglinicki T, Grune T. Protein oxidation and degradation during proliferative senescence of human MRC-5 fibroblasts. *Free Radic Biol Med* 2000;**28**:701–8.
131. Grune T, Reinheckel T, Joshi M, Davies KJA. Proteolysis in cultured liver epithelial cells during oxidative stress. Role of multicatalytic proteinase complex, proteasome. *J Biol Chem* 1995;**270**:2344–51.
132. Grune T, Reinheckel T, Davies KJA. Degradation of oxidized proteins in K562 human hematopoietic cells by proteasome. *J Biol Chem* 1996;**271**:15504–9.
133. Pacifici RE, Kono Y, Davies KJA. Hydrophobicity as the signal for selective degradation of hydroxyl radical modified hemoglobin by the multicatalytic protease complex, proteasome. *J Biol Chem* 1993;**268**:15405–11.
134. Grune T, Merker K, Jung T, Sitte N, Davies KJ. Protein oxidation and degradation during postmitotic senescence. *Free Radic Biol Med* 2005;**39**:1208–15.
135. Grune T, Blasig IE, Sitte N, Roloff B, Haseloff R, Davies KJ. Peroxynitrite increases the degradation of aconitase and other cellular proteins by proteasome. *J Biol Chem* 1998;**273**:10857–62.
136. Giulivi C, Pacifici RE, Davies KJ. Exposure of hydrophobic moieties promotes the selective degradation of hydrogen peroxide-modified hemoglobin by the multicatalytic proteinase complex, proteasome. *Arch Biochem Biophys* 1994;**311**:329–41.
137. Davies KJ, Shringarpure R. Preferential degradation of oxidized proteins by the 20S proteasome may be inhibited in aging and in inflammatory neuromuscular diseases. *Neurology* 2006;**66**:S93–6.
138. Yamanaka K, Ishikawa H, Megumi Y, Tokunaga F, Kanie M, Rouault TA, et al. Identification of the ubiquitin-protein ligase that recognizes oxidized IRP2. *Nat Cell Biol* 2003;**5**:336–40.
139. Lee BH, Lee MJ, Park S, Oh DC, Elsasser S, Chen PC, et al. Enhancement of proteasome activity by a small-molecule inhibitor of USP14. *Nature* 2010;**467**:179–84.

140. Lou MF, Dickerson Jr. JE. Protein-thiol mixed disulfides in human lens. *Exp Eye Res* 1992;**55**:889–96.
141. Chaves JM, Srivastava K, Gupta R, Srivastava OP. Structural and functional roles of deamidation and/or truncation of N- or C-termini in human alpha A-crystallin. *Biochemistry* 2008;**47**:10069–83.
142. Grey AC, Schey KL. Age-related changes in the spatial distribution of human lens alpha-crystallin products by MALDI imaging mass spectrometry. *Invest Ophthalmol Vis Sci* 2009;**50**:4319–29.
143. Hayes VH, Devlin G, Quinlan RA. Truncation of alphaB-crystallin by the myopathy-causing Q151X mutation significantly destabilizes the protein leading to aggregate formation in transfected cells. *J Biol Chem* 2008;**283**:10500–12.
144. Nakajima E, David LL, Riviere MA, Azuma M, Shearer TR. Human and monkey lenses cultured with calcium ionophore form alphaB-crystallin lacking the C-terminal lysine, a prominent feature of some human cataracts. *Invest Ophthalmol Vis Sci* 2009;**50**:5828–36.
145. Srivastava K, Gupta R, Chaves JM, Srivastava OP. Truncated human betaB1-crystallin shows altered structural properties and interaction with human betaA3-crystallin. *Biochemistry* 2009;**48**:7179–89.
146. Stella DR, Floyd KA, Grey AC, Renfrow MB, Schey KL, Barnes S. Tissue localization and solubilities of alphaA-crystallin and its numerous C-terminal truncation products in pre- and postcataractous ICR/f rat lenses. *Invest Ophthalmol Vis Sci* 2010;**51**:5153–61.
147. Su SP, McArthur JD, Truscott RJ, Aquilina JA. Truncation, cross-linking and interaction of crystallins and intermediate filament proteins in the aging human lens. *Biochim Biophys Acta* 2010;**1814**:647–56.
148. Wang Z, Schey KL. Phosphorylation and truncation sites of bovine lens connexin 46 and connexin 50. *Exp Eye Res* 2009;**89**:898–904.
149. Takemoto L, Fujii N, Boyle D. Mechanism of asparagine deamidation during human senile cataractogenesis. *Exp Eye Res* 2001;**72**:559–63.
150. Lampi KJ, Ma Z, Hanson SR, Azuma M, Shih M, Shearer TR, et al. Age-related changes in human lens crystallins identified by two-dimensional electrophoresis and mass spectrometry. *Exp Eye Res* 1998;**67**:31–43.
151. Gupta R, Srivastava OP. Effect of deamidation of asparagine 146 on functional and structural properties of human lens alphaB-crystallin. *Invest Ophthalmol Vis Sci* 2004;**45**:206–14.
152. Lampi KJ, Amyx KK, Ahmann P, Steel EA. Deamidation in human lens betaB2-crystallin destabilizes the dimer. *Biochemistry* 2006;**45**:3146–53.
153. Dudek EJ, Lampi KJ, Lampi JA, Shang F, King J, Wang Y, et al. Ubiquitin proteasome pathway-mediated degradation of proteins: effects due to site-specific substrate deamidation. *Invest Ophthalmol Vis Sci* 2010;**51**:4164–73.
154. Swamy MS, Tsai C, Abraham A, Abraham EC. Glycation mediated lens crystallin aggregation and cross-linking by various sugars and sugar phosphates in vitro. *Exp Eye Res* 1993;**56**:177–85.
155. Cheng R, Lin B, Ortwerth BJ. Rate of formation of AGEs during ascorbate glycation and during aging in human lens tissue. *Biochim Biophys Acta* 2002;**1587**:65–74.
156. Uchiki T, Weikel KA, Jiao W, Shang F, Caceres A, Pawlak D, et al. Glycation-altered proteolysis as a pathobiologic mechanism that links dietary glycemic index, aging, and age-related disease (in non diabetics). *Aging Cell* 2012;**11**(1):1–13.
157. Nagaraj RH, Linetsky M, Stitt AW. The pathogenic role of Maillard reaction in the aging eye. *Amino Acids* 2012;**42**:1205–20.
158. Takizawa N, Takada K, Ohkawa K. Inhibitory effect of nonenzymatic glycation on ubiquitination and ubiquitin-mediated degradation of lysozyme. *Biochem Biophys Res Commun* 1993;**192**:700–6.

159. Wong E, Cuervo AM. Integration of clearance mechanisms: the proteasome and autophagy. *Cold Spring Harb Perspect Biol* 2010;**2**:a006734.
160. Zheng Q, Su H, Tian Z, Wang X. Proteasome malfunction activates macroautophagy in the heart. *Am J Cardiovasc Dis* 2011;**1**:214–26.
161. Reneker LW, Chen H, Overbeek PA. Activation of unfolded protein response in transgenic mouse lenses. *Invest Ophthalmol Vis Sci* 2010;**52**:2100–8.
162. Bornheim R, Muller M, Reuter U, Herrmann H, Bussow H, Magin TM. A dominant vimentin mutant upregulates Hsp70 and the activity of the ubiquitin-proteasome system, and causes posterior cataracts in transgenic mice. *J Cell Sci* 2008;**121**:3737–46.
163. Guo W, Shang F, Liu Q, Urim L, Zhang M, Taylor A. Ubiquitin-proteasome pathway function is required for lens cell proliferation and differentiation. *Invest Ophthalmol Vis Sci* 2006;**47**:2569–75.
164. Awasthi N, Wagner BJ. Suppression of human lens epithelial cell proliferation by proteasome inhibition, a potential defense against posterior capsular opacification. *Invest Ophthalmol Vis Sci* 2006;**47**:4482–9.
165. Liu Q, Shang F, Whitcomb E, Guo W, Li W, Taylor A. Ubiquitin-conjugating enzyme 3 delays human lens epithelial cells in metaphase. *Invest Ophthalmol Vis Sci* 2006;**47**:1302–9.
166. Liu Q, Shang F, Zhang X, Li W, Taylor A. Expression of K6W-ubiquitin inhibits proliferation of human lens epithelial cells. *Mol Vis* 2006;**12**:931–6.
167. Kulka RG, Raboy B, Schuster R, Parag HA, Diamond G, Ciechanover A, et al. A Chinese hamster cell cycle mutant arrested at G2 phase has a temperature-sensitive ubiquitin-activating enzyme, E1. *J Biol Chem* 1988;**263**:15726–31.
168. Girard F, Strausfeld U, Fernandez A, Lamb NJ. Cyclin A is required for the onset of DNA replication in mammalian fibroblasts. *Cell* 1991;**67**:1169–79.
169. Sherr CJ, Roberts JM. CDK inhibitors: positive and negative regulators of G1-phase progression. *Genes Dev* 1999;**13**:1501–12.
170. Harper JW, Adami GR, Wei N, Keyomarsi K, Elledge SJ. The p21 Cdk-interacting protein Cip1 is a potent inhibitor of G1 cyclin-dependent kinases. *Cell* 1993;**75**:805–16.
171. Tvrdik D, Djaborkhel R, Nagy A, Eckschlager T, Raska I, Muller J. Cyclin D-cdk6 complex is targeted by p21(WAF) in growth-arrested lymphoma cells. *J Struct Biol* 2002;**140**:49–56.
172. Clarke DJ. Proteolysis and the cell cycle. *Cell Cycle* 2002;**1**:233–4.
173. Peters JM. The anaphase-promoting complex: proteolysis in mitosis and beyond. *Mol Cell* 2002;**9**:931–43.
174. DeSalle LM, Pagano M. Regulation of the G1 to S transition by the ubiquitin pathway. *FEBS Lett* 2001;**490**:179–89.
175. Yew PR. Ubiquitin-mediated proteolysis of vertebrate G1- and S-phase regulators. *J Cell Physiol* 2001;**187**:1–10.
176. Rape M, Kirschner MW. Autonomous regulation of the anaphase-promoting complex couples mitosis to S-phase entry. *Nature* 2004;**432**:588–95.
177. Rape M, Reddy SK, Kirschner MW. The processivity of multiubiquitination by the APC determines the order of substrate degradation. *Cell* 2006;**124**:89–103.
178. Song L, Rape M. Substrate-specific regulation of ubiquitination by the anaphase-promoting complex. *Cell Cycle* 2011;**10**:52–6.
179. Peters JM. The anaphase promoting complex/cyclosome: a machine designed to destroy. *Nat Rev Mol Cell Biol* 2006;**7**:644–56.
180. Skaar JR, Pagano M. Control of cell growth by the SCF and APC/C ubiquitin ligases. *Curr Opin Cell Biol* 2009;**21**:816–24.
181. Nakayama KI, Hatakeyama S, Nakayama K. Regulation of the cell cycle at the G1-S transition by proteolysis of cyclin E and p27Kip1. *Biochem Biophys Res Commun* 2001;**282**:853–60.

182. Petroski MD, Deshaies RJ. Mechanism of lysine 48-linked ubiquitin-chain synthesis by the cullin-RING ubiquitin-ligase complex SCF-Cdc34. *Cell* 2005;**123**:1107–20.
183. Jin L, Williamson A, Banerjee S, Philipp I, Rape M. Mechanism of ubiquitin-chain formation by the human anaphase-promoting complex. *Cell* 2008;**133**:653–65.
184. Matsumoto ML, Wickliffe KE, Dong KC, Yu C, Bosanac I, Bustos D, et al. K11-linked polyubiquitination in cell cycle control revealed by a K11 linkage-specific antibody. *Mol Cell* 2010;**39**:477–84.
185. Petroski MD, Deshaies RJ. Function and regulation of cullin-RING ubiquitin ligases. *Nat Rev Mol Cell Biol* 2005;**6**:9–20.
186. Kraft C, Herzog F, Gieffers C, Mechtler K, Hagting A, Pines J, et al. Mitotic regulation of the human anaphase-promoting complex by phosphorylation. *EMBO J* 2003;**22**:6598–609.
187. Pesin JA, Orr-Weaver TL. Regulation of APC/C activators in mitosis and meiosis. *Annu Rev Cell Dev Biol* 2008;**24**:475–99.
188. Hansen DV, Loktev AV, Ban KH, Jackson PK. Plk1 regulates activation of the anaphase promoting complex by phosphorylating and triggering SCFbetaTrCP-dependent destruction of the APC Inhibitor Emi1. *Mol Biol Cell* 2004;**15**:5623–34.
189. Eldridge AG, Loktev AV, Hansen DV, Verschuren EW, Reimann JD, Jackson PK. The evi5 oncogene regulates cyclin accumulation by stabilizing the anaphase-promoting complex inhibitor emi1. *Cell* 2006;**124**:367–80.
190. Moshe Y, Bar-On O, Ganoth D, Hershko A. Regulation of the action of early mitotic inhibitor 1 on the anaphase-promoting complex/cyclosome by cyclin-dependent kinases. *J Biol Chem* 2011;**286**:16647–57.
191. Liu Q, Shang F, Guo W, Hobbs M, Valverde P, Reddy V, et al. Regulation of the ubiquitin proteasome pathway in human lens epithelial cells during the cell cycle. *Exp Eye Res* 2004;**78**:197–205.
192. Whitcomb EA, Dudek EJ, Liu Q, Taylor A. Novel control of S phase of the cell cycle by ubiquitin-conjugating enzyme H7. *Mol Biol Cell* 2009;**20**:1–9.
193. Whitcomb EA, Taylor A. Ubiquitin control of S phase: a new role for the ubiquitin conjugating enzyme, UbcH7. *Cell Div* 2009;**4**:17.
194. Guo W, Shang F, Liu Q, Urim L, West-Mays J, Taylor A. Differential regulation of components of the ubiquitin-proteasome pathway during lens cell differentiation. *Invest Ophthalmol Vis Sci* 2004;**45**(4):1194–201.
195. Yang S, Wang-Su ST, Cai H, Wagner BJ. Changes in three types of ubiquitin mRNA and ubiquitin-protein conjugate levels during lens development. *Exp Eye Res* 2002;**74**:595–604.
196. Cai H, Singh I, Wagner BJ. Gene expression of the proteasome in rat lens development. *Exp Eye Res* 1998;**66**:339–46.
197. Caceres A, Shang F, Wawrousek E, Liu Q, Avidan O, Cvekl A, et al. Perturbing the ubiquitin pathway reveals how mitosis is hijacked to denucleate and regulate cell proliferation and differentiation in vivo. *PLoS One* 2010;**5**:e13331.
198. Zandy AJ, Bassnett S. Proteolytic mechanisms underlying mitochondrial degradation in the ocular lens. *Invest Ophthalmol Vis Sci* 2007;**48**:293–302.
199. Obin M, Mesco E, Gong X, Haas AL, Joseph J, Taylor A. Neurite outgrowth in PC12 cells. Distinguishing the roles of ubiquitylation and ubiquitin-dependent proteolysis. *J Biol Chem* 1999;**274**:11789–95.
200. Drinjakovic J, Jung H, Campbell DS, Strochlic L, Dwivedy A, Holt CE. E3 ligase Nedd4 promotes axon branching by downregulating PTEN. *Neuron* 2010;**65**:341–57.
201. Neves DD, Rehen SK, Linden R. Differentiation-dependent sensitivity to cell death induced in the developing retina by inhibitors of the ubiquitin-proteasome proteolytic pathway. *Eur J Neurosci* 2001;**13**:1938–44.

202. Galy A, Neron B, Planque N, Saule S, Eychene A. Activated MAPK/ERK kinase (MEK-1) induces transdifferentiation of pigmented epithelium into neural retina. *Dev Biol* 2002;**248**:251–64.
203. Xu W, Gong L, Haddad MM, Bischof O, Campisi J, Yeh ET, et al. Regulation of microphthalmia-associated transcription factor MITF protein levels by association with the ubiquitin-conjugating enzyme hUBC9. *Exp Cell Res* 2000;**255**:135–43.
204. Wilson R, Goyal L, Ditzel M, Zachariou A, Baker DA, Agapite J, et al. The DIAP1 RING finger mediates ubiquitination of Dronc and is indispensable for regulating apoptosis. *Nat Cell Biol* 2002;**4**:445–50.
205. Ditzel M, Wilson R, Tenev T, Zachariou A, Paul A, Deas E, et al. Degradation of DIAP1 by the N-end rule pathway is essential for regulating apoptosis. *Nat Cell Biol* 2003;**5**:467–73.
206. Wing JP, Schreader BA, Yokokura T, Wang Y, Andrews PS, Huseinovic N, et al. Drosophila Morgue is an F box/ubiquitin conjugase domain protein important for grim-reaper mediated apoptosis. *Nat Cell Biol* 2002;**4**:451–6.
207. Hays R, Wickline L, Cagan R. Morgue mediates apoptosis in the Drosophila melanogaster retina by promoting degradation of DIAP1. *Nat Cell Biol* 2002;**4**:425–31.
208. House CM, Frew IJ, Huang HL, Wiche G, Traficante N, Nice E, et al. A binding motif for Siah ubiquitin ligase. *Proc Natl Acad Sci USA* 2003;**100**:3101–6.
209. Bogdan S, Senkel S, Esser F, Ryffel GU, Pogge v Strandmann E. Misexpression of Xsiah-2 induces a small eye phenotype in Xenopus. *Mech Dev* 2001;**103**:61–9.
210. Li Y, Baker NE. The roles of cis-inactivation by Notch ligands and of neuralized during eye and bristle patterning in Drosophila. *BMC Dev Biol* 2004;**4**:5.
211. Deblandre GA, Lai EC, Kintner C. Xenopus neuralized is a ubiquitin ligase that interacts with XDelta1 and regulates Notch signaling. *Dev Cell* 2001;**1**:795–806.
212. Lai EC, Deblandre GA, Kintner C, Rubin GM. Drosophila neuralized is a ubiquitin ligase that promotes the internalization and degradation of delta. *Dev Cell* 2001;**1**:783–94.
213. Parks AL, Turner FR, Muskavitch MA. Relationships between complex Delta expression and the specification of retinal cell fates during Drosophila eye development. *Mech Dev* 1995;**50**:201–16.
214. Hicke L. A new ticket for entry into budding vesicles-ubiquitin. *Cell* 2001;**106**:527–30.
215. del Alamo D, Mlodzik M. Frizzled/PCP-dependent asymmetric neuralized expression determines R3/R4 fates in the Drosophila eye. *Dev Cell* 2006;**11**:887–94.
216. Lee JD, Amanai K, Shearn A, Treisman JE. The ubiquitin ligase Hyperplastic discs negatively regulates hedgehog and decapentaplegic expression by independent mechanisms. *Development* 2002;**129**:5697–706.
217. Ou CY, Lin YF, Chen YJ, Chien CT. Distinct protein degradation mechanisms mediated by Cul1 and Cul3 controlling Ci stability in Drosophila eye development. *Genes Dev* 2002;**16**:2403–14.
218. Ou CY, Wang CH, Jiang J, Chien CT. Suppression of Hedgehog signaling by Cul3 ligases in proliferation control of retinal precursors. *Dev Biol* 2007;**308**:106–19.
219. Bizzi A, Schaetzle B, Patton A, Gambetti P, Autilio-Gambetti L. Axonal transport of two major components of the ubiquitin system: free ubiquitin and ubiquitin carboxyl-terminal hydrolase PGP 9.5. *Brain Res* 1991;**548**:292–9.
220. Chen ST, von Bussmann KA, Garey LJ, Jen LS. Protein gene product 9.5-immunoreactive retinal neurons in normal developing rats and rats with optic nerve or tract lesion. *Brain Res Dev Brain Res* 1994;**78**:265–72.
221. Piccinini M, Merighi A, Bruno R, Cascio P, Curto M, Mioletti S, et al. Affinity purification and characterization of protein gene product 9.5 (PGP9.5) from retina. *Biochem J* 1996;**318**:711–6.

222. Liu Y, Fallon L, Lashuel HA, Liu Z, Lansbury Jr. PT. The UCH-L1 gene encodes two opposing enzymatic activities that affect alpha-synuclein degradation and Parkinson's disease susceptibility. *Cell* 2002;**111**:209–18.
223. Proctor CJ, Tangeman PJ, Ardley HC. Modelling the role of UCH-L1 on protein aggregation in age-related neurodegeneration. *PLoS One* 2010;**5**:e13175.
224. Wilkinson KD, Deshpande S, Larsen CN. Comparisons of neuronal (PGP 9.5) and non-neuronal ubiquitin C-terminal hydrolases. *Biochem Soc Trans* 1992;**20**:631–7.
225. Sano Y, Furuta A, Setsuie R, Kikuchi H, Wang YL, Sakurai M, et al. Photoreceptor cell apoptosis in the retinal degeneration of Uchl3-deficient mice. *Am J Pathol* 2006;**169**:132–41.
226. Poeck B, Fischer S, Gunning D, Zipursky SL, Salecker I. Glial cells mediate target layer selection of retinal axons in the developing visual system of Drosophila. *Neuron* 2001;**29**:99–113.
227. Chen X, Overstreet E, Wood SA, Fischer JA. On the conservation of function of the Drosophila fat facets deubiquitinating enzyme and Fam, its mouse homolog. *Dev Genes Evol* 2000;**210**:603–10.
228. Fischer-Vize JA, Rubin GM, Lehmann R. The fat facets gene is required for Drosophila eye and embryo development. *Development* 1992;**116**:985–1000.
229. Huang Y, Baker RT, Fischer-Vize JA. Control of cell fate by a deubiquitinating enzyme encoded by the fat facets gene. *Science* 1995;**270**:1828–31.
230. Fischer JA, Leavell SK, Li Q. Mutagenesis screens for interacting genes reveal three roles for fat facets during Drosophila eye development. *Dev Genet* 1997;**21**:167–74.
231. Overstreet E, Chen X, Wendland B, Fischer JA. Either part of a drosophila epsin protein, divided after the ENTH domain, functions in endocytosis of delta in the developing eye. *Curr Biol* 2003;**13**:854–60.
232. Chen H, Fre S, Slepnev VI, Capua MR, Takei K, Butler MH, et al. Epsin is an EH-domain-binding protein implicated in clathrin-mediated endocytosis. *Nature* 1998;**394**:793–7.
233. Carthew RW, Xu C. Endocytosis: why not wait to deubiquitinate? *Curr Biol* 2000;**10**:R532–4.
234. Wood SA, Pascoe WS, Ru K, Yamada T, Hirchenhain J, Kemler R, et al. Cloning and expression analysis of a novel mouse gene with sequence similarity to the Drosophila fat facets gene. *Mech Dev* 1997;**63**:29–38.
235. Taya S, Yamamoto T, Kanai-Azuma M, Wood SA, Kaibuchi K. The deubiquitinating enzyme Fam interacts with and stabilizes beta-catenin. *Genes Cells* 1999;**4**:757–67.
236. Le Bras S, Loyer N, Le Borgne R. The multiple facets of ubiquitination in the regulation of notch signaling pathway. *Traffic* 2010;**12**:149–61.
237. Weinmaster G, Fischer JA. Notch ligand ubiquitylation: what is it good for? *Dev Cell* 2011;**21**:134–44.
238. Tauriello DV, Maurice MM. The various roles of ubiquitin in Wnt pathway regulation. *Cell Cycle* 2010;**9**:3700–9.
239. Narimatsu M, Bose R, Pye M, Zhang L, Miller B, Ching P, et al. Regulation of planar cell polarity by Smurf ubiquitin ligases. *Cell* 2009;**137**:295–307.
240. Harhaj EW, Dixit VM. Deubiquitinases in the regulation of NF-kappaB signaling. *Cell Res* 2011;**21**:22–39.
241. Ungureanu D, Silvennoinen O. SLIM trims STATs: ubiquitin E3 ligases provide insights for specificity in the regulation of cytokine signaling. *Sci STKE* 2005;**2005**:pe49.
242. Carbia-Nagashima A, Arzt E. Intracellular proteins and mechanisms involved in the control of gp130/JAK/STAT cytokine signaling. *IUBMB Life* 2004;**56**:83–8.
243. Beinke S, Ley SC. Functions of NF-kappaB1 and NF-kappaB2 in immune cell biology. *Biochem J* 2004;**382**:393–409.
244. Adhikari A, Xu M, Chen ZJ. Ubiquitin-mediated activation of TAK1 and IKK. *Oncogene* 2007;**26**:3214–26.

245. Yamashita M, Fatyol K, Jin C, Wang X, Liu Z, Zhang YE. TRAF6 mediates Smad-independent activation of JNK and p38 by TGF-beta. *Mol Cell* 2008;**31**:918–24.
246. Kallio PJ, Wilson WJ, O'Brien S, Makino Y, Poellinger L. Regulation of the hypoxia-inducible transcription factor 1alpha by the ubiquitin-proteasome pathway. *J Biol Chem* 1999;**274**: 6519–25.
247. Salceda S, Caro J. Hypoxia-inducible factor 1alpha (HIF-1alpha) protein is rapidly degraded by the ubiquitin-proteasome system under normoxic conditions. Its stabilization by hypoxia depends on redox-induced changes. *J Biol Chem* 1997;**272**:22642–7.
248. Vallabhapurapu S, Karin M. Regulation and function of NF-kappaB transcription factors in the immune system. *Annu Rev Immunol* 2009;**27**:693–733.
249. Chen LF, Greene WC. Shaping the nuclear action of NF-kappaB. *Nat Rev Mol Cell Biol* 2004;**5**:392–401.
250. Scheidereit C. IkappaB kinase complexes: gateways to NF-kappaB activation and transcription. *Oncogene* 2006;**25**:6685–705.
251. Ghosh S, Hayden MS. New regulators of NF-kappaB in inflammation. *Nat Rev Immunol* 2008;**8**:837–48.
252. Sun SC, Ley SC. New insights into NF-kappaB regulation and function. *Trends Immunol* 2008;**29**:469–78.
253. Zandi E, Rothwarf DM, Delhase M, Hayakawa M, Karin M. The IkappaB kinase complex (IKK) contains two kinase subunits, IKKalpha and IKKbeta, necessary for IkappaB phosphorylation and NF-kappaB activation. *Cell* 1997;**91**:243–52.
254. Chen F, Bhatia D, Chang Q, Castranova V. Finding NEMO by K63-linked polyubiquitin chain. *Cell Death Differ* 2006;**13**:1835–8.
255. Brown K, Gerstberger S, Carlson L, Franzoso G, Siebenlist U. Control of IκB proteolysis by site-specific, signal-induced phosphorylation. *Science* 1995;**267**:1485–91.
256. Harper N, Hughes MA, Farrow SN, Cohen GM, MacFarlane M. Protein kinase C modulates tumor necrosis factor-related apoptosis-inducing ligand-induced apoptosis by targeting the apical events of death receptor signaling. *J Biol Chem* 2003;**278**:44338–47.
257. Micheau O, Tschopp J. Induction of TNF receptor I-mediated apoptosis via two sequential signaling complexes. *Cell* 2003;**114**:181–90.
258. Lee TH, Shank J, Cusson N, Kelliher MA. The kinase activity of Rip1 is not required for tumor necrosis factor-alpha-induced IkappaB kinase or p38 MAP kinase activation or for the ubiquitination of Rip1 by Traf2. *J Biol Chem* 2004;**279**:33185–91.
259. Ea CK, Deng L, Xia ZP, Pineda G, Chen ZJ. Activation of IKK by TNFalpha requires site-specific ubiquitination of RIP1 and polyubiquitin binding by NEMO. *Mol Cell* 2006;**22**:245–57.
260. Wu CJ, Conze DB, Li T, Srinivasula SM, Ashwell JD. Sensing of Lys 63-linked polyubiquitination by NEMO is a key event in NF-kappaB activation. *Nat Cell Biol* 2006;**8**:398–406.
261. Chen ZJ, Bhoj V, Seth RB. Ubiquitin, TAK1 and IKK: is there a connection? *Cell Death Differ* 2006;**13**:687–92.
262. Wang C, Deng L, Hong M, Akkaraju GR, Inoue J, Chen ZJ. TAK1 is a ubiquitin-dependent kinase of MKK and IKK. *Nature* 2001;**412**:346–51.
263. Chen ZJ, Parent L, Maniatis T. Site-specific phosphorylation of IκBα by a novel ubiquitination-dependent protein kinase activity. *Cell* 1996;**84**:853–62.
264. Alkalay I, Yaron A, Hatzubai A, Orian A, Ciechanover A, Ben-Neriah Y. Stimulation-dependent IκB phosporylkation marks NF-κB inhibitor for degradation via the ubiquitin-proteasome pathway. *Proc Natl Acad Sci USA* 1995;**92**:10599–603.
265. Gonen H, Bercovich B, Orian A, Carrano A, Takizawa C, Yamanaka K, et al. Identification of the ubiquitin carrier proteins, E2s, involved in signal-induced conjugation and subsequent degradation of IkappaBalpha. *J Biol Chem* 1999;**274**:14823–30.

266. Yaron A, Gonen H, Alkalay I, Hatzubai A, Jung S, Beyth S, et al. Inhibition of NF-kappa-B cellular function via specific targeting of the I-kappa-B-ubiquitin ligase. *EMBO J* 1997;**16**:6486–94.
267. Xiao G, Harhaj EW, Sun SC. NF-kappaB-inducing kinase regulates the processing of NF-kappaB2 p100. *Mol Cell* 2001;**7**:401–9.
268. Vallabhapurapu S, Matsuzawa A, Zhang W, Tseng PH, Keats JJ, Wang H, et al. Nonredundant and complementary functions of TRAF2 and TRAF3 in a ubiquitination cascade that activates NIK-dependent alternative NF-kappaB signaling. *Nat Immunol* 2008;**9**:1364–70.
269. Senftleben U, Cao Y, Xiao G, Greten FR, Krahn G, Bonizzi G, et al. Activation by IKKalpha of a second, evolutionary conserved. NF-kappa B signaling pathway. *Science* 2001;**293**:1495–9.
270. Rogers PR, Song J, Gramaglia I, Killeen N, Croft M. OX40 promotes Bcl-xL and Bcl-2 expression and is essential for long-term survival of CD4 T cells. *Immunity* 2001;**15**:445–55.
271. Angileri FF, Aguennouz M, Conti A, La Torre D, Cardali S, Crupi R, et al. Nuclear factor-kappaB activation and differential expression of survivin and Bcl-2 in human grade 2-4 astrocytomas. *Cancer* 2008;**112**:2258–66.
272. Matsuzawa A, Ichijo H. Molecular mechanisms of the decision between life and death: regulation of apoptosis by apoptosis signal-regulating kinase 1. *J Biochem (Tokyo)* 2001;**130**: 1–8.
273. Dudek EJ, Shang F, Taylor A. H(2)O(2)-mediated oxidative stress activates NF-kappaB in lens epithelial cells. *Free Radic Biol Med* 2001;**31**:651–8.
274. Fernandes AF, Guo W, Zhang X, Gallagher M, Ivan M, Taylor A, et al. Proteasome-dependent regulation of signal transduction in retinal pigment epithelial cells. *Exp Eye Res* 2006;**83** (6):1472–81.
275. Wu M, Bian Q, Liu Y, Fernandes AF, Taylor A, Pereira P, et al. Sustained oxidative stress inhibits NF-kappaB activation partially via inactivating the proteasome. *Free Radic Biol Med* 2009;**46**:62–9.
276. Parry GC, Martin T, Felts KA, Cobb RR. IL-1beta-induced monocyte chemoattractant protein-1 gene expression in endothelial cells is blocked by proteasome inhibitors. *Arterioscler Thromb Vasc Biol* 1998;**18**:934–40.
277. Wang GL, Jiang BH, Rue EA, Semenza GL. Hypoxia-inducible factor 1 is a basic-helix-loop-helix-PAS heterodimer regulated by cellular O2 tension. *Proc Natl Acad Sci USA* 1995;**92**:5510–4.
278. Semenza GL, Wang GL. A nuclear factor induced by hypoxia via de novo protein synthesis binds to the human erythropoietin gene enhancer at a site required for transcriptional activation. *Mol Cell Biol* 1992;**12**:5447–54.
279. Maxwell PH, Ratcliffe PJ. Oxygen sensors and angiogenesis. *Semin Cell Dev Biol* 2002;**13**:29–37.
280. Wenger RH. Cellular adaptation to hypoxia: $O_2$-sensing protein hydroxylases, hypoxia-inducible transcription factors, and $O_2$-regulated gene expression. *FASEB J* 2002;**16**:1151–62.
281. Manalo DJ, Rowan A, Lavoie T, Natarajan L, Kelly BD, Ye SQ, et al. Transcriptional regulation of vascular endothelial cell responses to hypoxia by HIF-1. *Blood* 2005;**105**:659–69.
282. Ivan M, Kondo K, Yang H, Kim W, Valiando J, Ohh M, et al. HIFalpha targeted for VHL-mediated destruction by proline hydroxylation: implications for O2 sensing. *Science* 2001;**292**:464–8.
283. Jaakkola P, Mole DR, Tian YM, Wilson MI, Gielbert J, Gaskell SJ, et al. Targeting of HIF-alpha to the von Hippel-Lindau ubiquitylation complex by O2-regulated prolyl hydroxylation. *Science* 2001;**292**:468–72.
284. Masson N, Willam C, Maxwell PH, Pugh CW, Ratcliffe PJ. Independent function of two destruction domains in hypoxia-inducible factor-alpha chains activated by prolyl hydroxylation. *EMBO J* 2001;**20**:5197–206.

285. Bento CF, Fernandes R, Ramalho J, Marques C, Shang F, Taylor A, et al. The chaperone-dependent ubiquitin ligase CHIP targets HIF-1alpha for degradation in the presence of methylglyoxal. *PLoS One* 2010;**5**:e15062.
286. Li Z, Wang D, Messing EM, Wu G. VHL protein-interacting deubiquitinating enzyme 2 deubiquitinates and stabilizes HIF-1alpha. *EMBO Rep* 2005;**6**:373–8.
287. Arjamaa O, Nikinmaa M. Oxygen-dependent diseases in the retina: role of hypoxia-inducible factors. *Exp Eye Res* 2006;**83**:473–83.
288. Arjamaa O, Nikinmaa M, Salminen A, Kaarniranta K. Regulatory role of HIF-1alpha in the pathogenesis of age-related macular degeneration (AMD). *Ageing Res Rev* 2009;**8**:349–58.
289. DeNiro M, Alsmadi O, Al-Mohanna F. Modulating the hypoxia-inducible factor signaling pathway as a therapeutic modality to regulate retinal angiogenesis. *Exp Eye Res* 2009;**89**:700–17.
290. Obin MS, Jahngen-Hodge J, Nowell T, Taylor A. Ubiquitinylation and ubiquitin-dependent proteolysis in vertebrate photoreceptors (rod outer segments). Evidence for ubiquitinylation of Gt and rhodopsin. *J Biol Chem* 1996;**271**:14473–84.
291. Obin M, Lee BY, Meinke G, Bohm A, Lee RH, Gaudet R, et al. Ubiquitylation of the transducin betagamma subunit complex. Regulation by phosducin. *J Biol Chem* 2002;**277**:44566–75.
292. Burns ME, Baylor DA. Activation, deactivation, and adaptation in vertebrate photoreceptor cells. *Annu Rev Neurosci* 2001;**24**:779–805.
293. Sokolov M, Lyubarsky AL, Strissel KJ, Savchenko AB, Govardovskii VI, Pugh Jr. EN, et al. Massive light-driven translocation of transducin between the two major compartments of rod cells: a novel mechanism of light adaptation. *Neuron* 2002;**34**:95–106.
294. Pirlich M, Muller C, Sandig G, Jakstadt M, Sitte N, Lochs H, et al. Increased proteolysis after single-dose exposure with hepatotoxins in HepG2 cells. *Free Radic Biol Med* 2002;**33**:283–91.
295. Pickering AM, Koop AL, Teoh CY, Ermak G, Grune T, Davies KJ. The immunoproteasome, the 20S proteasome and the PA28alphabeta proteasome regulator are oxidative-stress-adaptive proteolytic complexes. *Biochem J* 2010;**432**:585–94.
296. Zhang X, Zhou J, Fernandes AF, Sparrow JR, Pereira P, Taylor A, et al. The proteasome: a target of oxidative damage in cultured human retina pigment epithelial cells. *Invest Ophthalmol Vis Sci* 2008;**49**:3622–30.
297. Ethen CM, Hussong SA, Reilly C, Feng X, Olsen TW, Ferrington DA. Transformation of the proteasome with age-related macular degeneration. *FEBS Lett* 2007;**581**:885–90.
298. Obin M, Shang F, Gong X, Handelman G, Blumberg J, Taylor A. Redox regulation of ubiquitin-conjugating enzymes: mechanistic insights using the thiol-specific oxidant diamide. *FASEB J* 1998;**12**:561–9.
299. Yao D, Gu Z, Nakamura T, Shi ZQ, Ma Y, Gaston B, et al. Nitrosative stress linked to sporadic Parkinson's disease: S-nitrosylation of parkin regulates its E3 ubiquitin ligase activity. *Proc Natl Acad Sci USA* 2004;**101**:10810–4.
300. Ishii T, Sakurai T, Usami H, Uchida K. Oxidative modification of proteasome: identification of an oxidation-sensitive subunit in 26 s proteasome. *Biochemistry* 2005;**44**:13893–901.
301. Caballero M, Liton PB, Epstein DL, Gonzalez P. Proteasome inhibition by chronic oxidative stress in human trabecular meshwork cells. *Biochem Biophys Res Commun* 2003;**308**:346–52.
302. Okada K, Wangpoengtrakul C, Osawa T, Toyokuni S, Tanaka K, Uchida K. 4-Hydroxy-2-nonenal-mediated impairment of intracellular proteolysis during oxidative stress. Identification of proteasomes as target molecules. *J Biol Chem* 1999;**274**:23787–93.
303. Conconi M, Petropoulos I, Emod I, Turlin E, Biville F, Friguet B. Protection from oxidative inactivation of the 20S proteasome by heat-shock protein 90. *Biochem J* 1998;**333**(Pt 2):407–15.
304. Ding Q, Keller JN. Proteasome inhibition in oxidative stress neurotoxicity: implications for heat shock proteins. *J Neurochem* 2001;**77**:1010–7.

305. Shringarpure R, Grune T, Sitte N, Davies KJ. 4-Hydroxynonenal-modified amyloid-beta peptide inhibits the proteasome: possible importance in Alzheimer's disease. *Cell Mol Life Sci* 2000;**57**:1802–9.
306. Reinheckel T, Ullrich O, Sitte N, Grune T. Differential impairment of 20S and 26S proteasome activities in human hematopoietic K562 cells during oxidative stress. *Arch Biochem Biophys* 2000;**377**:65–8.
307. Scrofano MM, Shang F, Nowell Jr. TR, Gong X, Smith DE, Kelliher M, et al. Calorie restriction, stress and the ubiquitin-dependent pathway in mouse livers. *Mech Ageing Dev* 1998;**105**:273–90.
308. Scrofano MM, Shang F, Nowell Jr. TR, Gong X, Smith DE, Kelliher M, et al. Aging, calorie restriction and ubiquitin-dependent proteolysis in the livers of Emory mice. *Mech Ageing Dev* 1998;**101**:277–96.
309. Adamo AM, Pasquini LA, Moreno MB, Oteiza PI, Soto EF, Pasquini JM. Effect of oxidant systems on the ubiquitylation of proteins in the central nervous system. *J Neurosci Res* 1999;**55**:523–31.
310. Ramanathan M, Hassanain M, Levitt M, Seth A, Tolman JS, Fried VA, et al. Oxidative stress increases ubiquitin–protein conjugates in synaptosomes. *Neuroreport* 1999;**10**:3797–802.
311. Figueiredo-Pereira ME, Yakushin S, Cohen G. Accumulation of ubiquitinated proteins in mouse neuronal cells induced by oxidative stress. *Mol Biol Reports* 1997;**24**:25–38.
312. Shang F, Dudek E, Liu Q, Boulton ME, Taylor A. Protein quality control by the ubiquitin proteolytic pathway: roles in resistance to oxidative stress and disease. *Isr J Chem* 2006;**46**: 145–58.
313. Vernace VA, Arnaud L, Schmidt-Glenewinkel T, Figueiredo-Pereira ME. Aging perturbs 26S proteasome assembly in Drosophila melanogaster. *FASEB J* 2007;**21**:2672–82.
314. Ohtsuka H, Takahashi R, Goto S. Age-related accumulation of high-molecular-weight ubiquitin protein conjugates in mouse brains. *J Gerontol A Biol Sci Med Sci* 1995;**50**:B277–81.
315. Naash MI, Al-Ubaidi MR, Anderson RE. Light exposure induces ubiquitin conjugation and degradation activities in the rat retina. *Invest Ophthalmol Vis Sci* 1997;**38**:2344–54.
316. Bence NF, Sampat RM, Kopito RR. Impairment of the ubiquitin-proteasome system by protein aggregation. *Science* 2001;**292**:1552–5.
317. Bennett EJ, Bence NF, Jayakumar R, Kopito RR. Global impairment of the ubiquitin-proteasome system by nuclear or cytoplasmic protein aggregates precedes inclusion body formation. *Mol Cell* 2005;**17**:351–65.
318. Friguet B, Szweda LI. Inhibition of the multicatalytic proteinase (proteasome) by 4-hydroxy-2-nonenal cross-linked protein. *FEBS Lett* 1997;**405**:21–5.
319. Bulteau AL, Moreau M, Nizard C, Friguet B. Impairment of proteasome function upon UVA- and UVB-irradiation of human keratinocytes. *Free Radic Biol Med* 2002;**32**:1157–70.
320. Grune T, Jung T, Merker K, Davies KJ. Decreased proteolysis caused by protein aggregates, inclusion bodies, plaques, lipofuscin, ceroid, and 'aggresomes' during oxidative stress, aging, and disease. *Int J Biochem Cell Biol* 2004;**36**:2519–30.
321. Breusing N, Grune T. Regulation of proteasome-mediated protein degradation during oxidative stress and aging. *Biol Chem* 2008;**389**:203–9.
322. Villeneuve NF, Lau A, Zhang DD. Regulation of the Nrf2-Keap1 antioxidant response by the ubiquitin proteasome system: an insight into Cullin-ring ubiquitin ligases. *Antioxid Redox Signal* 2010;**13**(11):1699–712.
323. Nguyen T, Sherratt PJ, Huang HC, Yang CS, Pickett CB. Increased protein stability as a mechanism that enhances Nrf2-mediated transcriptional activation of the antioxidant response element. Degradation of Nrf2 by the 26 S proteasome. *J Biol Chem* 2003;**278**: 4536–41.

324. Ha KN, Chen Y, Cai J, Sternberg Jr. P. Increased glutathione synthesis through an ARE-Nrf2-dependent pathway by zinc in the RPE: implication for protection against oxidative stress. *Invest Ophthalmol Vis Sci* 2006;**47**:2709–15.
325. Mandal MN, Patlolla JM, Zheng L, Agbaga MP, Tran JT, Wicker L, et al. Curcumin protects retinal cells from light-and oxidant stress-induced cell death. *Free Radic Biol Med* 2009;**46**:672–9.
326. Pitha-Rowe I, Liby K, Royce D, Sporn M. Synthetic triterpenoids attenuate cytotoxic retinal injury: cross-talk between Nrf2 and PI3K/AKT signaling through inhibition of the lipid phosphatase PTEN. *Invest Ophthalmol Vis Sci* 2009;**50**:5339–47.
327. Venugopal R, Jaiswal AK. Nrf1 and Nrf2 positively and c-Fos and Fra1 negatively regulate the human antioxidant response element-mediated expression of NAD(P)H:quinone oxidoreductase1 gene. *Proc Natl Acad Sci USA* 1996;**93**:14960–5.
328. Wild AC, Moinova HR, Mulcahy RT. Regulation of gamma-glutamylcysteine synthetase subunit gene expression by the transcription factor Nrf2. *J Biol Chem* 1999;**274**:33627–36.
329. Nguyen T, Huang HC, Pickett CB. Transcriptional regulation of the antioxidant response element. Activation by Nrf2 and repression by MafK. *J Biol Chem* 2000;**275**:15466–73.
330. Dhakshinamoorthy S, Jaiswal AK. Functional characterization and role of INrf2 in antioxidant response element-mediated expression and antioxidant induction of NAD(P)H:quinone oxidoreductase1 gene. *Oncogene* 2001;**20**:3906–17.
331. Itoh K, Wakabayashi N, Katoh Y, Ishii T, Igarashi K, Engel JD, et al. Keap1 represses nuclear activation of antioxidant responsive elements by Nrf2 through binding to the amino-terminal Neh2 domain. *Genes Dev* 1999;**13**:76–86.
332. Cullinan SB, Gordan JD, Jin J, Harper JW, Diehl JA. The Keap1-BTB protein is an adaptor that bridges Nrf2 to a Cul3-based E3 ligase: oxidative stress sensing by a Cul3-Keap1 ligase. *Mol Cell Biol* 2004;**24**:8477–86.
333. Zhang DD, Lo SC, Cross JV, Templeton DJ, Hannink M. Keap1 is a redox-regulated substrate adaptor protein for a Cul3-dependent ubiquitin ligase complex. *Mol Cell Biol* 2004;**24**:10941–53.
334. Furukawa M, Xiong Y. BTB protein Keap1 targets antioxidant transcription factor Nrf2 for ubiquitination by the Cullin 3-Roc1 ligase. *Mol Cell Biol* 2005;**25**:162–71.
335. Sekhar KR, Yan XX, Freeman ML. Nrf2 degradation by the ubiquitin proteasome pathway is inhibited by KIAA0132, the human homolog to INrf2. *Oncogene* 2002;**21**:6829–34.
336. Stewart D, Killeen E, Naquin R, Alam S, Alam J. Degradation of transcription factor Nrf2 via the ubiquitin-proteasome pathway and stabilization by cadmium. *J Biol Chem* 2003;**278**:2396–402.
337. Kaspar JW, Niture SK, Jaiswal AK. Nrf2:INrf2 (Keap1) signaling in oxidative stress. *Free Radic Biol Med* 2009;**47**:1304–9.
338. Plafker KS, Nguyen L, Barneche M, Mirza S, Crawford D, Plafker SM. The ubiquitin-conjugating enzyme UbcM2 can regulate the stability and activity of the antioxidant transcription factor Nrf2. *J Biol Chem* 2010;**285**:23064–74.
339. Dreger H, Westphal K, Weller A, Baumann G, Stangl V, Meiners S, et al. Nrf2-dependent upregulation of antioxidative enzymes: a novel pathway for proteasome inhibitor-mediated cardioprotection. *Cardiovasc Res* 2009;**83**:354–61.
340. Dreger H, Westphal K, Wilck N, Baumann G, Stangl V, Stangl K, et al. Protection of vascular cells from oxidative stress by proteasome inhibition depends on Nrf2. *Cardiovasc Res* 2010;**85**:395–403.
341. Westphal K, Stangl V, Fahling M, Dreger H, Weller A, Baumann G, et al. Human-specific induction of glutathione peroxidase-3 by proteasome inhibition in cardiovascular cells. *Free Radic Biol Med* 2009;**47**:1652–60.

342. Chen J, Regan RF. Increasing expression of heme oxygenase-1 by proteasome inhibition protects astrocytes from heme-mediated oxidative injury. *Curr Neurovasc Res* 2005;**2**:189–96.
343. Meiners S, Ludwig A, Lorenz M, Dreger H, Baumann G, Stangl V, et al. Nontoxic proteasome inhibition activates a protective antioxidant defense response in endothelial cells. *Free Radic Biol Med* 2006;**40**:2232–41.
344. Yamamoto N, Sawada H, Izumi Y, Kume T, Katsuki H, Shimohama S, et al. Proteasome inhibition induces glutathione synthesis and protects cells from oxidative stress: relevance to Parkinson disease. *J Biol Chem* 2007;**282**:4364–72.
345. Kapeta S, Chondrogianni N, Gonos ES. Nuclear erythroid factor 2-mediated proteasome activation delays senescence in human fibroblasts. *J Biol Chem* 2010;**285**:8171–84.
346. Lee CS, Tee LY, Warmke T, Vinjamoori A, Cai A, Fagan AM, et al. A proteasomal stress response: pre-treatment with proteasome inhibitors increases proteasome activity and reduces neuronal vulnerability to oxidative injury. *J Neurochem* 2004;**91**:996–1006.
347. Cano M, Thimmalappula R, Fujihara M, Nagai N, Sporn M, Wang AL, et al. Cigarette smoking, oxidative stress, the anti-oxidant response through Nrf2 signaling, and age-related macular degeneration. *Vision Res* 2010;**50**:652–64.
348. Wei Y, Gong J, Yoshida T, Eberhart CG, Xu Z, Kombairaju P, et al. Nrf2 has a protective role against neuronal and capillary degeneration in retinal ischemia-reperfusion injury. *Free Radic Biol Med* 2011;**51**:216–24.
349. Zhao Z, Chen Y, Wang J, Sternberg P, Freeman ML, Grossniklaus HE, et al. Age-related retinopathy in NRF2-deficient mice. *PLoS One* 2011;**6**:e19456.
350. Ozawa Y, Nakao K, Kurihara T, Shimazaki T, Shimmura S, Ishida S, et al. Roles of STAT3/SOCS3 pathway in regulating the visual function and ubiquitin-proteasome-dependent degradation of rhodopsin during retinal inflammation. *J Biol Chem* 2008;**283**:24561–70.
351. Seufert W, Jentsch S. Ubiquitin-conjugating enzymes UBC4 and UBC5 mediate selective degradation of short-lived and abnormal proteins. *EMBO J* 1990;**9**:543–50.
352. Jahngen-Hodge J, Cyr D, Laxman E, Taylor A. Ubiquitin and ubiquitin conjugates in human lens. *Exp Eye Res* 1992;**55**:897–902.
353. Andersson M, SjOstrand J, Karlsson J. Proteolytic cleavage of N-Succ-Leu-Leu-Val-Tyr-AMC by the proteasome in lens epithelium from clear and cataractous human lenses. *Exp Eye Res* 1998;**67**:231–6.
354. Hosler MR, Wang-Su ST, Wagner BJ. Targeted disruption of specific steps of the ubiquitin-proteasome pathway by oxidation in lens epithelial cells. *Int J Biochem Cell Biol* 2003;**35**:685–97.
355. Andersson M, Sjostrand J, Karlsson JO. Differential inhibition of three peptidase activities of the proteasome in human lens epithelium by heat and oxidation. *Exp Eye Res* 1999;**69**:129–38.
356. Andersson M, Sjostrand J, Petersen A, Honarvar AK, Karlsson JO. Caspase and proteasome activity during staurosporin-induced apoptosis in lens epithelial cells. *Invest Ophthalmol Vis Sci* 2000;**41**:2623–32.
357. Mauger C, Del-Favero J, Ceuterick C, Lubke U, van Broeckhoven C, Martin J. Identification and localization of ataxin-7 in brain and retina of a patient with cerebellar ataxia type II using anti-peptide antibody. *Brain Res Mol Brain Res* 1999;**74**:35–43.
358. Naash M, Izbicka E, Anderson RE. Rat retina has an active and stable ubiquitin-protein conjugating system. *J Neurosci Res* 1991;**30**:433–41.
359. Yoshida T, Fukatsu R, Tsuzuki K, Aizawa Y, Hayashi Y, Sasaki N, et al. Amyloid precursor protein, A beta and amyloid-associated proteins involved in chloroquine retinopathy in ratsimmunopathological studies. *Brain Res* 1997;**764**:283–8.
360. Loeffler KU, Mangini NJ. Immunolocalization of ubiquitin and related enzymes in human retina and retinal pigment epithelium. *Graefes Arch Clin Exp Ophthalmol* 1997;**235**:248–54.

361. Anderson DH, Talaga KC, Rivest AJ, Barron E, Hageman GS, Johnson LV. Characterization of beta amyloid assemblies in drusen: the deposits associated with aging and age-related macular degeneration. *Exp Eye Res* 2004;**78**:243–56.
362. Leger F, Fernagut PO, Canron MH, Leoni S, Vital C, Tison F, et al. Protein aggregation in the aging retina. *J Neuropathol Exp Neurol* 2011;**70**:63–8.
363. Crabb JW, Miyagi M, Gu X, Shadrach K, West KA, Sakaguchi H, et al. Drusen proteome analysis: an approach to the etiology of age-related macular degeneration. *Proc Natl Acad Sci USA* 2002;**99**:14682–7.
364. Brett FM, Henson C, Staunton H. Familial diffuse Lewy body disease, eye movement abnormalities, and distribution of pathology. *Arch Neurol* 2002;**59**:464–7.
365. Louie JL, Kapphahn RJ, Ferrington DA. Proteasome function and protein oxidation in the aged retina. *Exp Eye Res* 2002;**75**:271–84.
366. Kapphahn RJ, Bigelow EJ, Ferrington DA. Age-dependent inhibition of proteasome chymotrypsin-like activity in the retina. *Exp Eye Res* 2007;**84**:646–54.
367. Uetama T, Ohno-Matsui K, Nakahama K, Morita I, Mochizuki M. Phenotypic change regulates monocyte chemoattractant protein-1 (MCP-1) gene expression in human retinal pigment epithelial cells. *J Cell Physiol* 2003;**197**:77–85.
368. Ambati J, Anand A, Fernandez S, Sakurai E, Lynn BC, Kuziel WA, et al. An animal model of age-related macular degeneration in senescent Ccl-2- or Ccr-2-deficient mice. *Nat Med* 2003;**9**:1390–7.
369. Arima K, Kinoshita A, Mishima H, Kanazawa N, Kaneko T, Mizushima T, et al. Proteasome assembly defect due to a proteasome subunit beta type 8 (PSMB8) mutation causes the autoinflammatory disorder, Nakajo-Nishimura syndrome. *Proc Natl Acad Sci USA* 2011;**108**:14914–9.
370. Hussong SA, Kapphahn RJ, Phillips SL, Maldonado M, Ferrington DA. Immunoproteasome deficiency alters retinal proteasome's response to stress. *J Neurochem* 2010;**113**:1481–90.
371. Ferrington DA, Hussong SA, Roehrich H, Kapphahn RJ, Kavanaugh SM, Heuss ND, et al. Immunoproteasome responds to injury in the retina and brain. *J Neurochem* 2008;**106**:158–69.
372. Hara T, Nakamura K, Matsui M, Yamamoto A, Nakahara Y, Suzuki-Migishima R, et al. Suppression of basal autophagy in neural cells causes neurodegenerative disease in mice. *Nature* 2006;**441**:885–9.
373. Mizushima N, Hara T. Intracellular quality control by autophagy: how does autophagy prevent neurodegeneration? *Autophagy* 2006;**2**:302–4.
374. Cuervo AM, Bergamini E, Brunk UT, Droge W, Ffrench M, Terman A. Autophagy and aging: the importance of maintaining "clean" cells. *Autophagy* 2005;**1**:131–40.
375. Mizushima N, Levine B, Cuervo AM, Klionsky DJ. Autophagy fights disease through cellular self-digestion. *Nature* 2008;**451**:1069–75.
376. Zhang C, Cuervo AM. Restoration of chaperone-mediated autophagy in aging liver improves cellular maintenance and hepatic function. *Nat Med* 2008;**14**:959–65.
377. Yang Z, Klionsky DJ. An overview of the molecular mechanism of autophagy. *Curr Top Microbiol Immunol* 2009;**335**:1–32.
378. Chen Y, Klionsky DJ. The regulation of autophagy—unanswered questions. *J Cell Sci* 2011;**124**:161–70.
379. Johansen T, Lamark T. Selective autophagy mediated by autophagic adapter proteins. *Autophagy* 2011;**7**:279–96.
380. Wang AL, Lukas TJ, Yuan M, Du N, Tso MO, Neufeld AH. Autophagy and exosomes in the aged retinal pigment epithelium: possible relevance to drusen formation and age-related macular degeneration. *PLoS One* 2009;**4**:e4160.
381. Wang AL, Lukas TJ, Yuan M, Du N, Tso MO, Neufeld AH. Autophagy, exosomes and drusen formation in age-related macular degeneration. *Autophagy* 2009;**5**:563–4.

382. Mitter SK, Rao HV, Qi X, Cai J, Sugrue A, Dunn Jr. WA, et al. Autophagy in the retina: a potential role in age-related macular degeneration. *Adv Exp Med Biol* 2012;**723**:83–90.
383. Chiu CJ, Taylor A. Dietary hyperglycemia, glycemic index and metabolic retinal diseases. *Prog Retin Eye Res* 2011;**30**:18–53.
384. Chiu CJ, Milton RC, Gensler G, Taylor A. Association between dietary glycemic index and age-related macular degeneration in nondiabetic participants in the Age-Related Eye Disease Study. *Am J Clin Nutr* 2007;**86**:180–8.
385. Chiu CJ, Milton RC, Klein R, Gensler G, Taylor A. Dietary carbohydrate and the progression of age-related macular degeneration: a prospective study from the Age-Related Eye Disease Study. *Am J Clin Nutr* 2007;**86**:1210–8.
386. Chiu CJ, Milton RC, Klein R, Gensler G, Taylor A. Dietary compound score and risk of age-related macular degeneration in the age-related eye disease study. *Ophthalmology* 2009;**116**:939–46.
387. Kaushik S, Wang JJ, Flood V, Tan JS, Barclay AW, Wong TY, et al. Dietary glycemic index and the risk of age-related macular degeneration. *Am J Clin Nutr* 2008;**88**:1104–10.
388. Weikel KA, Fitzgerald P, Shang F, Caceres MA, Bian Q, Handa JT, et al. Natural history of age-related retinal lesions that precede AMD in mice fed high or low glycemic index diets. *Invest Ophthalmol Vis Sci* 2012;**53**(2):622–32.
389. Marques C, Pereira P, Taylor A, Liang JN, Reddy VN, Szweda LI, et al. Ubiquitin-dependent lysosomal degradation of the HNE-modified proteins in lens epithelial cells. *FASEB J* 2004;**18**:1424–6.
390. Ryhanen T, Hyttinen JM, Kopitz J, Rilla K, Kuusisto E, Mannermaa E, et al. Crosstalk between Hsp70 molecular chaperone, lysosomes and proteasomes in autophagy-mediated proteolysis in human retinal pigment epithelial cells. *J Cell Mol Med* 2009;**13**:3616–31.
391. Viiri J, Hyttinen JM, Ryhanen T, Rilla K, Paimela T, Kuusisto E, et al. p62/sequestosome 1 as a regulator of proteasome inhibitor-induced autophagy in human retinal pigment epithelial cells. *Mol Vis* 2010;**16**:1399–414.

# Proteasome and Neurodegenerative Diseases

Betul Catalgol* and Tilman Grune†

*Department of Biochemistry, Faculty of Medicine, Genetic and Metabolic Diseases Research Center (GEMHAM), Marmara University, Haydarpasa, Istanbul, Turkey

†Institute of Nutrition, Friedrich Schiller University, Jena, Germany

I. Introduction ........ 397
II. Proteasome in Alzheimer's Disease ........ 399
III. Parkinson's Disease ........ 402
IV. Huntington's Disease ........ 403
V. Amyotrophic Lateral Sclerosis ........ 404
VI. Friedreich Ataxia ........ 405
References ........ 406

Proteasomal degradation of damaged proteins is involved in the pathogenesis of many neurodegenerative diseases including Alzheimer's, Parkinson's, Huntington's, stroke, and amyotrophic lateral sclerosis. A malfunction of the proteasomal activity may be the result or the consequence of protein aggregation, which is a key process for most neurodegenerative diseases. Because of the widespread aspects of the proteasomal involvement in the progression of these diseases, many studies are focused on this research.

## I. Introduction

The redox status of the brain is crucial in many neurodegenerative diseases. The accumulation of oxidized protein aggregates is higher and more pronounced in long-lived and postmitotic cells, which are devoid of cell divisions. So, protein aggregation is much more dramatic in postmitotic cells including neurons, as formed aggregates are not diluted into daughter cells upon cell division. So aged neurons may have 75% of their cellular volume filled with aggregated proteins.[1] This causes an increase in apoptotic susceptibility.[2] In addition, the brain is highly sensitive to oxidative damage due to its metabolic characteristics such as high oxygen consumption, high levels of peroxidizable polyunsaturated fatty acids, and high levels of iron in certain regions.[3] Furthermore, free radical-generating systems including mitochondria, nNOS, xanthine

Progress in Molecular Biology
and Translational Science, Vol. 109
DOI: 10.1016/B978-0-12-397863-9.00011-0




1877-1173/12 $35.00

oxidase, and monoamine oxidases show a very high activity in the brain. On the other hand, it was found that the antioxidant defense is poor due to low activities of antioxidant enzymes, such as superoxide dismutases, catalase, glutathione peroxidase, and glutathione reductase.[4–7] As a consequence of these properties, protein oxidation and proteasomal inhibition might be severe in the central nervous system and the viability of the neurons might also depend on the protein turnover.

The formation of protein aggregates is an important process in the aging brain and in neurodegenerative diseases (Fig. 1). The consequences of aggregate formation are dependent on the affected area and the cellular location and, of course, on the severity of the formation. Furthermore, metabolic disturbances might play a crucial role in protein accumulation. So a disturbance of proteolysis or of the protein substrate on the genetic level is certainly one of

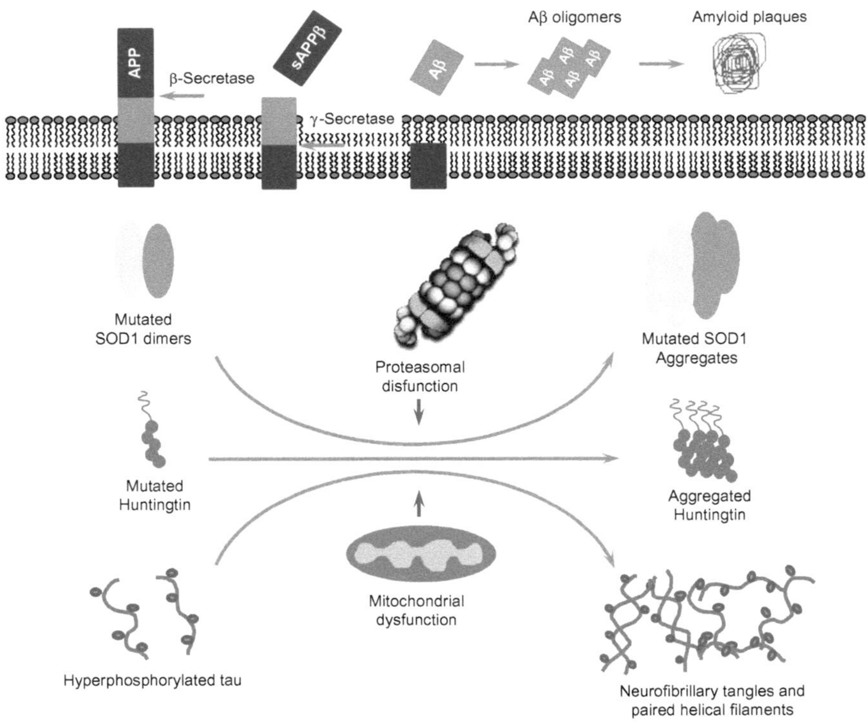


FIG. 1. Protein aggregates in neurodegeneration. Neurodegenerative diseases lead to extra- or intracellular protein aggregation. Several mutations and/or posttranslational modifications lead to this aggregation. Mitochondrial and proteasomal malfunctions promote this process. (See Color Insert.)

the possible factors. In addition, posttranslational modifications, including oxidative modifications, of proteins via metabolic and environmental factors consequently lead to changes in proteolytic susceptibility and perhaps aggregate formation.[8–11] Enzymatic posttranslational modification might also change the proteolytic susceptibility, as the phosphorylation of the tau protein, which causes reduced degradation and acceleration of tau aggregation.[12] In fact, the imbalance between the protein synthesis and degradation is always the reason for protein accumulation.[13] Extracellular protein deposits are degraded either by phagocytosis or by extracellular proteases, whereas intracellular aggregates are often autophagocytosed and accumulate in the lysosomes.[14,15] Lipofuscin, ceroid, and age-pigment-like fluorophores are the cross-linked modified proteins that are autophagocytosed and perhaps released into the cytosol, often undergoing such a cycle several times.[1,16]

## II. Proteasome in Alzheimer's Disease

Alzheimer's disease (AD) is one of the most widespread neurodegenerative diseases, with about 60% of cases overall. Alois Alzheimer was the physician/scientist who first described this disease, including the clinical symptoms, as short-term memory, sense of direction and orientation loss, and morphological hallmarks such as various protein aggregates. The disease is characterized by a severe disruption of long-term memory so that even close relatives are not recognized, accompanied by a progradient malfunction of motor neurons and the impairment of motor coordination. The changes include brain atrophy, loss of neurons, and synapses, often in temporal, frontal, and parietal cortex as well as in the hippocampus and amygdala. This disease is known to be age-dependent and the prevalence rises significantly with increasing age. While the prevalence is only between 1% and 3% in the age-range of 60–70, it is 25–35% in the age group of 85 or higher.[17]

The intracellular formation of the neurofibrillary tangles (NFTs) and extracellular accumulation of β-amyloid (Aβ) aggregates in senile plaques are pathophysiological hallmarks of the disease.[18] The main component of the senile plaques is the Aβ peptide, which is formed by the enzymatic cleavage of the amyloid precursor protein (APP) by β- and γ-secretases. In this so-called amyloidogenic APP-processing pathway, the β-secretase first cleaves the APP, forming an sAPPβ (s for soluble) and a smaller fragment called C99 remaining in the membrane. Further degradation by the γ-secretase results in the formation of $A\beta_{40}$ or $A\beta_{42}$, which are the major types found accumulated in the AD brain. The difference in the γ-secretase processing of Aβ is the result of the existence of two isoforms of native APP containing different carboxyl endings. In the non-amyloidogenic APP-processing pathway, an α-secretase cleaves the Aβ fragment, not allowing the formation of the critical $A\beta_{40}$ or $A\beta_{42}$. The transmembrane proteins presenilin-1 and -2, which are identified as parts of the γ-secretase enzyme, therefore, play a role in APP maturation. APPs are involved in many functions such as calcium homeostasis, cell growth, adhesion, axional vesicle transport, and free metal ion regulation.[19–22] The strong affinity of APP to $Cu^{2+}$ makes it likely to be an important basis for the Fenton reaction.[23] In addition, the formation of an $A\beta^{-}$ radical was observed, which further promotes lipid and protein oxidation. So the formation of the lipid peroxidation product hydroxynonenal (HNE) was shown to be increased in AD[24,25] along with an increase in the accumulation of protein oxidation products such as protein carbonyls in the frontal pole, hippocampus, and superior middle temporal gyrus,[26,27] or 3-nitrotyrosine in hippocampus and neocortical regions.[28,29]

The accumulation of hyperphosphorylated tau protein is another pathological marker in AD. The tau protein regulates the stability of microtubules in a way that non-phosphorylated tau stabilizes microtubules, whereas phosphorylation leads to a detachment of tau and microtubules and, therefore, to a destabilization of these structures. However, in AD the tau protein becomes hyperphosphorylated for unknown reasons. This enables the protein to form insoluble aggregates, which cannot be further degraded. Numerous kinases and phosphatases are involved in the regulation of tau phosphorylation, as the proline residue phosphorylation of tau is catalyzed via GSK3, cdk5, p38, and JNK whereas other residues are phosphorylated via PKA, PKC, CaMKII, MARK, or CKII.[30–35] As mentioned earlier, following their phosphorylation, tau proteins are dissociated from microtubules, which results in depolarization and a decrease in axonal transport in neuronal cells.[36] If hyperphosphorylated, the tau proteins form paired helical filaments (PHFs)[37] and NFTs.[38] The formation of NFTs was shown to be correlated with AD-related dementia.[39] However, contrary to the effects in AD, tau phosphorylation seems to have a protective effect in Huntington's disease (HD).[40]

There are many studies showing the role of proteasome in AD together with genetic and biochemical evidence.[41] First observations demonstrated that tau aggregates are known to be polyubiquitinated, but not degraded by proteasome. In the straight gyrus area of the AD brain, PHFs were shown to inhibit the proteasome[42] without affecting the proteasome amount. The remaining 20–50% of proteasomal activities were observed in the hippocampus and parahippocampal gyrus, the superior and middle temporal gyri, or the inferior parietal lobe.[43] Moreover, hyperphosphorylated tau was shown to be able to inhibit the proteasome following cross-linking with 4-HNE.[44] In a study carried out in HT22 mouse hippocampal tumor cells, recombinant tau was shown to be degraded by proteasome, and okadaic acid, as a protein phosphatase 1 and 2 inhibitor, was shown to block tau degradation.[12] Often ubiquitin in these structures is mutated to a ubiquitin-B mutant (UBB+1), which might cause the blockage of the degradation.[45] The carboxy terminus of the heat shock protein 70-interacting protein (CHIP) is a specific E3 protein possibly playing a role in the ubiquitination of tau protein.[46] Knockout of CHIP in mice was shown to lead to insoluble aggregate formation from hyperphosphorylated tau.[47] On the other hand, ubiquitination is not the only way to degrade tau, as it does not possess a complete folded structure and might be a substrate of the 20S proteasome without prior unfolding.[48] The ubiquitin carboxy-terminal hydrolase L1 (UCH-L1),[49,50] as a deubiquitinating protein, is shown to be oxidatively modified and downregulated in AD.[51,52]

Although Aβ is an extracellular peptide and accumulates extracellularly, the effect of this peptide on the proteasomal system was investigated in many studies. An inhibitory effect of Aβ on the proteasome was shown, which seems to be dose-dependent, and might reduce all three chymotrypsin-like, trypsin-like, and caspase-like proteolytic activities of the proteasome.[53–55] The binding to the inner chamber of the proteasome was proposed as a mechanism.[54] However, it was later proposed that perhaps cross-linking of the peptide has an important effect on the inhibition.[56] It seems that among the Aβ peptides, $A\beta_{40}$ is the most efficient inhibitor. But contrarily,[57] showed that $A\beta_{40}$ is less able to form aggregates compared to $A\beta_{42}$. N-terminal sequences $A\beta_{1-11}$, $A\beta_{1-16}$, and $A\beta_{25-35}$ also play an inhibitory role on the proteasome. Oh et al. have proved the inhibitory effect of $A\beta_{42}$ on the proteasomal activity.[58] Expression of an E2 ubiquitin-conjugating enzyme E2-25K (called Hip-2) was shown to be induced following the incubation of cells with $A\beta_{42}$ and this activated caspase-12, followed by induction of apoptosis.[59] Additionally, E2-25K downregulation in cortical neurons made them more resistant to Aβ-induced cytotoxicity and caspase-12-dependent cell death.

The turnover of several other AD-related proteins is regulated by proteasome such as PS1 and PS2[60,61]; nicastrin, which is in the complex of the γ-secretase[62]; anterior pharynx-defective 1 (Aph-1)[63]; presenilin enhancer-2 (Pen-2, which induces the cleavage of presenilin to its active N-terminal and C-terminal fragments)[64]; and TMP21.[65] A general increase in protein hydrophobicity was also shown to be an important factor for the proteasomal inhibition in AD.[66] Moreover, the proteasomal system responds to the changes during the disease, and therefore the immunoproteasome expression was shown to be increased in AD hippocampus and cerebellum in neurons and astrocytes when compared to age-matched controls.[67]

When all these results were considered, the inhibition of proteasome was shown to be sufficient for protein aggregation and AD-associated neuronal damage.[42]

## III. Parkinson's Disease

After AD, Parkinson's disease (PD) has a high prevalence among neurodegenerative disorders, and in the aged population, 1% is known to be affected by PD. Generally, the late-onset sporadic form is more common, but 5% of patients suffer from a familial form of PD.[68] Resting tremor, rigidity, and bradykinesia are among the main clinical symptoms, and they are mainly caused by the degeneration of nigrostriatal dopaminergic neurons located in the substantia nigra pars compacta.[68] So reduced dopamine levels are an important indicator of the disease.[69] Among the pathological hallmarks is the accumulation of the α-synuclein-related Lewy bodies. The protein generally accumulates in a polyubiquitinated form and is a histological marker for familial PD.[70,71] The parkin protein also plays an important role in the protein aggregation in PD.[72] Parkin functions as an E3-ligase with the RPN10 subunit of the 19S regulator in the 26S proteasome.[73] Another protein responsible for the aggregate formation in PD is again the UCH-L1. This is a deubiquitinating enzyme in neurons, which is able to catalyze the hydrolysis reaction of polyubiquitin chains to form single monomers.[74] Several genes have also been identified to play a role in familial PD including SNCA; PARK2; -3; -4; -5; -6; -7; -8; -9; -10; and NR4A2; and mutation in the PARK2 gene is the common form in familial PD.[70]

In a myriad of studies on PD, protein oxidation products have been shown to be increased along with the proposed increase in superoxide, peroxynitrite, and hydrogen peroxide radicals. Dopamine is also known to be a source of ROS due to its auto-oxidative properties. So the high iron concentrations in the substantia nigra might induce the dopamine-related superoxide formation.[75] 3-Nitrotyrosine was found to be high in cerebrospinal fluid and in nigral dopaminergic neurons in PD patients.[76,77] PD brains were shown to possess high protein carbonyl levels.[78] Furthermore, patients suffering from Lewy body dementia were shown to have high levels of protein carbonyls, too.[79] 3-Nitrotyrosine levels were demonstrated to be high not only directly in the Lewy bodies, but also in the amorphous deposits in neurons in the substantia nigra.[76] Interestingly, some antioxidative defense proteins seem to be damaged by this oxidative load in PD, as a nitration of manganese superoxide dismutase was found to be high in cerebrospinal fluid.[80] The aggregates of α-synuclein were shown to be formed following peroxynitrite formation and also dityrosine cross-linking.[81,82] Another marker of protein oxidation, glycosylation, was found to be increased in PD.[83] One of the main oxidant sources—the mitochondria—also seems to be an important source for the oxidative load in PD. Furthermore, the inhibition of mitochondrial complex I by the neurotoxin 1-methyl-4-phenylpyridinium ($MPP^+$) resulted in a decrease of the β5 and β1 activities of proteasome,[84] showing a direct link between oxidant formation and proteasomal malfunction. α-Subunits of the proteasome were shown to be affected in the substantia nigra in a rat model.[85] As the assembly of the proteasome requires α-subunits, this might cause a reduction in the levels of functional proteasome. Treatment of rats with the selective proteasome inhibitor, lactacystin, caused an increase in somatic symptoms, a more pronounced loss of dopaminergic neurons, and an increase in α-synuclein aggregate formation.[86] α-Synuclein aggregates are able to bind to the 20S proteasome and inhibit the chymotrypsin-like activity.[87] Moreover, monomeric and aggregated α-synucleins were shown to bind to the S6 subunit of the 19S regulator and were therefore able to inhibit 26S proteasomal activity.[88] Interestingly, in postmortem brain tissues of PD patients, PA700 and PA28 caps of the proteasome were shown to be downregulated by decreased expression, which might further contribute to the decrease in proteasomal activity.[89] The observed iron accumulation in PD was also shown to be involved in proteasome inhibition.[90] On the other hand, in cell culture, the proteasomal inhibitor lactacystin caused an increase in the labile iron pool and in ROS formation. This was accompanied by a ubiquitin-conjugated protein aggregate formation.[91]

In addition to the direct inhibition of proteasome, α-synuclein aggregates might cause an inhibition of some proteasome-related transcription factors. So, α-synuclein was shown to inhibit the NFκB cascade, which functions as an antiapoptotic pathway.[92] One possible explanation for this effect is the homology of α-synuclein with the chaperone 14-3-3, which binds to NFκB. On the other hand, GSK3β, as a proapoptotic protein, was shown to be triggered by α-synuclein and this protein might interact with antiapoptotic transcription factors β-catenin, CREB, STAT, NFAT, and c-jun and may also inhibit the phosphorylation of NFκB.

In the light of these results demonstrating the role of oxidative stress and proteasome in PD, several treatment approaches have been tried using GSH precursors (glutamyl cysteine ethyl ester, glutathione ethyl ester), thiol-containing antioxidants (α-lipoic acid),[93,94] monoamine oxidase inhibitors (depreneyl, rasagiline), or iron chelators (desferoxamine, V-28).[95,96]

## IV. Huntington's Disease

HD is an inherited, genetic disorder showing symptoms in early adulthood. The main genetic cause of this disease is the mutation of the gene IT15 on chromosome 4, resulting in a changed expression of the huntingtin protein.[97] Clinical symptoms are mostly motor dysfunctions, cognitive decline, and psychotic symptoms accompanied by subcortical dementia, depression, mania, personality changes, and memory defects.[98,99] During the progression of the disease, a neuronal loss in thalamus, hippocampus, spinal cord, and other brain regions is observed.[100] The polyubiquitinated huntingtin protein aggregates and inclusion bodies might be determined histologically. The huntingtin protein with its polyglutamine (polyQ) sequence at the N-terminus is the main protein that aggregates during the disease. A polyQ of more than 40 glutamine residues is accepted as required for aggregate formation and is, therefore, the starting point for the onset of clinical symptoms. Interestingly, the severity of the disease and the age of onset are dependent on the number of glutamine residues.[101] Possible functions of the huntingtin protein are the regulation of axional/vesicle transport and endocytosis.[19,102] The protein is primarily located in the nucleus but is also found in other parts of neurons.[97]

In addition to the mitochondria[103] and several transcription factors,[104] the proteasomal system is involved in the pathology of HD.[105,106] Huntingtin, in its fibrillary form, was shown to inhibit proteasomal activity; however, large polyglutamine aggregates show a less pronounced inhibitory effect.[107] This was explained with the role of the aggregate surface as a key factor of the inhibition process. In a study, the mechanism of the inhibition process was explained in detail as an interaction of polyQ sequences with α-rings of the 20S core proteasome. Such an interaction limits the substrate recognition or the binding of the 19S cap.[108] Furthermore, polyubiquitination of aggregates increases the inhibitory effect on the proteasome and, vice versa, the inhibition of proteasome contributes to the aggregate formation.[109] Inclusion bodies were also shown to interact with 20S, 19S, and ubiquitin and therefore decrease proteasome activity.[110,111]

## V. Amyotrophic Lateral Sclerosis

Amyotrophic lateral sclerosis (ALS) is a disease triggered by the loss of motor neurons in the motor cortex and the spinal cord and, therefore, leads to progressive loss of motor function. Protein aggregation is suggested to play a role in the loss of motor neurons, at least in some forms of the disease, where a form of the mutated Cu, Zn-superoxide dismutase (SOD1) is accumulating.[112,113] Chaperone activity and proteasomal degradation were shown to be decreased in the affected neurons of the spinal cord and these cells are recognized as being highly susceptible to aggregation. Maintenance of the chaperone system and, therefore, the native protein folding are important for ALS susceptibility.[114] Interestingly, SOD 1 aggregates were also found to be high in astrocytes in humans and rodent models.[115,116] As shown in general for all protein aggregates, these aggregates of SOD1 were also shown to affect cellular functions[117,118] and are a cause of decreased cell viability.[119,120]

In addition to the mutated SOD1, several different protein modification products were found to be increased in ALS patients including oxidized proteins, nitrated proteins, and protein adducts of HNE.[121,122] So, protein carbonyl levels are increased not only in frontal and motor cortex[123,124] but also in the spinal cord tissue of sporadic ALS patients.[125] The glial glutamate transporter protein was found to be modified by HNE.[126] 3-Nitrotyrosine levels were also found to be high in motor neurons[127–129] and in the cerebral fluid of ALS patients.[80,130,131] In a mouse model of ALS using SOD1$^{G93A}$ overexpression, creatine kinase was found to be inactivated via peroxynitrite formation.[132]

The proteasomal system plays an important role in the degradation of SOD as well as mutant SOD, which is most likely catalyzed by the HSP70–CHIP pathway.[133–135] Mutant SOD has been shown to be polyubiquitinated[136,137] and interestingly, only the mutant form tends to form polyubiquitinated aggregates, suggesting an involvement of the E3-ligases dorfin, CHIP, and NEDL1.[138] On the other hand, mutant SOD mediates the inhibition of the proteasome via aggregate formation.[114] Chymotrypsin-like proteasome activity was shown to decrease in the already mentioned experimental model using SOD1$^{G93A}$ transgenic mice. This inhibition was at least partially explained by the posttranslational modification of the β5-subunit. In the same study, the levels of α-subunits were also decreased, which causes a lack of cellular, available 20S proteasome in ALS.[139] Moreover, the $\beta 5_i$-subunit (the IFN-γ-inducible form) and the PA28 were shown to increase in astrocytes and glial cells of the spinal cord.[140] An interference of the proteasome with the mitochondrial homeostasis during ALS was also proposed.[141] The human ciliary neurotrophic factor (CNTF) is suggested to have the ability to increase the mRNA levels of the 20S proteasome. But the results obtained from clinical investigations did not show any correlation between CNTF and ALS progression.[142] However, the Keap1–Nrf2 pathway was induced in ALS, suggesting the attempt of an upregulation of antioxidative defense and proteasomal subunits.[143,144]

## VI. Friedreich Ataxia

Friedreich ataxia (FA) is an inherited neurodegenerative disease with onset of clinical symptoms before 25 years of age. The symptoms include hypertrophic cardiomyopathy, which is ultimately the main cause of death,[145] and a loss of sensoric system function and tendon reflexes.[146] The molecular basis of the disease is the mutation of the FRDA gene, which results in reduced expression of frataxin, a mitochondrial protein.[147] Frataxin is involved in cellular antioxidative defense,[148,149] iron storage,[150] and the buildup of mitochondrial iron–sulfur clusters.[151,152] In a yeast model expressing a mutant frataxin (ΔYFH1), which is used as a model for the mitochondrial failure comparable to FA, oxidative stress was found to be increased.[153] In addition to the brain, mitochondria-rich tissues such as heart, liver, and skeletal muscle are also affected by oxidative stress[154]in FA. The Lon protease, the mammalian protease catalyzing the removal of oxidized proteins in mitochondria,[155,156] or Pim1, the *Saccharomyces cerevisiae* homologue,[157] show an increased activity in FA models, while the proteasomal activity is decreased. However, the protein amounts of both enzymes do not change, suggesting metabolic regulation. In yeast, the levels of HNE- and HNE-modified proteins were shown to be increased.[153] However, contrary to the results obtained in the yeast model, the murine model did not show any evidence for the involvement of protein accumulation in FA pathology.[158] The reason for this is possibly the high Lon protease activity in the murine mitochondria compared to Pim-1 activity in the yeast.

## References


1. Yin D. Biochemical basis of lipofuscin, ceroid, and age pigment-like fluorophores. *Free Radic Biol Med* 1996;**21**:871–88.
2. Powell SR, Wang P, Divald A, Teichberg S, Haridas V, McCloskey TW, et al. Aggregates of oxidized proteins (lipofuscin) induce apoptosis through proteasome inhibition and dysregulation of proapoptotic proteins. *Free Radic Biol Med* 2005;**38**:1093–101.
3. Halliwell B. Reactive oxygen species and the central nervous system. *J Neurochem* 1992;**59**:1609–23.
4. Dringen R, Gutterer JM, Hirrlinger J. Glutathione metabolism in brain metabolic interaction between astrocytes and neurons in the defense against reactive oxygen species. *Eur J Biochem* 2000;**267**:4912–6.
5. Ernst A, Stolzing A, Sandig G, Grune T. Antioxidants effectively prevent oxidation-induced protein damage in OLN 93 cells. *Arch Biochem Biophys* 2004;**421**:54–60.
6. Ernst A, Stolzing A, Sandig G, Grune T. Protein oxidation and the degradation of oxidized proteins in the rat oligodendrocyte cell line OLN 93-antioxidative effect of the intracellular spin trapping agent PBN. *Brain Res Mol Brain Res* 2004;**122**:126–32.
7. Floyd RA, Carney JM. Free radical damage to protein and DNA: mechanisms involved and relevant observations on brain undergoing oxidative stress. *Ann Neurol* 1992;**32**(Suppl.):S22–7.
8. Giulivi C, Davies KJ. Dityrosine and tyrosine oxidation products are endogenous markers for the selective proteolysis of oxidatively modified red blood cell hemoglobin by (the 19 S) proteasome. *J Biol Chem* 1993;**268**:8752–9.
9. Grune T, Reinheckel T, Davies KJ. Degradation of oxidized proteins in K562 human hematopoietic cells by proteasome. *J Biol Chem* 1996;**271**:15504–9.
10. Grune T, Reinheckel T, Joshi M, Davies KJ. Proteolysis in cultured liver epithelial cells during oxidative stress. Role of the multicatalytic proteinase complex, proteasome. *J Biol Chem* 1995;**270**:2344–51.
11. Pacifici RE, Davies KJ. Protein degradation as an index of oxidative stress. *Methods Enzymol* 1990;**186**:485–502.
12. Poppek D, Keck S, Ermak G, Jung T, Stolzing A, Ullrich O, et al. Phosphorylation inhibits turnover of the tau protein by the proteasome: influence of RCAN1 and oxidative stress. *Biochem J* 2006;**400**:511–20.

13. Weimer JM, Kriscenski-Perry E, Elshatory Y, Pearce DA. The neuronal ceroid lipofuscinoses: mutations in different proteins result in similar disease. *Neuromolecular Med* 2002;**1**: 111–24.
14. Brunk UT, Terman A. The mitochondrial-lysosomal axis theory of aging: accumulation of damaged mitochondria as a result of imperfect autophagocytosis. *Eur J Biochem* 2002;**269**: 1996–2002.
15. Terman A, Gustafsson B, Brunk UT. The lysosomal-mitochondrial axis theory of postmitotic aging and cell death. *Chem Biol Interact* 2006;**163**:29–37.
16. Yin D. Studies on age pigments evolving into a new theory of biological aging. *Gerontology* 1995;**41**(Suppl. 2):159–72.
17. Evans DA, Funkenstein HH, Albert MS, Scherr PA, Cook NR, Chown MJ, et al. Prevalence of Alzheimer's disease in a community population of older persons. Higher than previously reported.. *JAMA* 1989;**262**:2551–6.
18. Lee VM, Goedert M, Trojanowski JQ. Neurodegenerative tauopathies. *Annu Rev Neurosci* 2001;**24**:1121–59.
19. Gunawardena S, Her LS, Brusch RG, Laymon RA, Niesman IR, Gordesky-Gold B, et al. Disruption of axonal transport by loss of huntingtin or expression of pathogenic polyQ proteins in Drosophila. *Neuron* 2003;**40**:25–40.
20. Mattson MP, Cheng B, Culwell AR, Esch FS, Lieberburg I, Rydel RE. Evidence for excitoprotective and intraneuronal calcium-regulating roles for secreted forms of the beta-amyloid precursor protein. *Neuron* 1993;**10**:243–54.
21. Rossjohn J, Cappai R, Feil SC, Henry A, McKinstry WJ, Galatis D, et al. Crystal structure of the N-terminal, growth factor-like domain of Alzheimer amyloid precursor protein. *Nat Struct Biol* 1999;**6**:327–31.
22. Smith DG, Cappai R, Barnham KJ. The redox chemistry of the Alzheimer's disease amyloid beta peptide. *Biochim Biophys Acta* 2007;**1768**:1976–90.
23. Barnham KJ, Haeffner F, Ciccotosto GD, Curtain CC, Tew D, Mavros C, et al. Tyrosine gated electron transfer is key to the toxic mechanism of Alzheimer's disease beta-amyloid. *FASEB J* 2004;**18**:1427–9.
24. Markesbery WR, Lovell MA. Four-hydroxynonenal, a product of lipid peroxidation, is increased in the brain in Alzheimer's disease. *Neurobiol Aging* 1998;**19**:33–6.
25. Sayre LM, Zelasko DA, Harris PL, Perry G, Salomon RG, Smith MA. 4-Hydroxynonenal-derived advanced lipid peroxidation end products are increased in Alzheimer's disease. *J Neurochem* 1997;**68**:2092–7.
26. Hensley K, Hall N, Subramaniam R, Cole P, Harris M, Aksenov M, et al. Brain regional correspondence between Alzheimer's disease histopathology and biomarkers of protein oxidation. *J Neurochem* 1995;**65**:2146–56.
27. Pamplona R, Dalfo E, Ayala V, Bellmunt MJ, Prat J, Ferrer I, et al. Proteins in human brain cortex are modified by oxidation, glycoxidation, and lipoxidation. Effects of Alzheimer disease and identification of lipoxidation targets.. *J Biol Chem* 2005;**280**:21522–30.
28. Hensley K, Maidt ML, Yu Z, Sang H, Markesbery WR, Floyd RA. Electrochemical analysis of protein nitrotyrosine and dityrosine in the Alzheimer brain indicates region-specific accumulation. *J Neurosci* 1998;**18**:8126–32.
29. Smith MA, Richey Harris PL, Sayre LM, Beckman JS, Perry G. Widespread peroxynitrite-mediated damage in Alzheimer's disease. *J Neurosci* 1997;**17**:2653–7.
30. Avila J, Santa-Maria I, Perez M, Hernandez F, Moreno F. Tau phosphorylation, aggregation, and cell toxicity. *J Biomed Biotechnol* 2006;**2006**:74539.
31. Robertson J, Loviny TL, Goedert M, Jakes R, Murray KJ, Anderton BH, et al. Phosphorylation of tau by cyclic-AMP-dependent protein kinase. *Dementia* 1993;**4**:256–63.

32. Sengupta A, Kabat J, Novak M, Wu Q, Grundke-Iqbal I, Iqbal K. Phosphorylation of tau at both Thr 231 and Ser 262 is required for maximal inhibition of its binding to microtubules. *Arch Biochem Biophys* 1998;**357**:299–309.
33. Singh TJ, Grundke-Iqbal I, Iqbal K. Phosphorylation of tau protein by casein kinase-1 converts it to an abnormal Alzheimer-like state. *J Neurochem* 1995;**64**:1420–3.
34. Taniguchi T, Kawamata T, Mukai H, Hasegawa H, Isagawa T, Yasuda M, et al. Phosphorylation of tau is regulated by PKN. *J Biol Chem* 2001;**276**:10025–31.
35. Yamamoto H, Hiragami Y, Murayama M, Ishizuka K, Kawahara M, Takashima A. Phosphorylation of tau at serine 416 by Ca2+/calmodulin-dependent protein kinase II in neuronal soma in brain. *J Neurochem* 2005;**94**:1438–47.
36. Cuchillo-Ibanez I, Seereeram A, Byers HL, Leung KY, Ward MA, Anderton BH, et al. Phosphorylation of tau regulates its axonal transport by controlling its binding to kinesin. *FASEB J* 2008;**22**:3186–95.
37. Mandelkow E, von BM, Biernat J, Mandelkow EM. Structural principles of tau and the paired helical filaments of Alzheimer's disease. *Brain Pathol* 2007;**17**:83–90.
38. Brion JP, Anderton BH, Authelet M, Dayanandan R, Leroy K, Lovestone S, et al. Neurofibrillary tangles and tau phosphorylation. *Biochem Soc Symp* 2001;**67**:81–8.
39. Caughey B, Lansbury PT. Protofibrils, pores, fibrils, and neurodegeneration: separating the responsible protein aggregates from the innocent bystanders. *Annu Rev Neurosci* 2003;**26**:267–98.
40. Arrasate M, Mitra S, Schweitzer ES, Segal MR, Finkbeiner S. Inclusion body formation reduces levels of mutant huntingtin and the risk of neuronal death. *Nature* 2004;**431**:805–10.
41. Oddo S. The ubiquitin-proteasome system in Alzheimer's disease. *J Cell Mol Med* 2008;**12**: 363–73.
42. Keck S, Nitsch R, Grune T, Ullrich O. Proteasome inhibition by paired helical filament-tau in brains of patients with Alzheimer's disease. *J Neurochem* 2003;**85**:115–22.
43. Keller JN, Hanni KB, Markesbery WR. Impaired proteasome function in Alzheimer's disease. *J Neurochem* 2000;**75**:436–9.
44. Hernandez F, Avila J. Tauopathies. *Cell Mol Life Sci* 2007;**64**:2219–33.
45. Lindsten K, de Vrij FM, Verhoef LG, Fischer DF, van Leeuwen FW, Hol EM, et al. Mutant ubiquitin found in neurodegenerative disorders is a ubiquitin fusion degradation substrate that blocks proteasomal degradation. *J Cell Biol* 2002;**157**:417–27.
46. Hatakeyama S, Matsumoto M, Yada M, Nakayama KI. Interaction of U-box-type ubiquitin-protein ligases (E3s) with molecular chaperones. *Genes Cells* 2004;**9**:533–48.
47. Dickey CA, Yue M, Lin WL, Dickson DW, Dunmore JH, Lee WC, et al. Deletion of the ubiquitin ligase CHIP leads to the accumulation, but not the aggregation, of both endogenous phospho- and caspase-3-cleaved tau species. *J Neurosci* 2006;**26**:6985–96.
48. Dorval V, Fraser PE. Small ubiquitin-like modifier (SUMO) modification of natively unfolded proteins tau and alpha-synuclein. *J Biol Chem* 2006;**281**:9919–24.
49. Luchansky SJ, Lansbury Jr. PT, Stein RL. Substrate recognition and catalysis by UCH-L1. *Biochemistry* 2006;**45**:14717–25.
50. Setsuie R, Wada K. The functions of UCH-L1 and its relation to neurodegenerative diseases. *Neurochem Int* 2007;**51**:105–11.
51. Castegna A, Aksenov M, Aksenova M, Thongboonkerd V, Klein JB, Pierce WM, et al. Proteomic identification of oxidatively modified proteins in Alzheimer's disease brain. Part I: creatine kinase BB, glutamine synthase, and ubiquitin carboxy-terminal hydrolase L-1. *Free Radic Biol Med* 2002;**33**:562–71.
52. Castegna A, Aksenov M, Thongboonkerd V, Klein JB, Pierce WM, Booze R, et al. Proteomic identification of oxidatively modified proteins in Alzheimer's disease brain. Part II:

dihydropyrimidinase-related protein 2, alpha-enolase and heat shock cognate 71. *J Neurochem* 2002;**82**:1524–32.

53. Gregori L, Fuchs C, Figueiredo-Pereira ME, Van Nostrand WE, Goldgaber D. Amyloid beta-protein inhibits ubiquitin-dependent protein degradation in vitro. *J Biol Chem* 1995;**270**:19702–8.
54. Gregori L, Hainfeld JF, Simon MN, Goldgaber D. Binding of amyloid beta protein to the 20 S proteasome. *J Biol Chem* 1997;**272**:58–62.
55. Lopez SM, Pasquini L, Besio MM, Pasquini JM, Soto E. Relationship between beta-amyloid degradation and the 26S proteasome in neural cells. *Exp Neurol* 2003;**180**:131–43.
56. Shringarpure R, Grune T, Sitte N, Davies KJ. 4-Hydroxynonenal-modified amyloid-beta peptide inhibits the proteasome: possible importance in Alzheimer's disease. *Cell Mol Life Sci* 2000;**57**:1802–9.
57. Crouch PJ, Harding SM, White AR, Camakaris J, Bush AI, Masters CL. Mechanisms of A beta mediated neurodegeneration in Alzheimer's disease. *Int J Biochem Cell Biol* 2008;**40**: 181–98.
58. Oh S, Hong HS, Hwang E, Sim HJ, Lee W, Shin SJ, et al. Amyloid peptide attenuates the proteasome activity in neuronal cells. *Mech Ageing Dev* 2005;**126**:1292–9.
59. Song S, Lee H, Kam TI, Tai ML, Lee JY, Noh JY, et al. E2-25K/Hip-2 regulates caspase-12 in ER stress-mediated Abeta neurotoxicity. *J Cell Biol* 2008;**182**:675–84.
60. Kim TW, Pettingell WH, Hallmark OG, Moir RD, Wasco W, Tanzi RE. Endoproteolytic cleavage and proteasomal degradation of presenilin 2 in transfected cells. *J Biol Chem* 1997;**272**:11006–10.
61. Marambaud P, Ancolio K, Lopez-Perez E, Checler F. Proteasome inhibitors prevent the degradation of familial Alzheimer's disease-linked presenilin 1 and potentiate A beta 42 recovery from human cells. *Mol Med* 1998;**4**:147–57.
62. He G, Qing H, Tong Y, Cai F, Ishiura S, Song W. Degradation of nicastrin involves both proteasome and lysosome. *J Neurochem* 2007;**101**:982–92.
63. He F, Stephens JM. Induction of SOCS-3 is insufficient to confer IRS-1 protein degradation in 3T3-L1 adipocytes. *Biochem Biophys Res Commun* 2006;**344**:95–8.
64. Bergman A, Hansson EM, Pursglove SE, Farmery MR, Lannfelt L, Lendahl U, et al. Pen-2 is sequestered in the endoplasmic reticulum and subjected to ubiquitylation and proteasome-mediated degradation in the absence of presenilin. *J Biol Chem* 2004;**279**:16744–53.
65. Liu S, Bromley-Brits K, Xia K, Mittelholtz J, Wang R, Song W. TMP21 degradation is mediated by the ubiquitin-proteasome pathway. *Eur J Neurosci* 2008;**28**:1980–8.
66. Dasuri K, Ebenezer P, Zhang L, Fernandez-Kim SO, Bruce-Keller AJ, Markesbery WR, et al. Increased protein hydrophobicity in response to aging and Alzheimer disease. *Free Radic Biol Med* 2010;**48**:1330–7.
67. Mishto M, Santoro A, Bellavista E, Sessions R, Textoris-Taube K, Dal PF, et al. A structural model of 20S immunoproteasomes: effect of LMP2 codon 60 polymorphism on expression, activity, intracellular localisation and insight into the regulatory mechanisms. *Biol Chem* 2006;**387**:417–29.
68. Chinta SJ, Andersen JK. Redox imbalance in Parkinson's disease. *Biochim Biophys Acta* 2008;**1780**:1362–7.
69. Bernheimer H, Birkmayer W, Hornykiewicz O, Jellinger K, Seitelberger F. Brain dopamine and the syndromes of Parkinson and Huntington. Clinical, morphological and neurochemical correlations. *J Neurol Sci* 1973;**20**:415–55.
70. Kawahara K, Hashimoto M, Bar-On P, Ho GJ, Crews L, Mizuno H, et al. alpha-Synuclein aggregates interfere with Parkin solubility and distribution: role in the pathogenesis of Parkinson disease. *J Biol Chem* 2008;**283**:6979–87.

71. Lee SJ. alpha-synuclein aggregation: a link between mitochondrial defects and Parkinson's disease? *Antioxid Redox Signal* 2003;**5**:337–48.
72. Moore DJ. Parkin: a multifaceted ubiquitin ligase. *Biochem Soc Trans* 2006;**34**:749–53.
73. Sakata E, Yamaguchi Y, Kurimoto E, Kikuchi J, Yokoyama S, Yamada S, et al. Parkin binds the Rpn10 subunit of 26S proteasomes through its ubiquitin-like domain. *EMBO Rep* 2003;**4**:301–6.
74. Wilkinson KD, Lee KM, Deshpande S, Duerksen-Hughes P, Boss JM, Pohl J. The neuron-specific protein PGP 9.5 is a ubiquitin carboxyl-terminal hydrolase. *Science* 1989;**246**:670–3.
75. Jenner P, Olanow CW. Oxidative stress and the pathogenesis of Parkinson's disease. *Neurology* 1996;**47**:S161–70.
76. Good PF, Hsu A, Werner P, Perl DP, Olanow CW. Protein nitration in Parkinson's disease. *J Neuropathol Exp Neurol* 1998;**57**:338–42.
77. Qureshi GA, Baig S, Bednar I, Sodersten P, Forsberg G, Siden A. Increased cerebrospinal fluid concentration of nitrite in Parkinson's disease. *Neuroreport* 1995;**6**:1642–4.
78. Alam ZI, Daniel SE, Lees AJ, Marsden DC, Jenner P, Halliwell B. A generalised increase in protein carbonyls in the brain in Parkinson's but not incidental Lewy body disease. *J Neurochem* 1997;**69**:1326–9.
79. Lyras L, Perry RH, Perry EK, Ince PG, Jenner A, Jenner P, et al. Oxidative damage to proteins, lipids, and DNA in cortical brain regions from patients with dementia with Lewy bodies. *J Neurochem* 1998;**71**:302–12.
80. Aoyama K, Matsubara K, Fujikawa Y, Nagahiro Y, Shimizu K, Umegae N, et al. Nitration of manganese superoxide dismutase in cerebrospinal fluids is a marker for peroxynitrite-mediated oxidative stress in neurodegenerative diseases. *Ann Neurol* 2000;**47**:524–7.
81. Paxinou E, Chen Q, Weisse M, Giasson BI, Norris EH, Rueter SM, et al. Induction of alpha-synuclein aggregation by intracellular nitrative insult. *J Neurosci* 2001;**21**:8053–61.
82. Souza JM, Giasson BI, Chen Q, Lee VM, Ischiropoulos H. Dityrosine cross-linking promotes formation of stable alpha -synuclein polymers. Implication of nitrative and oxidative stress in the pathogenesis of neurodegenerative synucleinopathies. *J Biol Chem* 2000;**275**:18344–9.
83. Castellani R, Smith MA, Richey PL, Perry G. Glycoxidation and oxidative stress in Parkinson disease and diffuse Lewy body disease. *Brain Res* 1996;**737**:195–200.
84. Caneda-Ferron B, De Girolamo LA, Costa T, Beck KE, Layfield R, Billett EE. Assessment of the direct and indirect effects of MPP+ and dopamine on the human proteasome: implications for Parkinson's disease aetiology. *J Neurochem* 2008;**105**:225–38.
85. McNaught KS, Bjorklund LM, Belizaire R, Isacson O, Jenner P, Olanow CW. Proteasome inhibition causes nigral degeneration with inclusion bodies in rats. *Neuroreport* 2002;**13**:1437–41.
86. Lev N, Melamed E, Offen D. Proteasomal inhibition hypersensitizes differentiated neuroblastoma cells to oxidative damage. *Neurosci Lett* 2006;**399**:27–32.
87. Lindersson E, Beedholm R, Hojrup P, Moos T, Gai W, Hendil KB, et al. Proteasomal inhibition by alpha-synuclein filaments and oligomers. *J Biol Chem* 2004;**279**:12924–34.
88. Snyder H, Mensah K, Theisler C, Lee J, Matouschek A, Wolozin B. Aggregated and monomeric alpha-synuclein bind to the S6' proteasomal protein and inhibit proteasomal function. *J Biol Chem* 2003;**278**:11753–9.
89. McNaught KS, Belizaire R, Isacson O, Jenner P, Olanow CW. Altered proteasomal function in sporadic Parkinson's disease. *Exp Neurol* 2003;**179**:38–46.
90. Zhu W, Li X, Xie W, Luo F, Kaur D, Andersen JK, et al. Genetic iron chelation protects against proteasome inhibition-induced dopamine neuron degeneration. *Neurobiol Dis* 2010;**37**: 307–13.
91. Li XP, Xie WJ, Zhang Z, Kansara S, Jankovic J, Le WD. A Mechanistic study of proteasome inhibition-induced iron misregulation in dopamine neuron degeneration. *Neurosignals* 2012.

92. Yuan Y, Jin J, Yang B, Zhang W, Hu J, Zhang Y, et al. Overexpressed alpha-synuclein regulated the nuclear factor-kappaB signal pathway. *Cell Mol Neurobiol* 2008;**28**:21–33.
93. Bharat S, Cochran BC, Hsu M, Liu J, Ames BN, Andersen JK. Pre-treatment with R-lipoic acid alleviates the effects of GSH depletion in PC12 cells: implications for Parkinson's disease therapy. *Neurotoxicology* 2002;**23**:479–86.
94. Karunakaran S, Diwakar L, Saeed U, Agarwal V, Ramakrishnan S, Iyengar S, et al. Activation of apoptosis signal regulating kinase 1 (ASK1) and translocation of death-associated protein, Daxx, in substantia nigra pars compacta in a mouse model of Parkinson's disease: protection by alpha-lipoic acid. *FASEB J* 2007;**21**:2226–36.
95. Fernandez HH, Chen JJ. Monamine oxidase inhibitors: current and emerging agents for Parkinson disease. *Clin Neuropharmacol* 2007;**30**:150–68.
96. Youdim MB, Lavie L. Selective MAO-A and B inhibitors, radical scavengers and nitric oxide synthase inhibitors in Parkinson's disease. *Life Sci* 1994;**55**:2077–82.
97. Bonilla E. Huntington disease. A review. *Invest Clin* 2000;**41**:117–41.
98. Haddad MS, Cummings JL. Huntington's disease. *Psychiatr Clin North Am* 1997;**20**:791–807.
99. Rosenblatt A, Leroi I. Neuropsychiatry of Huntington's disease and other basal ganglia disorders. *Psychosomatics* 2000;**41**:24–30.
100. Vonsattel JP, Myers RH, Stevens TJ, Ferrante RJ, Bird ED, Richardson Jr. EP. Neuropathological classification of Huntington's disease. *J Neuropathol Exp Neurol* 1985;**44**:559–77.
101. Andrew SE, Goldberg YP, Kremer B, Telenius H, Theilmann J, Adam S, et al. The relationship between trinucleotide (CAG) repeat length and clinical features of Huntington's disease. *Nat Genet* 1993;**4**:398–403.
102. Sugars KL, Rubinsztein DC. Transcriptional abnormalities in Huntington disease. *Trends Genet* 2003;**19**:233–8.
103. Acevedo-Torres K, Berrios L, Rosario N, Dufault V, Skatchkov S, Eaton MJ, et al. Mitochondrial DNA damage is a hallmark of chemically induced and the R6/2 transgenic model of Huntington's disease. *DNA Repair (Amst)* 2009;**8**:126–36.
104. Sugars KL, Brown R, Cook LJ, Swartz J, Rubinsztein DC. Decreased cAMP response element-mediated transcription: an early event in exon 1 and full-length cell models of Huntington's disease that contributes to polyglutamine pathogenesis. *J Biol Chem* 2004;**279**: 4988–99.
105. Diaz-Hernandez M, Valera AG, Moran MA, Gomez-Ramos P, Alvarez-Castelao B, Castano JG, et al. Inhibition of 26S proteasome activity by huntingtin filaments but not inclusion bodies isolated from mouse and human brain. *J Neurochem* 2006;**98**:1585–96.
106. Valera AG, Diaz-Hernandez M, Hernandez F, Ortega Z, Lucas JJ. The ubiquitin-proteasome system in Huntington's disease. *Neuroscientist* 2005;**11**:583–94.
107. Goswami A, Dikshit P, Mishra A, Mulherkar S, Nukina N, Jana NR. Oxidative stress promotes mutant huntingtin aggregation and mutant huntingtin-dependent cell death by mimicking proteasomal malfunction. *Biochem Biophys Res Commun* 2006;**342**:184–90.
108. Ortega Z, Diaz-Hernandez M, Lucas JJ. Is the ubiquitin-proteasome system impaired in Huntington's disease? *Cell Mol Life Sci* 2007;**64**:2245–57.
109. Waelter S, Boeddrich A, Lurz R, Scherzinger E, Lueder G, Lehrach H, et al. Accumulation of mutant huntingtin fragments in aggresome-like inclusion bodies as a result of insufficient protein degradation. *Mol Biol Cell* 2001;**12**:1393–407.
110. Davies SW, Turmaine M, Cozens BA, DiFiglia M, Sharp AH, Ross CA, et al. Formation of neuronal intranuclear inclusions underlies the neurological dysfunction in mice transgenic for the HD mutation. *Cell* 1997;**90**:537–48.
111. DiFiglia M, Sapp E, Chase KO, Davies SW, Bates GP, Vonsattel JP, et al. Aggregation of huntingtin in neuronal intranuclear inclusions and dystrophic neurites in brain. *Science* 1997;**277**:1990–3.

112. Cookson MR, Menzies FM, Manning P, Eggett CJ, Figlewicz DA, McNeil CJ, et al. Cu/Zn superoxide dismutase (SOD1) mutations associated with familial amyotrophic lateral sclerosis (ALS) affect cellular free radical release in the presence of oxidative stress. *Amyotroph Lateral Scler Other Motor Neuron Disord* 2002;**3**:75–85.
113. Jaiswal MK, Keller BU. Cu/Zn superoxide dismutase typical for familial amyotrophic lateral sclerosis increases the vulnerability of mitochondria and perturbs Ca2+ homeostasis in SOD1G93A mice. *Mol Pharmacol* 2009;**75**:478–89.
114. Kabashi E, Durham HD. Failure of protein quality control in amyotrophic lateral sclerosis. *Biochim Biophys Acta* 2006;**1762**:1038–50.
115. Durham HD, Roy J, Dong L, Figlewicz DA. Aggregation of mutant Cu/Zn superoxide dismutase proteins in a culture model of ALS. *J Neuropathol Exp Neurol* 1997;**56**:523–30.
116. Kato S, Horiuchi S, Liu J, Cleveland DW, Shibata N, Nakashima K, et al. Advanced glycation endproduct-modified superoxide dismutase-1 (SOD1)-positive inclusions are common to familial amyotrophic lateral sclerosis patients with SOD1 gene mutations and transgenic mice expressing human SOD1 with a G85R mutation. *Acta Neuropathol* 2000;**100**:490–505.
117. Matsumoto G, Stojanovic A, Holmberg CI, Kim S, Morimoto RI. Structural properties and neuronal toxicity of amyotrophic lateral sclerosis-associated Cu/Zn superoxide dismutase 1 aggregates. *J Cell Biol* 2005;**171**:75–85.
118. Sasaki S, Warita H, Murakami T, Shibata N, Komori T, Abe K, et al. Ultrastructural study of aggregates in the spinal cord of transgenic mice with a G93A mutant SOD1 gene. *Acta Neuropathol* 2005;**109**:247–55.
119. Bruening W, Roy J, Giasson B, Figlewicz DA, Mushynski WE, Durham HD. Up-regulation of protein chaperones preserves viability of cells expressing toxic Cu/Zn-superoxide dismutase mutants associated with amyotrophic lateral sclerosis. *J Neurochem* 1999;**72**:693–9.
120. Takeuchi H, Kobayashi Y, Yoshihara T, Niwa J, Doyu M, Ohtsuka K, et al. Hsp70 and Hsp40 improve neurite outgrowth and suppress intracytoplasmic aggregate formation in cultured neuronal cells expressing mutant SOD1. *Brain Res* 2002;**949**:11–22.
121. Perluigi M, Fai PH, Hensley K, Pierce WM, Klein JB, Calabrese V, et al. Proteomic analysis of 4-hydroxy-2-nonenal-modified proteins in G93A-SOD1 transgenic mice—a model of familial amyotrophic lateral sclerosis. *Free Radic Biol Med* 2005;**38**:960–8.
122. Poon HF, Hensley K, Thongboonkerd V, Merchant ML, Lynn BC, Pierce WM, et al. Redox proteomics analysis of oxidatively modified proteins in G93A-SOD1 transgenic mice—a model of familial amyotrophic lateral sclerosis. *Free Radic Biol Med* 2005;**39**:453–62.
123. Bowling AC, Barkowski EE, McKenna-Yasek D, Sapp P, Horvitz HR, Beal MF, et al. Superoxide dismutase concentration and activity in familial amyotrophic lateral sclerosis. *J Neurochem* 1995;**64**:2366–9.
124. Ferrante RJ, Browne SE, Shinobu LA, Bowling AC, Baik MJ, MacGarvey U, et al. Evidence of increased oxidative damage in both sporadic and familial amyotrophic lateral sclerosis. *J Neurochem* 1997;**69**:2064–74.
125. Shaw PJ, Ince PG, Falkous G, Mantle D. Oxidative damage to protein in sporadic motor neuron disease spinal cord. *Ann Neurol* 1995;**38**:691–5.
126. Pedersen WA, Fu W, Keller JN, Markesbery WR, Appel S, Smith RG, et al. Protein modification by the lipid peroxidation product 4-hydroxynonenal in the spinal cords of amyotrophic lateral sclerosis patients. *Ann Neurol* 1998;**44**:819–24.
127. Abe K, Pan LH, Watanabe M, Kato T, Itoyama Y. Induction of nitrotyrosine-like immunoreactivity in the lower motor neuron of amyotrophic lateral sclerosis. *Neurosci Lett* 1995;**199**: 152–4.
128. Abe K, Pan LH, Watanabe M, Konno H, Kato T, Itoyama Y. Upregulation of protein-tyrosine nitration in the anterior horn cells of amyotrophic lateral sclerosis. *Neurol Res* 1997;**19**:124–8.

129. Chou SM, Wang HS, Komai K. Colocalization of NOS and SOD1 in neurofilament accumulation within motor neurons of amyotrophic lateral sclerosis: an immunohistochemical study. *J Chem Neuroanat* 1996;**10**:249–58.
130. Beal MF, Ferrante RJ, Browne SE, Matthews RT, Kowall NW, Brown Jr. RH. Increased 3-nitrotyrosine in both sporadic and familial amyotrophic lateral sclerosis. *Ann Neurol* 1997;**42**:644–54.
131. Tohgi H, Abe T, Yamazaki K, Murata T, Ishizaki E, Isobe C. Remarkable increase in cerebrospinal fluid 3-nitrotyrosine in patients with sporadic amyotrophic lateral sclerosis. *Ann Neurol* 1999;**46**:129–31.
132. Wendt S, Dedeoglu A, Speer O, Wallimann T, Beal MF, Andreassen OA. Reduced creatine kinase activity in transgenic amyotrophic lateral sclerosis mice. *Free Radic Biol Med* 2002;**32**:920–6.
133. Choi JS, Cho S, Park SG, Park BC, Lee DH. Co-chaperone CHIP associates with mutant Cu/Zn-superoxide dismutase proteins linked to familial amyotrophic lateral sclerosis and promotes their degradation by proteasomes. *Biochem Biophys Res Commun* 2004;**321**:574–83.
134. Ishigaki S, Niwa J, Yamada S, Takahashi M, Ito T, Sone J, et al. Dorfin-CHIP chimeric proteins potently ubiquitylate and degrade familial ALS-related mutant SOD1 proteins and reduce their cellular toxicity. *Neurobiol Dis* 2007;**25**:331–41.
135. Urushitani M, Kurisu J, Tsukita K, Takahashi R. Proteasomal inhibition by misfolded mutant superoxide dismutase 1 induces selective motor neuron death in familial amyotrophic lateral sclerosis. *J Neurochem* 2002;**83**:1030–42.
136. Basso M, Massignan T, Samengo G, Cheroni C, De BS, Salmona M, et al. Insoluble mutant SOD1 is partly oligoubiquitinated in amyotrophic lateral sclerosis mice. *J Biol Chem* 2006;**281**:33325–35.
137. Ganesan S, Rohde G, Eckermann K, Sroka K, Schaefer MK, Dohm CP, et al. Mutant SOD1 detoxification mechanisms in intact single cells. *Cell Death Differ* 2008;**15**:312–21.
138. Miyazaki K, Fujita T, Ozaki T, Kato C, Kurose Y, Sakamoto M, et al. NEDL1, a novel ubiquitin-protein isopeptide ligase for dishevelled-1, targets mutant superoxide dismutase-1. *J Biol Chem* 2004;**279**:11327–35.
139. Kabashi E, Agar JN, Taylor DM, Minotti S, Durham HD. Focal dysfunction of the proteasome: a pathogenic factor in a mouse model of amyotrophic lateral sclerosis. *J Neurochem* 2004;**89**:1325–35.
140. Puttaparthi K, Elliott JL. Non-neuronal induction of immunoproteasome subunits in an ALS model: possible mediation by cytokines. *Exp Neurol* 2005;**196**:441–51.
141. Hervias I, Beal MF, Manfredi G. Mitochondrial dysfunction and amyotrophic lateral sclerosis. *Muscle Nerve* 2006;**33**:598–608.
142. Jho DH, Engelhard HH, Gandhi R, Chao J, Babcock T, Ong E, et al. Ciliary neurotrophic factor upregulates ubiquitin-proteasome components in a rat model of neuronal injury. *Cytokine* 2004;**27**:142–51.
143. Philbert MA, Beiswanger CM, Waters DK, Reuhl KR, Lowndes HE. Cellular and regional distribution of reduced glutathione in the nervous system of the rat: histochemical localization by mercury orange and o-phthaldialdehyde-induced histofluorescence. *Toxicol Appl Pharmacol* 1991;**107**:215–27.
144. Tanigawa S, Fujii M, Hou DX. Action of Nrf2 and Keap1 in ARE-mediated NQO1 expression by quercetin. *Free Radic Biol Med* 2007;**42**:1690–703.
145. Van Driest SL, Gakh O, Ommen SR, Isaya G, Ackerman MJ. Molecular and functional characterization of a human frataxin mutation found in hypertrophic cardiomyopathy. *Mol Genet Metab* 2005;**85**:280–5.
146. Ponka P. Hereditary causes of disturbed iron homeostasis in the central nervous system. *Ann N Y Acad Sci* 2004;**1012**:267–81.

147. Coppola G, Choi SH, Santos MM, Miranda CJ, Tentler D, Wexler EM, et al. Gene expression profiling in frataxin deficient mice: microarray evidence for significant expression changes without detectable neurodegeneration. *Neurobiol Dis* 2006;**22**:302–11.
148. Auchere F, Santos R, Planamente S, Lesuisse E, Camadro JM. Glutathione-dependent redox status of frataxin-deficient cells in a yeast model of Friedreich's ataxia. *Hum Mol Genet* 2008;**17**:2790–802.
149. Napoli E, Taroni F, Cortopassi GA. Frataxin, iron-sulfur clusters, heme, ROS, and aging. *Antioxid Redox Signal* 2006;**8**:506–16.
150. Gakh O, Adamec J, Gacy AM, Twesten RD, Owen WG, Isaya G. Physical evidence that yeast frataxin is an iron storage protein. *Biochemistry* 2002;**41**:6798–804.
151. Guillon B, Bulteau AL, Wattenhofer-Donze M, Schmucker S, Friguet B, Puccio H, et al. Frataxin deficiency causes upregulation of mitochondrial Lon and ClpP proteases and severe loss of mitochondrial Fe-S proteins. *FEBS J* 2009;**276**:1036–47.
152. Zhang S, Hacham M, Panepinto J, Hu G, Shin S, Zhu X, et al. The Hsp70 member, Ssa1, acts as a DNA-binding transcriptional co-activator of laccase in Cryptococcus neoformans. *Mol Microbiol* 2006;**62**:1090–101.
153. Bulteau AL, Dancis A, Gareil M, Montagne JJ, Camadro JM, Lesuisse E. Oxidative stress and protease dysfunction in the yeast model of Friedreich ataxia. *Free Radic Biol Med* 2007;**42**:1561–70.
154. Becker E, Richardson DR. Frataxin: its role in iron metabolism and the pathogenesis of Friedreich's ataxia. *Int J Biochem Cell Biol* 2001;**33**:1–10.
155. Bayot A, Basse N, Lee I, Gareil M, Pirotte B, Bulteau AL, et al. Towards the control of intracellular protein turnover: mitochondrial Lon protease inhibitors versus proteasome inhibitors. *Biochimie* 2008;**90**:260–9.
156. Ngo JK, Davies KJ. Importance of the lon protease in mitochondrial maintenance and the significance of declining lon in aging. *Ann N Y Acad Sci* 2007;**1119**:78–87.
157. Major T, von JB, Ruppert T, Mogk A, Voos W. Proteomic analysis of mitochondrial protein turnover: identification of novel substrate proteins of the matrix protease pim1. *Mol Cell Biol* 2006;**26**:762–76.
158. Seznec H, Simon D, Bouton C, Reutenauer L, Hertzog A, Golik P, et al. Friedreich ataxia: the oxidative stress paradox. *Hum Mol Genet* 2005;**14**:463–74.

# Index

Note: Page numbers followed by "*f*" indicate figures, and "*t*" indicate tables.


## A

Activator protein-1 (AP-1), 280
Age-related eye diseases
  cataract, 371
  immonoproteasome induction, 373
  lens architecture, 370–371
  lens transparency, 371
  Nakajo-Nishimura syndrome, 372–373
  protein control and signal transduction, 370
  RPE, 372
  VEGF expression, 372–373
Age-related macular degeneration
  NF-κB signaling pathway, 363–365
  signal transduction, 363
Aging. *See also* Ubiquitin-proteasome pathway (UPP)
  cellular proteome, 249–250
  intracellular protein degradation, 250
  lysosomal pathways, 251
  mechanism
    oxidative modification, 263–264
    phosphorylation, 264
    ROS, 261–262
    19S and 20S proteasome, 262
    S-glutathiolation, 263
    *in vitro* and *in vivo*, 262–263
  in organs and tissues
    adipose tissue, 255
    eye lens, 254
    muscle mass loss, 253
    proteasome activity loss, 252–253
    20S proteasome analysis, 254
    UV radiation, 254–255
  proteasome dysfunction
    ex vivo studies, 255–256
    immunosenescence, 256
    trabecular meshwork (TM), 256
    WI38 fibroblasts, 257
  proteasome function modulation
    AIRAP, 258
    E3 ubiquitin ligase, 258–259
    germ cell loss, 258
    macromolecular damage accumulation, 259–260
    in mammals, 259
    yeast cells, 258
  proteasome peptidase activity, 251
  proteostasis network, 249–250, 250*f*
  rejuvenating proteasome
    chemical/natural compounds, 264
    dose–response relationships, 267
    genetic approach, 264–265
    Nrf2 pathway, 266–267
    over expression, 265
    Rpn4-mediated, 266
Alzheimer's disease (AD)
  amyloid precursor protein (APP), 399–400
  β-amyloid (Aβ), 401
  characteristics, 399
  CHIP, 400–401
  Hsps, 137–138
  presenilin-1 and-2, 399–400
  proteasome role, 400–401
  tau protein accumulation, 400
Amyotrophic lateral sclerosis (ALS)
  characteristics, 404
  ciliary neurotrophic factor (CNTF), 405
  proteasomal system, 405
  SOD1, 404, 405
Angiogenic factors expression
  deubiquitinating enzymes, 366
  hypoxia-inducible factor (HIF), 365
  oxygen-regulated degradation, 366
Antiaging strategy. *See* Rejuvenating proteasome, aging
Antigen processing, 88–94
Apoptosis, 278–279
Atherosclerosis
  apoptosis, 301*f*
  complication stage, 306–308
  initiation stage, 301–303

Atherosclerosis (*cont.*)
  progression stage, 303–306
  proteasome inhibition, 308–311
Autophagic-lysosome pathway (ALP)
  lens protein quality control, 354–355
  *vs.* UPP, RPE
    autophagy, 373–374
    environmental stress, 375*f*
    glycemic diets, 374–375

## B

Bortezomib
  adverse effects, 171–173
  anti-HIV therapy, 171
  cancer, 284–286
  clinical trials, cancers, 167–169, 167*t*
  combination therapy, 169
  lupus-like disease, 170
  mantle cell lymphoma
    biomarkers, 166
    clinical trial, 166
    with rituximab, 166
  monotherapy, 168–169
  multiple myeloma
    anticancer effects, 163–164
    biotransformation pathway, 165
    chemosensitization, 164
    clinical trial, 164
    efflux, 165
    markers, 165–166
  myelofibrosis, 170

## C

Cancer
  activator protein-1 (AP-1), 280
  apoptosis, 278–279
  bortezomib, 284–286
  characteristics, 277
  IκB, 280–282
  metastasis and angiogenesis, 282
  p53, 279–280
  proteasome pathway, 283
  protein oxidation, 278
  protein regulation, 278*t*
  PSI, 283–284
  radiation therapy, 284
  retinoblastoma protein (RB)-E2F1 pathway, 283
  super-repressor, 280–282
  TRAIL, 278–279
  types, 277–278
  Wnt signaling, 282–283
Cardiac E3 ligases
  familial hypertrophic cardiomyopathy, 317–318
  ischemic heart disease, 315–317
  pathologic cardiac hypertrophy, 312–315
Cardiac proteasome regulation
  heterogeneity, 297–298
  posttranslational modification, 298–299
  20S proteasome regulation, 298*f*
Cardiovascular disease. *See* Ubiquitin-proteasome system (UPS)
Carfilzomib, 202–204
Cataract. *See* Ubiquitin-proteasome pathway (UPP)
Cellular antioxidant systems
  description, 369
  Nrf2, 369, 370
CEP-18770, 204–205
Chaperones interaction
  ERAD process, 115
  heat shock response
    classical Hsps regulation, 127–128
    heme oxygenase-1, 129–130
    high-molecular-weight, 115–119
    inhibition, proteasomal degradation, 131–133
    regulation, 121–123
    small-molecular-weight, 119–121
  important domains, 122*t*
  neurodegenerative diseases
    Alzheimer disease, 137–138
    description, 136–139
    Parkinson disease, 138–139
  oxidative stress and aging process, 133–136
  protein quality control, yeast, 139–150
    ERAD-L pathway, 144–146
    ERAD-M/C, 146–147
    ERAD pathway, 147
    mammalian ERAD process, 147
    proteasome inhibition, 148–150
  refolding/degradation
    BAG-1 and CHIP inhibition, 123–126
    co-chaperone-independent decision, 126–127
    hip and hop, 126
  stress, 114

## E

ERAD process. *See* Chaperones interaction
Eye diseases, pathogenesis. *See* Ubiquitin-proteasome pathway (UPP)

## F

Friedreich ataxia (FA)
characteristics, 405–406
Lon protease, 405–406
symptoms, 405–406

## G

Gene expression regulation
cell ability, 82
IFN-γ, 81
i-proteasome expression, 81
LMP2 and LMP7, 79
MECL-1 subunit, 80–81
promoter region, 79–80
stop codons, 80
thymoproteasome, 81
transcription factors, 81–82

## H

Heart failure
description, 325
early failure alterations, 325–326
late failure alterations, 326–327
Heat shock proteins (Hsps)
ERAD process, 115
heat shock response
classical Hsps regulation, 127–128
heme oxygenase-1, 129–130
high-molecular-weight
description, 115–116
Hsp40, 119
Hsp70, 117–118
Hsp90, 116–117
Hsp100, 116
important domains, 122*t*
inhibition, proteasomal degradation, 131–133
neurodegenerative diseases
Alzheimer disease, 137–138
description, 136–139
Parkinson disease, 138–139
oxidative stress and aging process, 133–136
protein quality control, yeast, 139–150
ERAD-L pathway, 144–146
ERAD-M/C, 146–147
ERAD pathway, 147
mammalian ERAD process, 147
proteasome inhibition, 148–150
refolding/degradation
BAG-1 and CHIP inhibition, 123–126
co-chaperone-independent decision, 126–127
hip and hop, 126
regulation, 121–123
small-molecular-weight, Hsp27, 119–121
20S proteasome, 5
stress, 114
Human disease, mutations and linkage
Arg/His, 104–105
autoinflammation, 103
CANDLE syndrome, 103
i-proteasome inhibitors, 102
JMP, 102–103
lipodystrophy, 103
LMP2 codon, 104
NNS patients, 103
PSMB8-associated syndrome, 102
SNPs, 104
Huntington's disease (HD)
characteristics, 403–404
polyubiquitination, 404

## I

Immunoproteasomes
description, 75–76
enzymatic activity
amino acids, 86
hydrolysis mechanism, 84–85
proteasome population, 86
subunits, 84, 85*f*
thymus-specific β5 subunit, 85–86
function
antigen processing, 88–94
endoproteolysis, 90
nonimmune roles, i-proteasome, 95–101
peptides, 88–90
protein degradation, 89*f*

Immunoproteasomes (*cont.*)
gene expression regulation
cell ability, 82
IFN-γ, 81
i-proteasome expression, 81
LMP2 and LMP7, 79
MECL-1 subunit, 80–81
promoter region, 79–80
stop codons, 80
thymoproteasome, 81
transcription factors, 81–82
human disease, mutations and linkage
Arg/His, 104–105
autoinflammation, 103
CANDLE syndrome, 103
i-proteasome inhibitors, 102
JMP, 102–103
lipodystrophy, 103
LMP2 codon, 104
NNS patients, 103
PSMB8-associated syndrome, 102
SNPs, 104
knockout (KO) mice
gene disruption, 86–87
i-proteasome subunits, 87–88
physiological role, 86
nomenclature, 76–77, 76*t*
20S core assembly
chaperone-mediated process, 82
i-proteasome, 83
knockout (KO) mice, 84
LMP2 subunit, 82–83
structure
catalytic subunit, 77, 78*f*, 78*t*
composition, 77
20S, 79
subtypes, 76
Inflammation, 365
I-proteasome inhibitors. *See* Immunoproteasomes

## K

Knockout (KO) mice
gene disruption, 86–87
i-proteasome subunits, 87–88
physiological role, 86

## L

Lens
cell proliferation regulation
cell cycle regulation, 356
E3, 356
lactacystin, 355
SCF and APC/C target cell, 356
Ubc3, 357
UbcH7, 357
fiber differentiation and lens development
characteristics, 357–359
K6W expression, 358*f*, 359*f*
mitochondria protein removal, 357–359
upregulation, 357–359
protein quality control
abnormal protein degradation, 355
autophagic-lysosome pathway (ALP), 354–355
calpain, 353–354
cell types, 350
characteristics, 351
deamidation, 354
disruption, 352
glutathiolation, 353
glycation, 354–355
modification types, 351
oxidative stress, 352
protein degradation, 352–353
role, 351–352

## M

Mantle cell lymphoma, bortezomib
biomarkers, 166
clinical trial, 166
with rituximab, 166
Marizomib, 200–202
Metastasis and angiogenesis, 282
Multiple myeloma, bortezomib
anticancer effects, 163–164
biotransformation pathway, 165
chemosensitization, 164
clinical trial, 164
efflux, 165
markers, 165–166
Myocardial ischemia
description, 327–328
IPC, 331–333

postischemic proteasome dysfunction, 328–329
proteasome inhibition, 333
11S-activated proteasome and myocardial ischemia, 329–331

## N

Neurodegenerative diseases
Alzheimer's disease
amyloid precursor protein (APP), 399–400
β-amyloid (Aβ), 401
characteristics, 399
CHIP, 400–401
presenilin-1 and-2, 399–400
proteasome role, 400–401
tau protein accumulation, 400
amyotrophic lateral sclerosis
characteristics, 404
ciliary neurotrophic factor (CNTF), 405
proteasomal system, 405
SOD1, 404, 405
brain, redox status, 397–398
Friedreich ataxia
characteristics, 405–406
Lon protease, 405–406
symptoms, 405–406
Hsps
Alzheimer disease, 137–138
description, 136–139
Parkinson disease, 138–139
Huntington's disease
characteristics, 403–404
polyubiquitination, 404
Parkinson's disease
characteristics, 402
Lewy body dementia, 402–403
oxidative stress role, 403
parkin protein, 402
proteasome inhibition, 403
protein aggregates formation, 398–399, 398*f*
NF-κB signaling pathway
AMD, 363–364
canonical and noncanonical, 364
IKK activation, 364
immune response, 365
inflammation, 365
NIK, 365
TRAF6, 364
Nutrition. *See* Autophagic-lysosome pathway (ALP)

## O

Oxidative stress. *See* Ubiquitin-proteasome pathway (UPP)

## P

Parkinson's disease (PD)
characteristics, 402
Hsps, 138–139
Lewy body dementia, 402–403
oxidative stress role, 403
parkin protein, 402
proteasome inhibition, 403
PQC and cardiac proteinopathy
cardiac, 319–321
genesis of, 322–323
potential mechanisms, 323–324
proteasome functional insufficiency, 321–322
treatment, 324–325
Proteasome inhibitors
aggregates formation. *See* (Neurodegenerative diseases)
bortezomib
adverse effects, 171–173
anti-HIV therapy, 171
clinical research, 163
clinical trials, cancers, 167–169, 167*t*
combination therapy, 169
FDA approval, 161–162
lupus-like disease, 170
mantle cell lymphoma, 166
MG-132, 162
monotherapy, 168–169
multiple myeloma, 161–162, 163–166
myelofibrosis, 170
PS-341, LDP-341, MLM341, 162
toxicological evaluation, 162–163
early drugs, 205–207
immunoproteasome inhibitors, 207–208
20S, second-generation, 173–205
carfilzomib, 202–204

Proteasome inhibitors (*cont.*)
CEP-18770, 204–205
marizomib, 200–202
UPD, 208
Protein damage. *See* 20S proteasome
Protein quality control
and signaling. *See* (Ubiquitin-proteasome pathway (UPP))
yeast, 139–150
ERAD-L pathway, 144–146
ERAD-M/C, 146–147
ERAD pathway, 147
mammalian ERAD process, 147
proteasome inhibition, 148–150
Protein triage. *See* Heat shock proteins (Hsps)

## R

Radiation therapy, 284
Redox regulation, eye
age-related inactivation, 368–369
capacity, 367
conjugating activity, 368
light exposure, 368–369
oxidative stress, 367–368
Refolding/degradation
BAG-1 and CHIP inhibition, 123–126
co-chaperone-independent decision, 126–127
hip and hop, 126
Rejuvenating proteasome, aging
chemical/natural compounds, 264
dose–response relationships, 267
genetic approach, 264–265
Nrf2 pathway, 266–267
over expression, 265
Rpn4-mediated, 266
Reperfusion injury. *See* Myocardial ischemia
Retinal development
apoptotic editing, 360–361
characteristics, 360
delta, 361
deubiquitinating enzymes, 362
drosophila IAP (DIAP), 360–361
fat facets, 363
Hedgehog and Decapentaplegic pathways, 361–362
liquid facets (Lqf), 363
SINA protein, 361
Retinoblastoma protein (RB)-E2F1 pathway, 283

## S

Signal transduction, 363

## T

Thymoproteasome, 19–20
20S core assembly
chaperone-mediated process, 82
i-proteasome, 83
knockout (KO) mice, 84
LMP2 subunit, 82–83
20S proteasome
adaptation, protein damage, 238–239
aging and age-related protein damage, 236–237
alternative forms
immunoproteasome and oxidative stress, 241–243
β-rings, 13*f*
β-subunits, 10–12
cardiac proteasome regulation, 298*f*
ClpQY, 5–7
degradation mechanism, 234–236
description, 2
Eubacteria, 5–7
forms of, 2
γ-interferon, 5–7
intracellular assembly, 12–15
mammalian cells, 7
modeling, 15
oxygen radicals
formation, 228–231
removal, 231–232
PA28/REG, 17*f*
PR39, 5
proteasomal regulator, 5
protein damage, 228
regulation, 15–25
hybrid proteasome, 21–23
immunoproteasome, 16–19
PA200 regulator protein, 23–25
11S regulator, 20–21

19S regulator, 16
thymus-specific proteasome, 19–20
removal, protein damage
description, 239–241
Pa200, 240–241
Pa28γ, 241
PARP, 240
*vs.* 26S proteasome, 232–234
11S regulator, 17*f*
structure, 2, 3*f*
subunits, 4*t*
α-subunits, 7–10

## U

Ubiquitin-proteasome pathway (UPP)
activation of, 349*f*
age-related eye diseases
cataract, 371
immonoproteasome induction, 373
lens architecture, 370–371
lens transparency, 371
Nakajo-Nishimura syndrome, 372–373
protein control and signal transduction, 370
RPE, 372
VEGF expression, 372–373
*vs.* ALP, RPE
autophagy, 373–374
environmental stress, 375*f*
glycemic diets, 374–375
angiogenic factors expression
deubiquitinating enzymes, 366
hypoxia-inducible factor (HIF), 365
oxygen-regulated degradation, 366
*vs.* cellular antioxidant systems
description, 369
Nrf2, 369, 370
cellular functions, 350
chain formation, 348–350
characteristics, 348
lens cell proliferation regulation
cell cycle regulation, 356
E3, 356
lactacystin, 355
SCF and APC/C target cell, 356
Ubc3, 357
UbcH7, 357
lens fiber differentiation and lens development
characteristics, 357–359
K6W expression, 358*f*, 359*f*
mitochondria protein removal, 357–359
upregulation, 357–359
lens protein quality control
abnormal protein degradation, 355
autophagic-lysosome pathway (ALP), 354–355
calpain, 353–354
cell types, 350
characteristics, 351
deamidation, 354
disruption, 352
glutathiolation, 353
glycation, 354–355
modification types, 351
oxidative stress, 352
protein degradation, 352–353
role, 351–352
lysine residues, 348–350
NF-κB signaling pathway
AMD, 363–364
canonical and noncanonical, 364
IKK activation, 364
immune response, 365
inflammation, 365
NIK, 365
TRAF6, 364
oxidative stress, vision function, 376
redox regulation, eye
age-related inactivation, 368–369
capacity, 367
conjugating activity, 368
light exposure, 368–369
oxidative stress, 367–368
retinal development
apoptotic editing, 360–361
characteristics, 360
delta, 361
deubiquitinating enzymes, 362
drosophila IAP (DIAP), 360–361
fat facets, 363
Hedgehog and Decapentaplegic pathways, 361–362
liquid facets (Lqf), 363
SINA protein, 361
signal transduction, 363
visual signaling

Ubiquitin-proteasome pathway (UPP) (*cont.*)
characteristics, 366
Tγ, 366
Ubiquitin-proteasome system (UPS)
atherosclerosis
apoptosis, 301*f*
complication stage, 306–308
initiation stage, 301–303
progression stage, 303–306
proteasome inhibition, 308–311
cardiac E3 ligases
familial hypertrophic cardiomyopathy, 317–318
ischemic heart disease, 315–317
pathologic cardiac hypertrophy, 312–315
cardiac proteasome regulation
heterogeneity, 297–298
posttranslational modification, 298–299
20S proteasome regulation, 298*f*
description, 296–297
deubiquitinates, 46–47
E3 ubiquitin–protein ligases, 44–46
heart failure
description, 325
early failure alterations, 325–326
late failure alterations, 326–327
modulation
in cancer and therapy, 59–62
E3 ligase role, 53
role, 52
19S and 11S complex induction, 58–59
20S complex induction, 54–58
Skp1-Cull-F-box E3 ligase, 52–53
transcriptional control, 51
von Hippel–Lindau (VHL) tumor, 54
WD40 protein, 53
myocardial ischemia/reperfusion injury
description, 327–328
IPC, 331–333
postischemic proteasome dysfunction, 328–329
proteasome inhibition, 333
11S-activated proteasome and myocardial ischemia, 329–331
PQC and cardiac proteinopathy
cardiac PQC, 319–321
genesis of, 322–323
potential mechanisms, 323–324
proteasome functional insufficiency, 321–322
treatment, 324–325
proteasome
Blm10/PA200, 50
PA28/REG/PA26, 50
19S complex, 49–50
20S proteasome, 47–49
ubiquitin, 42–44
UPP. *See* Ubiquitin-proteasome pathway (UPP)
UPS. *See* Ubiquitin-proteasome system (UPS)

## V

Velcade®. *See* Bortezomib
Visual signaling
characteristics, 366
Tγ, 366

## W

Wnt signaling, 282–283

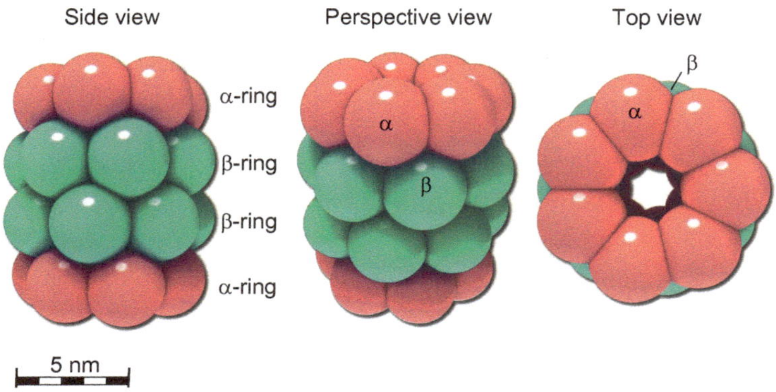


JUNG AND GRUNE, FIG. 1.

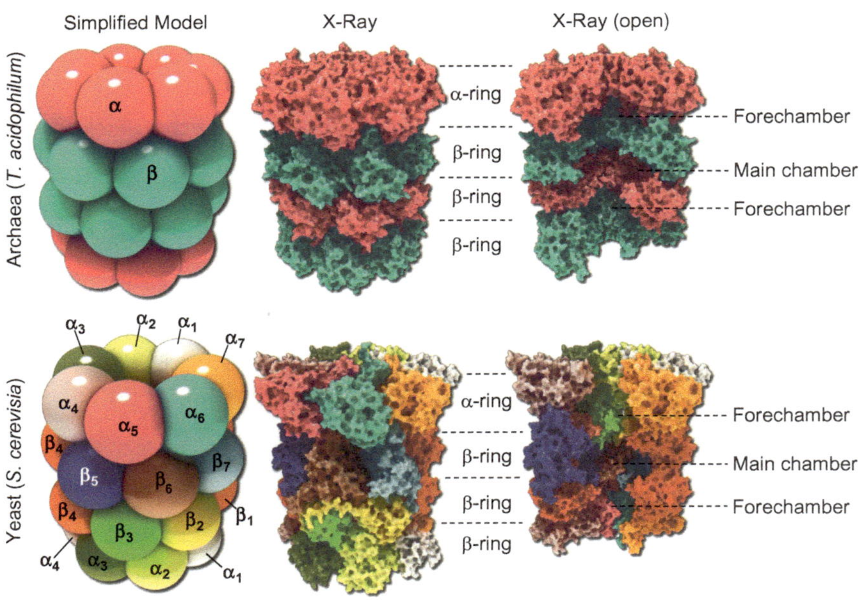


JUNG AND GRUNE, FIG. 2.

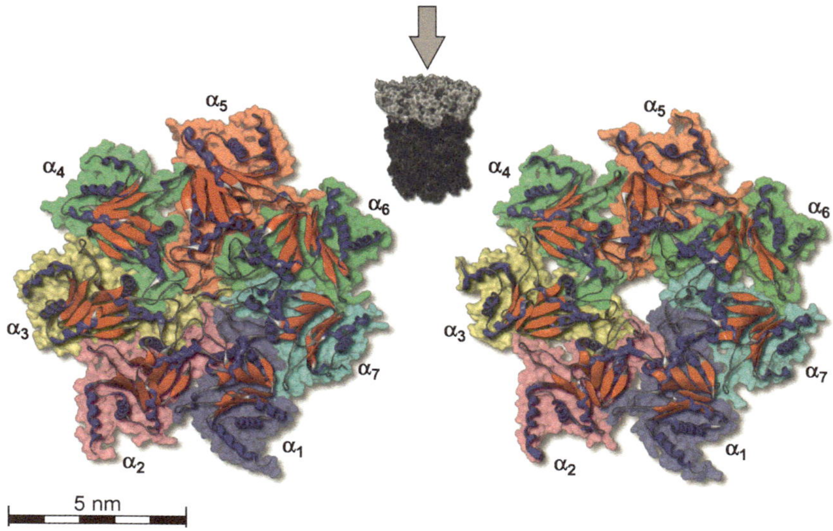


JUNG AND GRUNE, FIG. 3.

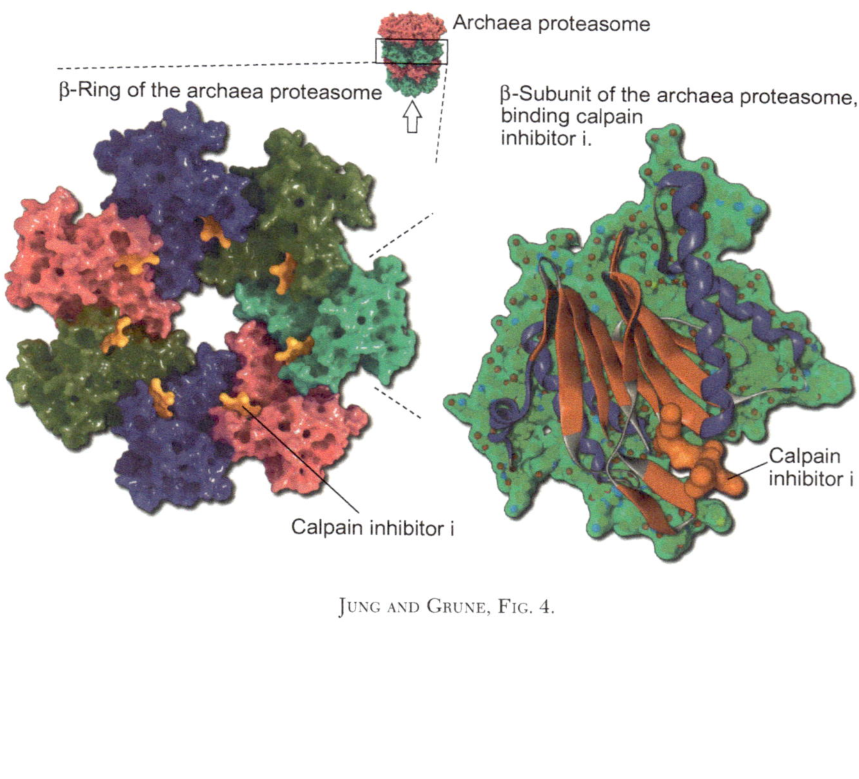


JUNG AND GRUNE, FIG. 4.

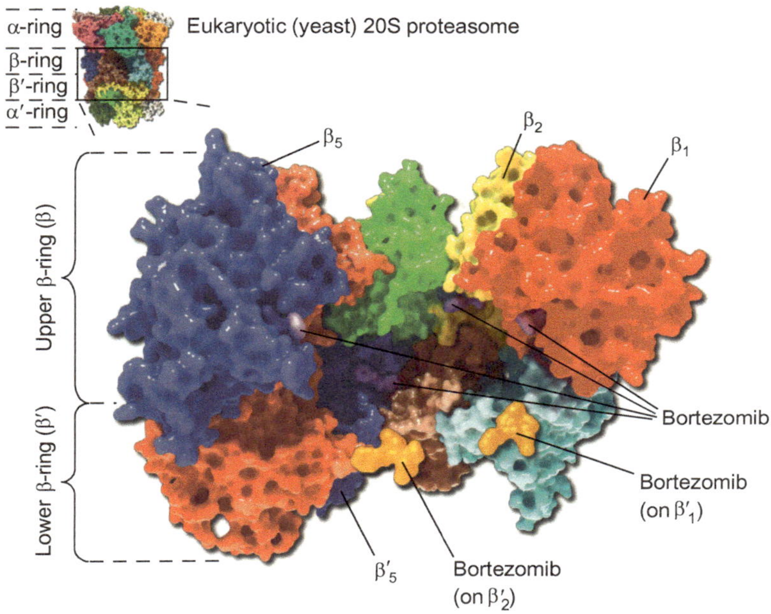


JUNG AND GRUNE, FIG. 5.

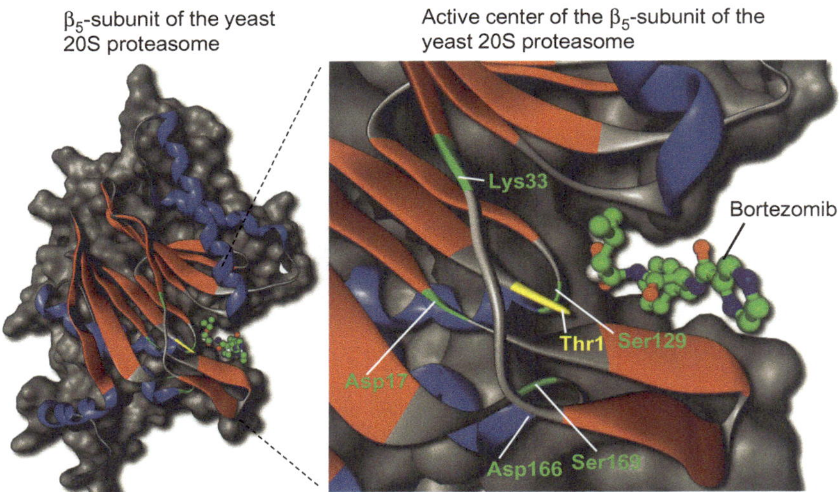


JUNG AND GRUNE, FIG. 6.

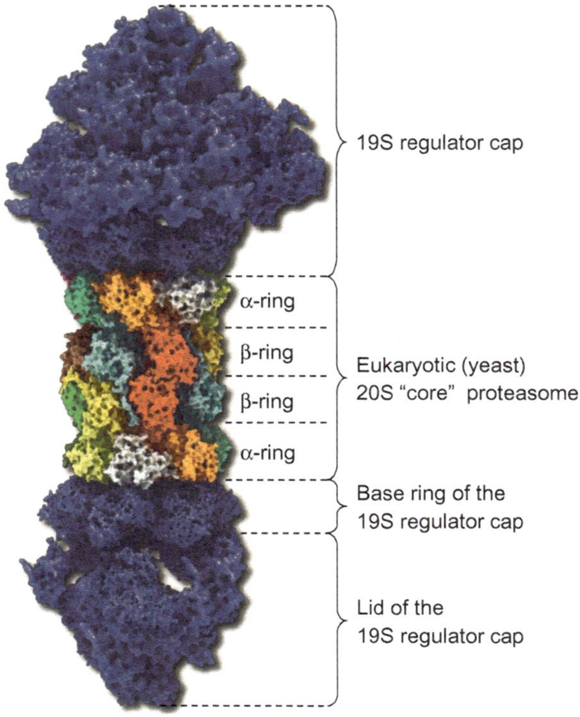


JUNG AND GRUNE, FIG. 7.

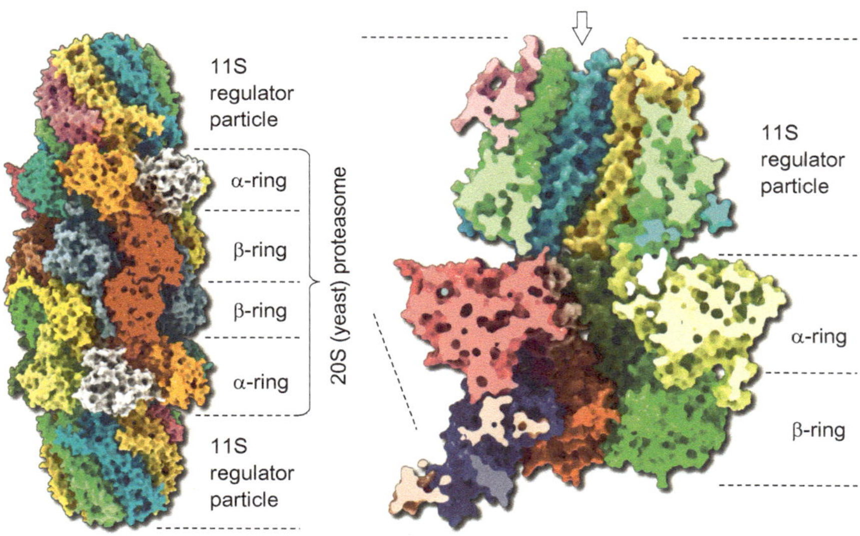


JUNG AND GRUNE, FIG. 8.

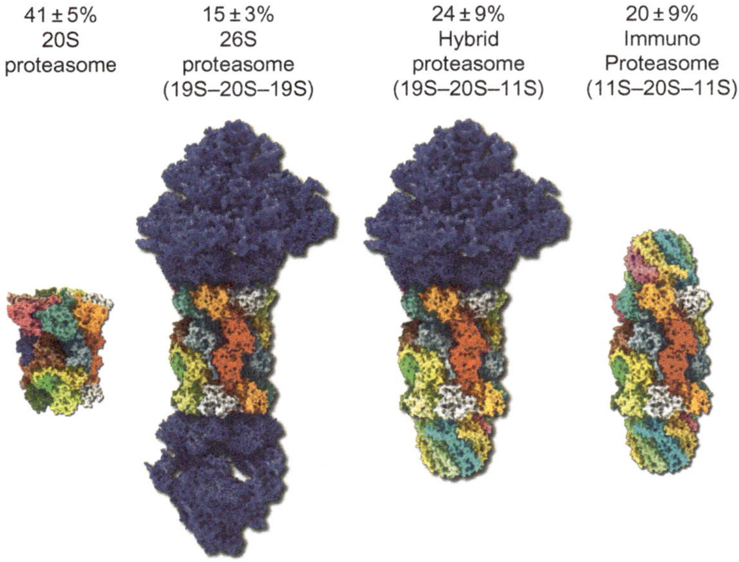


JUNG AND GRUNE, FIG. 9.

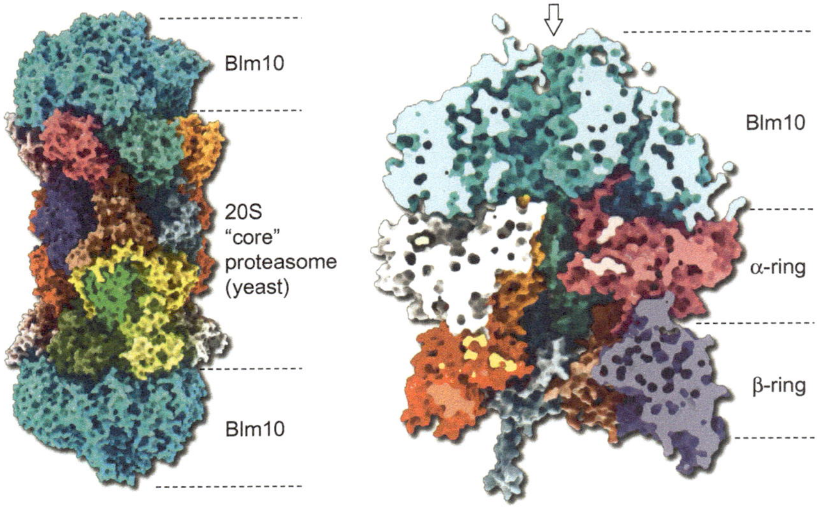


JUNG AND GRUNE, FIG. 10.

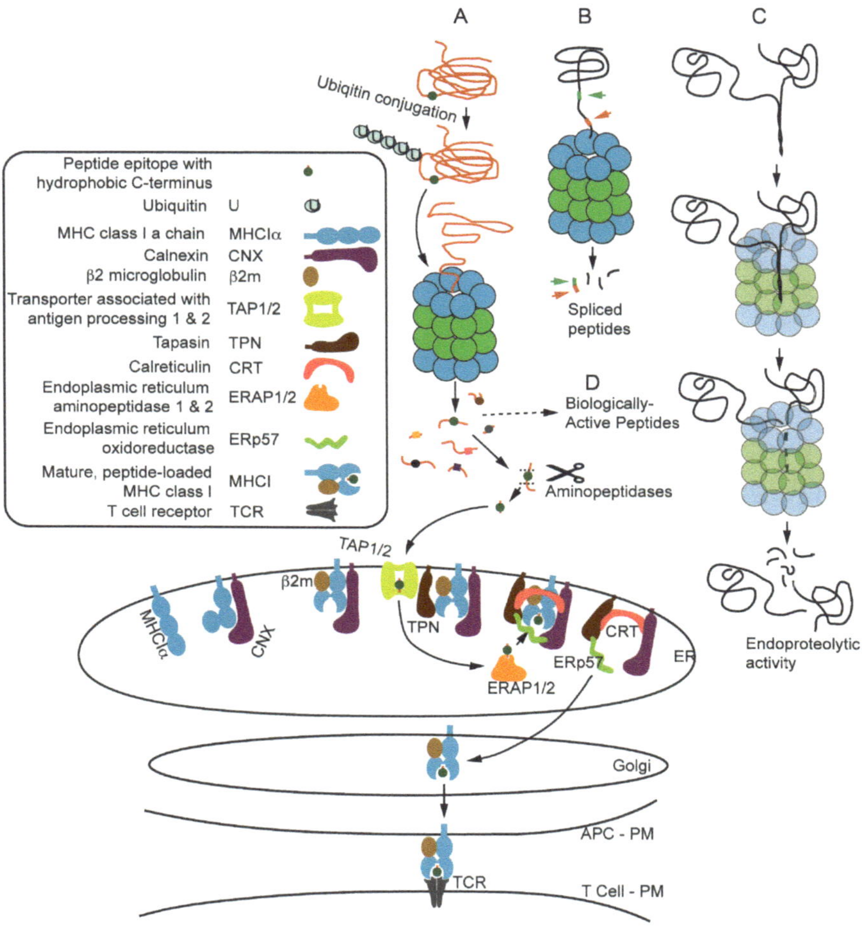


Ferrington and Gregerson, Fig. 3.

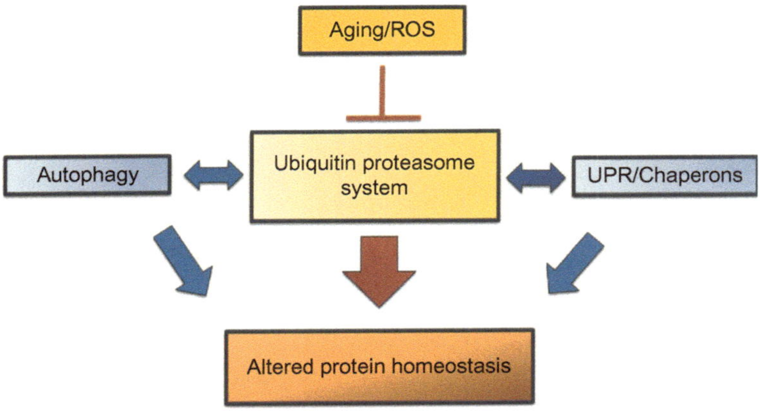


BARAIBAR AND FRIGUET, FIG. 1.

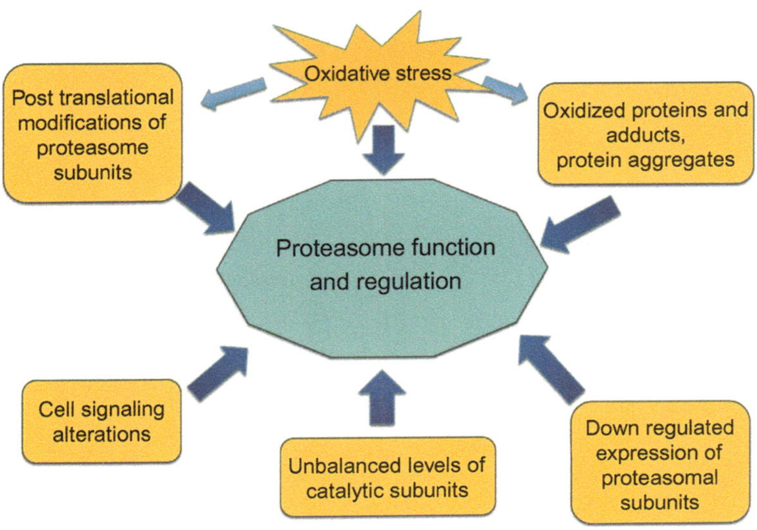


BARAIBAR AND FRIGUET, FIG. 2.

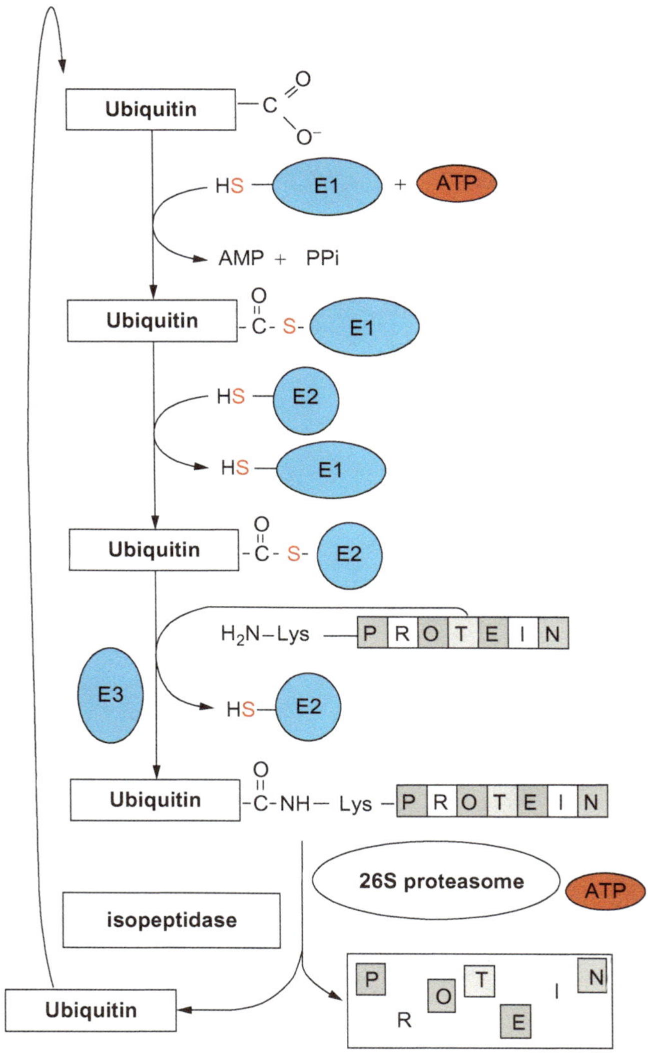


SHANG AND TAYLOR, FIG. 1.

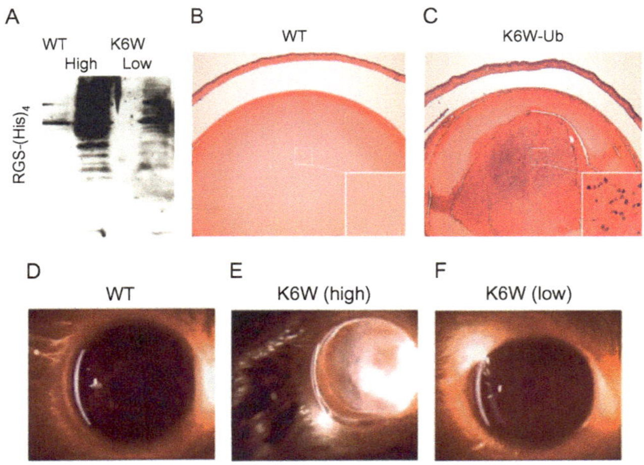


SHANG AND TAYLOR, FIG. 2.

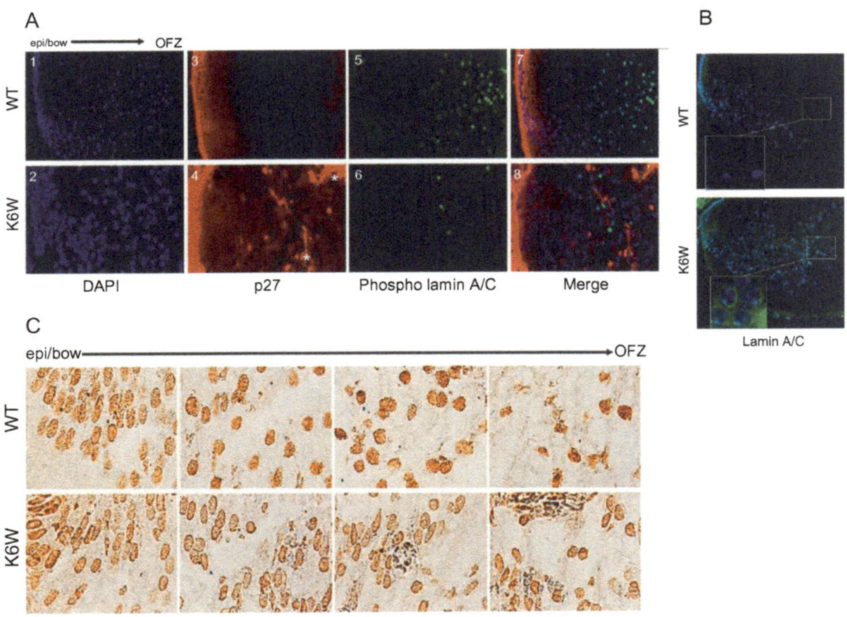


SHANG AND TAYLOR, FIG. 3.

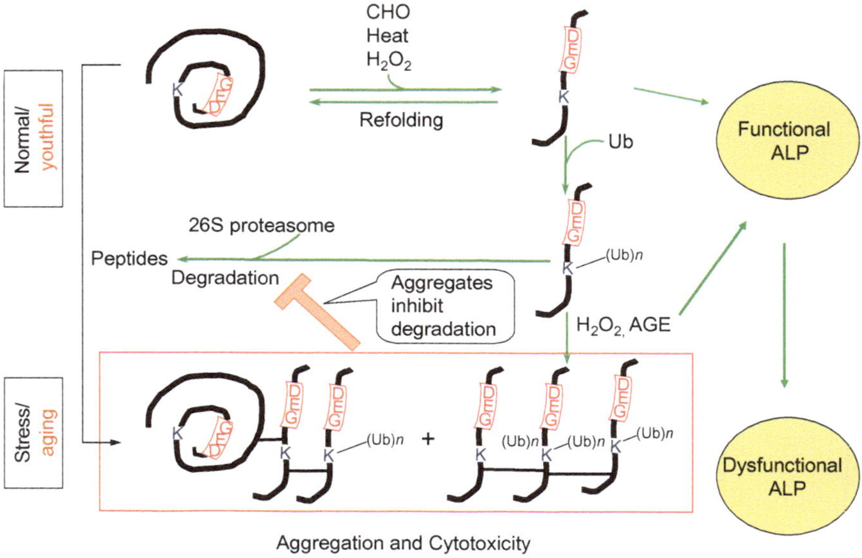


Shang and Taylor, Fig. 4.

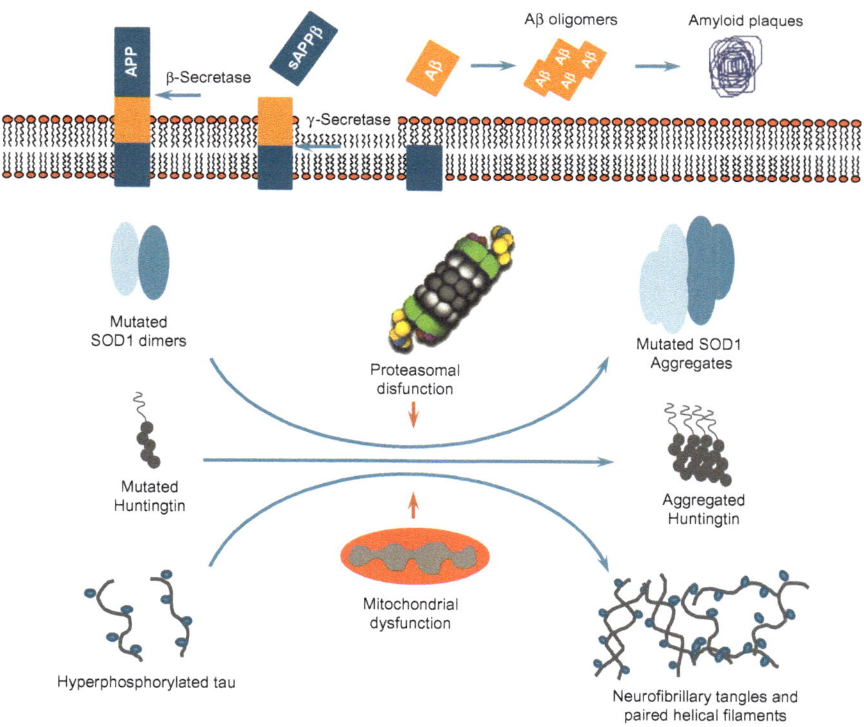


Catalgol and Grune, Fig. 1.

Kolibri 360 GmbH
Ferdinand-Jühlke-Straße 7
99095 Erfurt, Deutschland
produktsicherheit@kolibri360.de